Ultraviolet Spectroscopy and UV Lasers

T0265094

PRACTICAL SPECTROSCOPY

A SERIES

1. Infrared and Raman Spectroscopy (in three parts), *edited by Edward G. Brame, Jr., and Jeanette G. Grasselli*
2. X-Ray Spectrometry, *edited by H. K. Herglotz and L. S. Birks*
3. Mass Spectrometry (in two parts), *edited by Charles Merritt, Jr., and Charles N. McEwen*
4. Infrared and Raman Spectroscopy of Polymers, *H. W. Siesler and K. Holland-Moritz*
5. NMR Spectroscopy Techniques, *edited by Cecil Dybowski and Robert L. Lichter*
6. Infrared Microspectroscopy: Theory and Applications, *edited by Robert G. Messerschmidt and Matthew A. Harthcock*
7. Flow Injection Atomic Spectroscopy, *edited by Jose Luis Burguera*
8. Mass Spectrometry of Biological Materials, *edited by Charles N. McEwen and Barbara S. Larsen*
9. Field Desorption Mass Spectrometry, *László Prókai*
10. Chromatography/Fourier Transform Infrared Spectroscopy and Its Applications, *Robert White*
11. Modern NMR Techniques and Their Application in Chemistry, *edited by Alexander I. Popov and Klaas Hallenga*
12. Luminescence Techniques in Chemical and Biochemical Analysis, *edited by Willy R. G. Baeyens, Denis De Keukeleire, and Katherine Korkidis*
13. Handbook of Near-Infrared Analysis, *edited by Donald A. Burns and Emil W. Ciurczak*
14. Handbook of X-Ray Spectrometry: Methods and Techniques, *edited by René E. Van Grieken and Andrzej A. Markowicz*
15. Internal Reflection Spectroscopy: Theory and Applications, *edited by Francis M. Mirabella, Jr.*
16. Microscopic and Spectroscopic Imaging of the Chemical State, *edited by Michael D. Morris*
17. Mathematical Analysis of Spectral Orthogonality, *John H. Kalivas and Patrick M. Lang*
18. Laser Spectroscopy: Techniques and Applications, *E. Roland Menzel*
19. Practical Guide to Infrared Microspectroscopy, *edited by Howard J. Humecki*
20. Quantitative X-ray Spectrometry: Second Edition, *Ron Jenkins, R. W. Gould, and Dale Gedcke*
21. NMR Spectroscopy Techniques: Second Edition, Revised and Expanded, *edited by Martha D. Bruch*
22. Spectrophotometric Reactions, *Irena Nemcova, Ludmila Cermakova, and Jiri Gasparic*
23. Inorganic Mass Spectrometry: Fundamentals and Applications, *edited by Christopher M. Barshick, Douglas C. Duckworth, and David H. Smith*
24. Infrared and Raman Spectroscopy of Biological Materials, *edited by Hans-Ulrich Gremlich and Bing Yan*

Ultraviolet Spectroscopy and UV Lasers

edited by

Prabhakar Misra

**Howard University
Washington, D.C.**

Mark A. Dubinskii

**Magnon, Inc.
Reisterstown, Maryland**

CRC Press
Taylor & Francis Group
Boca Raton London New York

CRC Press is an imprint of the
Taylor & Francis Group, an **informa** business

First published 2002 by Marcel Dekker, Inc.

Published 2019 by CRC Press
Taylor & Francis Group
6000 Broken Sound Parkway NW, Suite 300
Boca Raton, FL 33487-2742

© 1991 by Taylor & Francis Group, LLC
CRC Press is an imprint of Taylor & Francis Group, an Informa business

First issued in paperback 2019

No claim to original U.S. Government works

ISBN-13: 978-0-367-44709-0 (pbk)
ISBN-13: 978-0-8247-0668-5 (hbk)

This book contains information obtained from authentic and highly regarded sources. Reasonable efforts have been made to publish reliable data and information, but the author and publisher cannot assume responsibility for the validity of all materials or the consequences of their use. The authors and publishers have attempted to trace the copyright holders of all material reproduced in this publication and apologize to copyright holders if permission to publish in this form has not been obtained. If any copyright material has not been acknowledged please write and let us know so we may rectify in any future reprint.

Except as permitted under U.S. Copyright Law, no part of this book may be reprinted, reproduced, transmitted, or utilized in any form by any electronic, mechanical, or other means, now known or hereafter invented, including photocopying, microfilming, and recording, or in any information storage or retrieval system, without written permission from the publishers.

For permission to photocopy or use material electronically from this work, please access www. copyright.com (http://www.copyright.com/) or contact the Copyright Clearance Center, Inc. (CCC), 222 Rosewood Drive, Danvers, MA 01923, 978-750-8400. CCC is a not-for-profit organiza-tion that provides licenses and registration for a variety of users. For organizations that have been granted a photocopy license by the CCC, a separate system of payment has been arranged.

Trademark Notice: Product or corporate names may be trademarks or registered trademarks, and are used only for identification and explanation without intent to infringe.

Visit the Taylor & Francis Web site at
http://www.taylorandfrancis.com

and the CRC Press Web site at
http://www.crcpress.com

Preface

Ultraviolet Spectroscopy and UV Lasers covers a range of subjects, from ultraviolet (UV) and vacuum ultraviolet (VUV) sources to the latest advances in instrumentation and techniques for absorption, emission, and fluorescence spectroscopy. The book will prove useful to scientists pursuing spectroscopy-related research in fields as varied and diverse as optical physics and engineering, analytical chemistry, biology, and laser technology. The book can also serve as a reference text for a two-semester special topics course for upper-level undergraduate and graduate students in these disciplines.

In the opening chapter, "Ultraviolet and Vacuum Ultraviolet Sources and Materials for Lithography," Liberman and Rothschild review radiation sources in the UV and VUV, followed by an appraisal of the key factors governing efficient detection of UV radiation. Their overview of sources and detectors, although presented in the context of lithographic applications, will prove useful for other UV spectroscopic applications as well. UV optical metrology is covered under two subheadings: measurements of bulk material properties and surfaces and measurement of thin film properties. Candidate optical materials for lithographic systems are reviewed in light of their index homogeneity, birefringent properties, and durability. An important consideration in the use of optical components for metrological applications at UV wavelengths is the issue of large absorption coefficients exhibited by vapors and surface adsorbates, which is the subject of discussion in the concluding segment of this chapter.

The second chapter, "Laser Optogalvanic Spectroscopy of Discharge Plasmas in the Ultraviolet Region" is coauthored by one of the editors (PM) of this book and provides an overview of the optogalvanic effect. Owing to its sensitivity and selectivity, the optogalvanic technique has been utilized for a wide variety of spectroscopic applications, including analytical flame spectrometry, atomic and molecular spectroscopy, laser stabilization and calibration, and plasma diagnostics. Haridass, Major, Misra, and Han provide both an experimental background and a theoretical framework for reporting an extensive array of optogalvanic spectral transitions and selective waveforms in the UV for both neon and argon. The mechanism associated with the generation of the optogalvanic signal is covered in depth, and a theoretical model is developed and used to analyze and interpret time-resolved waveforms of specific optogalvanic transitions of neon and argon in the context of electron collisional ionization and electron collisional transfer rates in a discharge medium.

In Chapter 3, "Spectra of the Isotopomers of CO^+, N_2^+, and NO in the Ultraviolet," Reddy and Haridass describe the experimental recording of data with a hollow cathode discharge tube, followed by a subsequent detailed analysis of the electronic spectra of these species. The ultraviolet spectroscopy of these molecules is important for the understanding and elucidation of significant astrophysical processes, as well as for improved understanding of associated terrestrial environmental and combustion phenomena. A review of the relevant electronic configurations and the general spectroscopic equations needed for a detailed spectral analysis and interpretation is provided. A comprehensive historical overview of past and present work is provided, in turn, for CO^+, N_2^+, and NO, followed by presentation and analysis of UV spectra for the isotopomers of all three molecular species. Schematic energy level diagrams illustrate the allowed spectral transitions and the associated branches in detail. Precise molecular parameters characterizing the various electronic states of the different species have been determined using rigorous nonlinear least-squares fitting routines and are tabulated for easy reference.

Photoabsorption cross-section measurements in the ultraviolet and vacuum ultraviolet are the subject of Chapter 4. Yoshino outlines the earlier cross-section measurements—prior to 1980—and discusses the current instrumentation available for precise cross-section measurements in the UV and VUV. A major uncertainty in cross-section measurements with a single path spectrometer is in estimating the background intensity level, since it may vary with wavelength and time during the course of the photoabsorption scan. It is also often difficult to maintain the gas pressure constant within the absorption cell, particularly for reactive and/ or absorbent species. Yoshino summarizes these difficulties in the context of cross-section measurements of the photoabsorption continuum under lower resolution, citing the specific example of the H_2O molecule. He then reviews measurements of cross-sections for molecular bands possessing associated fine rotational structure under superior resolution in relation to the Doppler widths of relatively simpler molecules. Line-by-line cross-section measurements of the Schumann-Runge bands of O_2 and their temperature dependence are then discussed. Line and band oscillator strengths are subsequently related via the Hönl-London factor, whereby once the band oscillator strengths are established, cross-sections for any temperature can in principle be determined.

In Chapter 5, "Ultraviolet and Vacuum Ultraviolet Laser Spectroscopy Using Fluorescence and Time-of-Flight Mass Detection," Lipson and Shi review the principles and techniques of ultraviolet lasers produced by frequency doubling in anisotropic crystals, nonresonant third harmonic generation, and two-photon resonant four-wave sum- and difference-frequency generation in isotropic gases. Their application focus is on single-photon excitations of supersonically jet-cooled gas-phase molecules that are detected either via laser-induced fluorescence or by ion detection via time-of-flight mass spectrometry. The authors provide a detailed discussion of supersonic jet expansions, which have now been established as a powerful tool for gas-phase spectroscopy, because a jet environment is highly conducive to the production and probing of rovibrationally cold molecules, free radicals, and clusters. This chapter provides practical examples of UV and VUV laser-induced fluorescence and time-of-flight excitation spectra in supersonic jets. The spectroscopy of several metal monohalide radicals and diatomic halogens is presented in some detail. A comprehensive catalog of atomic and molecular systems studied with VUV lasers is provided in tabular form. The table includes a summary of the electronic states investigated by specific spectroscopic methods for various atomic and molecular species, along with the pertinent literature references. The use of VUV lasers in tandem with mass spectrometry provides unique opportunities for detection and characterization of organic molecules. The practical use of these techniques is illustrated in the context of organic molecules (with several examples), and a detailed referenced tabulation of organic molecular systems detected by nonresonant VUV ionization has been included. The authors also provide a comprehensive bibliography of the literature associated with the twin techniques of fluorescence spectroscopy and time-of-flight mass detection.

In Chapter 6, Cefalas and Sarantopoulou review spectroscopy and applications of diatomic and triatomic molecules assisted by laser light at 157.6 nm. The authors begin with a description of the useful characteristics of the 157.6 nm radiation source—a molecular fluorine laser—and of its utility for photolithography, and then lead up to a summary of the laser's usage for elucidating the spectroscopy of such varied molecules as OH, O_2, HCl, DCl, CH_3Cl, H_2, OCS, H_2O, HDO, ICN, AsF_2, O_3, CO_2, and SiH. Spectroscopic investigations relating to the properties of neutral or charged clusters can be accomplished in the VUV via optical and mass spectroscopic techniques. Such approaches to study clusters in the VUV possess two distinct advantages, namely, that systems with a large band gap can be studied and that excitation of inner shell electrons allows element-specific information to be collected. In this context, the VUV spectroscopy of triatomic clusters is reviewed. It is followed by a discussion of the laser-induced fluorescence and mass spectroscopy of diatomic and triatomic clusters of mercury and a description of the interesting quantum beat phenomenon in mercury triatomic clusters, following de-excitation from the excited state to the ground state.

In "Tunable Solid-State Ultraviolet and Vacuum Ultraviolet Lasers" (Chapter 7), Ilev and Waynant comprehensively review the all-solid-state concept for producing widely tunable short-wavelength laser generation using nonlinear optical techniques and specific solid-state active materials (e.g., the rare-earth-doped wide band gap crystals). The authors provide an analytical outline of the general principles concerning basic nonlinear effects and then consider the primary all-solid-state nonlinear techniques and optical materials utilized for generating tunable UV/VUV radiation. Significant progress achieved by nonlinear frequency upconversion techniques is also presented. Optical fibers provide a good solid-state medium for broad-band optical frequency conversions, and when laser radiation is propagated in such fibers suitable conditions are realized for efficient nonlinear processes, which include, among others, stimulated Raman scattering, four-photon mixing, and self-phase modulation. Nonlinear effects in optical fibers are presented in considerable detail, especially those involving stimulated Raman scattering and its application for the design and development of tunable fiber Raman lasers in the UV. In addition to the all-solid-state nonlinear techniques for tunable UV/VUV laser generation, the use of rare-earth-activated wide band gap fluoride dielectric crystals for the development of solid-state UV/VUV lasers is reviewed, and leads naturally to the subject matter of the next chapter.

In Chapter 8, "VUV Laser Spectroscopy of Trivalent Rare-Earth Ions in Wide Band Gap Fluoride Crystals," methods and techniques used in vacuum ultraviolet spectroscopy are reviewed by Sarantopoulou and Cefalas, with special focus on the molecular fluorine laser and VUV absorption spectroscopy. Optical and electronic properties of wide band gap fluoride dielectric crystals doped with trivalent rare-earth ions are surveyed, followed by a section on the VUV absorption spectroscopy of these ions. Tables provide the energy positions of the $4f^{n-1}5d$ electronic configuration level onsets of the trivalent rare-earth ions in various crystal hosts, together with crystal field splitting of the main $4f^n$ electronic configuration levels in the presence of octahedral and tetragonal symmetry crystal fields. The laser-induced fluorescence spectroscopy of various rare-earth-activating ions excited at 157.6 nm is presented and discussed in relation to the active center formation with different site symmetries, direct emission and/or weak electron–phonon interactions, emission due to repopulation via phonon trapping and reabsorption of lattice vibrations, and spin-forbidden transitions.

In Chapter 9, "Spectroscopy of Broad-Band UV-Emitting Materials Based on Trivalent Rare-Earth Ions" by Moncorgé, the latest developments in broad-band UV laser systems are discussed and the pros and cons of using specific rare-earth ions for different applications are evaluated. Data summarizing the positions and fluorescence lifetimes of metastable levels in different hosts doped with rare-earth ions are presented, along with the related excited-state absorption spectra. Color center formation, also known as solarization, following excitation by the UV light, can be a critical problem in UV-emitting materials with direct excitation. Such solarization effects are discussed with respect to polarization of the optical transitions and in relation to the purity and dopant substitution processes associated with the host crystals used.

Oldenburg and Eden, in ''Generation of Coherent Ultraviolet and Vacuum Ultraviolet Radiation by Nonlinear Processes in Intense Optical Fields'' (Chapter 10) focus on three important processes: continuum generation, four-wave mixing, and high-order harmonic generation. Researchers in the past decade and a half have observed a variety of nonlinear effects that are a direct result of intense optical fields available to them. The generation of odd harmonics when an intense optical field interacts with an atom or molecule in the gas phase offers the potential for the development of a table-top source emitting soft X-ray radiation or coherent XUV. In addition, harmonic generation is also an exceptional tool for investigating the interaction between an atomic system and a strong optical field. The authors explore all of the above issues and their ramifications.

In Chapter 11, Liu, Sarukura, and Dubinskii (a coeditor of this volume) comprehensively review the all-solid-state, short-pulse, tunable, ultraviolet laser sources based on Ce^{3+}-activated fluoride crystals. Rare earth ions incorporated in appropriate host crystals exhibit vibronic lasing and can be implemented for the development of compact and widely tunable ultrashort-pulse laser sources in the UV. Prior to detailed discussions of newer solid-state laser materials, the authors review the existing UV laser systems that use efficient frequency conversion and also those that allow direct generation of coherent UV radiation. The spectroscopic and laser properties of Ce:LLF and Ce:LiCAF crystals are reviewed in depth, followed by a discussion on the generation of subnanosecond pulses from Ce^{3+}-doped fluoride lasers. Short-cavity subnanosecond pulse generation, combined with regenerative amplification, has been used in the design of the novel so-called self-injection-seeded pulse train lasers. The authors introduce a generalized passive version of this device, followed by detailed descriptions of the self-injection-seeded pulse train Ce:LLF and Ce:LiCAF lasers. Tunability data for the short-cavity Ce:LLF and Ce:LiCAF lasers are presented in the near-UV, followed by a demonstration of tunable subnanosecond pulse generation around 230 nm by sum-frequency generation.

Fluorescence lifetime spectroscopy requires UV pulses with pulse durations shorter than the fluorescence lifetime of a large number of molecules. ''Diode-Pumped Picosecond UV Lasers by Nonlinear Frequency Conversions'' (Chapter 12) addresses the critical issues related to the development of compact, efficient, and reliable diode-pumped short-pulsed solid-state lasers that can be effectively utilized for such fluorescence lifetime measurements. Balembois et al. present several diode-pumped sources that emit picosecond pulses in the UV. Three kinds of techniques have been used to generate the picosecond pulses (followed by appropriate multipass or regenerative amplification): active mode locking of a laser cavity, gain-switching of a laser diode, and passive Q-switching of an ultra-short-cavity microchip laser. All the sources developed use diode-pumped laser crystals that emit in the near infrared, and the emission is subsequently converted to UV wavelengths utilizing nonlinear crystals for harmonic generation. A pair of laser sources has been developed, based on a Ti:sapphire regenerative amplifier seeded by either an actively mode-locked diode-pumped Cr^{3+}:$LiSrAlF_6$ picosecond oscillator or a gain-switched picosecond laser diode. These lasers produced an average power of 1 mW in the wavelength region 270–285 nm. Another efficient laser source described here is based on a passive Q-switched Nd:YAG laser that seeds a multipass (specifically 2 or 4) Nd:YVO_4 amplifier. Such potentially portable systems hold promise for optical sampling and for the semiconductor industry, besides their obvious utility for fluorescence spectroscopy.

Ultraviolet spectroscopy can provide a wealth of information about the Earth's atmosphere via remote sensing techniques. In "Atmospheric Ultraviolet Spectroscopy," Chance and Rothman survey the field of UV remote sensing measurements, with a special focus on satellite-based measurements. They provide an overview of the underlying physics, followed by a description of the data analysis techniques used to process raw data and extract meaningful atmospheric parameters. The HITRAN spectroscopic database has become the standard reference for absorption cross-section information related to atmospheric measurements. The authors report on the progress and status of the extension of the HITRAN database into the ultraviolet regime, and point out that the UV data are stored in a separate directory in the HITRAN compilation and further subdivided into cross-sectional and line parameter data. The concluding segment of Chapter 13 presents an overview of satellite UV spectrometers that have been used for atmospheric investigations followed by a survey of current and planned instrumentation for atmospheric ultraviolet spectroscopy.

In the final chapter of the book, "Ultraviolet Spectroscopy in Astronomy," Carruthers provides an account of the early measurements and the difficulties encountered in recording UV spectra of astronomical objects. He then describes the characteristics required of materials for transmission optics and those of coatings for reflective optics used in designing efficient UV spectroscopic instrumentation. Typical spectrograph configurations and designs are reviewed in the context of space measurements and stellar observations. The evolution of ultraviolet detectors is reviewed, covering photographic films, photomultipliers, channel electron multipliers, and electronic imaging devices. Examples of current UV spectrographic instruments and their applications to UV space astronomy are presented, which include measurements with the Hubble Space Telescope, the Hopkins UV Telescope, the Interstellar Medium-Absorption Profile Spectrometer, and the Far-Ultraviolet Spectroscopic Explorer, as well as Extreme-UV (below about 90 nm) Spectroscopy Missions. A summary of the scientific results from various UV space astronomy missions is provided. Different types of measurements are summarized: those related to the solar atmosphere and solar activity, those involving temperatures and luminosities of hot stars, as well as measurements of the composition and properties of the interstellar medium and studies relating to extragalactic objects and cosmology.

The editors wish to acknowledge all of the contributors in putting the book together, despite their busy schedules. We would also like to acknowledge Ms. Barbara Mathieu's tireless efforts in this endeavor and appreciate Mr. Russell Dekker's patience and support.

Professor Misra wishes to dedicate this book in fond and cherished memory of his father, Shri Prem Krishna Misra, who always valued a good book. Dr. Dubinskii would like to dedicate his efforts on this book with gratefulness to the memory of his teacher, Professor S. A. Altshuller, one of the true pioneers of electronic paramagnetic resonance and nuclear magnetic resonance.

Prabhakar Misra
Mark A. Dubinskii

Contents

Contributors

François Balembois, Ph.D.   Non Linear Optics Group, Laboratoire Charles Fabry de l'Institut d'Optique, Orsay, France

Alain Brun, Ph.D.   Non Linear Optics Group, Laboratoire Charles Fabry de l'Institut d'Optique, Orsay, France

George R. Carruthers, Ph.D.   Space Science Division, Naval Research Laboratory, Washington, D.C.

Alciviadis-Constantinos Cefalas, Ph.D.   National Hellenic Research Foundation, Theoretical and Physical Chemistry Institute, Athens, Greece

Kelly Chance, Ph.D.   Atomic and Molecular Physics Division, Harvard–Smithsonian Center for Astrophysics, Cambridge, Massachusetts

Frédéric Druon, Ph.D.   Non Linear Optics Group, Laboratoire Charles Fabry de l'Institut d'Optique, Orsay, France

Mark A. Dubinskii, Ph.D.   Magnon, Inc., Reisterstown, Maryland

J. G. Eden, Ph.D.   Department of Electrical and Computer Engineering, and Office of Vice-Chancellor for Research, University of Illinois, Urbana, Illinois

Patrick Georges, Ph.D.   Non Linear Optics Group, Laboratoire Charles Fabry de l'Institut d'Optique, Orsay, France

Xianming L. Han, Ph.D.   Department of Physics and Astronomy, Butler University, Indianapolis, Indiana

C. Haridass, Ph.D.*   Department of Physics and Astronomy, Howard University, Washington, D.C.

Ilko K. Ilev, Ph.D.   U.S. Food and Drug Administration, Rockville, Maryland

Vladimir Liberman, Ph.D.   Lincoln Laboratory, Massachusetts Institute of Technology, Lexington, Massachusetts

R. H. Lipson, Ph.D.   Department of Chemistry, University of Western Ontario, London, Ontario, Canada

Zhenlin Liu, Ph.D.   Hosono Transparent Electro-Active Materials Project, Exploratory Research for Advanced Technology, Japan Science and Technology Corporation, Kawasaki, Kanagawa, Japan

H. Major, Ph.D.   Department of Physics and Astronomy, Howard University, Washington, D.C.

Prabhakar Misra, Ph.D.   Department of Physics and Astronomy and the Center for the Study of Terrestrial and Extraterrestrial Atmospheres, Howard University, Washington, D.C.

Richard Moncorgé, Ph.D.   Centre Interdisciplinaire de Recherche Ions-Lasers, UMR 6637 CNRS-CEA-ISMRA, Université de Caen, Caen, France

A. L. Oldenburg, Ph.D.   Department of Electrical and Computer Engineering, University of Illinois, Urbana, Illinois

S. Paddi Reddy, D.Sc.   Department of Physics and Physical Oceanography, Memorial University of Newfoundland, St. John's, Newfoundland, Canada

Laurence S. Rothman, Ph.D.   Atomic and Molecular Physics Division, Harvard–Smithsonian Center for Astrophysics, Cambridge, Massachusetts

Mordechai Rothschild, Ph.D.   Submicrometer Technology Group, Lincoln Laboratory, Massachusetts Institute of Technology, Lexington, Massachusetts

---

* *Current affiliation*: Department of Physical Sciences, Belfry School, Belfry, Kentucky.

Evangelia Sarantopoulou, Ph.D. National Hellenic Research Foundation, Theoretical and Physical Chemistry Institute, Athens, Greece

Nobuhiko Sarukura, Ph.D. Institute for Molecular Science, Okazaki National Research Institutes, Myodaiji, Okazaki, Japan

Y. J. Shi Department of Chemistry, University of Western Ontario, London, Ontario, Canada

Ronald W. Waynant, Ph.D. Electro-Optics Branch, U.S. Food and Drug Administration, Rockville, Maryland

Kouichi Yoshino, Ph.D. Atomic and Molecular Physics Division, Harvard–Smithsonian Center for Astrophysics, Cambridge, Massachusetts

1

Ultraviolet and Vacuum Ultraviolet Sources and Materials for Lithography

Vladimir Liberman and Mordechai Rothschild
Massachusetts Institute of Technology, Lexington, Massachusetts

I. MOTIVATION FOR ULTRAVIOLET LITHOGRAPHY

The relentless drive of the semiconductor industry to smaller device dimensions continues to present new challenges for all the enabling fabrication technologies, and in particular to microlithography. At present, the most advanced microelectronic devices have dimensions of 0.18 μm, and it is expected that these will be reduced further within the next few years to 100 nm and below. The mainstay of mass production lithography has been projection optical lithography, in which the pattern on a photomask is imaged in reduction onto the silicon wafer. From fundamental principles we know that the resolution in such a configuration is directly proportional to the wavelength. Thus, the most straightforward way of reducing printed dimensions is to use radiation sources at shorter wavelengths. Several years ago the industry made the transition from a continuous-wave mercury discharge lamp at 365 nm to a pulsed excimer laser operating at 248 nm (KrF lasers). It is widely expected that in the near future a similar transition will take place to 193 nm (ArF lasers). Beyond that, 157 nm (F_2 lasers) is being explored for sub-100-nm lithography. This shift in lithographic wavelengths from the near-ultraviolet (UV) to deep-UV and eventually to vacuum-UV (VUV) poses new challenges related to optical materials, optical coatings, detectors, and ambient control. This chapter summarizes the key issues encountered at the shorter lithographic wavelengths, and the present state of the art in addressing them.

The definition of the terms "deep-UV" and "vacuum-UV" is not always unambiguous. In lithography "deep-UV" has been applied to 248 and 193 nm, and "VUV" to 157 nm. This chapter will follow this convention: wavelengths in the region 250–190 nm are designated as "deep-UV," and those between 190 and 100 nm as "VUV." The name "vacuum-UV" for wavelengths below 190 nm is a misnomer, since the radiation can propagate in purged ambients as well, such as nitrogen, argon, or helium. The main restriction is that molecular oxygen be absent, since it absorbs strongly in this spectral region (see discussions in Sec. VII). However, even at 193 nm one should avoid the presence of significant amounts of oxygen, because discrete absorption bands do exist there and the atmospheric attenuation can vary from wavelength to wavelength.

II. RADIATION SOURCES

In the deep-UV and VUV spectral regions there exist several radiation sources that can be adapted to specific applications. The main use of such sources in lithography is, of course, for the actual lithographic process (i.e., exposure of resist-coated wafers). However, there are other needs as well: for interferometry, absorption measurements, and at-wavelength inspection of photomasks, among others. The lithographic radiation sources must have high power in order to facilitate a high throughput of the projection systems, and they must be robust, with a low cost of ownership.

The sources that have become the de facto industry standard are excimer lasers. These gas lasers rely on electric discharges to generate the requisite laser population inversion. At 248 nm the radiating species are krypton fluoride molecules. These are formed in their electronically excited state by gas-phase recombination of krypton and fluorine atoms in the presence of a buffer gas (helium or neon). At 193 nm the lasing species are ArF molecules, and at 157 nm they are F_2 molecules. The ground states of the KrF and ArF molecules are dissociative, and therefore a population inversion (and amplified emission) is established as soon as the upper state is populated via collisions. Although this feature makes excimer lasers efficient, it also causes the natural linewidth to be quite broad (a few hundred picometers). In lithography the upper limit on acceptable laser linewidth is determined by the dispersion of the lens materials and by the allowable chromatic aberration of the optical system. For high-numerical aperture optics, the laser linewidth must be narrowed to less than 1 pm [1]. This task is achieved with intracavity optical components such as prisms and gratings. Line-narrowed excimer lasers at 248 and 193 nm are available commercially from several vendors. Their pulse repetition rates exceed 1 kHz and are expected to approach 5 kHz in the near future. Their average power is ~10 W [2,3]. Excimer lasers have a high gain and, therefore, the pulse duration is quite short, typically ~10–20 ns full-width at half maximum. Only few cavity round trips can be generated, and the spatial coherence is consequently low (less than 100 μm correlation length) [4].

The 157 nm laser is slightly different. The lower electronic state of the lasing transition is nondissociative, and, therefore, the natural output spectrum has a discrete structure. It consists of two peaks, each ~1 pm wide, separated by ~100 pm [5]. It is relatively easy to select one of the lines, using intracavity optics, but it is quite difficult to narrow the width of the remaining line to less than ~0.5 pm. In addition, the 157 nm emission of molecular fluorine is accompanied by a red emission (~700 nm) of fluorine atoms generated in the discharge [6]. This red emission is especially noticeable when the buffer gas is helium, and amounts to 3–5% of the total laser power.

Solid-state coherent radiation sources in this spectral range lag significantly behind excimer lasers with regard to power, cost, and reliability. There are no lasers operating in the deep-UV and VUV, but there are schemes for frequency upconversion using Nd:YAG or Ti:sapphire lasers [7]. For instance, the fifth harmonic of a 1.06 μm Nd:YAG laser emits at 213 nm. The nonlinear media required for upconversion are usually crystals such as barium beta-borate (BBO) or lithium borate (LBO). They are transparent to ~190 nm, but their exact transmission at 193 nm, as well as their resistance to laser damage at this wavelength, are strongly determined by crystalline impurities and defects. The efficiency of the frequency upconversion process into these short wavelengths is less than 10%. Other upconversion schemes do not rely on nonlinear crystals but on the generation of anti-Stokes lines in a gaseous Raman-active medium, or on near-resonant nonlinear processes in certain vapors [8]. At present none of these techniques has been turned into commercially available systems. All of the approaches relying on nonlinear optical processes require high peak powers of a pump laser, and are therefore inherently pulsed sources. They also hold the promise of narrow-linewidth, highly coherent radiation, in contrast with the excimer lasers.

Incoherent sources in the deep-UV and VUV are based on electrical discharges of gases, whether continuous wave (cw) or pulsed. Weak lamps, but covering a broad spectral range, employ low-pressure hydrogen or deuterium molecules. They emit mostly at ~140–170 nm in several rotational–vibrational bands, and they also have weak continua at wavelengths below and above this range [9]. Deuterium lamps are frequently used in spectrophotometers. More intense lamps have limited spectral range. For example, mercury lamps radiate at 254 and 185 nm, corresponding to atomic transitions in mercury. The relative intensity of the 185 nm line varies, since it may be strongly attenuated by absorption in the window material used for the discharge envelope. Other commercially available lamps rely on discharges in xenon (172 nm), krypton (145 nm), or argon gas (126 nm) [10]. The spectra of these are broad, 5–10 nm, and are due to radiative transitions in the respective dimer molecules. The lamp efficiency and its reliability decrease rapidly with wavelength. Therefore, xenon lamps are much easier to obtain and maintain than argon lamps. Their power is nevertheless low compared to that of excimer lasers: it is measured in milliwatts per square centimeter, as opposed to the watts one encounters with lasers. One more wavelength should be mentioned: 121.6 nm. This wavelength of an atomic transition in hydrogen atoms, the so-called Lyman-alpha line, can be obtained quite efficiently in discharges containing hydrogen. Several schemes have been proposed for higher-power 121.6 nm lamps, but they have not yet been demonstrated experimentally.

III. DETECTION OF UV RADIATION

The choice of the detector for the UV region is dictated by several factors related to the incident radiation:

1. Temporal shape of incident radiation and required temporal resolution
2. Incident level of radiation and required dynamic range of the detector
3. Incident wavelength of radiation and required wavelength sensitivity of the detector
4. Durability expectations of the detector

Four types of detectors are commonly used: pyroelectric detectors, thermopiles, photodiodes, and photomultiplier tubes.

The operation of pyroelectric devices is based on a temperature dependence of the electric polarization in certain classes of materials, such as ferroelectrics. Thus, a change in temperature resulting from an incident laser pulse leads to a voltage signal [11]. Pyroelectric detectors are well suited for direct energy and power measurements of excimer laser radiation. For pulsed energy detection, pyroelectric detectors can accommodate fluences as high as hundreds of $mJ/cm^2/$ pulse and as low as $\mu J/cm^2/$pulse with fast amplifier circuitry. Within this range, three to four orders of magnitude of dynamic range for pulse energy detection can be commonly obtained. For a flat spectral response and efficient energy coupling, the front detector surface may be coated with an absorbing organic paint or a baked-on black ceramic coating. As an alternative, uncoated pyroelectric detectors can be used, whose front surface electrode is a thin layer of chromium. The uncoated design offers the advantage of a much faster detector response for resolving individual pulses at repetition rates above a few hundred hertz; however, the wavelength response is determined by the spectral reflectivity of chromium, and thus can vary by 15–20% across the UV spectrum. When operating pyroelectric detectors in fluence regimens below the material damage threshold, the detector durability is determined by the lifetime of the front coating or a front surface electrode. For instance, in recent tests performed at the Massachusetts

Institute of Technology (MIT) Lincoln Laboratory at 157 nm, a commercial pyro-electric detector was observed to fail after 3×10^9 pulses at an incident fluence of 0.5 mJ/cm²/pulse. The mode of failure appeared to be delamination of the front surface electrode [12].

The pyroelectric response of a detector may be a function of many factors, including crystal size and type, the value of the pyroelectric coefficient, and its dependence on temperature. Therefore, the absolute calibration of pyroelectric detectors is not always straightforward. At present, the National Institute of Standards and Technology (NIST) offers calibration services for pyroelectric detectors at 248 and 193 nm with an accuracy of $\pm$ 1% [13].

Thermopile devices are useful for measuring continuous wave (CW) radiation sources or average power of pulsed UV radiation. A thermopile consists of an array of thermocouple or thermistor devices that convert incident power into a temperature rise. Because of their simplicity, thermopile probes are consider-ably less expensive than pyroelectric devices. They can measure intensities from several kilowatts to milliwatts per cm². The front surface of a thermopile is typi-cally a black, highly absorptive coating, with a spectral response constant within a few percent over the UV spectral region [14]. The main utility of thermopile probes is the direct determination of incident laser power and laser stability. With NIST calibration, the measurement accuracy can be better than 2% [13]. The most common damage mechanism involves ablation of the black coating. The coating can be reapplied by a detector manufacturer at a nominal cost.

Photodiodes are an excellent choice for detecting highly attenuated pulsed excimer laser radiation when high temporal resolution (better than 1 ns) and wide dynamic range are required. Photodiodes operate on the principle of converting incident photons into electron–hole pairs, which are swept to the opposing ends of a p–n junction, generating current at the device terminals. Diodes for UV applications can be made from a variety of semiconducting materials, including Si, PtSi, and GaP [15]. Photodiodes exhibit excellent linearity over several orders of magnitude: pulsed UV laser energies from under 1 nJ/cm² to hundreds of µJ/cm² can be measured. The external wavelength sensitivity of the photodiode is dependent on the surface reflectance, which is a function of the substrate's optical constants. In addition, the internal quantum efficiency of the device increases with photon energy. The combined effect is a varying spectral responsivity. For instance, for a typical Si-based p–n junction device, a 50% increase in overall sensitivity may be expected as the wavelength is reduced from 248 to 157 nm. Furthermore, as the device ages under UV laser exposure, its wavelength respon-sivity may change [16].

For metrology applications requiring extremely low-level signal detection, such as transmittance, reflectance, and ellipsometry measurements in the ultravio-let, photomultiplier tubes (PMTs) offer the best solution. In a popular configura-tion the PMT window is coated with a fluorescing material, such as sodium salycilate. Ultraviolet light striking the sodium salycilate will cause it to fluoresce at a wavelength of approximately 400 nm, which can in turn be detected by a PMT designed for the visible spectral region [17]. For ultimate detection of low signals, UV-sensitive solar-blind PMTs are used. These have photocathodes insensitive to visible radiation, such as CsI or CsTe. PMTs have constant responsivities over thousands of hours when used with low-level CW radiation. They are not generally used for measuring pulsed UV laser radiation since both sodium salycilate and photocathode materials can be easily damaged with high peak powers.

IV. UV OPTICAL METROLOGY

A. Bulk Materials

In discussing UV optical metrology relevant to microlithography, it is convenient to classify the topic into two parts: measurement of bulk material properties and surfaces, and measurement of thin film properties. Bulk materials relevant to UV microlithography would include lens materials, such as fused silica, single crystal fluorides (calcium fluoride, barium fluoride, and others), and photomask materials, such as fused silica. For lens materials, important optical properties include transmission and loss at the lithographic wavelength, and highly accurate refractive index values. The transmission and optical losses of the lens materials have immediate impact on the overall system transmission and throughput. Furthermore, residual absorption causes lens heating, which, through the temperature dependence of the index of refraction, can cause unacceptable lens aberrations. An accurate knowledge of the index of refraction, its temperature dependence, and its dispersion is essential to the design and tolerancing of near-diffraction limited optics such as those used in lithographic systems.

The transmission and maximum loss requirements are a sensitive function of a particular lens design. For 193 nm and 157 nm applications, a consensus [12] has been developed that acceptable losses in a bulk material should not exceed 0.5%/cm at the wavelength of use, which is expressed as a maximum bulk absorption coefficient of 0.002/cm, base 10.

When prescreening optical materials, one must separate bulk losses, such as bulk absorption and scatter, which are intrinsic to the material, from surface losses, which are a function of surface polishing and ambient conditions. For a sample of length l the internal transmission is:

$$T_{int} = 10^{-[2\beta+(\alpha+\gamma)l]} \tag{1}$$

where β refers to the loss of a single surface, including absorption and scatter; α is the internal absorption coefficient per unit length; and γ is the bulk scatter coefficient per unit length [18]. Experimentally, one measures the external transmission, which includes the internal transmission as well as surface reflections due to index mismatch (Fresnel losses). The internal transmission can then be approximated by:

$$T_{int} = T_{ext}(n + 1)^4/(16n^2) \tag{2}$$

where n is the refractive index of the material. The procedure for obtaining bulk losses involves measuring the transmission of several samples of different lengths, all with the same nominal surface finish. Then, applying Eqs. (1) and (2) above, one derives $\alpha + \gamma$. In order to obtain the bulk absorption coefficient, separate bulk scatter measurements must be performed. Additional metrology complications may arise because the material characteristics can change in the course of laser irradiation: surface losses may change due to laser cleaning or contamination and bulk losses may vary due to laser induced bleaching or absorption. Thus, when long-term irradiation experiments requiring a high degree of accuracy are needed, it is desirable to irradiate several samples of the same grade and different lengths simultaneously.

Because of the expected high initial transmission of lens materials, samples at least several centimeters in length need to be measured to obtain appreciable loss signals. Commercial spectrophotometers are not well suited for measuring longer samples. This is due in part to the limited length of the sample compartment, and in part is a result of the high divergence of the spectrometer beam. From our experience at MIT Lincoln Laboratory, the above issues can cause transmission measurement errors >2% for an 8 cm-long sample at wavelengths below 200 nm. This measurement discrepancy of >0.25%/cm is a significant fraction of the specification of initial material transmission.

An alternative, more accurate, and precise technique of transmission measurements of long bulk samples utilizes an excimer laser light source. This technique is especially useful in conjunction with long-term materials durability studies, since transmission measurements can be performed in situ [18]. A typical measurement setup (Fig. 1) involves using a beamsplitter before the sample to reflect part of the incident beam onto a reference detector. The main part of the laser beam is incident on the sample detector, which is positioned directly behind the sample station. Measurements are obtained ratiometrically. For the incident laser fluences of interest (from 0.1 to 10 mJ/cm²/pulse), pyroelectric detectors are well suited to the task of transmission measurements. Most material evaluation studies for UV lithographic lens applications are performed using the laser ratiometry method described above, whether they are carried out by the material suppliers, by independent laboratories, or by lithographic lens manufacturers. In using pulsed laser ratiometry for in situ single-wavelength transmission measure-

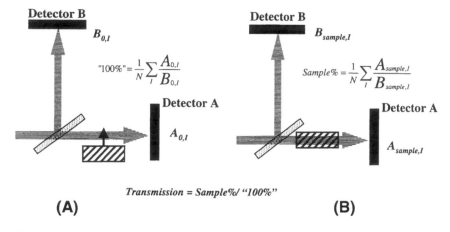

Figure 1 Schematic representation of an in-situ laser-based transmission measurement scheme. A. Measurement configuration for obtaining "100%" transmission. B. Measurement configuration for obtaining sample transmission.

ments, one must be careful to separate potential two-photon absorption from linear absorption. The total absorption coefficient is

$$\alpha_{tot} = \alpha_0 + \beta I \tag{3}$$

where α_0 is the linear absorption coefficient, β is the two-photon absorption coefficient, and I is the peak intensity of the laser pulse. The two-photon absorption coefficient can be obtained by plotting the total absorption coefficient as a function of incident peak intensity and obtaining the slope of the linear regression. Typical two-photon absorption measurements for either fused silica or calcium fluoride at 193 nm yield $\beta = 0.5 - 1$ cm/GW (base 10) [19–22]. For a pulse width on the order of 10 ns and a fluence of 10 mJ/cm^2/pulse, $\beta I = 0.0005-0.001$/cm (base 10), compared to the specification of α_{tot} less than 0.002/cm (base 10). Thus, in order to keep the effect of two-photon absorption below ~25% of the total measured value, fluences below 10 mJ/cm^2/pulse should be used in laser ratiometry.

Microlithographic projection systems are designed to have near-diffraction limited performance, with very small wavefront aberrations. Their design is sensitive to the exact value of the lens material's index of refraction, which must be known to within 10 ppm. The most accurate measurements to date of the index of refraction of materials transparent in the UV region have been obtained by the minimum deviation method [23].

B. Thin Films

Accurate knowledge of the optical properties of thin films is important to many areas of microlithography. For instance, a lithographic lens manufacturer may want to develop a variety of multilayer thin film coatings applied to lens surfaces that will either eliminate surface reflections or allow the element to reflect a prescribed amount, depending on the particular lens design. Although the principles of designing such thin film coatings are well understood, the optical properties of the constituent layers often are not. This lack of information has hampered the development of high-performance coatings in the UV wavelength range.

Many lithographic applications require knowledge of the refractive index and absorption coefficients of thin films, in addition to reflection and transmission data. For instance, knowledge of the refractive index of thin films for lens coatings is needed for precise coating designs. Another application is the tailoring and modeling of photoresist performance, which require knowledge of the optical constants of resists and their constituents (such as photoacid generators). Similar data are needed for the optimization of bottom antireflectance coatings (BARCs), which are applied to the wafer surface prior to the deposition of photoresists to suppress back reflections into the resist. The extraction of optical constants of thin films is best performed with ellipsometry, as discussed in Sec. IV.B.2 below.

1. Transmittance and Reflectance Measurements

For the lifetime testing of optical coatings, accurate spectrophotometric reflectance and transmittance measurements need to be performed periodically. The stringent demands of the material's durability impose strict requirements on the measurement reproducibility. For instance, for durability studies of antireflectance coatings and high reflector coatings, both transmission and reflectance over the UV spectral range may need to be recorded over periods of months with repeatability around 0.1%. In addition, it is not sufficient to record single-wavelength data that are available with a laser-based measurement (as is the case with bulk materials), since useful information about degradation mechanisms of coatings can be learned only from their full spectral behavior.

Transmission and reflection measurements in the near-UV spectrum can be performed to high accuracy and precision, but special challenges arise for metrology below 200 nm. Both water vapor and oxygen absorb strongly in this wavelength region and therefore the instruments must either be purged with a transparent gas or operated under vacuum. The ambient constraints complicate the instrument design, and significantly increase the times needed for sample transfer and measurement. All the transmitting windows, bulb envelopes, and photomultiplier windows must be made either of UV-grade synthetic fused silica or, for operation below 180 nm, of magnesium or calcium fluoride. If polarizing elements are needed in the VUV, the most common option is Rochon prisms made of optically contacted magnesium fluoride elements. They are designed to deflect the extraordinary ray by several degrees while letting the ordinary ray pass straight through.

A simple VUV spectrometer design utilizes a ratiometric method to normalize bulb intensity fluctuations during a wavelength scan. In that configuration, light from a monochromator is incident on a beamsplitter, where a part of the beam is deflected onto a reference detector and the rest of the beam is used for sample transmission measurement. By contrast, in a dual beam–single detector system the two beams are formed using a chopper wheel. During a wavelength scan, for every chopper cycle both a reference and a signal scan are obtained. The technique produces a superior signal-to-noise ratio. The dual beam–single detector designs are quite common for near-UV and visible spectrometers, but they have only recently been incorporated into VUV spectrometers.

In addition to the challenges inherent in obtaining precise readings in the deep-UV, accuracy is difficult to achieve since there are no primary transmission or reflection standards in this wavelength range. The difficulty in developing such standards in the deep-UV stems from the fact that the optical properties of materials are not known with sufficient accuracy, and surface losses due to polishing or surface contamination can significantly influence measurements below 200 nm (see Sec. VII) [24]. Therefore, standards for sample preparation and surface treatment need to be established as part of every measurement procedure. The impact of surface cleaning on the transmission and reflection measurements has only recently been assessed, and a considerable amount of work must be performed before reliable transferable standards are developed.

Several specular reflectance configurations for high accuracy and precision measurements have been successfully implemented in the deep-UV and VUV. Integrating spheres, while popular at longer wavelengths, are not practical below 200 nm because of the lack of a suitable coating reflector material. In the absence of reflectance standards for the VUV range, one needs to use self-referencing schemes, including goniometer-based measurements and use of absolute reflectance accessories, such as V–W and V–N methods. In the latter, the nomenclature describes the beam path traced by a light beam in the baseline and reflectance measurement positions, respectively. The advantage of goniometer over fixed-angle setups is an ability to measure a wide range of specular reflectance signals. However, this arrangement is not compatible with so-called straight-through single-detector spectrometers. For the single-detector schemes, V–W and V–N methods can be used (Fig. 2). Both methods utilize movable mirrors to ensure that, for both baseline and reflectance measurements, the same number of mirrors direct the light beam and the path length is identical for both baseline and sample measurements.

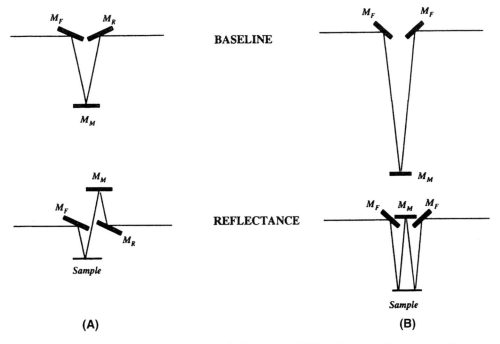

Figure 2   Geometric representations of V-N (A) and V-W reflectance (B) methods. M_F, fixed mirror; M_R, rotatable mirror; M_M, movable mirror.

2. Ellipsometry

Although accurate measurements of reflection and transmission of thin films, as described above, can be used to extract the material optical constants [25], variable angle spectroscopic ellipsometry (VASE) is better suited for such purposes. Ellipsometric parameters relate to a ratio of two reflectivities of different polarizations; thus, the measurements are self-normalizing. Additionally, the phase change upon reflection, obtained in ellipsometric measurements, is sensitive to the presence of very thin films and to the interface structure and therefore can yield valuable information in these areas [26,27]. The extension of VASE to VUV encounters certain practical challenges, similar to those discussed above in the context of spectrophotometry: sources, coating and window materials, polarizers and detectors need to be engineered or optimized for this wavelength range. Ellipsometers must work in a purged or vacuum ambient; thus, instrument access is limited, and automation of the alignment procedure must be engineered. Significant progress has been made recently towards the design of purged UV

VASE, and commercial systems are now available with comparable performance to that obtained in the near-UV [28,29].

V. OPTICAL MATERIALS

There is a limited choice of optical materials in the deep-UV and VUV that would both meet the initial requirements of the excimer-laser based lithographic systems and exhibit the required lifetime durability. For instance, some materials with excellent damage resistance, such as crystalline quartz, magnesium fluoride, and sapphire, are intrinsically birefringent and are thus unsuitable for a lithographic system that is polarization-sensitive. Coupled with the initial requirements of low birefringence, candidate materials must exhibit excellent index homogeneity and must be available in sufficient quantity to meet the demand of the semiconductor industry. The only practical materials that could meet the preceding requirements are amorphous fused silica and crystalline fluorides, primarily calcium and barium fluorides.

A. Fused Silica

The fused silica used in UV lithography applications is of very high purity. All the trace element impurities are in the sub-100 parts per billion (ppb) level. Hydroxyl groups and hydrogen are present in various amounts in 193 nm and 248 nm grade fused silica, because they apparently improve the resistance of fused silica to laser induced damage. Fused silica for 157 nm applications is somewhat different: since OH absorbs at this wavelength, the material must be "dried." Instead of the hydroxyl groups that are removed, fluorine atoms are added to the network. This process extends the transmission edge of the fused silica to below 157 nm, since fluorine titrates absorbing defects and dangling bonds.

The high-purity fused silica represents an atomic structure very close to the ideal continuous random network, comprised of perfect $Si(O_{1/2})_4$ tetrahedra (Fig. 3A) joined at the corners with statistically determined Si–O–Si and dihedral angles [30,31]. In an amorphous fused silica the tetrahedra form closed rings with varying numbers of members, from three tetrahedra and up (Fig. 3B).

Modeling the absorption band-edge of fused silica by an Urbach rule for near band-edge absorption by excitons [32] predicts a lower absorption limit of $\sim 10^{-5}$/cm at 193 nm. However, the most transparent fused silica that has been measured exhibits much larger absorption coefficients, in the range of 10^{-3}/cm. This discrepancy between expectation and experiment indicates that the transparency of fused silica in the deep UV is limited by defects, whether intrinsic or radiation-induced, by OH impurities, and by bulk scattering (Fig. 4) [30].

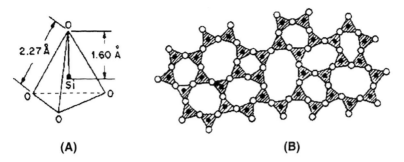

Figure 3 A. Basic structural unit of silicon dioxide. B. Two-dimensional representation of the amorphous fused silica. (From Refs. 30, 31.)

Several types of lattice defects contribute to the deep-UV absorption of fused silica. One is the nonbridging oxygen hole center, the spectral signature of which is an absorption band at ~4.8 eV (260 nm). This defect can be represented by the notation $\equiv$Si–O·, where the three parallel lines represent three separate bonds to oxygen atoms, and the dot represents an unpaired electron. The width of the band is broad enough to overlap the 248 nm excimer laser line. However, this defect does not contribute to absorption at 193 nm and below.

The second defect, the E′ center, can be represented by the notation $\equiv$Si·, and refers to an unpaired electron on a silicon atom. The absorption band caused by this defect is centered at 5.8 eV (215 nm) and is broad enough that it overlaps the 193 nm laser line. The laser-induced formation of E′ centers is believed to be the dominant cause of the observed degradation in deep-UV transmission when fused silica is irradiated with 193 nm lasers.

Scattering-induced transmission losses in fused silica arise from statistical density fluctuations. The general form of this scattering loss, γ, can be expressed by [33]

$$\gamma \propto (n^8/\lambda^4)p^2 T_f \beta_T \tag{4}$$

where n is the refractive index, λ is the wavelength, p is the photoelastic constant, T_f is the fictive temperature, and β_T is the isothermal compressibility at T_f. The contribution of the bulk scattering losses in the deep-UV can be extrapolated from total integrated scatter measurements performed in the near-UV and visible, using the expression above to correct for wavelength and dispersion effects. From those measurements, a bulk scatter loss of $\gamma \approx 10^{-3}$/cm at 193 nm has been derived [18]. Thus, for the high-purity fused silica, bulk scatter is comparable to absorption from intrinsic defects.

The most critical optical elements in a lithographic system are those forming the projection optics. It is expected that in a manufacturing environment they

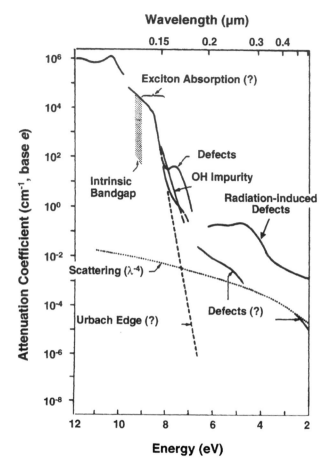

Figure 4 Attenuation coefficients of amorphous fused silica as a function of wavelength and energy. (From Ref. 30.)

will be exposed to fluences up to ~0.1 mJ/cm²/pulse for 4–10 × 10⁹ pulses per year. The fluences are low enough that the electric fields are orders of magnitude below the threshold for dielectric breakdown in fused silica. They are also low enough that, coupled with the low absorption coefficient, the induced temperature rise is only a few degrees even at high pulse repetition rates. Therefore, any laser-induced changes are not caused by either breakdown or thermal effects. Instead, there may be small photo-induced processes whose cumulative effect is notice-able only after large pulse counts. Several laser-induced phenomena in fused silica have been reported, but their mechanistic details and even their scaling laws are not yet fully understood. It has been previously observed that 248 nm irradiation at higher fluences of initially transparent fused silica causes a sudden drop in transmission after a few million pulses, a phenomenon termed strong absorption transition (SAT) [34]. This effect has not been documented with lasers below 248 nm, nor has its fluence scaling been reported. Other changes that occur in fused silica after prolonged irradiation are a gradual loss in transmission and a laser-induced structural rearrangement of the fused silica network, which can manifest itself as densification. All the above phenomena adversely affect the performance of optical systems by changing either their transmission or their effective optical path length.

Our description of irradiation-induced damage of fused silica will be structured in two parts: high-dose damage studies of fused silica for 193 nm lens materials applications, and lower-dose irradiation studies of modified (fluorine-doped) fused silica for 157 nm photomasks. The behavior of fused silica at 248 nm is qualitatively similar to that at 193 nm but the damage rates are about one order of magnitude smaller.

1. Fused Silica for 193 nm and 248 nm Applications

Most earlier reports of UV laser-induced damage to bulk fused silica are based on studies performed at fairly high fluences (10^2 mJ/cm^2/pulse) pulse counts (on the order of 10×10^6 pulses) [31,35–46]. More recent damage studies of fused silica are specifically aimed at understanding degradation effects relevant to microlithography (i.e., those observed in the lower-fluence, higher pulse count regimes) [18,47–51]. At MIT Lincoln Laboratory we surveyed 10 grades of fused silica from six commercial suppliers designed for 193 nm [50]. All were designated as lithographic grade. To increase the pulse accumulation rate the irradiation test bed used 193 nm radiation from two free-running 400 Hz excimer lasers, achieving an effective 800 Hz pulse repetition rate. The incident laser beams were temporally interleaved and spatially overlapped by a 50/50 beam splitter and subsequently distributed along 12 beamlines to deliver an incident laser fluence in the range 0.25–4 mJ/cm^2/pulse. More than 100 different samples were tested. For the material with the highest transmission, lifetime durability tests were extended to pulse counts on the order of 10^9.

The details of material growth were not made available, in order to protect the suppliers' confidentiality. The study aimed instead to assess the state of fused silica as it applies to serving the needs of the lithographic community by deriving general trends in degradation behavior and analyzing the spread among various grades.

Induced Absorption. For the samples tested, the initial 193 nm bulk absorption coefficient was found to vary from 0.001 to 0.01/cm (base 10), depending on the grade tested. The majority of the samples had initial absorption coefficients below 0.002/cm (base 10), which is a target value for maximum acceptable absorption derived in consultation with lithographic lens manufacturers. Upon irradiation, several different trends were observed (Fig. 5). Most samples displayed varying degrees of transmission recovery at the onset of irradiation. Depending on the amount of recovery, this phenomenon could be either laser-induced surface cleaning (1–2% total per sample, length-independent) or bulk bleaching (up to 2% per cm, scaling with sample length). Thereafter, the 193 nm absorption coefficients were found either to reach a plateau or exhibit slow continuous growth. No SAT phenomenon was observed for any sample for the largest doses exposed (5 × 10⁹ pulses at 1 mJ/cm²/pulse).

In general, the details of the degradation mechanism of fused silica under low-level irradiation can be a fairly complex function of small amounts of impurities in the glass, such as OH and H_2. A recent study has correlated the degree of damage under similar irradiation condition to the amount of hydrogen dissolved in the glass network [48]. The time-dependent absorption behavior has been modeled as an interplay of two competing processes: activation of preexisting defects into E′ centers by incident photons and quenching of these defects by dissolved hydrogen to form nonabsorbing SiH bonds [48].

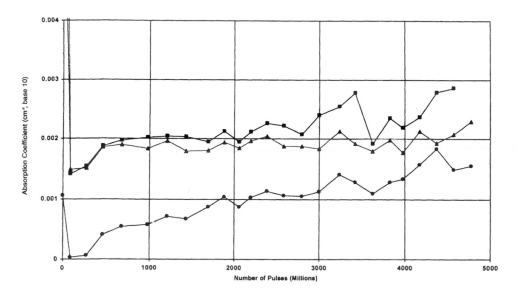

Figure 5   193 nm absorption coefficients of several grades of fused silica as a function of a number of laser pulses (in millions). The incident fluence is 1 mJ/cm²/pulse. (From Ref. 50.)

The absorption measurements shown in Figure 5 represent a permanently induced absorption. However, fused silica also exhibits a temporary reduction in transmission following initial exposure to laser irradiation, and a partial recovery in transmission once the laser is turned off [40,48,52]. Typical time resolved absorption data for a fused silica sample are shown in Figure 6. [53]

Laser-Induced Structural Changes: Densification and Rarefaction. It has been reported for several decades that, when exposed to prolonged low levels of ionizing and nonionizing radiation, fused silica undergoes densification, as manifested by an increase in its refractive index in the irradiated area. This phenomenon has been observed to occur also with the subbandgap irradiation of excimer lasers. The somewhat indirect observations of densification are small, typically a few parts per million (ppm). Because of their magnitude it has not been possible to obtain direct evidence of the structural rearrangement of the lattice or the density increase. Interferometric measurements of wavefront retardation or measurements of stress-induced birefringence through the photoelastic effect are used instead to measure the amount of densification. Nevertheless, the impact of material densification, even in the ppm range, on the performance of a high-quality lithographic system can be devastating, causing unacceptable wavefront aberrations. Over a lens lifetime, it is estimated that a maximum of 1 ppm densification may be tolerated.

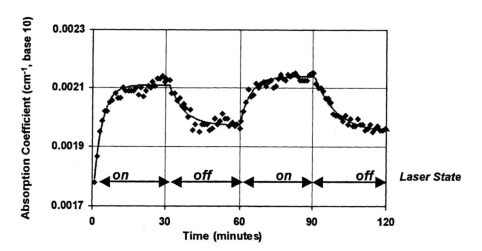

Figure 6 Temporal behavior of 193 nm fused silica absorption as the incident laser is turned on and off. Data shown are obtained for an 8-cm long sample at an incident fluence of 0.25 mJ/cm^2/pulse. (From Ref. 53.)

It has been proposed that UV laser-induced densification follows a power law. If N is the pulse count and the fluence is Φ, then the fractional densification δ is

$$\delta = k(N\Phi^2/\tau)^x \tag{5}$$

where k and x are empirically derived constants, with x typically between 0.5 and 0.8, and τ is a measure of effective laser pulsewidth, assuming that a two-photon effect is operative [54]. This power law has been confirmed experimentally in long-term studies performed at MIT Lincoln Laboratory and elsewhere [19,39,50,51,55,56]. The irradiation was typically with unpolarized light, at fluences of $<0.5–100$ mJ/cm^2/pulse, and for pulse counts up to 5×10^9. Some of these studies reported that k and x as defined in Eq. (5) varied somewhat between grades and with incident fluence levels.

To extend the applicability of the densification scaling to the lifetime doses of lithographic tools ($40–100 \times 10^9$ pulses at 0.1 mJ/cm^2/pulse), an ultramarathon irradiation study has been recently performed [57]. It used time-delay pulse multiplexing of a 2 kHz lithography-grade 193 nm laser beam for an effective operating rate of 8 kHz. The study used polarized light and periodically assessed both interferometric wavefront distortion and stress-induced birefringence of the irradiated fused silica samples. Rather than showing conventional densification through increase in the optical pathlength, the study revealed the opposite effect: rarefaction. The magnitude of this effect was about 0.5 ppm/cm after 40×10^9 pulses. Early studies of fused silica did detect expansion under ionizing irradiation [58–60], but this is the first demonstration of such a phenomenon with UV laser light. The accompanying birefringence signal did not suggest a development of a symmetrical stress field that would be induced by a round laser spot, but exhibited anisotropy that seemed to be induced by the linear polarization of the laser. While at present the laser-induced rarefaction phenomenon is not well understood, it is clearly relevant to attempts to predict the practical lifetime of lithographic lens optics.

2. Modified Fused Silica

Even the highest-quality 193 nm fused silica is not transmissive enough at 157 nm. However, in the last year significant progress has been made to improve its transmission substantially in the 155–175 wavelength region by modifying its composition [61,62]. The general approach consists of two steps: reduce the hydroxyl content of fused silica, since the OH moieties absorb at wavelengths above about 175 nm; and introduce fluorine into the glassy network, in order to titrate all the unpaired electrons on the silicon atoms. Figure 7 shows the transmission of conventional UV-grade fused silica and modified fused silica [63]. The transmission of a 1 cm thick modified fused silica is about 82% at 157 nm, as measured

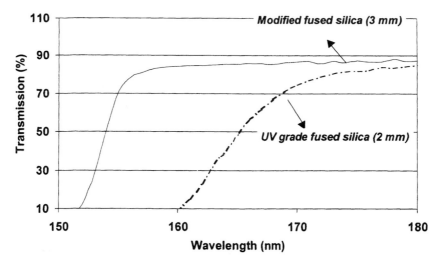

Figure 7   Transmission comparison of standard UV grade fused silica (dot-dash) and a modified fused silica (solid line) samples.

in a vacuum-UV spectrometer. This value should be compared to the theoretical maximum of 88%, obtained from the index of refraction and by allowing for only Fresnel reflection losses. It thus appears that there still is a residual absorption of ~6% in the sample measured. In fact, 3–4% losses can be attributed to surface losses that are removed under laser radiation. The intrinsic absorption coefficient of the modified fused silica can be as low as 0.02/cm (base 10) at 157 nm.

The primary application of modified fused silica is as a transmissive substrate for photomasks at 157 nm. The requirements for good mask material are more relaxed than those for lens materials. High initial transmission needs to be ensured only for a 6 mm thickness (the standard mask thickness), and lifetime requirements must be met only for ~60 × 10⁶ pulses at 0.1 mJ/cm²/pulse. Several modified fused silica samples were evaluated at MIT Lincoln Laboratory, and some met the criteria for low initial absorption as well as for stable transmission at the pulse counts and fluences listed above [12].

B.   Calcium Fluoride

The lattice structure of crystalline calcium fluoride is shown in Figure 8. Each fluoride ion is tetrahedrally bound to four calcium ions. Calcium fluoride has excellent transmission down to 130 nm. It is inert, has very low water solubility, and can be grown in large-diameter boules. Because of its excellent resistance to high-power laser irradiation, the material has found extensive use in illuminator

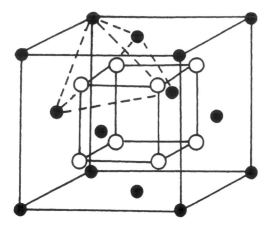

Figure 8   A unit cell of calcium fluoride. Light balls, F⁻ ions; dark balls, Ca²⁺ ions.

designs for 193 nm lithographic systems. The growth in material processing and polishing infrastructure has positioned CaF_2 as the primary candidate for lens materials for 157 nm lithography as well.

Early studies assessing transmission and radiation-induced color centers in undoped calcium fluoride had difficulties separating impurity-induced from intrinsic defects [64–66]. Nevertheless, the most often seen intrinsic color centers are the F center, a single electron trapped in a fluorine vacancy; and the M center, an aggregate of two nearest F centers. The absorption peak of the F band is at ~380 nm, while the M center is dichroic with the two main peaks at 360 nm and ~550–600 nm (see Fig. 9). Since the intrinsic color centers are sufficiently far from 193 nm, the transmission properties at that wavelength and at 157 nm are limited by impurities in the material. In fact, long-term low-fluence 193 nm durability studies of calcium fluoride showed that the initial absorption first increases and then saturates [18]. Similar long-term studies under 157 nm irradiation revealed no added 157 nm absorption for pulse counts in excess of 1×10^9 [12,63]. Although considerable variations exist among calcium fluoride suppliers, total absorption coefficients (initial plus laser induced) of less than 0.003 cm⁻¹ (base 10) can be readily obtained for a number of materials at both 193 and 157 nm wavelengths.

Due to its single-crystal nature, calcium fluoride does not undergo radiation-induced density changes such as those observed in fused silica. Furthermore, bulk scattering in the material is at least an order of magnitude less than that of fused silica. The main challenges in producing CaF_2 for lithographic lens applications include the ability to grow large-diameter crystals (>200 mm) with no dislocations, perfect crystallinity over the full open aperture, low stress-induced

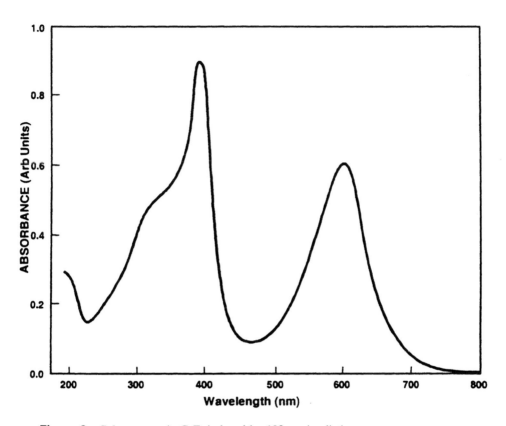

Figure 9   Color centers in CaF$_2$ induced by 193 nm irradiation.

birefringence (<1 nm/cm at 633 nm), and good index homogeneity (better than 1 ppm at 663 nm).

C.   Other Fluorides

For projection lens systems utilizing an all-refractive optical design, a second lens material would be required to provide achromatization across the spectral bandwidth of the 157 nm excimer laser ($\approx$1 pm). The second material option is being pursued in addition to engineering the lasers to bandwidths below $\approx$0.5 pm. Such second refractive lens materials must meet the same stringent requirements as CaF$_2$: transmission, durability, index homogeneity, low stress birefringence, among others. The material should also have good mechanical, thermal, and environmental stability.

Table 1 summarizes key properties of several potential material candidates.

Table 1 Relevant Properties of Potential Lens Materials for 157-nm Lithography

Material	Refractive index at 20°C	Index dispersion, $dn/d\lambda$ (/nm)	dn/dT in N_2 (ppm per °C)	OPD ($\times 10^{-6}$/deg)	Thermal Conductivity (W/m/K)	Comments
CaF_2	1.559	−0.0026	8	12	9.7	Primary choice for lens material.
Modified SiO_2	1.654	−0.0050	39	24	1.4	Marginal absorption (>0.02/cm), long-term radiation durability appears satisfactory.
LiF	1.486	NA	0–4	12–15	11.7	Optical quality, radiation durability, polishing may pose challenges.
SrF_2	1.576	−0.0031	9	12	8.5	Index dispersion is marginally different from that of CaF_2.
BaF_2	1.657	−0.0044	10.6	13.6	7	Index and dispersion complement CaF_2. Transmission and laser damage must be further tested. Prone to thermal shock.

All wavelength-dependent data are at 157.6 nm. All material properties are from Ref. 67, with the exception of LiF, which is from Ref. 69.

All the materials listed are nonbirefringent and are expected to have good transmission properties at 157 nm. The optical pathlength difference (OPD) shown in Table 1 is defined as

$$\text{OPD} = 1/n[(n - 1)\alpha + dn/dT] \tag{6}$$

where α is the coefficient of thermal expansion. The highest-performing modified fused silica has transmission somewhat inferior to that of single crystal fluorides. It also has a thermal conductivity about an order of magnitude lower than that of fluorides, resulting in a higher temperature rise per unit incident laser flux. Its OPD per degree is the highest of the materials listed because of its high thermo-optic coefficient dn/dT at 157 nm. These disadvantages, in addition to marginal initial transmission, would limit the use of modified fused silica to very few elements in an optical train. Of the remaining materials listed, barium fluoride is the best candidate. It has sufficiently different dispersion from that of calcium fluoride to enable achromatization over ~1 pm [67], high thermal conductivity, and good radiation durability when exposed to gamma rays [68], and in initial 157 nm durability testing [12,63]. However, an infrastructure for growing large sizes of this material still needs to be developed. Techniques for polishing and figuring of this material to the required tolerances have not yet reached maturity, nor have low-stress birefringence and index inhomogeneity been demonstrated.

Strontium fluoride has optical properties intermediate between those of calcium fluoride and barium fluoride. In recent 157 nm tests it showed promising initial transmission and laser durability results [12]. However, refractive index and dispersion measurements performed at NIST [67] suggest that the index dispersion of SrF_2 may not be sufficiently different from that of CaF_2 to warrant its use as a second refractive material. Thus, there is not likely to be a strong demand for strontium fluoride for UV lithographic applications.

VI. OPTICAL COATINGS

Dielectric coatings applied to lithographic elements can be classified into three main categories: antireflective designs for throughput enhancement and ghost image suppression, high reflectors for beam steering purposes; and partial reflectors/beamsplitters employed primarily in catadioptric designs. The coating stacks will consist of layers of alternating high- and low-index materials, as compared to the substrate index. The layers must be transparent enough at the wavelength of use to enable an efficient design and to ensure good damage resistance. For 157 nm lithography the choice of coating materials will be primarily limited to fluorides, since most oxide compounds are too absorptive. For instance, low-index materials may include magnesium and aluminium fluorides, while high-index materials may include lanthanum and gadolinium fluorides. For 193 nm the choice of materials is more relaxed; oxides may also be used, in addition to fluorides. Therefore, the oxides of aluminum, hafnium and zirconium may also be considered as high-index candidates, while for low-index materials silicon dioxide is a possible choice.

In addition to the optical properties of the materials used in thin film coatings, the microscopic structure of the films plays an important role. Thin films tend to be more porous than bulk materials, and airborne molecules may adsorb onto the surfaces of the micropores. It was previously observed that water can incorporate into thin films, thus changing their effective index of refraction and thereby shifting the optical response of multilayers. This effect occurs even at wavelengths at which water is transparent (400–700 nm) [70], and it would be even more pronounced below 180 nm, where water also absorbs radiation. This effect may be mitigated by increasing the film deposition temperature, since the packing density changes from ~70% at room temperature to over 90% at 300°C [71]. However, higher deposition temperatures are not always practical because of thermally induced stresses at the substrate–coating interface. For UV applications, both water and hydrocarbon contamination need to be considered in their effect on coating layers. Although condensed water is nearly transparent above 200 nm, it does absorb significantly in the VUV range. The absorption coefficient of condensed water is 7/μm (base 10) at 157 nm. Hydrocarbons have strong absorption throughout the UV range, depending on their molecular structure.

Even small amounts of water adsorption can have significant effects on the performance of multilayer coatings in the VUV range. By way of example, we have modeled the effect of 1% moisture incorporation into coating layers for a LaF_3/MgF_2 system designed for 157 nm [63]. The simulations (Fig. 10) show that even such a modest amount of water can cause a drop of 2% in the transmittance of an antireflective design and in the reflectance of a high-reflectance system. By comparison, for a design centered at 193 nm, where the absorption coefficient of water is $<1 \times 10^{-5}$/μm, a change of less than 0.2% is observed for either an antireflective or a full reflector stack.

A. COATINGS DURABILITY STUDIES

Laser-induced coating damage has received a large amount of attention in recent years because of high-power laser optics applications [42,72–75]. Most studies involved the determination of laser-induced damage thresholds (LIDTs) using so-called one-on-one or S-on-one measurement techniques. The former method is a measurement of energy required to produce damage to a coating in a single laser pulse. The latter involves a determination of a maximum laser pulse energy that a coating can withstand for a fixed number of pulses, usually a few tens to a few hundreds. Typical one-on-one LIDT thresholds for dielectric coatings in

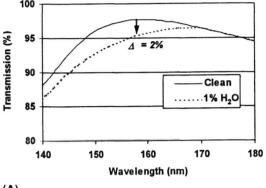

(A)

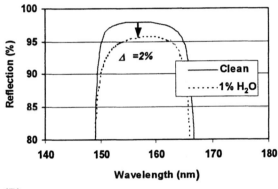

(B)

Figure 10 Effects of 1% moisture incorporation into thin film dielectric layers on (A) transmission of a three-layer antireflectance coating and (B) reflection of a 20-layer full reflector. (From Ref. 63.)

the UV are several J/cm^2 and S-on-one thresholds are two to four times lower [42].

A recent investigation of 248 nm laser damage of multilayer dielectric stacks employed photothermal displacement and mirage techniques to assess 1-on-1 LIDTs in both fluoride and oxide systems [72]. The study found that LIDT values were dominated by defect absorption within the dielectric layers for a LaF$_3$/MgF$_2$ system, and by interface absorption for Al$_2$O$_3$/SiO$_2$ layers. In the case of interface absorption, thermal transport of the energy away from the absorbing region was found to be a critical factor in determining the damage threshold.

Mann et al. [74] also found that the substrate's thermal conductivity was a critical parameter in determining LIDT values of dielectric 248 nm antireflective coatings. The work was extended to 193 nm studies of dielectric-based reflectors. It found significant effects due to hydrocarbon contamination and diffusion into the dielectric layers. Laser conditioning was found to drive out the contaminants and to increase the 193 nm LIDT values.

Few studies have addressed long-term, low-fluence degradation of optical coatings. Two previous studies analyzed UV degradation of dielectric optics at 248 and 308 nm, for exposures up to $\approx 10^8$ pulses and moderately high laser fluences (up to 0.5 J/cm^2/pulse) [74,75]. Both studies reported no visible damage to the coatings, only a slight degradation in optical properties and a possible increase in the absorption coefficient of the dielectric layers.

Within the lifetime of a lithographic tool an average antireflective coating will be subjected to up to 10^{11} pulses at laser repetition rates of 1 kHz and fluences ranging from 0.1 mJ/cm^2/pulse to a few mJ/cm^2/pulse. Since these fluence values are three orders of magnitude below typical single-pulse damage thresholds, one would not expect the coating performance to be determined by LIDT values. Instead, long-term cumulative exposure effects may result either in subtle degradation of optical properties or, perhaps, the eventual failure of the coating. Thus, systematic experiments designed to identify the extent and mechanisms of laser-induced degradation of optical coatings under realistic exposure conditions are needed.

Long-term damage assessment of antireflective coatings for 193 nm lithography was performed at MIT Lincoln Laboratory using incident fluences of <15 mJ/cm^2/pulse and pulse counts of up to 10^9 pulses [50]. Antireflective coatings from nine different commercial suppliers were evaluated under identical conditions. Upon initial irradiation all the coatings underwent a certain amount of surface cleaning, which resulted in higher transmission. The magnitude of this recovery was up to 2%, and its magnitude was found to vary across coatings. This so-called laser cleaning is observed also with bulk materials, but it is more pronounced in coatings, presumably because it is related to surface roughness and/or porosity.

A variety of catastrophic failures of coatings were also observed, although these tended to occur early in the course of the irradiation studies, at pulse counts under 100×10^6. The failure phenomena included high losses, either in the layers or at the interfaces, peeling of the coating, or an apparent thinning of the layers.

It was also noted in these studies that antireflective coatings on fused silica had ~1% higher 193 nm transmission than coatings deposited on calcium fluoride substrates. In the course of 193 nm laser irradiation these losses decreased to the level of the losses on the SiO_2 substrates. This cleaning was very slow, and could take place over 10^9 pulses. No correlation was found between surface roughness and the amount of optical losses. The losses are believed to be an interface phenomenon, and the ''laser cleaning'' perhaps a slow, laser-induced out-diffusion of contaminants through the coating layers.

Thin-film coatings at 157 nm are only now beginning to emerge with the performance and durability comparable to that observed in 193 nm lithography. The main difficulties are qualitatively the same as at longer wavelengths, but quantitatively more pronounced: a limited choice of materials, larger impact of contaminants, as well as the lack of a robust metrology infrastructure.

VII. ATMOSPHERIC ABSORPTION AND SURFACE CONTAMINATION

A critical issue for the operation of optical components and for metrology at short wavelengths is the high absorption coefficient of vapors and surface adsorbates. As can be seen from Figure 11, propagation of UV radiation becomes progressively difficult below 200 nm due to gas-phase absorption by O_2 and H_2O molecules [76]. Even at 193 nm, purging the laser beamline with nitrogen is recommended to suppress ozone formation. By comparison, 157 nm radiation can propagate in an air atmosphere only over distances on the order of a millimeter before being completely absorbed.

Beam delivery systems in the VUV spectral region may be purged with an inert gas or with nitrogen, which is transparent to wavelengths below 120 nm. As an alternative, experiments may be performed under vacuum. However, great

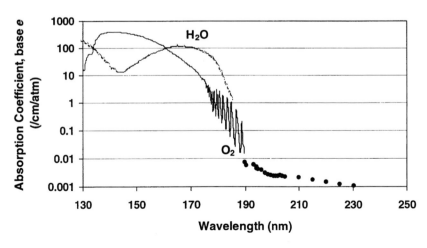

Figure 11 Absorption coefficients of water vapor and oxygen throughout the ultraviolet range. (From Ref. 76.)

care must be taken that the vacuum is established by an oil-free pump and that ultrahigh vacuum techniques are practiced to prevent surface contamination of samples from gas-phase adsorbates. Figure 12 shows that 30 min exposure of a precleaned CaF_2 sample to a 5×10^{-5} torr of vacuum resulted in a transmission drop of several percent throughout the VUV range [24]. In that example, the vacuum was maintained by an oil-free turbomolecular pump backed by an oil-based roughing pump. The contamination can be explained by backstreaming of hydrocarbon-based oils into the vacuum chamber. Note that the original sample transmission could be completely restored by in situ 157 nm laser cleaning.

Ambient contamination and sample packaging can drastically affect measurements in the VUV region. To put the absorptivity of contaminants in perspective, Figure 13 shows the absorption of liquid water and solid polyethylene, a representative material for modeling common organic contaminants such as alkanes [77]. Even a very thin film of 1 nm organic material can cause a 1% transmission drop at 157 nm. For a 1 nm layer of water, an almost 2% drop in transmission is expected. In fact, in one set of experiments, in the course of sample exchange with a supplier of optical coatings, several antireflectance coatings inadvertently came in contact with a clean polyethylene packaging membrane. All

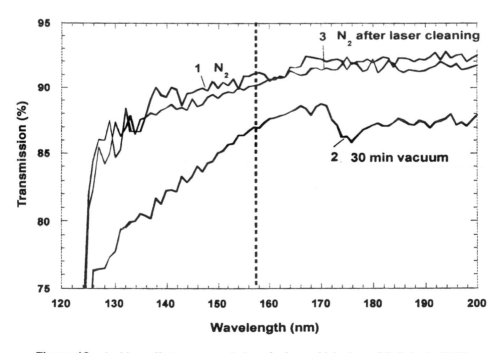

Figure 12   Ambient effects on transmission of a 2 mm thick piece of CaF_2 in the VUV range. (From Ref. 24.)

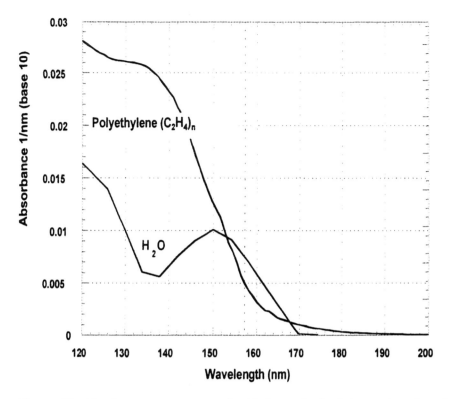

Figure 13   Absorbance per nanometer of a thin layer of polyethylene and condensed water. (From Ref. 77.)

these samples consistently exhibited transmission around 70% when measured in the VUV spectrometer prior to laser exposure. However, upon brief exposure to 157 nm irradiation, dramatic transmission recovery was observed (>25% increase in transmission). By comparison, coating samples shipped in such a way that only an edge contact was made inside the shipping container exhibited only a modest transmission increase (about 1–2%). The effect of surface contamination on samples may be expected to be more dramatic than on uncoated ones because of the possibility of detuning of the dielectric stack and because of the inherent porosity of some dielectric layers, as has been discussed in Sec. VI.

ACKNOWLEDGMENTS

The authors thank all the people and organizations that over the last few years have provided us with samples to be tested and have shared with us their vast amounts of technical insight. Special thanks to our colleagues at MIT Lincoln Laboratory, without whom much of the data reported here would have been impossible to obtain. This work was sponsored in part by the Advanced Lithography Program of the Defense Advanced Research Projects Agency, under Air Force contract F19628-95-C-0002 and in part under a Collaborative Research and Development Agreement between MIT Lincoln Laboratory and SEMATECH. Opinions, interpretations, conclusions, and recommendations are those of the authors and are not necessarily endorsed by the United States Government.

REFERENCES

1. AI Ershov, T Hofmann, WN Partlo, IV Fomenkov, G Everage, PP Das, D Myers. Proc SPIE 3334: 1021–1030, 1998.
2. U Stamm, R Patzel, J Kleinschmidt, K Vogler, W Zschocke, I Bragin, and D Basting. Proc SPIE 3334: 1010–1013, 1998.
3. TP Duffey, T Embree, T Ishihara, R Morton, WN Partlo, T Watson, R Sandstrom. Proc SPIE 3334: 1014–1020, 1998.
4. R Sandstrom. Proc SPIE 1264: 505–519, 1990.
5. K Vogler, U Stamm, I Bragin, F Voss, SV Govorkov, G Hua, J Kleinschmidt, R Paetzel. Proc SPIE 4000: 1515–1528, 2000.
6. CB Collins, FW Lee, JM Carrol. Appl Phys Lett 37: 857–859, 1980.
7. U Stamm, W Zschocke, B Nikolaus, P Genter, D Basting. Proc SPIE 2991: 194–200, 1997.
8. T Uchimura, T Onoda, C-H Lin, T Imasaka. Rev. Sci. Instrum. 70: 3254–3258, 1999.
9. Oriel Company. http://www.oriel.com/homepage/down/pdfla.htm
10. Ushio Semiconductor Company http://www.ushiosemi.com/excimer.htm
11. W Tiffany. http://www.molectron.com/AppNote005a.html
12. V Liberman, M Rothschild, JHC Sedlacek, RS Uttaro, AK Bates, K Orvek. Proc SPIE 4000: 488–495, 2000.
13. ML Dowell, CL Cromer, RW Leonhardt, TR Scott. AIP Proc 449, 539–541, 1998.
14. JD Buck. http://www.molectron.com/AppNote002a.html
15. Hamamatsu Company. http://usa.hamamatsu.com/opto-semi/photodiodes/default.htm
16. R Gupta, KR Lykke, P-S Shaw, JL Dehmer. Proc SPIE 3818: 27–33, 1999.
17. Mc Pherson Inc. http://www.mcphersoninc.com/detectors/model650detectorassembly.htm
18. V Liberman, M Rothschild, JHC Sedlacek, RS Uttaro, A Grenville, AK Bates, C Van Peski Opt Lett 24: 58–60, 1999.
19. NV Morozov. Proc SPIE 2428: 153–169, 1995.
20. KR Mann, E Eva. Proc SPIE 3334: 1055–1061, 1998.
21. R DeSalvo, AA Said, DJ Hagan, EW Van Stryland, M Sheik-Bahae. IEEE J Quantum Electron. 32: 1324–33, 1996.
22. E Eva, K Mann. Appl Surf Sci 109–110, 52–57, 1997.
23. R Gupta, JH Burnett, U Griesmann, M Walhout. Appl Opt 37: 5964–5968, 1998.

24. TM Bloomstein, V Liberman, M Rothschild, DE Hardy, RB Goodman. Proc SPIE 3676: 342–349, 1999.
25. V Liberman, TM Bloomstein, M Rothschild. Proc SPIE 3998: 480–491, 2000.
26. JA Woollam, B Johs, CM Herzinger, J Hilfiker, R Synowicki, CL Bungay. Proc SPIE CR72, 3–28, 1999.
27. B Johs, JA Woollam, CM Herzinger, J Hilfiker, R Synowicki, CL Bungay. Proc SPIE CR72, 29–57, 1999.
28. JN Hilfiker, B Sing, RA Synowicki, CL Bungay. Proc SPIE 3998: 390–398, 2000.
29. P Boher, JP Piel, P Evrard, C Defranoux, M Espinosa, J Stehle. Proc SPIE 3998: 379–389, 2000.
30. DL Griscom. J Ceram Soc Jpn 99: 923–942, 1991.
31. DL Griscom. Proc SPIE 541: 38–59, 1985.
32. IT Godmanis, AN Trukhin, K Hubner. Physica Status Solidi B 116: 279–287, 1983.
33. S Sakaguchi, S Todoroki, S Shibata. J Am Ceram Soc 79: 2821–2824, 1996.
34. DJ Krajnovich, IK Pour, AC Tam, WP Leung, MV Kulkarni. Opt Lett 18: 453–455, 1993.
35. DR Sempolinski, TP Seward, C Smith, N Borrelli, C Rosplock. J Non Crystal Solids 203: 69–77, 1995.
36. P Schermerhorn. Proc SPIE 1835: 70–79, 1992.
37. R Schenker, L Eichner, H Vaidya, S Vaidya, WG Oldham. Proc SPIE 2440: 118–25, 1995.
38. S Thomas, B Kuhn. Proc SPIE 2966: 56–64, 1996.
39. RE Schenker, WG Oldham. J Appl Phys 82: 1065–1071, 1997.
40. C Pfleiderer, N Leclerc, KO Greulich. J Non Crystal Solids 159: 145–153, 1993.
41. E Eva, K Mann, S Thomas. Proc SPIE 2966: 72–99, 1997.
42. SS Wiseall, DC Emmony. Proc SPIE 369: 521–526, 1983.
43. E Eva, K Mann. Appl Phys A 62: 143–149, 1996.
44. JH Stahis, MA Kastner. Philos Mag 49: 357–362, 1984.
45. V Uhl, KO Greulich, S Thomas. Appl Phys A 65: 457–462, 1997.
46. TE Tsai, DL Griscom. Phys Rev Lett 67: 2517–20, 1991.
47. C Smith, NF Borrelli, DC Allan, TP Seward. Proc SPIE 3051: 116–21, 1997.
48. RJ Araujo, NF Borrelli, C Smith. Proc SPIE 3424: 2–9, 1998.
49. V Liberman, M Rothschild, JHC Sedlacek, RS Uttaro, A Grenville, AK Bates, CK Van Peski. Proc SPIE 3427: 411–419, 1998.
50. V Liberman, M Rothschild, JH Sedlacek, RS Uttaro, AK Bates, CK Van Peski. Proc SPIE 3578: 2–15, 1999.
51. V Liberman, M Rothschild, JHC Sedlacek, RS Uttaro, A Grenville. J Non Cryst Solids 244: 159–171, 1999.
52. N Leclerc, C Pfleiderer, H Hitzler, J Wolfrum, KO Greulich, S Thomas, W English. J Non Crystal Solids 149: 115–121, 1992.
53. V Liberman, M Rothschild, JHC Sedlacek, RS Uttaro, AK Bates, C Van Peski. Proc SPIE 3679: 1137–1142, 1999.
54. TP Duffey, T Embree, T Ishihara, R Morton, WN Partlo, T Watson, R Sandstrom. Proc SPIE 3334: 1014–1020, 1998.
55. DC Allan, C Smith, NF Borrelli, TP Seward III. Opt Lett 21: 1960–1962, 1996.
56. NF Borrelli, C Smith, DC Allan, TP Seward, III. J Opt Soc Am B 14: 1606–15, 1996.

57. RG Morton, RL Sandstrom, GM Blumenstock, Z Bor, CK Van Peski. Proc SPIE 4000: 496–510, 2000.
58. CB Norris, EP EerNisse. J Appl Phys 45: 3876–3882, 1974.
59. JE Shelby. J Appl Phys 50: 3702–3706, 1979.
60. JA Ruller, EJ Friebele. J Non Crystal Solids 136: 163–72, 1991.
61. CM Smith, LA Moore. Proc SPIE 3676: 834–841, 1999.
62. Y Ikuta, S Kikugawa, T Kawahara, H Mishiro, N Shimodaira, A Masui, S Yoshizawa. Proc SPIE 4000: 1510–1514, 2000.
63. V Liberman, TM Bloomstein, M Rothschild, JHC Sedlacek, RS Uttaro, AK Bates, CV Peski, K Orvek. J Vac Sci Technol B 17: 3273–3279, 1999.
64. BG Ravi, S Ramasamy. Int J Mod Phys B 6: 2809–2836, 1992.
65. DR Rao, HN Bose. J Phys Soc Jpn 28: 152–7, 1970.
66. F Abba, Y Marqueton, EA Decamps. Physica Status Solidi A 35: 129–132, 1969.
67. JH Burnett, R Gupta, U Griesmann. Proc SPIE 4000: 1503–1509, 2000.
68. S Majewski, MK Bentley. Nucl Instrum Methods Phys Res A, Accel 260: 373–376, 1987.
69. P Laporte, JL Subtil, M Courbon, M Bon, L Vincent. J Opt Soc Am 73: 1062–1069, 1983.
70. S Ogura, N Sugawara, R Hiraga. Thin Solid Films 30: 3–10, 1975.
71. U Kaiser, N Kaiser, P Weissbrodt, U Mademann, E Hacker, H Muller. Thin Solid Films 217: 7–16, 1992.
72. E Welsch, K Ettrich, H Blaschke, P Thomsen-Schmidt, D Schafer, N Kaiser. Opt Eng 36: 504–14, 1997.
73. F Rainer, WH Lowdermilk, D Milam, CK Carniglia, TT Hart, TL Lichtenstein. Appl Opt 24: 496–500, 1985.
74. K Mann, B Granitza, E Eva. Proc SPIE 2966: 496–504, 1997.
75. DJ Krajnovich, M Kulkarni, W Leung, AC Tam, A Spool, B York. Appl Opt 31: 6062–75, 1992.
76. H Okabe, Photochemistry of Small Molecules. New York: Wiley Interscience, 1978.
77. ED Palik, ed. Handbook of Optical Constants of Solids. vol. 2. San Diego, CA: Academic Press, 1991.

2

Laser Optogalvanic Spectroscopy of Discharge Plasmas in the Ultraviolet Region

C. Haridass,* H. Major, and Prabhakar Misra
Howard University, Washington, D.C.

Xianming L. Han
Butler University, Indianapolis, Indiana

I. INTRODUCTION

The optogalvanic effect (OGE) is due to a change (increase or decrease) in the electrical properties (conductivity) within a self-sustained gaseous discharge when illuminated by radiation that is resonant with an atomic or molecular transition of the element within the discharge. The main principle of this effect is that the absorption of radiation causes a redistribution of populations in the atomic or molecular energy levels. Under steady-state discharge conditions, there exists a dynamic equilibrium (as a result of various radiative and/or collisional and/or collective processes) between the various plasma species causing a well-defined impedance to the flow of current. At dynamic equilibrium, changes in the electron/ion densities and/or mobilities, with a concomitant change in the electrical impedance of the plasma, are caused by optical perturbations. Such a change in impedance alters the current in the plasma and can be either real for direct current (DC) discharges or complex for alternating current (AC) discharges.

The optogalvanic effect has been used for many years as a sensitive and reliable method for recording calibration spectra when tunable lasers are employed for monitoring species (e.g., the hydroxyl radical, OH) that are of primary

---

* *Current affiliation*: Belfry School, Belfry, Kentucky.

importance to stratospheric photochemistry or tropospheric air quality. Optogalvanic spectroscopic/detection methods are well-established techniques, although the optogalvanic effect in discharges is not completely well understood. Optogalvanic spectroscopy in a gas discharge plasma is differentiated from other charged particle detection techniques, such as space–charge-limited diodes [1] and gas-filled proportional counters [2], by the presence of a sustained discharge involving an electron gas at a relatively high temperature (0.5–10 eV) [3]. Laser optogalvanic (LOG) spectroscopy is performed by directing a tunable dye laser beam into a plasma (such as that generated in a hollow cathode discharge lamp). When the wavelength of the laser beam coincides with the absorption of a species in the plasma, the rate of ionization of that species changes momentarily because of laser-perturbed collisional ionization. The associated impedance change can be detected as a voltage drop across a ballast resistor in the lamp feeding circuit.

If the laser photon flux is high enough, ionization can occur due to the laser photoionization of excited-state species produced by stepwise resonant or multiphoton laser excitation (whereby the collisional step is replaced by photoionization). The term resonance ionization spectroscopy (RIS) or multiphoton ionization (MPI) is then used. MPI may also be referred to as resonance multiphoton ionization or (REMPI). RIS and MPI differ from the optogalvanic effect in that while they both occur in the plasma/flame environment, they do not necessarily require such a medium [4].

II. OVERVIEW OF PAST AND PRESENT WORK

The earliest observation related to the OGE was made by Foote and Mohler [5]. They used a tungsten lamp dispersed by a monochromator while studying the photoionization of Cs vapor in a thermionic diode tube, in an effort to detect small levels of ionization. These authors also observed ionization of the Cs occurring when the sample was irradiated by wavelengths longer than 318 nm, which corresponded to wavelengths of the atomic Cs spectrum. They postulated that radiation at these wavelengths produced excited-state atoms, which upon collision with other atoms acquired sufficient additional energy to become ionized. Penning [6] observed the OGE for atoms when he detected a variation in the impedance of a neon discharge irradiated by emission from an adjacent neon discharge. Terenin [7] observed the phenomena in molecules when he studied photoionization of salt vapors. Garscadden et al. [8] observed the OGE brought about by a change in the discharge current of a gas-discharge laser when the laser came over threshold.

It was not until the development of tunable dye lasers, however, that the OGE became widely used as a spectroscopic tool. Green et al. [9] demonstrated that high-sensitivity spectra of the species in a discharge could be obtained with a tunable dye laser. The electronic excitation of the atoms in the discharge allowed for the observation of transitions starting from the metastable or excited states, while the use of a hollow-cathode discharge made it possible to perform spectroscopy on a gas-phase sample of refractory elements produced by cathodic sputtering. Optogalvanic transitions were observed for various species sputtered from hollow cathodes [10–13] and from gas-discharge flashlamps filled with neon and argon [9,14–20]. Optogalvanic signals from commercial neon lamps, extensively used as spectral light sources, have been reported in the 540–750 nm [9,14,16,21] and 2440–2780 nm [15] regions. Gusev and Kompancts [22] pointed out the difficulty of obtaining OG resonances with neon at wavelengths shorter than 580 nm, owing to low oscillator strengths of the resonances. Zhu et al., [23] however, observed laser-assisted OG signals with neon in the 337–598 nm region using a commercial iron-neon hollow cathode lamp. These authors identified 223 OG transitions associated with neon energy levels and found that there were more neon OG lines in the near UV region than in the yellow and red region and that these UV transitions possessed fairly strong intensities. The identified lines in the study by Zhu et al. [23] were calibrated with interference fringes from the two surfaces of an etalon recorded simultaneously with the OG signal. OG transitions of argon have also been observed in the following wavelength regions: 367–422 nm [20], 415–670 nm [17], 425–700 nm [22], 420–740 nm [18], 555–575 nm [24], 727–772 nm [16], 360–740 nm [19], and 2440–2780 nm [15].

Most of the work cited above, for both neon and argon, describes the general nature of the optogalvanic signals in various wavelength regimens. However, very few studies [25,26,27] reported in the literature identify and quantitatively characterize the dominant physical processes contributing to the production of the optogalvanic signals in a discharge plasma. Han et al., [25] using a simple mathematical rate equation model, with only a few parameters characterizing the atomic state population, produced a good simulation for the observed time-dependent OG waveform for a specific neon transition. In contrast, a model given by Stewart et al. [27] used nearly 150 parameters for the same transition of the neon atom and still failed to describe the observed waveform satisfactorily.

Owing to its sensitivity and selectivity, the optogalvanic technique has been used successfully for a wide variety of laser spectroscopic experiments, including analytical flame spectroscopy/spectrometry [28–30], laser calibration [11,14,19], laser stabilization [31,32], plasma diagnostics [33–34], atomic spectroscopy [4,21], molecular spectroscopy [35–38], and dynamics [39–41].

Turk et al. [28] introduced analytical flame spectrometry using laser-enhanced ionization (the optogalvanic effect in flames). They compared their detection limits to other methods of the time and found competitive sensitivity for a number of elements with a much simpler system. While aspirating a dilute sodium solution into the flame, a measurable increase in the current through the flame was observed when the laser wavelength was tuned to the sodium resonance lines at 589.0 nm and 589.6 nm. The signal responded linearly to Na concentrations in the range 2–1000 ng/ml [29].

Skolnick [31] used the OGE as a method to stabilize the frequency of a single J value CO_2 laser to the center of its output power. Skolnick's method of frequency stabilization was based on modulation of the laser frequency and simultaneous sensing of the resulting perturbation of the laser plasma tube's impedance. Such stabilization resulted in a simple frequency-control system without using a 10 μm light detector or any ancillary optical component [31].

Ausschnitt et al. [33] demonstrated the application of multiphoton optogalvanic spectroscopy to hydrogen discharge plasma diagnostics. The ground and excited-state densities and the translational temperatures of hydrogen and deuterium atoms within the plasmas were probed [33]. The use of two unequal energy photons resonant with an intermediate state enabled three-dimensional probing of the ground state and discrimination between hydrogen and deuterium.

The first LOG detection of molecular ions (N_2^+ and CO^+) was reported by Walkup et al. [38] Their results indicated an optogalvanic mechanism that involves direct alteration of ion mobility by laser excitation due to a difference in charge-exchange collision rates for excited versus ground-state ions. This technique proved to be a sensitive probe of ions in the cathode dark space [38]. The LOG spectroscopic technique may be regarded as a complement to the traditional absorption or fluorescence methods rather than a replacement.

Intermodulated optogalvanic spectroscopy (IMOGS), a newer method of Doppler-free saturation spectroscopy, was described by Lawler et al. [42] Closely related to intermodulated fluorescence spectroscopy, IMOGS, uses the detection scheme of optogalvanic spectroscopy. The authors compared the sensitivity of IMOGS to that of Doppler-free saturated absorption spectroscopy. Using appropriate parameters in their experiment, they estimated that $(S/N)_{IMOGS}/(S/N)_{SAS} = 100$. Since other spectroscopic techniques have been devised that can improve the sensitivity of saturated interference spectroscopy, they concluded that IMOGS would be most useful in the region of the spectrum where low-noise detectors, interferometric-quality optics, or high-quality polarizers were unavailable.

In order to comply with environmentally mandated regulations, the US Department of Energy is attempting to implement online, real-time monitors (with limits of detection of 1 pCi l^{-1}) to measure the concentration of radionuclides in the off-gases of mixed hazardous waste and radioactive waste thermal treatment systems [43–44]. A detection limit of 1 pCi l^{-1} corresponds to 465 μg m^{-3} for U-235 and 2980 μg m^{-3} for U-238. Abhilasha et al. [43] developed a LOGS system to measure the concentration of uranium present in the off-gases of mixed hazardous waste thermal treatment systems. Their technique yielded limits several orders of magnitude below that attainable by current real-time online radioactive disintegration counting methods and that sought by US Department of Energy (4.6 µg m^{-3} corresponding to a detection limit of 0.0099 pCi l^{-1} for U-235 or 0.0016 pCi l^{-1} for U-238). It was noted, however, that the bandwidth of their current laser system (nominally 0.4 cm^{-1} in the UV) was insufficient to resolve the uranium isotopically.

A field-deployable LOGS system was developed by Monts et al. [44] more recently to measure the concentration of metal species present in thermal treatment systems. LOGS is an ultrasensitive diagnostic tool for online real-time measurements of the concentration of volatile toxic trace metals, radionuclides, and other gas-phase pollutants present in off-gases of mixed-waste thermal treatment systems, where metal species exist primarily as airborne metal oxides and/or metallic particles rather than as free single atoms. Real-time monitoring of airborne contaminants plays a vital role in the protection of personnel and in safeguarding the environment [44]. Since the amplitude of the LOGS signal can be related to species concentration, LOGS has the capability of monitoring, in real time, the concentration of a pollutant of interest.

Miron et al. [39] studied the dynamics of the OGE in hollow cathode lamps by recording time-dependent OG signals and treating the hollow cathode lamp (HCL) as a circuit element. They found that a stable operation of the HCL does not require a high-impedance source, like that needed for a longitudinal discharge tube (for pressure ranges of 1–10 torr). The investigators reported that the duration of the OG signal (a few microseconds) is independent of the lifetime (10–500 ns) of excited levels populated by the laser. They concluded that it was in fact the impulse response of the electronic circuit (which includes the HCL). The variation of the response time with current implied that the discharge recovered more quickly with increased electron density.

This chapter will fill a gap that exists in the optogalvanic spectral map in the near ultraviolet (UV) region, and study the time-dependent OG waveforms of neon and argon, in order to determine the dominant process or processes affecting the OGE and thereby quantify important associated rate parameters. Argon and neon are excellent choices for especially accomplishing the first objective, because they are commonly used buffer gases within hollow cathode lamps and provide an acceptable density of lines in the visible and near (UV) regions. The near UV region is of importance because in this wavelength domain fall the electronic spectra of important atmospheric and combustion species, which include chemical intermediates such as the hydroxyl and alkoxy radicals. The second and more ambitious objective was complicated because the environment of a discharge is too complex to be described by a single general theoretical model; and a quantitative understanding of the OGE is possible only if a reasonable model of the discharge can be made and the cross sections of the involved processes are well known. An effective use of the rate equation formalism to describe the different discharge processes can lead to a very large number of equations for which cross-section values are unavailable. Thus, any knowledge of the properties of a particular discharge is usually largely empirical. An extended and refined version of a mathematical rate equation model for the population distribution of the atomic species in a commercial hollow cathode discharge lamp, set forth by Han et al. [25] in a relatively recent paper on collisional ionization in a discharge plasma, was used in this investigation. The time-resolved profiles of LOG waveforms of argon and neon were analyzed using commercial hollow cathode discharge lamps, with an emphasis on the low-discharge current regimen (0.2–3.0 mA), since it had not been covered in previous studies and promised to provide new insight into multilevel atomic systems in a gas discharge plasma. The physics of time-resolved OG waveforms provided quantitative information on the rates of excited-state collisional processes in a gas discharge plasma.

Collisional ionization is a primary chemical process occurring in the upper atmospheres and in ionospheres of planets. The flux of energetic photons in the upper atmosphere can induce chemical reactions that are extremely difficult to reproduce under similar conditions in the laboratory. An understanding of the ionization processes and the subsequent chemistry is essential for answering questions such as the phenomenon responsible for the existence of sharp ionization layers (so-called ledges), or to the ionic contributions of various physical phenomena, such as thermospheric heating and auroral excitations. The study of the optogalvanic effect may provide some critical insights into more effective parameterization in models of ionospheric excitation. Neon and argon are good candidates for prototype studies because they do not posses rotational or vibrational degrees of freedom. However, neon is isoelectronic with HF, CH_4, and other trace gases, while argon is isoelectronic with HCl, F_2, and other species. By studying the collisional excitation of these noble gases under controlled conditions, suitable approximations can be made to explain the behavior of important trace gases in the upper atmosphere. By fitting the OG waveforms, parameters associated with rates of decay and those proportional to the cross sections of electron collisional ionization are obtained. Allowances have to be made for rotational and vibrational degrees of freedom, and decay rates obtained would probably not be very accurate, but there should be many similarities between the neon/argon prototype and a so-called unified atom framework of the trace gases.

III. EXPERIMENTAL WORK

The experimental arrangement used for laser optogalvanic spectroscopy is shown in Figure 1. In this figure, a dye laser (DL) is pumped by the second harmonic of a Nd:YAG-laser (YL) running at 10 Hz. The output beam had a pulse duration of about 20 ns and a nominal line width of 0.07 cm^{-1} (without any intracavity etalon). The tuning range in the near UV (and visible) was covered by several laser dyes. One of the beams passed through a 1 mm aperture and entered either

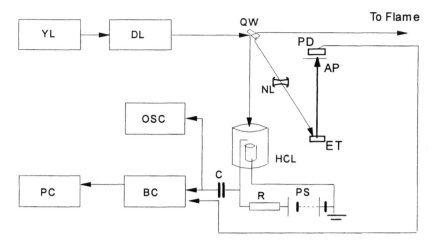

Figure 1 Schematic experimental arrangement for laser optogalvanic spectroscopy YL, Nd:YAG laser; DL, dye laser; QW, quartz wedge; PD, photodiode; AP, aperture; NL, negative lens; ET, etalon; HCL, hollow cathode discharge lamp; PS, power supply; R, resistor; C, capacitance; BC, boxcar; OSC, oscilloscope; PC, personal computer.

a commercial (Perkin-Elmer) iron–neon HCL or a Laser Galvatron (L2783-26ANE-FE, Hamamastu Co.), which is also an HCL containing argon (2 Torr), neon (3 Torr), and trace amounts of iron vapor. The second split beam can be directed to record simultaneously the laser-induced fluorescence (LIF) spectrum of a free radical (e.g., OH) in a propane–air flame. An interference fringe pattern was also simultaneously recorded using the etalon for calibration purposes. A high-voltage power supply (PS) and a ballast resistor (R) of 20 KΩ for the iron–neon lamp (and 30KΩ for the Ar-containing commercial Laser Galvatron) were used to record the OG spectra. When the laser pulse was resonantly absorbed by the discharge medium, the voltage across the lamp varied, and these variations were suitably coupled via a capacitor (C) to a boxcar (BC) integrator. Temporal evolution of the signal was recorded using a digital oscilloscope (OSC). Outputs of the boxcar and the photodiode were recorded with a computer-aided (PC) data acquisition system.

IV. THEORY

A. Mechanism

The mechanism of the optogalvanic effect involves laser enhancement or suppression of the ionization rates of a particular species present in the plasma. Several models have been developed, each appropriate for a given type of plasma, but they are essentially phenomenological and do not account for many of the complex processes occurring in ionized gases that could, in principle, be involved in a laser-induced impedance change. Yet the fact that a number of different plasma sources exhibit the OGE (flames, positive column DC discharges, hollow cathode DC discharges, radiofrequency (RF) discharges) attests to its generality [37].

Illumination of a gas discharge with radiation at a wavelength corresponding to an atomic transition of a species in the discharge causes perturbations to the steady-state population of two or more levels. This, in turn, causes a change in the electrical properties of the discharge. The optogalvanic effect can correspond to an increase or decrease in discharge current depending on the kinetics of the levels whose populations are perturbed by the laser. When the laser frequency is tuned to the transitions $E_{L1} \rightarrow E_{L2}$ between two levels of atoms or ions in the discharge, the laser initially transfers some population in E_{L1} to E_{L2}, which subsequently relaxes to E_{L3} by radiative and/or collision processes in a time frame of tens of nanoseconds. Owing to the different ionization probabilities from E_{L1} and E_{L3}, such a population change will result in a change in the discharge current, ΔI, which is then detected as a voltage change $\Delta V = R \, \Delta I$ across the ballast resistor R. In general, positive and negative signals are observed, depending on the characteristics of levels E_{L1} and E_{L3} involved in the transition $E_{L1} \rightarrow E_{L3}$. If IP_{L1} is the total ionization probability of an atom in level E_{L1}, the voltage change ΔV produced by the laser-induced change $\Delta n_{L1} = n^f_{L1} - n^i_{L1}$ is given by: [45]

$$\Delta V = R \, \Delta I = a[\Delta n_{L1} IP_{L1} + \Delta n_{L3} \, IP_{L3}] \tag{1}$$

where Δn_{L1} is the laser-induced change in the population of level 1 (final $-$ initial) and Δn_{L3} is the change in the population of level 3. If the laser excites atoms from a level E_{L1}, with a small probability of ionization, to a level E_{L3}, which has a larger probability of ionization, the discharge current will increase. If on the other hand, the net effect of the laser takes atoms from a level with a large probability of ionization to a level with a small probability of ionization, the discharge current will decrease. The latter often arises when the atom is excited out of a long-lived metastable state into a higher-lying state that is not metastable, which results in rapid decay to a state lower than the original metastable state. Some of the lowest excited levels in noble gases are metastable, having values of $J = 0$ or 2. Metastable atoms exist in high concentrations in glow discharges and thus can have a great impact on the optogalvanic effect.

The convention used throughout the present study is based on an external measuring circuit that takes the OG signal off the cathode end of the discharge lamp. The traditional *positive* OG signal is one in which there is a net increase in ionization or increase in current ($\Delta I > 0$). It is the result of a decrease in impedance, which appears as a decrease in voltage ΔV across the HCL (or an increase in voltage across the ballast resistor) that causes the waveform to appear with an initial negative peak. A *negative* OG signal results from the depletion of the metastable population. The net effect is an increase in impedance and therefore an increase in ΔV, whereby $\Delta I < 0$. (Note: When signals are measured from the anode, all signs are reversed.)

Several different ionization reactions can contribute to the discharge current. Electron impact ionization, collisions among the discharge species, and Penning ionization are perhaps the most important and likely candidates. We use the notation: B = buffer gas atom, S = sputtered atom and B* = an excited state of B, so that various ionization processes may be represented as follows.

Electron impact ionization occurs from excited levels as well as from the ground state, and although it is not always the dominant ionizing reaction, it usually contributes to the effect to some degree. It may result from a one-step ionization defined as

$$B + e^- \rightarrow B^+ + 2\,e^- \tag{2}$$

Such direct ionization dominates at low pressures, since the electrons can acquire enough energy between collisions to ionize an atom. A two-step, or multistep ionization, is also possible and defined by

$$B + e^- \rightarrow B* + e^- \tag{3}$$
$$B* + e^- \rightarrow B^+ + 2\,e^-$$

This kind of process is important in noble gas discharges and has a significant impact on the optogalvanic effect.

Collisions among the excited atoms of the discharge species can be represented by

$$B* + B* \rightarrow B^+ + B + e^- \tag{4}$$

These types of collisions are mostly seen in the case of noble gas atoms and are very effective because of the high energies (on the order of 16.6 eV for neon and 11.6 eV for argon) of the metastable levels involved. The buffer gas ions liberate atoms from the surface of the cathode by colliding with the cathode material. Thus, the liberated atoms form a minor constituent of the discharge species.

Penning ionization can be represented by

$$B* + S \rightarrow S^+ + B + e^- \tag{5}$$

In Penning ionization, the excited buffer gas atoms may ionize the sputtered atoms. This kind of process is important for discharges in noble gases and influences the characteristics of the observed discharge current. In general, at higher discharge currents, the mechanism of electron impact ionization dominates over Penning ionization, leaving the latter hardly noticeable [46,47].

Broglia et al. [48] demonstrated that photoionization can influence the discharge current in hollow cathode DC lamps. The researchers showed that the dark space of hollow-cathode discharges exhibits signals much more rapidly than the conventional OG signals and related them to direct laser photoionization. Since our work was based on results in which the laser beam is collimated to fill the region between the electrodes, it was first assumed that we might be able to see this feature; however, the energy flux was so weak in the low-current region used in this study that we could not do so. Broglia et al. [49] also admitted that experimental evidence showed that in a low-power regimen only the OG signal characterized by the μs time scale arises.

It was still evident, however, that the induced current resulting from the OGE could, in principle, be due to many factors. One of the main objectives of the present work was to study extensive sets of OG transitions of neon and argon. As expected, the largest identifiable contribution of the OGE waveform turned out to be the electron collisional ionization of the buffer gas, and contributions such as Penning ionization and others only appeared as either slight additions or perturbations.

A nonlinear least-squares fitting program written by Han [50] of Butler University in Fortran code was used and was capable of fitting multiple exponential functions to experimental data of the form:

$$W = ae^{-bt} + ce^{-dt} + ee^{-ft} + \cdots$$

The fit could run up to 200 iterations or until a precision of 10^{-6} was reached. W is the waveform intensity; $a, c, e, \ldots$, etc., are coefficients related to effective collisional cross sections; and $b, d, f, \ldots$, etc., are exponential parameters that allow quantification of the decay rates and thus the effective lifetime of the states involved in the OG transition.

B. Theoretical Model

A model involving only buffer gas (neon or argon) transitions, developed by Han et al., [25] was modified and used in this study to account for the results associated with time-resolved waveforms of LOG transitions. Neon or argon atoms are detected by the optogalvanic effect (via increased or decreased discharge current) only as they are ionized. When the laser is tuned to a transition of the buffer gas, the number of atoms in each state is determined mainly by three processes electron collisional excitation, radiative depopulation, and electron collisional ionization. Certain other nonnegligible processes can contribute to the OG effect: in particular, the redistribution of excited buffer gas atoms due to atomic and electronic collisions. Subsequent relaxation of these atoms to the corresponding lower states may be subject to further ionization. In terms of the Paschen notation, for

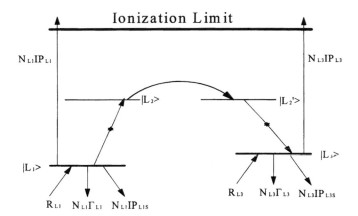

Figure 2 A simplified two-level energy diagram illustrates the dynamics involved in the generation of the optogalvanic signal (* indicates laser excitation and radiative decays). In the case of neon $1s_5-2p_9$ transition, $|L_1\rangle$ represents $|1s_5\rangle$, $|L_2\rangle$ represents $|2p_9\rangle$, and $|L_2'\rangle$ represents other $|2p\rangle$ energy states. For other optogalvanic transitions, $|L_2\rangle$ and $|L_2'\rangle$ could represent the same energy states.

the neon $1s-2p$ transition, there are four $1s$ states and ten $2p$ states, and therefore a total of 30 allowed radiative transitions [51,52].

Prior to laser excitation, there exists a state of dynamic equilibrium in the atomic distribution in each energy level. Upon laser excitation, a fraction of the excited buffer gas atoms is collisionally transferred from $|L_2\rangle$ to other $|L_2'\rangle$ states and then radiatively decays back to the $|L_3\rangle$ state, which should happen within the radiative lifetime of the $|L_2\rangle$ state. To a first-degree approximation, let us treat the other three $1s$ states as a collective $|L_3\rangle$ state for the sake of simplicity of mathematical derivation. We also make the approximation that, after one collision, the atoms transferred out of states $|L_1\rangle$ and $|L_3\rangle$ reach the steady-state distribution, and therefore we neglect the possibility of collisional transfer between $|L_1\rangle$ and $|L_3\rangle$. The preceding above approximations were validated by the observed experimental data. If these approximations were invalid, all four $1s$ states would have to be treated separately. The solution for the time-dependent OG signal would consist of four exponential functions instead of the two exponential functions considered by Han et al. [25] in treating the visible neon transition $1s_5-2p_9$.

The population change over time (as illustrated in Fig. 2) can be expressed as

$$\frac{dN_{L1}}{dt} = R_{L1} - N_{L1}(\Gamma_{L1} + IP_{L1} + IP_{L1s}) \qquad (6)$$

$$\frac{dN_{L3}}{dt} = R_{L3} - N_{L3}(\Gamma_{L3} + IP_{L3} + IP_{L3s}) \tag{7}$$

where R_{L1} and R_{L3} are the rate constants of collisional excitation to the states $|L_1\rangle$ and $|L_3\rangle$, respectively, from all lower states. N_{Li} and Γ_{Li} are, respectively, the neon population and effective decay rates of the state $|L_i\rangle$. IP_{Li} is the total probability of ionization of the state $|L_i\rangle$, I is the current, and P_{Li} is related to the collisional ionization cross section of the state $|L_i\rangle$. P_{L1s} and P_{L3s} are the electron collisional transfer rate parameters to 1s steady-state distribution from $|L_1\rangle$ and $|L_3\rangle$ states, respectively. At steady state, $dN_{L1}/dt = dN_{L3}/dt = 0$. Therefore, the steady-state population of each state can be expressed as

$$n_{L1} = \frac{R_{L1}}{\Gamma_{L1} + IP_{L1} + IP_{L1s}} \tag{8}$$

and

$$n_{L3} = \frac{R_{L3}}{\Gamma_{L3} + IP_{L3} + IP_{L3s}} \tag{9}$$

Solving Eqs. (6) and (7) gives the change in population of both $|L_1\rangle$ and $|L_3\rangle$ states as a function of time:

$$N_{L1}(t) = n_{L1} + c_{L1} \exp[-(\Gamma_{L1} + IP_{L1} + IP_{L1s})t] \tag{10}$$

$$N_{L3}(t) = n_{L3} + c_{L3} \exp[-(\Gamma_{L3} + IP_{L3} + IP_{L3s})t] \tag{11}$$

where c_{L1} and c_{L3} coefficients that will be determined by the initial conditions. Thus, the population changes caused by the laser excitation are

$$\Delta N_{L1}(t) = N_{L1}(t) - n_{L1} = c_{L1} \exp[-(\Gamma_{L1} + IP_{L1} + IP_{L1s})t] \tag{12}$$

$$\Delta N_{L3}(t) = N_{L3}(t) - n_{L3} = c_{L3} \exp[-(\Gamma_{L3} + IP_{L3} + IP_{L3s})t] \tag{13}$$

for $|L_1\rangle$ and $|L_3\rangle$ states, respectively. After laser excitation, the neon population redistribution in the 1s states gives the initial conditions as

$$\Delta N_{L1}(0) = -\Delta N_0 \quad \text{and} \quad \Delta N_{L3}(0) = +\Delta N_0 \tag{14}$$

Using the initial conditions given in Eq. (14), the change in populations given in Eq. (12) and (13) can be shown to be

$$\Delta N_{L1}(t) = -\Delta N_0 \exp[-(\Gamma_{L1} + IP_{L1} + IP_{L1s})t] \tag{15}$$

$$\Delta N_{L3}(t) = \Delta N_0 \exp[-(\Gamma_{L3} + IP_{L3} + IP_{L3s})t] \tag{16}$$

Therefore, the OGE signal due to ionization of $|L_1\rangle$ and $|L_3\rangle$ states can be expressed as

$$
\begin{aligned}
\Delta w(t) &= \Delta N_{L1} IP_{L1} + \Delta N_{L3} IP_{L3} \\
&= \Delta N_0 \{ IP_{L3} \exp[-(\Gamma_{L3} + IP_{L3} + IP_{L3s})t] \\
&\quad - IP_{L1} \exp[-(\Gamma_{L1} + IP_{L1} + IP_{L1s})t] \}
\end{aligned}
\tag{17}
$$

Equation (17) is recognized as a form composed of two exponential decays associated with $|L_1\rangle$ and $|L_3\rangle$ states and gives the experimental time-resolved OGE signal as

$$
w(t) = \frac{\Delta w(t)}{\Delta N_0} = a e^{-bt} - c e^{-dt}
\tag{18}
$$

where

$$
a = IP_{L3}
\tag{19}
$$

$$
b = \Gamma_{L3} + I(P_{L3} + P_{L3s})
\tag{20}
$$

$$
c = IP_{L1}
\tag{21}
$$

$$
d = \Gamma_{L1} + I(P_{L1} + P_{L1s})
\tag{22}
$$

By varying the discharge current, and therefore changing the evolution rate, a set of time-resolved OGE waveforms can be compiled to reveal the effects of varying discharge currents. Based on the theoretical model described above, each spectrum can then be fitted by a nonlinear least-squares method to obtain the parameters associated with the optical energy states. The effect of the instrumental time constant τ is crucial in determining the fast time region of the OGE waveform and can only be determined experimentally [53,54]. The instrumental response becomes less significant as the OGE signal evolves in the time domain. When τ is included, Eq. (18) can be rewritten as:

$$
S(t) = \frac{a}{1 - b\tau}(e^{-bt} - e^{(t/\tau)}) - \frac{c}{1 - d\tau}(e^{-dt} - e^{(t/\tau)})
\tag{23}
$$

where a and c are the positive and negative amplitudes, b and d are proportional to the decay rates of the 1s states, and τ is the instrumental time constant. Collisional ionization parameters are proportional to the electron collisional ionization cross sections associated with the $|L1\rangle$, $|L3\rangle$ states. Han et al. [25] were successful in fitting the waveforms resulting from the neon transition at 640.22 nm in the 4–14 mA current range. The present investigation extended the limits of this same visible transition down to 0.2 mA. Other neon transitions and several argon transitions (both in the visible and UV range) were subsequently analyzed to expand the model further and test the limits of the basic theory.

V. IDENTIFICATION AND ASSIGNMENT OF OPTOGALVANIC TRANSITIONS USING THE J–L COUPLING SCHEME

Precise wavelength calibration of tunable lasers is critical when performing LIF spectroscopy of atoms and molecules. Although some OG spectral wavelength tables do exist, very few of the compilations are either readily available or complete enough for calibrating tunable laser output in the near-UV region. The laser-assisted optogalvanic effect provides a good solution for existing wavelength calibration inadequacies in these regions, which proves vital for combustion and atmospheric monitoring purposes.

OG spectroscopy can occasionally provide high levels of intensity for transitions of some species that produce very weak signals or those that are undetectable by more traditional methods. The differences in OG and fluorescence spectra can be particularly dramatic for transitions arising from metastable levels in noble gases [26], which is not surprising since the intensity of the OG signal is not an absolute measure of the oscillator strength of the transition. The mechanism responsible for the OG signal is complex but it is clearly dependent upon the discharge current.

A. Wavelength Calculations and Edlen's Formula

In order to accurately identify and assign the new lines of the optogalvanic spectra of neon and argon, the positions of the rotational lines of the OH radical were also simultaneously measured and their bin numbers d recorded. The air wavelengths λ_{air} of the reference OH lines were taken from Dieke and Crosswhite [55], and their corresponding bin numbers, were fitted to the least-square polynomial:

$$\lambda_{air} = \sum_{i=0}^{N} a_i(d - d_0)^i \qquad (24)$$

where d_0 is the bin number reading of the first reference line and the a_i is the least-square coefficient associated with the polynomial fit of order N. The wavelengths in air (in Å) were then converted to vacuum wavenumbers ν (in cm^{-1})

$$\nu_{air} = 10^8/(\eta \times \lambda_{vac}) \quad \text{in cm}^{-1} \qquad (25)$$

where λ_{vac} is in 10^{-8} cm and the index of refraction in air η is given by Edlen's formula [56]:

$$\eta = 1 + 6432.8 \times 10^{-8} + \frac{2,942,810}{146 \times 10^8 - v^2} + \frac{25,540}{41 \times 10^8 - v^2} \qquad (26)$$

Vacuum wavenumbers of the energy levels for neon and argon were taken from Bashkin and Stoner [57], who have compiled the atomic energy levels and the pertinent Grotrian diagrams. The source for neon and argon air wavelengths was a set of tables of spectral lines for neutral and ionized atoms by Striganov and Sventitskii [58], which we have referred to as the *atlas*. Additional argon energy-level tables used were supplied by Sugar [59].

B. Spectroscopy of the Noble Gases and Racah Notation

Once the neon and argon spectra were recorded and precisely cross-calibrated using the rotationally resolved LIF spectrum of the OH radical, the OG transitions were assigned using the J–L coupling scheme represented in the Racah notation [60]. Although cumbersome, Racah's notation provides the best model of the physical situation of the excited states of noble gases. Pashen notation, most commonly used, especially for neon, is simply a system of shorthand symbols that must be treated as arbitrary names given to the energy levels [60]. Even though the familiar letters s, p, and d are used, we cannot assume that a Paschen symbol with the letter s, for example, refers to an outer electron in an s orbit. To find out which quantum-mechanical model represents a specific level, we must consult a list of the symbols that relates them to the model, such as Moore's atomic-energy-level tables [61]. Figure 3 and 4 illustrate these ideas for the lowest excited states of neon and argon atoms, respectively.

The p-shell of noble gases, being completely filled, ends with the electron configurations $2p^6$, $3p^6$, $4p^6$, and so on. The total angular momentum **J**, the orbital angular momentum **L**, and the total spin of such closed shells are all zero. When excited, one of the p-electrons moves out of its closed shell leaving a five-electron core behind. In the case of neon, for example, the excited configuration is of the form $2p^5 3s$, $2p^5 3p$, and so on (omitting the preceding closed shells, $1s^2$, $2s^2$, etc.).

J–L coupling (using the Racah notation) is the best approximation of the coupling mechanism in an excited noble gas atom. In **J–L** coupling, the orbital angular momentum of the outer electron **l** is coupled to the total angular momentum of the core $\mathbf{J_c}$, and a resultant vector **K** is obtained by $\mathbf{K} = \mathbf{J_c} + \mathbf{l}$. Total angular momentum **J** of the atom is obtained by adding (vectorially) the spin of the outer electron to **K**. Racah notation consists of the outer electron configuration followed by [K], that is, 3s[3/2] or 3s'[1/2], where the prime is an indication of the antiparallel combination of core orbital momentum and spin. Thus, for the lowest excited states of neon, two possibilities are obtained from 3s[3/2], where $J = 2$ or 1, and can be written $3s[3/2]_2$, $3s[3/2]_1$; while $3s'[1/2]_0$, $3s'[1/2]_1$ are the other two possibilities. Two of these levels are metastable with long radiative

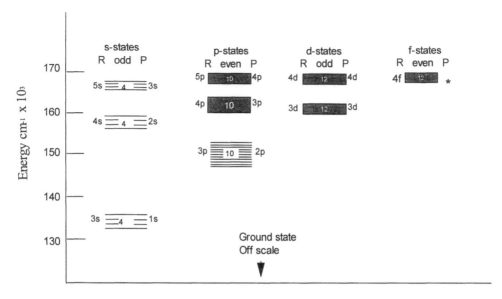

Figure 3 Plot of the lowest excited states of the neon atom displaying relative energy levels, the number of terms, and both Racah (R) and Paschen (P) notations. *, Some terms in this group have special Paschen symbols.

lifetimes, since they have $J = 0$ and $J = 2$, and cannot undergo an electric dipole transition to the ground state with $J = 0$, because of the selection rule ($\Delta J = J' - J = 0, \pm 1$; excluding 0-0). The 3p[K] configurations ($J_c = 3/2, 1/2$) lead to 10 possible states.

The change in the angular momentum of the core from $J_c = 1/2$ to 3/2 affects the energy levels of neon by relatively small amounts, and thus the levels for different J_c are not shown separately on the neon chart (Fig. 3). The primes of the Racah symbols are also omitted. The separation of the energy levels with different J_c increases progressively with increasing atomic number and, therefore, is already noticeable in the case of argon. Note the grouping of the levels according to J_c values as well as the primes are shown in the argon chart (Fig. 4). The columns in the charts contain infinitely many energy levels with the Racah configuration numbers ns, ns′, np, np′, and so on. All unprimed sequences tend to the same limit; the energy of ionization for a singly ionized noble gas atom in the $^2P_{3/2}^0$ state. All primed sequences tend to a different limit, which represents the ionization energy required to produce an ion in the $^2P_{1/2}^0$ state. The latter energy is greater, according to Hund's rules, and the energy difference increases with increasing atomic number.

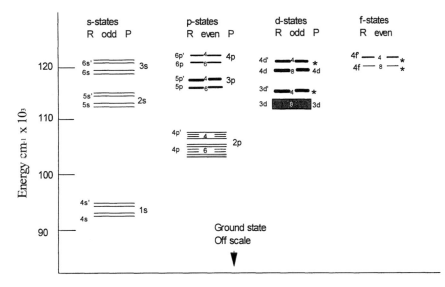

Figure 4 Plot of the lowest excited states of the argon atom displaying relative energy levels, the number of terms and both Racah (R) and Paschen (P) notations. *, Miscellaneous or improvised symbols.

Table 1 provides a guideline for the values of the angular momentum of the core, J_c; the resultant vector, K; the total angular momentum, J; and the number of terms in the J–L coupling scheme.

C. Assigned Neon Transitions

A series of measurements has previously been carried out to observe the laser-assisted OG signals of neon in an Fe–Ne hollow cathode lamp in the 337–598 nm wavelength interval [62]. In the present work, a total of 220 OG transitions were observed. Some of the observed neon transitions were revisited and looked at from a theoretical perspective. The approach was to confirm the earlier optogalvanic assignments by comparing the experimental wavenumbers with calculated wavenumbers and to extend the earlier analysis by identifying and assigning new transitions. The transitions that were experimentally observed were analyzed using the approach described below.

The vacuum wavenumbers of the energy levels for the neon atom were taken from Bashkin and Stoner [57] for ns, n′s′, np, n′p′, nd and n′d′; and the vacuum wavenumbers for the allowed transitions ($\Delta J = 0, \pm 1$) were determined

Table 1 Calculation of Quantum Numbers and the Numbers of Terms in a $J-L$ Coupling Scheme for the Noble Gases

Excited state	J_c ($l_c = 1$, $s_c = 1/2$)	l value (outer)	Allowed K values	$J = K \pm 1/2$	Number of terms
$2p^53s$	$J_c = 3/2, 1/2$	0	3/2, 1/2	2,1; 1,0	4
$2p^53p$	$J_c = 3/2, 1/2$	0, 1	5/2, 3/2, 1/2	3,2; 2,1; 1,0	$6 + 4 = 10$
			3/2, 1/2	2,1; 1,0	
$2p^53d$	$J_c = 3/2, 1/2$	0, 1, 2	7/2, 5/2, 3/2, 1/2	4,3; 3,2; 2,1; 1,0	$8 + 4 = 12$
			3/2, 1/2	2,1; 1,0	

by subtracting the energies of the respective levels, $s-p$, $s' - p$, . . . and so on. These vacuum wavenumbers were then converted into air wavelengths using Edlen's formula (Eq. 26) for the refractive index, η. The Striganov and Sventitskii [58] atlas was used as reference for the neon air wavelengths. The calculated vacuum wavenumbers, the air refractive indices, the calculated air wavelengths, together with the air wavelengths reported in the atlas, and the explicitly assigned optogalvanic transitions have been summarized in Table 2.

Based on Table 2, it is evident that the calculated values agree well with those reported in the literature, thereby confirming the accuracy of the present theoretical treatment. All but the last three transitions have been assigned in our

Table 2 Calculated and Observed Neon Optogalvanic Spectral Lines and Their Assignments

Calculated (ν_{vac} in cm^{-1})	η	Calculated[a] (λ_{air} Å)	Atlas[b] (λ_{air} Å)	Observed[c] (λ_{air} Å)	Transition
29665.8806	1.00028726	3369.9078	3369.9069	—	$3s[1.5]^0 2\text{-}4p'[0.5]1$
29615.4275	1.88828720	3375.6490	3375.6489	—	$3s[1.5]^0 2\text{-}4p'[1.5]1$
29249.3118	1.00028681	3417.9035	3417.9031	—	$3s[1.5]^0 1\text{-}4p'[1.5]2$
29248.4335	1.00028681	3418.0062	3418.007	—	$3s[1.4]^0 1\text{-}4p'[0.5]1$
29197.9804	1.00028676	3423.9126	3423.9120	—	$3s[1.5]^0 1\text{-}4p'[1.5]1$
28996.5111	1.00028655	3447.7028	3447.7022	—	$3s[1.5]^0 2\text{-}4p\ [1.5]2$
28970.7809	1.00028652	3450.7650	3450.7641	—	$3s[1.5]^0 2\text{-}4p\ [1.5]1$
28942.0143	1.00028649	3454.1949	3454.1942	—	$3s[1.5]^0 2\text{-}4p\ [0.5]0$
28889.0801	1.00028644	3460.5243	3460.5235	—	$3s'[0.5]^0 0\text{-}4p'[0.5]1$
28857.2727	1.00028640	3464.3387	3464.3385	—	$3s[1.5]^0 2\text{-}4p\ [2/5]2$
28838.6270	1.00028638	3466.5787	3466.5781	—	$3s'[0.5]^0 0\text{-}4p'[1.5]1$
28788.8632	1.00028633	3472.5711	3472.5706	—	$3s[1.5]^0 2\text{-}4p\ [2.5]3$
28579.0640	1.00028612	3498.0640	3498.0632	—	$3s[1.5]^0 1\text{-}4p\ [1.5]2$
28553.3338	1.00028609	3501.2163	3501.2154	—	$3s[1.5]^0 1\text{-}4p\ [1.5]1$
28476.0316	1.00028601	3510.7212	3510.7207	—	$3s[1.5]^0 2\text{-}4p\ [0.5]1$
28439.8256	1.00028597	3515.1907	3515.1900	—	$3s[1.5]^0 1\text{-}4p\ [2.5]2$
28397.1683	1.00028593	3520.4712	3520.4714	—	$3s'[0.5]^0 1\text{-}4p'[0.5]0$
27819.8852	1.00028535	3593.5257	3593.5263	—	$3s'[0.5]^0 1\text{-}4p'[1.5]2$
27768.5538	1.00028530	3600.1687	3600.1694	—	$3s'[0.5]^0 1\text{-}4p'[1.5]1$
27699.2311	1.00028524	3609.1790	3609.1787	—	$3s'[0.5]^0 0\text{-}4p[0.5]1$
23469.0079	1.00028148	4259.7395	—	4259.77	$3p[2.5]3\text{-}13s[1.5]^0 1$
23435.2085	1.00028146	4265.8832	—	4265.89	$3p[2.5]2\text{-}10s'[0.5]^0 1$
20288.6972	1.00027915	4927.4772	—	4927.48	$3p[0.5]0\text{-}10s'[0.5]1$

[a] Calculated using the energy levels in Ref. 57 and Edlen's formula [56].
[b] From Ref. 58.
[c] Observed air wavelength in our laboratory not reported in the atlas [58].

Table 3 Optogalvanic Spectral Lines of Ar Atoms in a Laser Galvatron (301.800–320.800 nm)

Observed λ (nm)	Atlas λ (nm)	Calculated λ (nm)[a]	Δλ (nm)[b]	Transition
301.907	—	301.902	—	$4s[3/2]^0_2$-$13p[3/2]_2$
301.944	—	301.940	—	$4s[3/2]^0_2$-$13p[5/2]_2$
303.486	—	303.483	—	$4s[3/2]^0_2$-$12p[3/2]_1$
303.495	—	303.501	—	$4s[3/2]^0_2$-$12p[5/2]_3$
303.536	—	303.534	—	$4s[3/2]^0_2$-$12p[5/2]_2$
303.592	—	303.597	—	$4s[3/2]^0_2$-$12p[1/2]_1$
303.938	—	303.938	—	$4s[3/2]^0_1$-$17p[5/2]_2$
304.564	—	304.563	—	$4s[3/2]^0_1$-$16p[5/2]_2$
305.292	—	305.292	—	$4s[3/2]^0_1$-$15p[3/2]_2$
305.327	—	305.327	—	$4s[3/2]^0_1$-$15p[5/2]_2$
305.627	305.628	—	0.001	$4s[3/2]^0_2$-$11p[3/2]_1$
305.696	—	305.696	—	$4s[3/2]^0_2$-$11p[5/2]_2$
305.717	—	305.716	—	$4s[3/2]^0_2$-$11p[1/2]_1$
306.207	306.206	—	−0.001	$4s[3/2]^0_2$-$8p'[3/2]_2$
306.285	306.282	—	−0.003	$4s[3/2]^0_2$-$8p'[3/2]_1$
306.347	306.344	—	−0.003	$4s[3/2]^0_2$-$8p'[1/2]_1$
307.538	—	307.538	—	$4s[3/2]^0_1$-$13p[3/2]_2$
308.647	308.647	—	0.0	$4s[3/2]^0_2$-$10p[3/2]_1$
308.781	308.781	—	0.0	$4s[3/2]^0_2$-$10p[5/2]_3$
308.917	308.917	—	0.0	$4s[3/2]^0_2$-$10p[1/2]_1$
309.018	309.018	—	0.0	$4s[3/2]^0_1$-$12p[1/2]_0$
309.130	309.132	—	0.002	$4s[3/2]^0_2$-$6f'[5/2]_3$
309.175	—	309.177	—	$4s[3/2]^0_1$-$12p[3/2]_2$
309.181	—	309.178	—	$4s[3/2]^0_1$-$12p[3/2]_1$
310.012	310.010	—	−0.002*	$4s[3/2]^0_1$-$8f[5/2]_{2,3}$
311.066	311.066	—	0.0	$4s[3/2]^0_1$-$11p[1/2]_0$
311.406	—	311.405	—	$4s[3/2]^0_1$-$11p[3/2]_1$
311.497	311.497	—	0.0	$4s[3/2]^0_1$-$11p[1/2]_1$
311.620	311.622	—	0.002	$4s[3/2]^0_1$-$8p'[1/2]_0$
312.005	312.006	—	0.001	$4s[3/2]^0_1$-$8p'[3/2]_2$
313.080	313.080	—	0.0	$4s[3/2]^0_2$-$9p[3/2]_2$
313.106	313.104	—	−0.002	$4s[3/2]^0_2$-$9p[3/2]_1$
313.230	313.231	—	0.001	$4s[3/2]^0_2$-$9p[5/2]_2$
313.286	313.287	—	0.001	$4s[3/2]^0_2$-$9p[5/2]_3$
313.429	313.427	—	−0.002	$4s[3/2]^0_2$-$9p[1/2]_1$
314.264	314.260	—	−0.004*	$4s[3/2]^0_1$-$10p[1/2]_0$
314.540	314.542	—	0.002	$4s[3/2]^0_1$-$10p[3/2]_2$
314.560	314.563	—	0.001	$4s[3/2]^0_1$-$10p[3/2]_1$
314.823	314.820	—	−0.003	$4s[3/2]^0_1$-$10p[1/2]_1$
315.040	315.042	—	0.002	$4s[3/2]^0_1$-$6f'[5/2]_2$

Table 3 Continued

Observed λ (nm)	Atlas λ (nm)	Calculated λ (nm)[a]	Δλ (nm)[b]	Transition
315.152	315.152	—	0.0	$4s[3/2]^{0}_{2}$-$7f[5/2]_{3,2}$
315.229	315.229	—	0.0	$4s[3/2]^{0}_{2}$-$7f[3/2]_{1,2}$
315.957	315.955	—	−0.002	$4s[3/2]^{0}_{1}$-$8f[5/2]_{2}$
316.005	316.006	—	0.001	$4s[3/2]^{0}_{1}$-$8f[5/2]_{1,2}$
317.222	317.218	—	−0.004	$4s[3/2]^{0}_{2}$-$7p'[3/2]_{2}$
317.296	317.296	—	0.0	$4s[3/2]^{0}_{2}$-$7p'[1/2]_{1}$
317.371	317.371	—	0.0	$4s[3/2]^{0}_{2}$-$7p'[3/2]_{1}$
318.663	318.663	—	0.0	$4s[3/2]^{0}_{1}$-$9p[1/2]_{0}$
319.145	319.150	—	0.005	$4s[3/2]^{0}_{1}$-$9p[3/2]_{2}$
319.173	319.172	—	−0.001	$4s[3/2]^{0}_{1}$-$9p[3/2]_{1}$
319.492	319.493	—	0.001	$4s'[1/2]^{0}_{0}$-$11p[1/2]_{1}$
319.510	319.512	—	0.002[c]	$4s[3/2]^{0}_{1}$-$9p[1/2]_{1}$
320.036	320.039	—	0.003	$4s[3/2]^{0}_{2}$-$8p[3/2]_{2}$
320.086	320.084	—	−0.002	$4s[3/2]^{0}_{2}$-$8p[3/2]_{1}$
320.112	320.112	—	0.0	$4s'[1/2]^{0}_{0}$-$8p'[3/2]_{1}$
320.285	320.285	—	0.0	$4s[3/2]^{0}_{2}$-$8p[5/2]_{2}$
320.368	320.366	—	−0.002	$4s[3/2]^{0}_{2}$-$8p[5/2]_{3}$
320.749	320.750	—	0.001	$4s[3/2]^{0}_{2}$-$8p[1/2]_{1}$

[a] New OG spectral lines of Ar-atom assigned based on calculations using Eq. (26).
[b] Δλ = (Atlas-observed) value.
[c] MIT atlas [63].

laboratory earlier, but the wavelengths determined were either atlas or observed values. We have treated these same transitions theoretically using the method described above to calculate their wavelengths. During this process, to the best of our knowledge, three new neon OG transitions were identified and assigned and are presented in Table 2.

D. Assigned Argon Transitions

The OG spectrum of argon, recorded using a laser galvatron, was recorded simultaneously with the LIF spectrum of the OH radical in a propane–air flame. The spectral region 290.886–320.750 nm was covered by using several dyes. Equation (24) was used to calculate the wavelengths of the argon lines in air. These values were then converted to vacuum wavenumbers v (cm^{-1}) using Eqs. (25) and (26). Table 3 contains a summary of all the argon transitions observed with the laser galvatron in the 301.800–320.800 nm wavelength region. The transitions were calculated based on the energy levels of the ns, n's', np . . . atomic levels taken from the Grotrian diagram compliation [57] and subsequently compared with the values of the transitions given in the atlas [58] and the NIST database compilation [59].

VI. MODELING OF OPTOGALVANIC WAVEFORMS

A. Optogalvanic Waveforms Associated with the Atomic Neon Transitions in a Hollow Cathode Discharge

For the neon 1s–2p transition, using the Paschen notation, there are four 1s states and ten 2p states, which yield a total of 30 allowed radiative transitions. These allowed transitions are shown in Figure 5. Of the four 1s states, two (1s$_3$ and 1s$_5$) are metastable with radiative lifetimes on the order of 1s, while the other two (1s$_2$ and 1s$_4$) decay to the ground state within 1 ns. Lawler [64] has shown that the optogalvanic effect in the positive column could be modeled in terms of the perturbed rate equations, including the dynamic impedance. His theoretical model showed excellent agreement with the experimental results for helium. Doughty and Lawler [65] extended the model containing terms of the perturbed-rate equations, including the dynamic impedance proposed by Lawler, [64] to neon in the visible region and obtained good agreement with a simplified description of its fairly complex energy level system. Kane [66] has modeled the saturated laser optogalvanic effect in neon and explained the variation in the optogal-

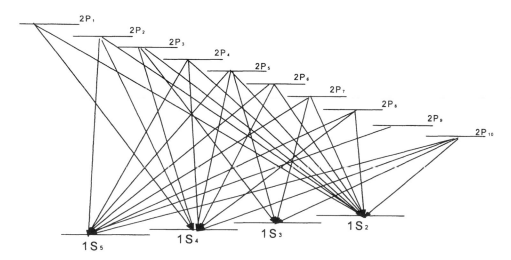

Figure 5 Energy level diagram shows all allowed neon transitions from the first to the second excited state. 1S–2P, Paschen notation.

vanic line profile with laser power. The improved theoretical understanding of the optogalvanic effect in a variety of discharges and under different conditions has made it an important quantitative diagnostic technique. However, as models have progressed from a simple basic structure to a more comprehensive inclusion of important discharge processes, it is apparent that some of the descriptions did not agree well with the experimental observations [67]. Such discrepancies led Stewart et al. [68] to develop a model that included and identified specifically and quantitatively the dominant effects of electron collisional transfer and resonance radiation wall losses. The authors included separate rate equations (with about 150 parameters) for neon atoms, in each of the four 1s states, and discussed the agreement of the model with their measured optogalvanic signals for a number of transitions in the visible region involving $1s_4$ and $1s_5$ states, and also over a wide range of discharge gas pressures and tube currents. Han et al. [25] subsequently used a theoretical model involving a handful of parameters (four) to study the optogalvanic effect associated with the signal of neon that Stewart et al. [68] had analyzed earlier. They clearly identified and quantitatively characterized the dominant physical processes that contributed to the production of the OG signal. Literature survey revealed, to the best of our knowledge, that so far this type of modeling has not been performed systematically for the atomic neon transitions that fall in the ultraviolet spectral region. This prompted us to carry out a comprehensive study of the time-resolved waveforms of neon transitions in the ultraviolet region and thereby attempt to identify and qualitatively characterize the dominant physical processes contributing to the production of the OG signal.

To illustrate our ideas, we have chosen the neon $3s[3/2]_1^0-5p[5/2]2$ ($1s_4-4p_8$ in Paschen notation) transition at 301.735 nm recorded at various discharge currents: 0.6, 0.8, 1.0, 1.2, 1.4, and 1.6 mA. In case of the $1s_4 \rightarrow 4p_8$ transition for neon, there are a total of 30 optically allowed transitions, one of which involves the $1s_4$ and $4p_8$ states and is presented in the energy level diagram shown in Figure 6. On laser excitation from the $1s_4$ to the $4p_8$ state, the excited neon atoms relax to optically allowed $1s_2$, $1s_3$, and $1s_5$ states. The hollow cathode lamp we have used for recording the neon OG spectra presented here is a commercial Fe–Ne hollow cathode lamp. The wavelengths of the OG-spectral lines of neon were calibrated using an interference pattern recorded with an etalon. Our main aim has been to fit the signal independently and to obtain parameters that yield information about the decay rates, the amplitudes associated with the energy states involved in the transition, and also to determine the instrumental time constant.

To perform a fit of the time-resolved OG waveforms, we have to consider (based on Fig. 6) the involvement of four ($1s_4$, $1s_2$, $1s_3$, and $1s_5$) optically decaying states. According to this scheme, the observed waveform can be fitted to the following expression

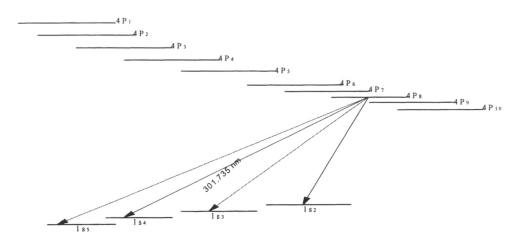

Figure 6 Schematic representation of the energy level diagram shows the neon ($1s_4$–$4p_8$) transition at 301.735 nm and the associated optically allowed transitions.

$$s(t) = \frac{a}{1 - b\tau} (e^{-bt} - e^{-t/\tau}) + \frac{c}{1 - d\tau} (e^{-dt} - e^{-t/\tau})$$
$$+ \frac{e}{1 - f\tau} (e^{-ft} - e^{-t/\tau}) + \frac{g}{1 - h\tau} (e^{-ht} - e^{-t/\tau})$$

$$(27)$$

where a, c, e, and g are the positive, negative, positive, and negative amplitudes, respectively; and b, d, f, and h are the decay rates for the states involved in the transition, respectively. τ is the instrumental time constant of the waveform representing the OG signal. Although, in principle, one can include all the four terms to fit the observed optogalvanic signal, in the present work we have used only the first two terms in Eq. (27) to fit the observed neon OG waveforms satisfactorily for three (0.6, 0.8, and 1.0 mA) currents. A nonlinear least-squares program was used to fit Eq. (27). The observed and fitted time-resolved optogalvanic waveforms for 0.6 mA, 0.8 mA, and 1.0 mA currents, respectively, are shown in Figure 7. The values of the fitted constants along with their standard deviations for the three currents are given in Table 4. Owing to a high correlation between the amplitudes a and c, we have considered the value of c to be a constant in the least-squares fit for the three currents.

The observed time-resolved OG waveform for 1.2 mA current was likewise fitted to the first two terms in Eq. (27), while the remaining two OG waveforms recorded at 1.4 and 1.6 mA currents, respectively, were satisfactorily fitted to the first three terms in Eq. (27). The observed and fitted time-resolved optogalvanic waveforms for 1.2, 1.4, and 1.6 mA currents, respectively, are shown in Figure 8. The values of the fitted constants, along with their standard deviations, are given in Table 5.

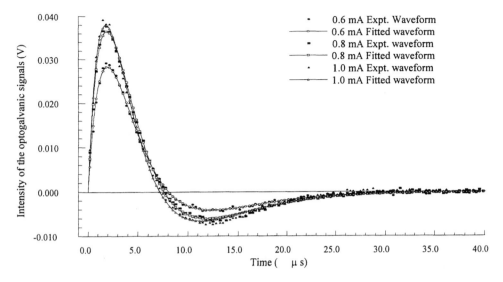

Figure 7 Fitted and observed optogalvanic waveforms of the Ne-transition ($1s_4$–$4p_8$) at 301.735 nm for three (0.6, 0.8, and 1.0 mA) different currents.

Owing to a high correlation between the amplitudes a and c for the waveform recorded at 1.2 mA, and that among the amplitudes a, c, and e for the waveforms recorded at currents of 1.4 and 1.6 mA, in the present work the values of c and e were fixed in the least-squares fit. Based on Tables 4 and 5, it is evident that the parameters b and d (representing the decay rates) show increasing and decreasing trends, respectively, as the discharge current is increased. However, the decay rate parameter f remains constant as the current is increased.

Table 4 Fitted Parameters Obtained from a Nonlinear Least-Squares Fit of the Observed Optogalvanic Waveforms of the Neon Transition at 301.735 nm for 0.6, 0.8, and 1.0 mA Currents

Parameter	Value (for 0.6 mA)	Value (for 0.8 mA)	Value (for 1.0 mA)
τ	5.47(11) (μs)	5.11(11) (μs)	5.13(14) (μs)
a	8.855(49) $\times$ 10^{-1} (V)	6.705(63) $\times$ 10^{-1} (V)	8.196(91) $\times$ 10^{-1} (V)
b	4.640(44) $\times$ 10^{-1} (μs^{-1})	5.133(63) $\times$ 10^{-1} (μs^{-1})	5.318(81) $\times$ 10^{-1} (μs^{-1})
c	-6.78099×10^{-1} (V)[a]	-4.11148×10^{-1} (V)[a]	-5.35445×10^{-1} (V)[a]
d	3.698(16) $\times$ 10^{-1} (μs^{-1})	3.361(12) $\times$ 10^{-1} (μs^{-1})	3.654(18) $\times$ 10^{-1} (μs^{-1})

[a] Fixed in the least-squares fit.

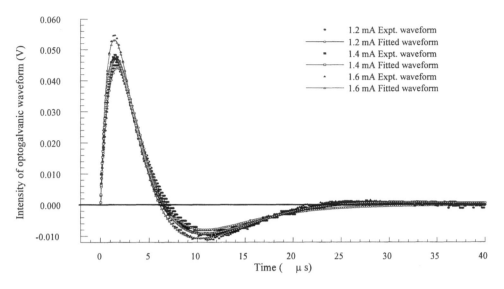

Figure 8 Fitted and observed optogalvanic waveforms of the Ne-transition ($1s_4$–$4p_8$) at 301.735 nm for three (1.2, 1.4, and 1.6 mA) different currents.

The average value of τ obtained from the data set associated with the neon 301.735 nm transition is 5.15 μs $\pm$ 5%. A good agreement between different waveform fits is not always possible because the discharge lamp is part of the electrical circuit and has a variable nature. The slightest change in conditions, such as temperature, current setting, laser power setting, or wavelength tuning,

Table 5 Fitted Parameters Obtained from a Nonlinear Least-Squares Fit of the Observed Optogalvanic Waveforms of the Neon Transition at 301.735 nm for 1.2, 1.4, and 1.6 mA Currents

Parameter	Value (for 1.2 mA)	Value (for 1.4 mA)	Value (for 1.6 mA)
τ	4.87(32) (μs)	5.87(15) (μs)	5.40(14) (μs)
a	8.60(25) $\times$ 10^{-1} (V)	6.63(17) $\times$ 10^{-1} (V)	7.22(21) $\times$ 10^{-1} (V)
b	5.69(23) $\times$ 10^{-1} (μs^{-1})	9.20(22) $\times$ 10^{-1} (μs^{-1})	1.008(26) (μs^{-1})
c	-5.29626×10^{-1} (V)[a]	-4.35617×10^{-1} (V)[a]	-4.36331×10^{-1} (V)[a]
d	3.699(43) $\times$ 10^{-1} (μs^{-1})	1.791(27) $\times$ 10^{-1} (μs^{-1})	1.799(29) $\times$ 10^{-1} (μs^{-1})
e		2.68471 $\times$ 10^{-1} (V)[a]	2.68447 $\times$ 10^{-1} (V)[a]
f		1.487(29) $\times$ 10^{-1} (μs^{-1})	1.488(32) $\times$ 10^{-1} (μs^{-1})

[a] Fixed in the least-squares fit.

can cause a large change in the plasma response, and thus experimental time constants can be far from constant within data sets.

Figure 9 represents a plot of the parameter b versus current (I) in mA. It indicates the linear relationship between relaxation rate and discharge current. Recalling Eq. (20), $b = \Gamma_{L3} + I(P_{L3} + P_{L3s})$, where Γ_{L3} is the decay rate of the other s states, I is the current, and P is the rate parameter related to the total electron collisional ionization cross section. The intercept (Γ_{L3}) of the line is the effective decay rate (0.37/μs), implying an effective lifetime of 2.8 μs for the upper states ($1s_2$, $1s_3$, or $1s_5$). The parameter d associated with the lower state ($1s_4$) has a relatively constant value of 0.36/μs ± 5% and implies an intercept (Γ_{LI}) of 0.36/μs corresponding to a lifetime of 2.8 μs. Thus, within our experimental uncertainty, the effective decay rates of both the lower and upper states have the same value: 2.8 μs. The effective lifetime derived from the s states is much longer than the radiative lifetime of the $4p_8$ state (~2 ns).

The values of the electron ionization rate parameters P_{L1} and P_{L3} are related to electron ionization cross sections in complicated ways. These depend upon electron collisional ionization cross sections, the electron energy distributions in

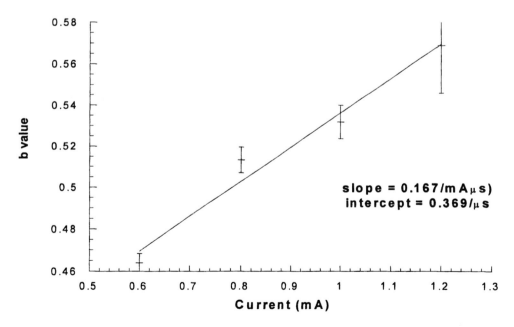

Figure 9 A plot of the parameter b for the Ne transition at 301.735 nm as a function of discharge current I exhibits direct proportionality between the relaxation rate and the current.

the discharge lamp, and the discharge geometry. In cases where redistribution of the 1S block occurs, they are the combination electron ionization rate parameters plus electronic collisional transfer rate parameters. The slope of the line in Figure 9, therefore, cannot be directly converted into the electron collisional ionization cross section associated with the upper state. However, the slope does contain information about the electron collisional ionization and electronic collisional transfer rates that can aid theoretical calculations in discharge physics.

B. Optogalvanic Waveforms Associated with the Atomic Argon Transitions in a Hollow Cathode Discharge

The optogalvanic transitions of atomic argon prove very useful, especially in the UV, in calibrating tunable dye laser wavelengths because often weak emission lines could be excited via the optogalvanic effect. Argon and neon are both attractive choices for use in wavelength calibration of spectra because gaseous or volatile elements are present in a sample discharge volume at higher concentrations and hence the associated OG signals are easier to measure than those of less volatile elements. Furthermore, commercial hollow cathode lamps are generally filled with argon or neon as buffer gas and therefore these atoms are readily available. The optogalvanic spectrum of atomic argon in the near-ultraviolet region plays an important role in our understanding of combustion processes because it overlaps with the electronic spectrum of the hydroxyl radical. OG signals from argon have been reported in a fragmented manner in the literature. Yet so far (to the best of our knowledge) there is no systematic study of the time-resolved waveforms of argon in the UV using a model that includes and identifies specifically and quantitatively the dominant effects of electron collisional transfer and resonance radiation wall losses.

C. Argon UV Transition at 297.905 nm

In the present work, we have chosen the argon II $4s^2P - 4p'^2P^\circ$ ($J = 1/2-1/2$) transition at 297.905 nm and recorded its time-resolved OGE waveform at two (0.25 and 0.5 mA) different currents in order to understand the collisional ionization of the excited state of argon atoms in a gas discharge plasma. A typical time-resolved experimental OG waveform extending from 0 to 250 μs for 0.25 mA and from 0 to 400 μs for 0.5 mA, together with the corresponding fitted curves, are shown in Figures 10 and 11. Unlike the optogalvanic signals of neon transitions produced in the hollow cathode tube, which have less noise, the optogalvanic signals of argon have oscillations associated with it. It is believed to be due to the electrical resistance of the lamp and its associated circuit, which, when combined with the negative resistance, creates characteristic oscillations. Such oscillations have been found by Lee et al. [69] while studying the optogalvanic signals of tin with commercial hollow-cathode lamps. The observed waveforms

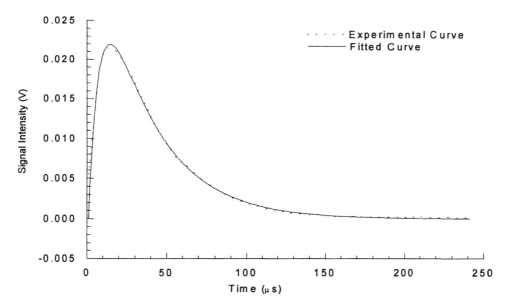

Figure 10 Fitted and observed waveforms of the argon II $4s^2P-4p'^2P°$ transition at 297.905 nm for 0.25 mA.

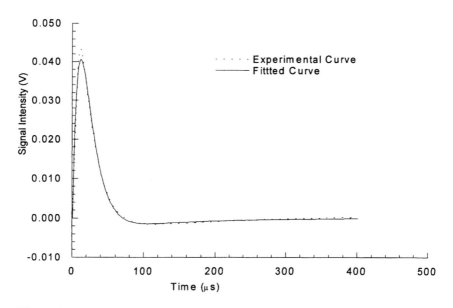

Figure 11 Fitted and observed waveforms of the argon II $4s^2P-4p'^2p°$ transition at 297.905 nm for 0.50 mA.

were fitted to the expression given in Eq. (27). For the optogalvanic signal recorded at a discharge current of 0.25 mA, only the first term in Eq. (27) was
used in the nonlinear least-squares fit to obtain the parameters τ (instrumental
time constant), a (negative amplitude), and b (the decay rate). Curve fitting with
these three parameters gave excellent results, as shown in Figure 10. The fitted
parameters, along with their standard deviations, are given in Table 6. For the
OG signal recorded at a discharge current of 0.5 mA, the two terms in Eq. (27)
were likewise used to obtain a satisfactory fit between the observed and fitted
waveforms. Owing to high correlation between the amplitudes (a and c) and the
decay rates (b and d), only the values of three parameters (τ, a and b), along
with their standard deviations, are given in Table 6. A plot of the values of the
parameter b vs. current would not be convincing evidence for linearity, since
there are only two points (which would result in an automatic straight line), but
the increase in amplitude with increasing current is obvious by nothing the a
parameter values in Table 6.

D. Argon UV Transition at 320.366 nm

One more OG waveform study of the argon $4s[3/2]^{0}_{2}-8p[5/2]_{3}$ transition at
320.366 nm has been chosen in the present chapter to allow us to understand the
collisional ionization of a different excited state in a gas discharge plasma. Time-
resolved OG waveforms of three different current (0.2, 0.4, and 0.6 mA) are
shown in Figure 12. The observed waveforms were fitted to the expression given
in Eq. (27). The first two terms were used in the nonlinear least-squares fit to
obtain the relevant parameters, namely, τ (instrumental time constant), a (positive

Table 6 Fitted Parameters Obtained from a Least-
Squares Fit of the Observed Optogalvanic Waveforms
of the Argon Transition at 297.905 nm for Currents
0.25 and 0.5 mA

	Current	
Parameter	0.25 mA	0.50 mA
$\tau(\mu s)$	3.44(23)	4.39(59)
$a(V)$	0.03322(97)	0.091(8)
$b(\mu s^{-1})$	6.11(23) $\times$ 10^{-2}	1.28(18) $\times$ 10^{-1}
$c(V)^a$		$-0.004(1)$
$d(\mu s^{-1})$		1.81(54) $\times$ 10^{-2}

[a] Fixed in the least-square fit.

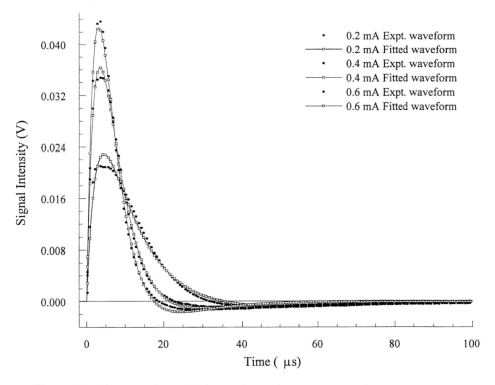

Figure 12 Fitted and observed OG waveforms of the argon $4s[3/2]_2^0-8p[5/2]_3$ transition at 320.266 nm for three (0.2, 0.4, and 0.6 mA) different currents.

Table 7 Fitted Parameters Obtained from a Nonlinear Least-Squares Fit of the Observed Optogalvanic Waveforms of the Argon Transition at 320.366 nm for Currents 0.2, 0.4, and 0.6 mA

Parameter	Current		
	0.2 mA	0.4 mA	0.6 mA
$\tau(\mu s)$	19(2)	8.33 (75)	4.97 (65)
a (V)	0.326(45)	0.294 (27)	0.207 (30)
b (μs^{-1})	0.419(44)	0.416 (30)	0.406 (53)
c (V)[a]	−0.039	−0.053	−0.021
d (μs^{-1})	0.078(64)	0.1336 (74)	0.1052 (83)

[a] Fixed in the least-square fit.

amplitude), and b and d (the decay rate constants). The fitted parameters, along with their standard deviations, are given in Table 7.

VII. SUMMARY AND CONCLUSIONS

The optogalvanic effect gives rise to efficient, sensitive, and economical spectroscopic and detection methods. The OGE complements other optical techniques, such as traditional absorption and fluorescence spectroscopy and optoacoustic spectroscopy. It is intrinsically more sensitive than absorption spectroscopy, since the OGE is based upon a signal whose background is zero, while absorption spectroscopy records a small variation superimposed on a large signal. When compared to LIF, the OGE has the advantage of not being affected either by the background signal due to the discharge luminosity or by the scattering of the excitation signal. This becomes particularly important when the fluorescence is detected at the same wavelength as the absorbed radiation.

Of the various laser-assisted spectroscopic and detection methods, the optogalvanic effect is economical, since commercially available lamps can be used. A glow discharge is an inexpensive way to obtain large densities of excited states in volatile elements, especially those involving metastable states. Gaseous states of refractory elements are easily produced in hollow cathode lamps via sputtering. Rydberg states can also be studied by the OG technique, since these can be reached by exciting radiative transitions starting from metastable levels that are well populated in discharges, particularly in the noble gases. In general, it has been demonstrated that OG techniques perform favorably compared to traditional optical techniques, with the extra advantage of inherent simplicity. However, they cannot provide the maximum signal-to-noise ratio in detecting spectra.

The present investigation began by recording extensive LOG spectra of neon and argon in the near-ultraviolet region, where the known and well-documented neon lines were used to calibrate the rotationally resolved LIF spectrum of the hydroxyl radical, which in turn was used to calibrate unknown argon OG spectral transitions in the near UV. By adopting this method, several new argon OG transitions were identified and assigned using the J–L coupling scheme. A dependable set of OG transitions suitable for calibration of tunable lasers in the visible and near-UV regions are now available, which is so vital for combustion and atmospheric monitoring purposes. To the best of our knowledge, all the experimentally observed argon OG transitions in the present study, spanning the wavelength region 301.800–320.800 nm (a total of 58 lines), have been documented optogalvanically in our laboratory.

The detailed investigation presented here was centered around the optogalvanic effect associated with a DC hollow cathode lamp, which is a special form of DC discharge characterized by two main regions: the cathode dark region (where 90% of the voltage drop takes place within a region <1 mm) and an expanded negative glow region (showing ~5% voltage drop). It was also convenient to use commercial hollow cathode lamps because they are relatively inexpensive, and can be found containing many elements in the periodic table, and also possess very low noise levels (on the order of shot noise). By irradiating the entire cathode cavity, the signals produced were in fact a combination of many of the different processes described in the theory section. As expected, the prominent process determining the OGE signal, in the current regimen of the study, was consistent with the electron collisional ionization of the buffer gas (either neon or argon). As previously stated, either superposition or perturbation was used to account for other processes evident in the recorded waveforms.

We found the following general trends:

The amplitude of the OG signal increased with increasing discharge current until saturation due to an increase in electron population was reached.

The amplitude of the OG signal increased with increasing laser power (for a fixed discharge current) until optical saturation was reached.

The relaxation rates depended linearly on the current.

Saturation due to increased laser power for a fixed current can be explained as optical saturation. For each neon and argon transition studied, whenever there were enough data recorded for a detailed analysis, the linear relationship between relaxation rates and the discharge current was confirmed.

The results of this investigation support the theory that collisional ionization is the dominant factor in producing the OGE signal in the low-current regimen (0.2–3.0 mA). This conclusion is based on the fact that the ionization stage of the OG process is consistent with the collisional ionization model and the fit between the data and the mathematical model was impressively close.

In addition, the analysis performed to understand time-dependent OG waveforms generated at different currents yielded quantitative information about relaxation rates and also determined parameters related to the electron collisional cross sections of the states involved in the transitions.

Perhaps one of the greatest disadvantages of the OGE is that the useful regimen of a discharge is limited to a narrow range of parameters. The condition becomes particularly delicate when the current has to be continually increased (as is necessary for sputtering in hollow cathode lamps). When performing OG experiments in commercial lamps, each lamp has a small current and voltage range, with only minimal changes in pressure possible.

There are times, however, when the OG technique is the best detection method available. One example would be when absorption spectroscopy is not efficient, because the sample is optically thin and there is no fluorescence signal available, as is often the case in molecular infrared spectroscopy.

The greatest difficulty encountered in the present investigation was the interference from the electrical resonance oscillations while performing waveform studies in the low-current regimen (0.2–3.0 mA). Working around the inconvenience caused by these oscillations led to new insights into the electrical properties of the hollow cathode lamp. Our study of the oscillations themselves provided a method of determining the current limits within which dependable OG data could be taken. It is envisioned that a more specific study of the oscillations themselves will not only produce more information about the OG phenomenon but also yield new ways in which the effect may be turned into an advantage to allow for a better understanding of discharge phenomena. This specific goal was outside the scope of the present investigations, however, and is recommended as a subject for further study. It is hoped that the new quantitative information pertaining to the optogalvanic effect provided here can be used to gain better understanding of plasma and gaseous media.

ACKNOWLEDGMENT

We acknowledge the many helpful discussions and critique of the manuscript provided by Dr. J.C. Travis of the National Institute of Standards and Technology (Gaithersburg, MD).

REFERENCES

1. D Popescu, ML Pascu, CB Collins, GW Johnson, I Popescu. Phys Rev A 8: 1666, 1973.
2. GS Hurst, MG Payne, SD Kramer, JP Young. Rev Mod Phys 51: 767, 1979.
3. JEM Goldsmith, JE Lawler. Optogalvanic spectroscopy. Contemp phys 11(2): 235–248, 1981.
4. JC Travis, GC Turk, JR DeVoe, PK Schenck, and CA van Dijk. Prog Analyt Atom Spectrosc 7: 199, 1984.
5. PD Foote, FL Mohler. Photo-electric ionization of caesium vapor. Phys Rev 26: 195, 1925.
6. FM Penning, Physica 8: 137, 1928.
7. A Terenin, Photoionization of salt vapors. Phys Rev 36: 147–148, 1930.
8. A Garscadden, P Bletzinger, EM Friar. J Appl Phys 35: 3432, 1964.
9. RB Green, RA Keller, GG Luther, PK Schenck, JC Travis. Galvanic detection of optical absorptions in a gas discharge. Appl Phys Lett 29: 727–729, 1976.
10. RA Keller, R Engleman, Jr., BA Palmer. Appl Opt 19: 836, 1980.

11. NJ Dovichi, DS Moore, RA Keller, Appl Opt 21: 1468, 1982.
12. BM Suri, R Kapoor, GD Saksena, PRK Rao. Opt Commun 52: 315, 1985.
13. F Babin, P Camus, JM Gagne, P Pillet, J Boulmer. Opt Lett 12: 468, 1987.
14. DS King, PK Schenck, KC Smyth, JC Travis. Appl. Opt 16: 2617, 1977.
15. MH Begemann, RJ Saykally. Opt Commun 40: 277, 1982.
16. JR Nestor, Appl Opt 21: 4154, 1982.
17. BR Reddy, P Venkteswarlu, MC George. Opt Commun 75: 267, 1990.
18. M Duncan, Devonshire. Inst Phys Conf Ser 113(6): 207, 1990.
19. Y Oki, T Izuha, M Maeda, C Honda, Hasegawa, H Futani, J Izumi, K Masuda. Jpn J Appl Phys 30: L1744, 1991.
20. M Hippler, J Pfab. Opt Commun 97: 347, 1993.
21. KC Smyth, PK Schenck. Chem Phys Lett 55: 466, 1978.
22. VM Gusev, ON Kompanets. Sov J Quantum Electron 17: 1515, 1978.
23. X Zhu, AH Nur, P Misra. J Quant Spectrosc. Radiat Transfer 52: 167, 1994.
24. M-C Su, RR Ortiz, DL Monts. Opt Commun 61: 257, 1987.
25. XL Han, W Wisehart, SE Conner, M-C Su, DL Monts. Collisional ionization of excited state neon in a gas discharge plasma. Contrib Plasma Phys 34: 439–452, 1995.
26. B Barbieri, N Beverini, A Sasso. Optogalvanic spectroscopy. Rev Mod Phys 62(3): 612, 1990.
27. RS Stewart, KI Hamad, KW McKnight. In: RS Stewart, JE Lawler, eds. Optogalvanic Spectroscopy. Institute of Physics Conference Series no 113. Bristol, UK: Institute of Physics, p 89, 1990.
28. GC Turk, JC Travis, JR DeVoe. Anayltical flame spectrometry with laser enhanced ionization. Anal Chem 50(6): 817–820, 1978.
29. GC Turk. The application of laser enhanced ionization to analytical flame spectrometry, Ph.D. dissertation, University of Maryland, College Park, MD, 1978.
30. PK Schenck, JC Travis, GC Turk, TC O'Haver. Laser-enhanced ionization flame velocimeter. Appl Spectrosc 36(2): 168, 1982.
31. ML Skolnick. Use of plasma tube impedance variations to frequency stabilize a CO_2 laser. IEEE J Quant Electron QE(6): 139–140, 1970
32. RB Green, RA Keller, GG Luther, PK Schenck, JC Travis. Use of an opto-galvanic effect to frequency-lock a continuous wave dye laser, IEEE J Quant Electron 13: 63, 1977.
33. CP Ausschnitt, GC Bjorklund, RR Freeman. Hydrogen plasma diagnostics by resonant multiphoton optogalvanic spectroscopy. Appl Phys Lett 33: 851–853, 1978.
34. RW Shaw, CM Barshick, LW Jennings, JP Young, JM Ramsey. Mass Spectrum. 10: 316, 1996.
35. CR Webster, CT Rettner. Laser optogalvanic spectroscopy of molecules. Laser Focus 19: 41–51, 1983.
36. CR Webster, RT Menzies. Infrared laser optogalvanic spectroscopy of molecules. J Chem Phys 78(5): 2121, 1983.
37. HH Telle. Optogalvanic spectroscopy of molecules and complexes. In: RS Stewart, JE Lawler, eds. Proceedings of the Second International Meeting on Optogalvanic Spectroscopy and Allied Topics. Glasgow: Strathclyde University, 1990.

38. R Walkup, RW Dreyfus, P Avouris. Laser optogalvanic detection of molecular ions. Phys Rev Let 50(23): 1846, 1983.
39. E Miron, I Smilanski, J Liran, S Lave, G Erez. Dynamic optogalvanic effect in rare gases and uranium. IEEE J Quan Electron QE-15, 194–196, 1979.
40. N Matsuno. Time-resolved spectroscopy of collisional ionization dynamics in neon discharge plasma,'' Master's Thesis, Butler University, 1997.
41. R Shuker, M Hakham-Itzhaq. Optogalvanic spectroscopy of plasma processes and autoinization levels. In: RS Stewart, JE Lawler, eds. Optogalvanic Spectroscopy. Institute of Physics Conference Series 113. New York: Institute of Physics, 263, 1990.
42. JE Lawler, AI Lerence Ferguson, JEM Goldsmith, DJ Jackson, SL Schawlow. Doppler-free intermodulated optogalvanic spectroscopy. Phys Rev Let 42(16): 1046–1049, 1979.
43. Abhilasha, S Qian, DL Monts. Laser optogalvanic detection of airborne uranium particles. Appl Phys B 65: 625–632, 1997.
44. DL Monts, Abhilasha, S Qian, D Kumar, X Yao, SP McGlynn. Comparison of atomization sources for a field-deployable laser optogalvanic spectrometry system. J Thermophys. Heat Transf 12(1): 66, 1998.
45. Demtroder, W. Laser Spectroscopy, 2nd ed. New York: Springer-Verlag, 405–407, 1996.
46. C Haridass, Y-B She, H Major, P Misra. Proceeding of the International Conference on Laser '97, New Orleans, LA. McLean, VA: STS Press, 514–521, 1998.
47. P Misra, C Haridass, H Major. Laser optogalvanic spectroscopy of neon and argon in a discharge plasma and its significance for microgravity combustion. Proceedings of the 5th International Microgravity Combustion Workshop, Cleveland, OH, 205, 1999.
48. MF Broglia, F Catoni, A Montone, P Zampetti. Galvanic detection of laser photoionization in hollow-cathode discharges: experimental and theoretical study. Phy Rev A 36(2): 705–714, 1987.
49. M Brogia, F Catoni, P Zampetti. Temporal behavior of the optogalvanic signal in a hollow cathode lamp. J Phys (Paris), C7–44, 479, 1983.
50. XL Han, Private communication. Butler University, 1999.
51. JJ Brehm, WJ Mullin. Introduction to the Structure of Matter. New York: John Wiley & Sons, 528, 1989.
52. HE White. Introduction to Atomic Spectra. New York: McGraw-Hill, 203–205, 274–277, 1934.
53. R Shuker, A Ben-Amar, G Erez. Theoretical and experimental study of the resonant optogalvanic effect in neon discharge. J Phys (Paris), C7–44: 35, 1983.
54. A Ben-Amar, G Erez, RJ Shuker. J Appl Phys 54: 3688, 1983.
55. DH Dieke, HM Crosswhite. J Quant Spectrosc. Radiat. Transfer. 2: 97–199, 1961.
56. BJ Edlen. J Opt Soc Am 43: 339–344, 1953.
57. S Bashkin, JO Stoner, Jr. Atomic Energy Levels and Grotrian Diagrams, vols. 1, 2. New York: North-Holland, 1978.
58. AR Striganov, NS Sventitskii. Tables of Spectral Lines of Neutralized and Ionized Atoms. New York: Plenum, 1968.
59. Sugar, J. Private communication, NIST, 1998.

60. BA Lengyel. Introduction to Laser Physics. New York: John Wiley & Sons, 177, 1967

61. CE Moore. Atomic energy levels. National Standards Reference Data Service, National Bureau of Standards, vol 1: 1971.

62. AH Nur. Laser optogalvanic spectroscopy and laser-induced chemical kinetic studies pertaining to the methoxyl radical. Ph.D. dissertation, Howard University, Washington, DC, 1994.

63. FM Phelps, III. MIT Wavelengths Tables, vol. 2, Cambridge, UA: MIT Press, 1982.

64. JE Lawler. Phys Rev A, 22: 1025, 1980.

65. DK Doughty, JE Lawler. Phys Rev A, 28: 773, 1983.

66. DM Kane. J Appl Phys 56: 1267, 1984.

67. A Sasso, M Ciocca, E Arimondo, J Opt Soc Am B 5: 1984, 1988.

68. Stewart, RS, RW McKnight, KI Hamad. J Phys D, 23: 832, 1990.

69. SP Lee, EW Rothe, GP Rothe. J Appl Phys 61(1): 109, 1987.

3
Spectra of the Isotopomers of CO⁺, N₂⁺, and NO in the Ultraviolet

S. Paddi Reddy
Memorial University of Newfoundland, St. John's,
Newfoundland, Canada

C. Haridass
Howard University, Washington, D.C.*

I. INTRODUCTION

This chapter describes the experimental results obtained from the ultraviolet spectra of the isotopomers of CO^+, N_2^+, and NO excited in a hollow-cathode discharge tube of special design, developed in our laboratory. Spectra of these molecules are of considerable importance in the understanding of several aspects of molecular spectroscopy and their applications to astrophysical phenomena.

A study of the electronic spectra of the molecular ion CO^+ enables us to understand the chemical and physical processes that take place in the solar and stellar atmospheres, comet tails, and the interstellar space. The molecular ion has also been used as a monitoring probe for chemical dynamics in the investigation of the environmental research and combustion processes. The electronic spectrum of CO^+ consists of four band systems: the comet tail ($A\,^2\Pi_i$–$X\,^2\Sigma^+$), the First Negative ($B\,^2\Sigma^+$–$X\,^2\Sigma^+$), the Baldet-Johnson ($B\,^2\Sigma^+$–$A\,^2\Pi_i$), and the Marchand-D'Incan-Janin ($C\,^2\Delta_r$–$A\,^2\Pi_i$) systems occurring in the region 1800–8500 Å. Pluvinel and Baldet [1] first observed the comet-tail system in the tail of the comet Morehouse 1908c. Dixon and Woods [2] observed the first terrestrial molecular ion CO^+ by microwave spectroscopy. Strong evidence of the presence of CO^+ in Orion molecular cloud 1 was provided by Erickson and coresearchers [3]. The comet-tail system of CO^+ that occurs in the spectral region 3080–8500 Å has

---

* *Current affiliation*: Belfry School, Belfry, Kentucky.

applications in the study of radiative heating of hypersonic spacecrafts at escape velocity and in the atmospheric fringe of the planet Venus, which contains considerable amounts of CO_2. Spectroscopic observations of these objects by rockets and satellites have increased their importance in astrophysics. The previous work done on the spectra of the $^{12}C^{16}O^+$ ion by numerous researchers has been reviewed by Krupenie [4], Huber and Herzberg [5], Jakubek and coresearchers [6], and Haridass and colleagues [7,8]. Prasad and Reddy [9], Reddy and Prasad [10], and Prasad and Reddy [11] have reviewed the work done on the A–X, B–A, and B–X systems of $^{13}C^{18}O^+$, respectively.

Spectra of neutral molecular nitrogen N_2 and its ion N_2^+ play a significant role in the atmospheric and astrophysical phenomena. For example, the Meinel (A $^2\Pi_{i,u}$–X $^2\Sigma_g^+$) and the First Negative (B $^2\Sigma_u^+$–X $^2\Sigma_g^+$) band systems of $^{14}N_2^+$ were observed in the auroral emission; the First Positive (B $^3\Pi_g$–A $^3\Sigma_u^+$), Second Positive (C $^3\Pi_u$–B $^3\Pi_g$), Vegard-Kaplan (A $^3\Sigma_u^+$–X $^1\Sigma_g^+$) and Lyman-Birge-Hopefield (A $^1\Pi_g$–X $^1\Sigma_g^+$) band systems of $^{14}N_2$ were also observed in the auroral emissions; the First Negative system of $^{14}N_2^+$ was also found in the spectra of comet tails (see the review article by Lofthus and Krupenie [12]). From radio astronomical measurements the abundance ratios of isotopes such as $^{15}N/^{14}N$ were estimated for lunar samples, solar wind, and meteoritic samples (see Clayton [13]). The study of molecular spectra of celestial sources provides the abundance ratios of isotopes such as $^{15}N/^{14}N$, which, in turn, gives a better understanding of the nuclear processes through which the energy is generated in the celestial objects.

The laboratory data on the spectra of $^{14}N_2$, $^{14}N_2^+$, $^{15}N_2$, and $^{15}N_2^+$ would be very useful in the study of their spectra resulting from the atmospheric and astrophysical phenomena mentioned above. The spectra of $^{14}N_2$ and $^{14}N_2^+$ have been thoroughly investigated in the laboratory. However, the spectroscopic data on $^{15}N_2$ and $^{15}N_2^+$ are either fragmentary or nonexistent. In an attempt to obtain comprehensive laboratory data on the spectra of $^{15}N_2^+$, its First Negative (B $^2\Sigma_u^+$–X $^2\Sigma_g^+$) system has been investigated in considerable detail for the first time by Reddy and Prasad [14] in our laboratory.

Nitric oxide (NO) is produced in the earth's atmosphere mainly by the oxidation of N_2O (which is a by-product of microbial metabolism) by the excited atomic $O(^1D)$ and to some extent by high-altitude aircrafts, nuclear blasts, volcanoes, lightning, and other phenomena [15]. It also plays a role in the destruction of ozone (O^3) but is involved in the photochemical production of O_3 in the troposphere by reaction with smog; it is a major product of internal combustion engines and combustion power plants [16]. The gamma (γ) (A $^2\Sigma^+$–X $^2\Pi_r$) system of NO has important applications in atmospheric science. For example, the bands of this system are used to measure the NO column densities in the mesosphere in order to interpret emission in aurora [17], and are also found to be potentially attractive for an optically pumped ultraviolet laser [18].

II.   THE HOLLOW-CATHODE DISCHARGE TUBE

The details of the design of the hollow-cathode discharge tube and the associated
gas-handling system are shown schematically in Figure 1. The hollow cathode
(F) made from a copper cylinder is 75 mm long, 17 mm in outer diameter, and
1.5 mm in wall thickness. It is sliver-soldered (H) to the lower section of a 19
mm inner diameter Kovar-Pyrex seal (G). The upper end of this seal is joined
to the main Pyrex glass body (D), 140 mm long and 46 mm in outer diameter,
of the discharge tube. The side arm (E), 17 mm in outer diameter, branches out
from the main body. A tungsten anode (A) is fused into the branch of the side
arm. Two quartz (SI-UV) windows (B and C) (ESCO Products Inc., Oak Ridge,
NJ), 1.5 mm thick, are attached to the ground end-surfaces of the anode and
cathode branches of the discharge tube with low vapor pressure Torr Seal (Varian
Associated Inc., Lexington, MA). Ball and socket arrangements (not shown in
Fig. 1) are provided, one for each of the reservoir sections (R_1 and R_2) and the
discharge tube to facilitate their easy connection to and disconnection from the
pumping system. The discharge tube and the secondary reservoir R_2 are thoroughly evacuated first. A small quantity of the experimental gas N_2 or CO is then
admitted from the primary reservoir R_1 into R_2 through stopcock S_1. By closing
the stopcock S_3, a small amount of the gas from R_2 is admitted into the discharge
tube through stopcock S_2. Under normal operating conditions, an applied direct
current (DC) voltage of 1100 V from a power supply unit rated at 2000 V and
250 mA gives a discharge current of 65 mA. The discharge is operated when the
gas is at rest. In the best operating conditions the pressure inside the discharge

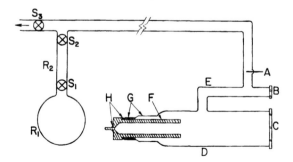

Figure 1   Schematic diagram of the hollow-cathode discharge tube and the associated
gas-handling system. A. tungsten anode; B, C. quartz windows; D. Pyrex glass body; E.
anode branch; F. copper hollow-cathode; G. Kovar-Pyrex seal; H. silver-soldering; R_1, R_2,
primary and secondary gas reservoirs; S_1, S_2, S_3: stopcocks. Arrow shows the connection to
the vacuum system.

tube is ~100 Pa as read on a thermocouple gauge. The hollow-cathode discharge tube described here is similar in some respects to the one used by Herzberg and coresearchers [19], who excited the emission spectra of triatomic H_3 and D_3 in the cathode column of a hollow-cathode discharge tube cooled with liquid nitrogen and operated in a vertical configuration with flowing H_2 and D_2, respectively. These spectra correspond to the transitions between their Rydberg states.

The emissions from the anode column and cathode glow with the carbon monoxide gas were photographed on a 2 m Bausch and Lomb dual grating spectrograph in the first order of a 1200 grooves/mm grating blazed at 1.0 μm. Those with the molecular nitrogen gas were photographed on the same spectrograph in the first order of a 600 grooves/mm grating blazed at 2.5 μm. Kodak Spectrum Analysis No 1 and 103-F plates were used to photograph the spectra of N_2 and N_2^+ from 334 to 523 nm. Kodak SWR plates were used to record the spectra of CO, CO^+, and CO_2 from 210 to 300 nm. Exposure times varied from 15 to 30 min. A DC Fe arc was used as a source for the reference spectra. Details of the hollow-cathode discharge tube are given by Reddy and Prasad [20].

As an example to illustrate the characteristics of the hollow-cathode discharge tube, the spectra of the anode column and cathode glow obtained with the N_2 gas from 334 to 523 nm are shown in juxtaposition in Figure 2. The

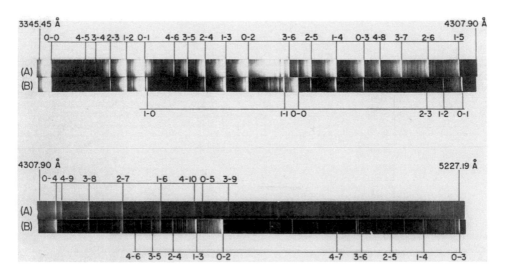

Figure 2   Spectra excited in the hollow-cathode discharge tube with the N_2 gas from 3340 to 5230 Å. Top. Spectrum of the anode column shows the Second Positive ($C\,^3\Pi_u - B\,^3\Pi_g$) system of the neutral $^{14}N_2$ molecule. Bottom. Spectrum of the cathode glow shows the First Negative System ($B\,^2\Sigma_u^+ - X\,^2\Sigma_g^+$) of $^{14}N_2^+$. Excitation conditions for both spectra are identical.

spectrum of the anode column (indicated by (A)) contains almost exclusively the Second Positive (C $^3\Pi_u$–B $^3\Pi_g$) system of the neutral N_2 molecule. The only exception is the appearance of the prominent 0-0 band of the First Negative system of N_2^+ with low intensity. The spectrum of the cathode glow (indicated by (B) in Fig. 2) consists of the First Negative (B $^2\Sigma_u^+$–X $^2\Sigma_g^+$) system of N_2^+. Also seen in Figure 2B is the occurrence of some normally very intense bands of the Second Positive system of N_2 in the cathode glow with comparatively lower intensity. The vibrational assignments of the bands of N_2 and N_2^+ are identified in Figure 2 and their wavelength data can be obtained from Lofthus and Krupenie [12].

III. ELECTRONIC CONFIGURATIONS

The type of molecular binding and the nature of the electronic states of a molecule are determined by electrons in the outermost shells of the constituent atoms of the molecule. The electronic configurations (Herzberg [21]) of the ground states of the C, N, and O atoms are:

C: K $2s^22p^2$; N: K $2s^2\ 2p^3$; O: K $2s^2\ 2p^4$.

The electronic configurations of the ground (X $^2\Sigma^+$) and the first two excited states (A $^2\Pi_i$ and B $^2\Sigma^+$) of CO^+ are:

K	K	$(z\sigma)^2$	$(y\sigma)^2$	$(w\pi)^4$	$(x\sigma)$:	X $^2\Sigma^+$
...	...	...	$(y\sigma)^2$	$(w\pi)^3$	$(x\sigma)^2$:	A $^2\Pi_i$
...	...	...	$(y\sigma)$	$(w\pi)^4$	$(x\sigma)^2$:	B $^2\Sigma^+$

The electronic configuration of the ground X $^2\Sigma_g^+$ state and the first excited state A $^2\Pi_i$ of homonuclear N_2^+ are given by

K	K	$(z\sigma_g)^2$	$(y\sigma_u)^2$	$(w\pi_u)^4$	$(x\sigma_g)$:	X $^2\Sigma^+$
...	...	...		$(w\pi_u)^3$	$(x\sigma_g)^2$:	A $^2\Pi_{i,u}$

The next excited states B $^2\Sigma_g^+$ and C $^2\Sigma_u^+$ of N_2^+ arise from the mixture of the following configurations:

K K $(z\sigma_g)^2$ $(y\sigma_u)$ $(w\pi_u)^4$ $(x\sigma_g)^2$

and

... ... $(y\sigma_u)^2$ $(w\pi_u)^3$ $(x\sigma_g)$ $(v\pi)$

The configurations of the ground state X $^2\Pi_r$ and the first excited doublet state A $^2\Sigma^+$ of NO are:

K	K	$(z\sigma)^2$	$(y\sigma)^2$	$(w\pi)^4$	$(x\sigma)^2$	$v\pi$:	X$^2\Pi_r$
...	...	...	...	...	...	$u\sigma$:	A $^2\Sigma^+$

In these configurations the electron from the antibonding $v\pi$ orbital is excited to the $u\sigma$ Rydberg orbital.

IV. GENERAL SPECTROSCOPIC EQUATIONS

The total energy E (in ergs) of a diatomic molecule (excluding translational and nuclear spin energies) is expressed by [21]:

$$E = E_e + E_v + E_r \tag{1}$$

and the term value T (in cm^{-1}) in terms of the vibrational and rotational quantum numbers v and J respectively is written as

$$T = T_e + G(v) + F_v(J) \tag{2}$$

where T_e, $G(v)$ and $F_v(J)$ are the electronic, vibrational, and rotational terms, respectively. The electronic terms of different multiple components of an electronic state (such as $^2\Pi$, etc.) are

$$T_e = T_0 + A\Lambda\Sigma_s \tag{3}$$

where T_0 is the electronic term, A and Σ_s are the quantized projections along the internuclear axis of the electron orbital angular momentum **L** and its spin angular momentum **S**. A is the spin orbit coupling constant and is positive or negative depending on whether the electronic state is regular or inverted. For Σ and Π states A = 0 and 1 respectively. The electronic state Σ is designated as Σ^+ or Σ^- depending on whether the wave function Ψ_e remains unchanged or changes sign upon reflection at the plane passing along the internuclear axis. The multiplicity of an electronic state is given by 2S + 1, which is the number of components along the internuclear axis. The vibrational term $G(v)$ is expressed as

$$G(v) = \omega_e(v + 1/2) - \omega_e x_e(v + 1/2)^2 + \omega_e y_e(v + 1/2)^3 + \cdots \tag{4}$$

where ω_e is the vibrational frequency and $\omega_e x_e$, $\omega_e y_e$, and so on are the anharmonic terms of the vibrational motion. The wave number v_0 of the band origin is given by

$$v_0 = v_e + G'(v') - G''(v'') \tag{5}$$

The isotopic shift Δv of a vibrational band is given by the expression

$$\begin{aligned}
\Delta v = {}& (1 - \rho) \, [\omega_e'(v' + 1/2) - \omega_e''(v'' + 1/2)] \\
& -(1 - \rho^2) \, [\omega_e' x_e'(v' + 1/2)^2 - \omega_e'' x_e''(v'' + 1/2)^2] \\
& +(1 - \rho^3)[\omega_e' y_e'(v' + 1/2)^3 - \omega_e'' y_e''(v'' + 1/2)^3]
\end{aligned} \tag{6}$$

where $\rho = (\mu_1/\mu_2)^{1/2}$, μ_1 and μ_2 being the reduced masses of the isotopomers. The rotational and stretching constants B_v and D_v are expressed as

$$B_v = B_e - \alpha_e(v + 1/2) + \gamma_e(v + 1/2)^2 + \delta_e(v + 1/2)^3 + \cdots \qquad (7)$$

and

$$D_v = D_e + \beta_e(v + 1/2) \, \varepsilon_e(v + 1/2)^2 + \cdots \qquad (8)$$

V. $^{12}C^{16}O^+$

A. Earlier Work on $^{12}C^{16}O^+$

1. The Comet-Tail System

After the earlier laboratory work of Fowler [22,23] on $^{12}C^{16}O^+$, several researchers observed the bands of the comet-tail (A–X) system (3080 Å–8500 Å) of this molecular ion [4,5]. Baldet [24] observed most of the known bands of this system and Birge [25] identified the transition as A $^2\Pi_r$–$^2\Sigma^+$. Coster and colleagues [26], Schmid and Gerö [27], and Bulthuis [28] performed the rotational analysis of several of these bands ($v' = 2$ to 11 and $v'' = O$–2). However, Rao [29] did the rotational analysis of the 0–2, 0–3, and 0–4 bands of this system and identified the upper state A as an inverted $^2\Pi$ (i.e., $^2\Pi_i$) state. Rao revised the vibrational assignments by lowering the previous assignments by three units. These new assignments were later confirmed by Asundi and colleagues [30] and Dhumwad et al. from the observed isotope shifts of the bands of the A–X system and the B–A system of $^{13}C^{16}O^+$ and $^{12}C^{18}O^+$, respectively, from the corresponding bands of $^{12}C^{16}O^+$. Gagnaire and Goure [32] reanalyzed the 2–0 band, but their data were found to be erroneous by Coxon and Foster [33]. Katayama and Welsh [34] and Brown et al. [35] reinvestigated the 0–0 band observed in laser-induced fluorescence. Coxon and Foster [33] carried out the deperturbation analysis of the experimental data of four bands of earlier researchers: 5-0 (i.e., 8–0 band) [26], 10–1 and 10–2 (i.e., 13–1 and 13–2) bands [28], and 0-2 band [29] and concluded that the A, $v = 0$, 5, and 10, vibrational levels are perturbed. They found that the perturbation shifts up to $\sim$14.5 cm^{-1} in the A $^2\Pi_{1/2} - X\ ^2\Sigma^+$ subband at $J = 14.5$ and 22.5 cm^{-1} in the two doublet components of A $^2\Pi_{1/2}$, $v = 0$, $2.5 \le J \le 7.5$ in the R_{21} and Q_{32} branches. They suggested that this perturbation was caused by the mixing of the A $^2\Pi_{1/2}$, $v = 0$ level, with X $^2\Sigma^+$, $v = 10$ level, which is associated with the unusual double crossing between the e levels of X $^2\Sigma^+$. The perturbations in the A $^2\Pi$, $v = 0$ level, relative to those in the A $^2\Pi$, $v = 5$ and 10 levels, are weak. As the off-diagonal spin-orbit matrix elements of the A, $v = 0$–X, $v = 0$, interaction were obtained incorrectly in [34]. Coxon and Foster [33] refitted the data [34] for the 0–0 band of the A–X system. Vujisic and Pesic [36] reinvestigated the rotational structure of the 4–0, 3–0, and 0–2 bands.

2.   The First Negative (B $^2\Sigma^+$–X $^2\Sigma^+$) System

The vibrational assignments of this system were given by Blackburn [37] and
Biskamp [38]. Coster et al. [26] and Schmid [39] performed the rotational analysis of a few bands of this system prior to Biskamp [38] without providing reliable
rotational constants. Schmid and colleagues [27] did the rotational analysis of
the 0–0 band of this system and established that the ground state X is the common
lower state of the comet-tail and first negative systems. Woods [40] studied the
splitting of B $^2\Sigma^+$ and X $^2\Sigma^+$ states. Rao [29] analyzed 15 bands of this system:
the $|\gamma'-\gamma''|$ value reported by him is one-half of that reported by Woods. Recently
Misra and co-researchers [41] re-examined the 0–0, 0–1, and 0–2 bands of the
B–X system and reported the spin-splitting constants of several vibrational levels
of the X and B states by combining the earlier experimental data of Rao [29].
They supported the latter's conclusions regarding the existing controversy about
the magnitude of the $|\gamma' - \gamma''|$ differences.

3.   The Baldet-Johnson (B $^2\Sigma^+$–A $^2\Pi_i$) System

After the initial observations of this system by Baldet [42], Johnson [43] and
Bulthuis [44] performed the rotational analysis of the 0–0 and 0–1 bands. Rao
and Sarma [45] summarized the results of the A–X, B–X, and B–A systems of
CO^+. Conkić and coresearchers [46] reinvestigated the rotational structure of the
0–1 band, but others [35] who attempted to reanalyze the former's wave-number
data claimed that their results were questionable because of possible errors in
calibration and measurements. Jakubek and colleagues [6] performed the rotational analysis of the 1–0, 0–0, and 0–1 bands of this system under higher resolution and obtained molecular constants of the B $^2\Sigma^+$, $v = 0$ and 1 and A $^2\Pi_i$, $v =
0$ and 1 states.

4.   The Marchand-D'Incan-Janin (C $^2\Delta_r$–$^2\Pi_i$) System

This system, which is inherently weak, was first observed at lower resolution by
Marchand et al. [47], who assigned the vibrational quantum numbers for the
observed bands. Cossart and Cossart-Magos [48] observed several bands of this
system in the presence of Ne and did the rotational analysis of the 0–2, 1–1, and
2–2 bands, photographed under higher resolution. These authors confirmed the
vibrational assignments of Marchand et al. [47] by observation of isotopic shifts
of the $^{13}C^{16}O^+$ bands from those of $^{12}C^{16}O^+$. They obtained molecular constants
of C $^2\Delta_r$ by holding the constants for the A $^2\Pi$, $v = 2$, level, determined from
Gagnaire and Goure [32], fixed in the fit, and those of the A, $v = 1$ by interpola-

tion. However, data from Gagnaire and Goure [32] were found erroneous by Coxon and Foster [33].

5. Microwave and Infrared Studies

The $^{12}C^{16}O^+$ ion has a strong electric dipole moment and hence is expected to have strong microwave and infrared spectra. Its pure rotational spectrum is the $X\,^2\Sigma^+$, $v = 0$ level, which was observed in the laboratory by Sastry and core-searchers [49] and Piltch and coresearchers [50] in the millimeter region, and by van den Heuvel and colleagues [51] in the submillimeter region. Bogey and co-workers [52,53] reported observing rotational transitions in the X, $v = 0$–4 levels. Davies and Rothwell [54] measured the fundamental band in the X state of CO^+ in emission using velocity modulation technique. Haridass et al. [7] reviewed the earlier work done on the four electronic band systems of $^{12}C^{16}O^+$ and its infrared and microwave spectra. They have also analyzed the rotational structure of the individual bands of the A–X system and reanalyzed the data of the B–X [55,41] and B–A [6] systems, and the infrared [54] and microwave data of the $v = 0$, 1, and 2 levels [49,53].

The molecular constants and band origins obtained from these analyses were combined and all multiple estimates were reduced to a single set of values using a grand merge procedure. It was believed that the nonsmooth variations of molecular constants in the A $^2\Pi_i$ state, especially the higher-order constants such as D_v, A_{Dv}, p_v, and q_v, are attributed to the neglect of perturbations. Bembe-nek and coresearchers [56] recently recorded new bands in the Baldet-Johnson (B $^2\Sigma^+$–A $^2\Pi_i$) system of $^{12}C^{16}O^+$ and performed rotational analysis. They have excluded the perturbed lines of the bands with a common lower $v'' = 0$ level and claimed that the molecular constants of the B $^2\Sigma^+$ and A $^2\Pi_i$ states thus obtained show a smooth variation with the vibrational quantum number ($v + 1/2$). They pointed out that the Λ-doubling parameters p_v and q_v of the A $^2\Pi_i$ state obtained by Haridass and coresearchers [7] do not obey traditionally recognized polynomial dependence on the vibrational quantum number ($v + 1/2$). Weak perturbations in the A $^2\Pi_i$, $v = 0$, state at $2.5 \le J \le 6.5$ due to X $^2\Sigma^+$, $v = 10$ state result in extra lines of the subband of A $^2\Pi_{1/2}$–X $^2\Sigma^+$, which have been observed by Katayama and Welsh [34] in the laser-induced spectra of $^{12}C^{16}O^+$. Deperturbation analysis for the A $^2\Pi_i$ state of CO^+ from the available spectro-scopic data [26,28,34,55] of the perturbed levels A, $v = 0$, 5, and 10 using a modern direct fitting approach was carried out by Coxon and Foster [33]. These authors have also calculated an accurate Rydberg, Klein, Rees (RKR) potential curve for the X $^2\Sigma^+$ state and tabulated the term values and the RKR turning points for high-lying vibrational levels. Remeasurement and subsequent analysis of the 0–0 band of the A–X system of CO^+ using laser-induced fluorescence excitation were also carried out by Brown and coresearchers [35]. From the time-resolved laser-induced fluorescence of CO^+, radiative decay rates and the velocity-averaged cross sections for the A $^2\Pi_i$ ($v = 0$–3) levels have been obtained by Miller and colleagues [57,58].

B. Experimental Details

The molecular ion CO^+ was excited in the cathode column of the hollow-cathode discharge tube. The high-resolution spectra of seven bands were photographed on the 3.4 m Jarrell-Ash spectrograph equipped with a 1200 grooves/mm grating blazed at 1.4 µm: five bands (4–0, 3–0, 2–0, 1–0, and 2–1) in the third order and two bands (1–1 and 0–2) in the second order. Three bands were photographed on the 2.0-m Bausch and Lomb spectrograph, two bands (0–1 and 0–3) in the second order of a 1200 grooves/mm grating blazed at 1.0 µm, and one band (0–4) in the third order of a 600 grooves/mm grating blazed at 2.5 µm. The slit width was maintained at 30 µm for the former and 20 µm for the latter. The reciprocal dispersions of the spectra photographed varied from 0.60 Å/mm at 3820 Å in the third order to 1.05 Å/mm at 7200 Å in the second order. Overlapping orders of the spectra were eliminated using Corning and Hoya glass filters. The exposure times on the photographic plates (103 a-o, 103-F, and 1-N) varied from 45 min to 12 h. An Fe–Ne hollow-cathode lamp was used as source for the reference spectra whose wavelengths were taken from work by Crosswhite [59].

C. Analysis of the Spectra

The vacuum wavenumbers of the band heads of 17 bands of the comet-tail (A $^2\Pi_i$–X $^2\Sigma^+$) system of CO^+, their relative intensities, and the vibrational assignments are given in Table 1. In this system, for the bands 8–1, 7–1, 5–0, 3–3, and 1–2, for which the rotational analysis was not carried out, the estimated band origins ($T'_\pi - T''_\Sigma$) using the derived molecular constants (see below) are also included in Table 1. Each of the bands of this system gives rise to four heads formed by the R_{21ee}, Q_{21fe}/R_{22ff}, R_{11ee}, and Q_{11fe}/R_{12ff} branches in the order of decreasing wavenumber. The two heads formed by the first two branches are from the $^2\Pi_{1/2}$–$^2\Sigma^+$ component and the other two heads formed by the latter two branches are from the $^2\Pi_{3/2}$–$^2\Sigma^+$ component. The head due to the R_{11ee} branch of the 8–1 band and the head due to the Q_{21fe}/R_{22ff} branches of the 7–1 band could not be measured. The upper A $^2\Pi_i$ state of the comet-tail system belongs to Hund's case (a) and its lower X $^2\Sigma^+$ state belongs to Hund's case (b). The rotational structure of a band arising from a $^2\Pi_i$–$^2\Sigma^+$ transition gives rise to 12 branches designated R_{21ee}, R_{22ff}, Q_{21ef}, Q_{22ef}, P_{21ee}, and P_{22ff} from $^2\Pi_{1/2}$–$^2\Sigma^+$, and R_{11ee}, R_{12ff}, Q_{11fe}, Q_{12ef}, P_{11ee}, and P_{12ff} from $^2\Pi_{3/2}$–$^2\Sigma^+$, as shown schematically in Figure 3. The parity of the levels shown in the e/f notation is according to Brown

Table 1 Comet-Tail (A $^2\Pi_i$–X $^2\Sigma^+$) Band System of $^{12}C^{16}O^+$ Molecule

Band head[a] (cm^{-1})	Band origin[b] $T'_\pi - T''_\Sigma$ (cm^{-1})	Band origin[c] $T'-T''$ (cm^{-1})	Relative intensity[d]	Assignment $v'-v''$
29826.11[e]	29750.58[f]		vw	8–1
29817.05[e]				
29692.28[e]				
29288.41	29212.897(2)	29212.896(2)	w	6–0
29276.55				
29163.94				
29153.11				
28484.47[e]	28403.09[f]		w	7–1
28367.51[e]				
28347.42[e]				
27891.72	27812.16[f]		m	5–0
27878.69				
27763.28				
27751.66				
26462.51	26384.722(1)	26384.720(1)	s	4–0
26449.35				
26336.33				
26324.26				
25009.13	24930.349(1)	24930.349(2)	vs	3–0
24994.99				
24882.58				
24870.02				
23529.01	23449.235(1)	23449.253(1)	vs	2–0
23514.03				
23402.28				
23388.91				
22022.07	21941.223(1)	21941.215(1)	m	1–0
22006.17				
21895.08				
21880.93				
21345.58	21265.241(1)	21265.239(1)	m	2–1
21330.09				
21218.80				
21204.94				
19839.33	19757.564(2)	19757.565(1)	m	1–1
19822.62				
19712.07				
19697.34				

Table 1 Continued

Band head[a] (cm^{-1})	Band origin[b] $T'_\pi - T''_\Sigma$ (cm^{-1})	Band origin[c] $T' - T''$ (cm^{-1})	Relative intensity[d]	Assignment $v' - v''$
18550.12	18469.59[f]		vw	3–3
18535.64^2				
18424.15[e]				
18409.14[e]				
18305.32	18222.331(2)	18222.289(1)	s	0–1
18287.64				
18177.66				
18162.05				
17685.77[e]	17603.70[f]		vw	1–2
17674.91[e]				
17559.40[e]				
17542.94[e]				
17070.47	16988.472(2)	16988.480(1)	vw	2–3
17053.51				
16943.22				
16928.20				
16152.55	16068.569(2)	16068.578(1)	s	0–2
16133.97				
16024.72				
16008.42				
14030.53	13945.476(1)	13945.473(1)	m	0–3
14011.01				
13902.46				
13885.49				
11938.78	11852.560(1)	11852.557(1)	w	0–4
11918.22				
11810.37				
11792.68				

[a] The four heads identified for each band are formed by the R_{21ee}, Q_{21fe}/R_{22ff}, R_{11ee}, and Q_{11fe}/R_{12ff} branches in the order of decreasing wavenumber.
[b] The number in parentheses indicates the uncertainty in the last digit and corresponds to one standard deviation.
[c] Band origins obtained from the global fit.
[d] Abbreviations for the relative intensities vs, s, m, w, and vw represent very strong, strong, medium, weak, and very weak, respectively.
[e] Measured from the medium dispersion spectrum.
[f] Estimated value from the derived molecular constants.

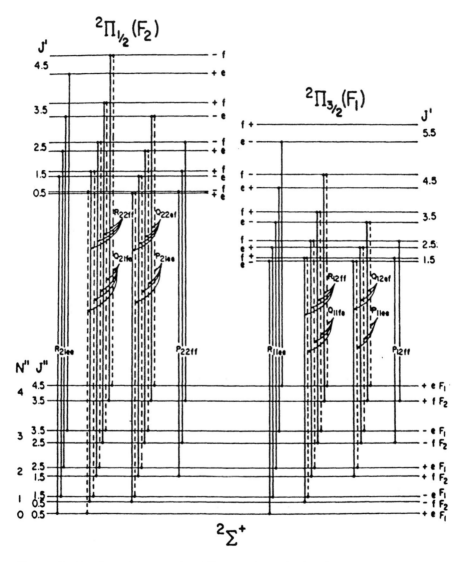

Figure 3 A schematic energy level diagram shows the first few rotational transitions for all 12 branches of a band of a $^2\Pi_i$–$^2\Sigma^+$ system.

Reddy and Haridass

Table 2 Matrix Elements of the Hamiltonian
for States $^2\Sigma^+$ and $^2\Pi$

Molecular constant	Labeling[a]	Matrix element[b,c]
T_v	1, 1	1
	2, 2	1
	3, 3	1
B_v	1, 1	$x(x \mp 1)$
	2, 2	$x^2 - 1$
	3, 3	$x^2 + 1$
	2, 3	$-(x^2 - 1)^{1/2}$
D_v	1, 1	$x^2(x \mp 1)^2$
	2, 2	$-x^2(x^2 - 1)$
	3, 3	$-x^2(x^2 + 3)$
	2, 3	$2x^2(x^2 - 1)^{1/2}$
γ^v	1, 1	$0.5(x \mp 1)$
A_v	2, 2	0.5
	3, 3	-0.5
A_{Dv}	2, 2	$0.5\,(x^2 - 1)$
	3, 3	$-0.5(x^2 + 1)$
p_v	3, 3	$\mp 0.5x$
q_v	3, 3	$\mp x$
	2, 3	$\pm 0.5x(x^2 - 1)^{1/2}$

[a] Labels 1, 2, and 3 refer to states $^2\Sigma^+$, $^2\Pi^{3/2}$, and $^2\Pi_{1/2}$, respectively.
[b] $x = J + 0.5$.
[c] In the notation $\pm$ and $\mp$ the upper and lower signs refer to the e and f levels, respectively.
Source: Ref. 60.

and coresearchers [60]. As the bands of this system are degraded to longer wave-
lengths, the R_{21ee}, Q_{21fe}/R_{22ff}, R_{11ee}, and Q_{11fe}/R_{12ff} branches form four different
heads in the rotational structure of a band. The characteristic band origin
$T'_\pi - T''_\Sigma$ for a band is estimated from the analysis (see below). The band origins
of the $^2\Pi_{1/2}-^2\Sigma^+$ and $^2\Pi_{3/2}-^2\Sigma^+$ subbands of a band are obtained from $(T'_\pi - T''_\Sigma)$
$\pm (1/2)A_v$, A_v being the spin–orbit coupling constant of the $^2\Pi$ state. The effective
Hamiltonian for the $^2\Pi$ and $^2\Sigma^+$ states of diatomic molecules, discussed in detail
by Brown and colleagues [61], was used to analyze the rotational structure of
the comet-tail bands of $^{12}C^{16}O^+$. A complete list of the matrix elements of this
Hamiltonian for the $^2\Pi$ and $^2\Sigma^+$ states used by Haridass et al. [7,8] are taken
from Amiot and coresearchers [62] and Douay and co-workers [63] and are listed
in Table 2. The relevant matrix elements of the Hamiltonian are:

$$^2\Sigma^{e,f}: T_v + B_v x\,(x \mp 1) - D_v x^2 (x \mp 1)^2 + 0.5\gamma_v (x \mp 1)$$
$$^2\Pi_{1/2}^{e,f}: T_v - 0.5A_v - 0.5A_{Dv}[x^2 + 1] + B_v[x^2 + 1]$$
$$- D_v x^2 [x^2 + 3] \mp 0.5p_v x \mp q_v x$$
$$^2\Pi_{3/2}^{e,f}: T_v - 0.5A_v - 0.5A_{Dv}[x^2 - 1] + B_v[x^2 - 1] \qquad (9)$$
$$- D_v x^2 [x^2 - 1]$$
$$\{^2\Pi_{1/2}, {}^2\Pi_{3/2}\}^{e,f}: - B_v[x^2 - 1]^{1/2} + 2.0D_v x^2 [x^2 - 1]^{1/2}$$
$$\pm 0.5q_v x[x^2 - 1]^{1/2},$$

where B_v and D_v are the rotational constants, γ_v is the spin-rotation constant, A_v and A_{Dv} are the spin-orbit constants, p and q are the Λ-doubling constants, and $x = (J + 0.5)$. The upper and lower state signs $\pm$ or $\mp$ in the above matrix elements refer to the e/f levels, respectively. The rotational structure of a section of the 0–2 band of the comet-tail system photographed on the 3.4 m Jarrell-Ash spectrograph in the second order of the 1200 grooves/mm grating is shown in Figure 4. The four heads and the 10 branches (out of the expected 12) of their rotational structure are clearly identified. Each of the branches, Q_{22ef}/P_{21ee} in the $^2\Pi_{1/2}-^2\Sigma^+$ subband and R_{12ff}/Q_{11fe} in the $^2\Pi_{3/2}-^2\Sigma^+$ subband, forms two close pairs and appears as one branch. The intensities of the lines of the resolved pairs (R_{22ff}, Q_{21fe}) and (Q_{12ef}, P_{11ee}) are found to be approximately equal. These pairs of branches give directly the spin-splitting in the ground $^2\Sigma^+$ state. To verify the observed intensity pattern of a band in the $^2\Pi_i-^2\Sigma^+$ transition of CO^+, intensities of the branches were calculated using expressions given by Earls [64]. The calculated intensities are displayed in Figures 5 and 6. In these figures, it is clearly seen that the ratio of the intensities of the unresolved pairs (Q_{22ef}/P_{21ee} and Q_{11fe}/R_{12ff}) is approximately 3:1 and the ratio of the resolved pairs (Q_{22ef} and P_{21ee} and Q_{12ef} and P_{11ee}) is approximately 1:1. In the case of the 1–0 and 1–1 bands (i.e., bands with a common upper level), it was possible to observe 12 branches since the bands were strong and the rotational lines were resolved. To rule out the possibility of systematic shifts in the observed spectra due to various experimental procedures, the spectra can in principle be calibrated with the atomic lines of carbon and oxygen if they appear in the CO^+ spectrum itself. However, these atomic lines were not observed [7,8]. The emission spectrum has the advantage of displaying several bands involving the same upper vibrational levels and therefore provides a critical check of the assignments and systematic shifts. Since the ground ($^2\Sigma^+$) state rotational levels of CO^+ are known accurately, it is possible to calculate the shifts for the bands arising from the same v' and different v'', by combining the observed wavenumbers of the rotational levels with the corresponding term values of the ground state. The ground state term values for $v = 0, 1, 2, 3,$ and 4 for both e and f levels are calculated from the Dunham parameters and the spin-rotation constant γ_v given by Bogey et al. [53] and the G(v) values calculated by Coxon and Foster [33]. For example, the rovibrational term values

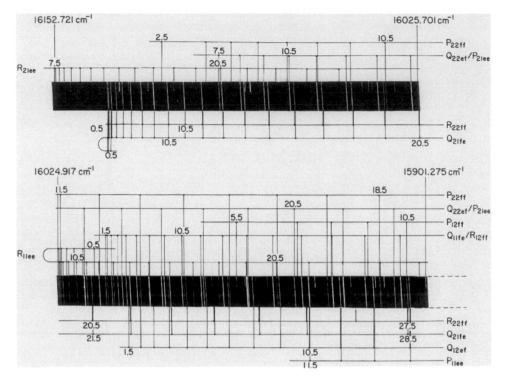

Figure 4 Rotational structure of a section of the 0–2 band of the comet-tail (A $^2\Pi_i$–X $^2\Sigma^+$) system of $^{12}C^{16}O^+$, photographed on the 3.4 m Jarrell-Ash spectrograph in the second order of a 1200 grooves/mm grating.

of A $^2\Pi_i$, $v = 0$, e and f levels are obtained by combining the observed wavenumbers of the 0–1, 0–2, 0–3, and 0–4 bands of the A–X system to the corresponding term values of the $v = 1$, 2, 3, and 4 levels of the X $^2\Sigma^+$ state. The term values of the $v' = 0$ level calculated from the 0–2 band are found to be consistently lower by 0.197 cm^{-1} than those calculated from the 0–1, 0–3, and 0–4 bands. Using a similar procedure, the rovibronic term values of the A $^2\Pi_i$, $v = 1$ and 2 levels are calculated by combining the wavenumbers of the 1–0 and 1–1, 2–0, and 2–1 bands, respectively, of the A–X system to the corresponding term values of the X $^2\Sigma^+$, $v'' = 0$ and 1 as mentioned above. The calculated term values of the $v' = 1$ level from the 1–0 band are found to be consistently lower by 0.265 cm^{-1} than those calculated from the 1–1 band. A systematic shift was observed for the term values of the $v' = 2$ level calculated from the 2–1 band with those calculated from the 2–0 band and is found to be -0.113 cm^{-1}. The corrected vacuum wavenumbers and the rotational quantum numbers of the spec-

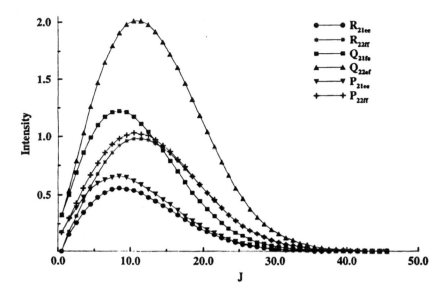

Figure 5 Intensity distribution of the six branches in the 0–2 band of the comet-tail $(^2\Pi_{1/2}-^2\Sigma^+)$ subsystem of $^{12}C\,^{16}O^+$.

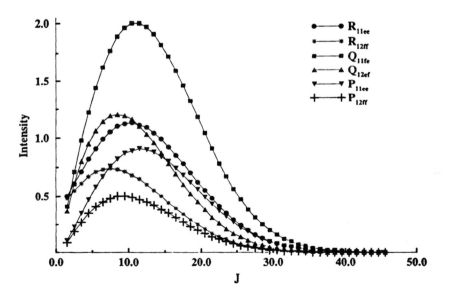

Figure 6 Intensity distribution of the six branches in the 0–2 band of the comet-tail $(^2\Pi_{3/2}-^2\Sigma^+)$ subsystem of $^{12}C^{16}O^+$.

tral lines of all the 10 bands (4–0, 3–0, 2–0, 1–0, 2–1, 1–1, 0–1, 0–2, 0–3, and 0–4) are deposited in the Electronic Depository of the Supplementary Material of the *Journal of Molecular Spectroscopy* [8].

D. RESULTS AND DISCUSSION

The wavenumber data of the 10 bands, the infrared data of the rovibronic spectra [54], and the microwave data of the $v = 0, 1, 2, 3$, and 4 levels for the X $^2\Sigma^+$ state [49,53]—a total of 2513 transitions—were used simultaneously in a nonlinear least-squares fit [8]. The weights were calculated from the expression $w = 1/\sigma^2$, where σ is the uncertainty for the spectral positions of the lines. The values of σ are in the range 0.01–0.1 cm^{-1} for the optical data, 0.001–0.01 cm^{-1} for the infrared data, and 0.2–0.6×10^{-5} cm^{-1} for the microwave data. The residuals between the calculated and observed wavenumber data of the optical spectra for all 10 bands are also given in the Electronic Depository of Supplementary Material [8]. The dimensionless variance in the fit is 2.5 and the root-mean-square deviation is ~0.016 cm^{-1}. A set of 55 molecular parameters for the $v = 0$ to 4 levels for both states A $^2\Pi_i$ and X $^2\Sigma^+$ thus obtained is provided in Table 3. From this table the value of the Λ-doubling parameters p_v and q_v do not show irregularities in their v-dependence. The B_0 value of the A $^2\Pi_i$ state obtained by Haridass et al. [8] is 1.5796211(84) cm^{-1}. The high precision of the B_0 value up to the sixth decimal place is due to the inclusion of the microwave data. This value is in agreement in the fourth decimal place within the error limits to the values 1.579629(102) and 1.579710(104) cm^{-1} [6] reported from analysis of the individual bands 0–0 and 1–0 of the B $^2\Sigma^+$–A $^2\Pi_i$ system. Reanalyzed data of the 1–0 band [6] show that the value for B_0 of the A state is 1.579606(74) cm^{-1}. Using the value of 1.579496(26) cm^{-1} and the values of the other molecular parameters reported by Bembenek and co-workers [56], the wavenumbers of the 0–v″, bands with v″ = 1 to 4 of the A–X system can be reproduced up to $J = 22.5$ with an accuracy of 0.01 cm^{-1}. Beyond this J value there is a deviation in the $(v_{obs}-v_{cal})$ values from -0.028 to -0.195 cm^{-1} for $J = 24.5$ to $J = 33.5$. Thus we conclude that the molecular constants given in Table 3 can be used to reproduce the wavenumbers up to the J levels [8]. Since least-squares technique is only an interpolation method, extrapolation of the data to higher rotational quantum numbers will not give accurate values. From Table 3 we claim that the values of the Λ-doubling parameters p_v and q_v do not show irregularities in their v-dependence. The B_v values listed in Table 3 were fitted to the standard spectroscopic relation given by Eq. (7), and the resulting equilibrium molecular constants B_e, α_e, γ_e, and δ_e along with their standard deviations are presented in Table 4. The D_v values were fitted to Eq. (8) and the resulting values of D_e and β_e for the A state, and D_e, β_e, and ε_e for the X state are also listed in Table 4. The value

Table 3 Molecular Constants[a] (in cm^{-1}) of the A $^2\Pi_i$ and X $^2\Sigma^+$ States of $^{12}C^{16}O^+$

Molecular constant	$v = 0$	$v = 1$	$v = 2$	$v = 3$	$v = 4$
A $^2\Pi_i$					
T_v	20406.2166(18)	21941.4793(12)	23449.2278(14)	24930.3444(18)	26384.7229(17)
B_v	1.5796211(84)	1.5600719(74)	1.5407277(96)	1.521349(11)	1.5019686(98)
$D_v \times 10^6$	6.735(11)	6.4641(96)	6.661(12)	6.690(12)	6.600(10)
$-A_v$	122.0513(18)	121.9829(20)	121.8874(23)	121.8126(30)	121.8720(29)
$-A_{Dv} \times 10^4$	2.179(55)	1.357(63)	1.467(70)	1.546(80)	0.804(81)
$-q_v \times 10^4$	2.619(75)	2.439(93)	2.61(11)	2.21(10)	3.16(12)
$p_v \times 10^2$	1.549(14)	1.293(17)	1.259(18)	1.054(21)	0.943(23)
X $^2\Sigma^+$					
T_v	0	2183.92064(36)	4337.4571(19)	6460.7397(27)	8553.6771(28)
B_v	1.967462265(60)	1.94843818(64)	1.92935252(90)	1.910180(16)	1.890985(16)
$D_v \times 10^6$	6.3170(18)	6.3674(47)	6.4278(81)	6.399(22)	6.475(20)
$\gamma_v \times 10^3$	9.10556(50)	9.0495(34)	8.9729(38)	8.972(65)	8.766(63)

[a] Number in parentheses is the uncertainty in the last digit and corresponds to one standard deviation.

Table 4   Equilibrium Molecular Constants [a, b] (in cm^{-1})
for the A $^2\Pi_i$ and $X\,^2\Sigma^+$ States of $^{12}C^{16}O^+$

Molecular constants	$X\,^2\Sigma^+$	A $^2\Pi_i$
T_e		19628.298(69)
$G(0)$	1103.23(21)	
ω_e	2214.15(15)	1562.79(18)
$\omega_e X_e$	15.150(25)	13.926(89)
$\omega_e y_e$		0.65(12)
B_e	1.97695082(80)	1.589482(20)
$\alpha_e \times 10^2$	1.89615(19)	1.9806(36)
$\gamma_e \times 10^4$	−3.018(70)	1.66(18)
$\delta_e \times 10^4$		−1.92(24)
$D_e \times 10^6$	6.2856(44)	6.752(12)
$\beta_e \times 10^8$	6.52(81)	−3.01(36)
$\varepsilon_e \times 10^9$	−5.5(2.2)	
$p_e \times 10^2$		1.631(22)
$\alpha_p \times 10^3$		1.99(23)
$\beta_p \times 10^4$		1.08(49)

[a] Weighted least-squares fits were used to obtain the molecular constants here.
[b] Number in parentheses is the uncertainty in the last digit and corresponds to one standard deviation.

of $D_1 = 6.4641(96) \times 10^{-6}$ cm^{-1} of the A state is not included in determining the parameters D_e and β_e. The equilibrium molecular constants p_e, α_p, and β_p are determined by fitting the values of the Λ-doubling parameter p_v given in Table 3 to the expression

$$p_v = p_e - \alpha_p(v + 1/2) + \beta_p(v + 1/2)^2 \qquad (10)$$

and their values are also given in Table 4. The term values T_v for both A and X states given in Table 3 were fitted to the expression:

$$T_v = T_e + \omega_e(v + 1/2) - \omega_e x_e(v + 1/2)^2 + \omega_e y_e(v + 1/2)^3 \qquad (11)$$

and the vibrational constants ω_e, $\omega_e x_e$, and $\omega_e y_e$ and T_e thus obtained along with their standard deviations are also listed in Table 4. In fitting the Eqs. (7), (8), (10), and (11), a weighted least-squares fit was used in each case. The $^2\Pi_i$ state of the comet-tail system of CO$^+$ belongs neither to Hund's case (a) nor to Hund's case (b), but to intermediate case (case (a) for small rotation and case (b) for large rotation). To confirm this a Fortrat diagram for the 0–2 band of the comet-

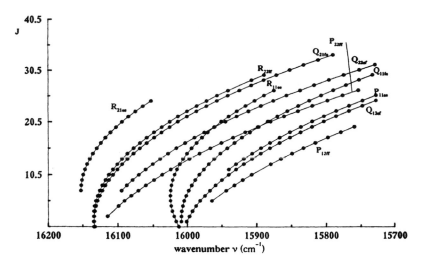

Figure 7 Fortrat diagram of the 0–2 band of the A $^2\Pi_i$–X $^2\Sigma^+$ system of $^{12}C^{16}O^+$.

tail system of $^{12}C^{16}O^+$ is constructed and shown in Figure 7. In this figure, the branches R_{22ff} and R_{11ee}, Q_{22ef} and Q_{11fe}, and P_{22ff} and P_{11ee} draw closer together with increasing J, as expected for the intermediate case [see, for example, Herzberg [21]. For the $^2\Pi_i$ state, the value of A/B is found to be ≈ -77. Jakubek and coresearchers [6] did the rotational analysis of the 0–0, 1–0, and 0–1 bands of the Baldet-Johnson (B $^2\Sigma^+$–A $^2\Pi_i$) system of $^{12}C^{16}O^+$ and obtained the merged molecular constants of the B and A states. Their merged molecular constants of the B state are given in Table 5. The controversy over the $|\gamma_v'-\gamma_v''|$ difference of the bands of the First Negative (B $^2\Sigma^+$–X $^2\Sigma^+$) system existing in the literature was reexamined by Haridass and coresearchers [7]. The values of $|\gamma_v'-\gamma_v''|$ obtained by Haridass et al. [7] together with those of Woods [40] and Rao [55] are

Table 5 Merged Molecular Constants of the B $^2\Sigma^+$ State of $^{12}C^{16}O^+$

	$v = 0$	$v = 1$
B_v	1.784564(54)	1.754520(56)
$D_v \times 10^6$	7.941(78)	8.153(78)
$\gamma_v \times 10^2$	2.116(11)	2.065(15)
T_v	25227.6135(24)	26906.9837(29)

Table 6 Comparison of the Values of $|\gamma_v'-\gamma_v''|$ (in cm^{-1})

	$|\gamma_v'-\gamma_v''|$		
$v'-v''$	Woods (40)	Rao (55)	Haridass and co-researchers (7)
0–1	0.0180	0.009	0.0092
0–2	0.0159	0.008	0.0092
0–3	0.0153	0.010	0.0093
1–2	0.0172	0.009	0.0092
1–3	0.0158	0.008	0.0092
1–4	0.0150	0.010	0.0067
1–5	0.0159	0.009	0.0079
2–4	0.0149	0.009	0.0085
2–5	0.0153	0.009	0.0096
2–6	0.0150	0.008	0.0069
3–5	0.0153	0.007	0.0071

presented in Table 6. The values obtained by Haridass and colleagues [7] are closer to those of Rao [55], except for the 1–4 band, but differ a great deal from those of Woods [40].

VI. $^{13}C^{18}O^+$

A. Earlier Work on the Spectra of Other Isotopomers of CO

From a study of the comet-tail system (A $^2\Pi$–X $^2\Sigma^+$) of $^{12}C^{16}O$, Rao [29] identified the state A to be an inverted $^2\Pi_i$ (Sec. V). The new vibrational assignments in [29] were confirmed from the observed isotope shifts of the band heads of the $^{13}C^{16}O^+$ [30] and those of the Baldet-Johnson system (B $^2\Sigma^+$–A $^2\Pi_i$) of $^{12}C^{18}O^+$ [31]. The 2–0 and 0–0 bands, respectively, of the A–X system of the $^{13}C^{16}O^+$ molecule have been investigated [65,66]. The rotational analysis of five bands (4–0, 3–0, 2–0, 1–0, and 2–1) of this system of the $^{14}C^{16}O^+$ molecule has been performed [67]. The rotational structure of four bands (4–0, 3–0, 2–0, and 0–2) of this system of the $^{12}C^{18}O^+$ molecule has been analyzed [68]. In our laboratory several investigations [69–72] were made on the electronic spectra of the neutral $^{13}C^{18}O$ molecule. Prasad and Reddy [9] have reported the first observation of nine bands (5–0, 4–0, 3–0, 2–0, 1–0, 2–1, 1–1, 0–1, and 0–2) of the comet-tail system of the $^{13}C^{18}O^+$ molecule and the results obtained from the rotational analysis of eight of them (except the very weak 1–0 band).

Regarding the work done on the Baldet-Johnson system of the isotopically substituted carbon monoxide ions, Conkić and coresearchers [73] analyzed the 1–0 and 0–0 bands of $^{12}C^{18}O^+$, the 1–0, 0–0, and 0–1 bands of $^{13}C^{16}O^+$, and the 1–0 band of $^{12}C^{16}O^+$. Brown and colleagues [74], who attempted to reanalyze the wave number data of Conkić et al. [73], claimed that the data are questionable because of possible errors in calibration and measurements. Jakubek et al. [75] performed the rotational analysis of the 1–0, 0–0, and 0–1 bands of this system in the $^{14}C^{16}O^+$ ion. Reddy and Prasad [10] observed the 1–0, 0–0, and 0–1 bands of this system of $^{13}C^{18}O^+$ and analyzed this rotational structure for the first time.

Misra and coresearchers [41] performed a partial analysis of the 0–0, 0–1, and 0–2 bands of the First Negative system of $^{13}C^{16}O^+$, and reported a few rotational constants and $|\gamma' - \gamma'|$ differences for the 0–1 and 0–2 bands. Shvangiradze and colleagues [76] and Pešić and coresearchers [77] studied the isotope shifts of the bandheads of this system of $^{13}C^{16}O^+$ relative to those of $^{12}C^{16}O^+$. Regarding the work done on this system of $^{12}C^{18}O^+$, the isotope shifts of the bandheads have been studied by Pešić et al. [77]; Janjić and Pešić [78] recorded the spectrum at moderate dispersion and performed the rotational analysis of 10 bands. The spin splitting of the spectral lines was not resolved. Twelve bandheads of the First Negative system and reported isotope shifts of these bands relative to those of $^{13}C^{16}O^+$ have been reported [77]. Pešić and coresearchers [79] recorded six bands of this system at a moderate dispersion. The spin splitting was not observed in any of these bands, and their assignment of rotational quantum numbers was incorrect, particularly for the 0–1 band. Prasad and Reddy [11] observed a total of 22 bands with $v' = 0-4$ and $v'' = 0-7$ of the First Negative system and analyzed the rotational structure of 13 of these bands with $v' = 0-2$ and $v'' = 0-5$, photographed under high resolution.

B. Experimental Details

The A–X, B–A, and B–X systems of $^{13}C^{18}O^+$ were excited in the cathode column of the cathode of the hollow-cathode discharge tube (Sec. III). Carbon-13 and oxygen-18 gas with a rated purity of 99% ^{13}C and 95% ^{18}O (Merck Sharpe and Dohme) was used. A DC voltage of 1100 V applied between the electrodes maintained the discharge at a current of ~65 mA when the pressure inside the discharge tube was ~0.8 torr. An Fe–Ne hollow-cathode lamp was used as a source of reference spectra [59] for the three band systems. Nine bands of the comet-tail system of $^{13}C^{18}O^+$ occurring in the spectral region 3165–6165 Å were first photographed under medium dispersion on a 2 m Bausch and Lomb dual grating spectrograph, in the first order of a 600 grooves/mm grating, blazed at 2.5 μm. Two of these bands (0–1 and 0–2) were photographed under high resolution in the second order of a 1200 grooves/mm grating blazed at 1.0 μm on the Bausch and Lomb spectrograph. Of the remaining seven bands (5–0, 4–0, 3–0, 2–0, 1–

0, 2–1, and 1–1), which were photographed under high resolution on a 3.4 m Jarrell-Ash spectrograph, the 1–1 band was recorded in the second order and the rest were in the third order of a 1200 grooves/mm grating blazed at 1.4 μm. The exposure times for the high-resolution spectra on the Bausch and Lomb spectrograph were about 1 h, whereas on the Jarrell-Ash spectrograph the times varied from 45 min (for the 3–0 band) to 8.5 h (for the 1–1 band). Kodak Spectrum Analysis No. 1, 103 a-O, and 103-F plates were used to photograph the spectra. The slit widths were maintained at 20 and 30 μm on the Bausch and Lomb and the Jarrell-Ash spectrographs, respectively. The reciprocal dispersions of the spectra are ~0.56 Å/mm at 3950 Å in the third order on the Jarrell-Ash spectrograph and about 1.42 Å/mm at 6200 Å in the second order on the Bausch and Lomb spectrograph.

The 1–0, 0–0, and 0–1 bands of the Baldet-Johnson system of $^{13}C^{18}O^+$ occurring in the spectral region 3700–4225 Å were photographed on Kodak Spectrum Analysis No. 1 plates under high resolution on a 3.4 m Jarrell-Ash Ebert grating spectrograph in the third order of a 1200 grooves/mm grating blazed at μm. The exposure time was 1 h for the stronger 1–0 and 0–0 bands and about 2 h for the 0–1 band. A slit width of 30 μm gave a reciprocal dispersion of 0.58 Å/mm at 3960 Å. The accuracy of the measurements was ±0.003 Å.

A total of 22 bands of the B–X system of $^{13}C^{18}O^+$, degraded to longer wavelengths and occurring in the spectral region 2110–2755 Å, were initially photographed in the first order of a 1200 grooves/mm grating blazed at 1.0 μm, on a 2 m Bausch and Lomb dual-grating spectrograph. Thirteen of these bands were then photographed under high resolution in the fifth and sixth orders of a 1200 grooves/mm grating blazed at 1.4 μm on a 3.4 m Jarrell-Ash Ebert grating spectrograph using Kodak SWR plates. The 0–3, 1–4, and 2–5 bands were recorded in the fifth order and the remaining 10 bands (1–0, 2–1, 0–0, 1–1, 0–1, 1–2, 2–3, 0–2, 1–3, and 2–4) were recorded in the sixth order. The slit width was maintained at 20 μm for the strong bands and at 30 μm for the weak bands to photograph the high-resolution spectra. An exposure time of 30 min was required to photograph the weak bands under medium dispersion on the Bausch and Lomb spectrograph. The exposure times for the high-resolution spectra varied from 5 min for the 0–0 band to 2 h for the 2–4 band. The measured reciprocal dispersions are about 0.2 Å/mm at 2450 Å for high-resolution spectra on the Jarrell-Ash spectrograph and 4.1 Å/mm for the medium-resolution spectrum on the Bausch and Lomb spectrograph. The accuracy of the measurements was ±0.002 Å for the high-resolution spectra and ±0.02 Å for the medium-resolution spectrum.

C. Results and Discussion

The rotational quantum numbers and the vacuum wavenumbers of the spectral lines of eight bands (5–0, 4–0, 3–0, 2–0, 2–1, 1–1, 0–1 and 0–2) of the comet-tail system of $^{13}C^{18}O^+$ are listed as a supplementary publication of the *Journal of Chemical Physics* [9]. The data for the spectral lines of 1–0, 0–0 and 0–1 bands of the Baldet-Johnson system of $^{13}C^{18}O^+$ are listed similarly [10]. The rotational quantum numbers and the vacuum wavenumbers of 13 bands out of total of 22 of the First Negative system, which are rotationally analyzed, are deposited in the Editor's Office of the *Journal of Molecular Spectroscopy*, [11]. The authors [11] may also be contacted for the spectral data. In the rest of this subsection we describe the analysis of the First Negative system of $^{13}C^{18}O^+$.

1. Spectrum of CO$^+$ in the Region 2113 Å–2656 Å

A spectrogram of the First Negative (B $^2\Sigma^+$–X $^2\Sigma^+$) system of $^{13}C^{18}O^+$ photographed under medium dispersion is shown in Figure 8. Of a total of 22 bands observed, 18 are shown in this spectrogram. Because the commercial sample of $^{12}C^{18}O$ used in the present work contained some traces of ^{12}C and ^{16}O, a few bandheads of $^{12}C^{18}O^+$ and $^{13}C^{16}O^+$ were also observed on the photographic plates. In addition, a few bands of the fourth positive (A $^1\Pi$–X $^1\Sigma^+$) system of $^{13}C^{18}O$ can also be seen in Figure 8. The vacuum wavenumbers of the bandheads of $^{13}C^{18}O^+$, their relative intensities, and the vibrational quantum numbers are given by Prasad et al. [11].

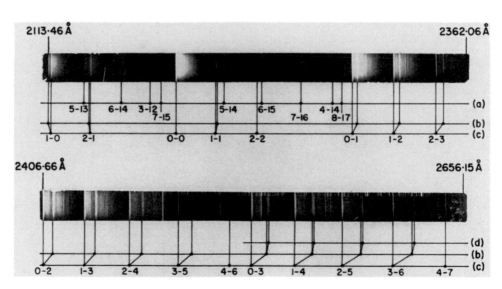

Figure 8 The First Negative System (B $^2\Sigma^+$–X $^2\Sigma^+$) of the $^{13}C^{18}O^+$, photographed on a 2 m Bausch and Lomb dual grating spectrograph in the first order of a 1200 grooves/mm grating in the region 2110–2660 Å. (a) Fourth Positive System (A $^1\Pi$–X $^1\Sigma^+$) of $^{13}C^{18}O$; (b), (c), (d) First Negative System of $^{12}C^{18}O^+$, $^{13}C^{18}O^+$, and $^{13}C^{16}O^+$, respectively.

2. Rotational Analysis of the First Negative System

The First Negative (B $^2\Sigma^+$–X $^2\Sigma^+$) system of $^{13}C^{18}O^+$ represents a transition between two states that belong to Hund's case (b). The rotational structure of a band arising from such a transition consists of four main branches (R_{11ee}, R_{22ff}, P_{11ee}, and P_{22ff}) and two satellite branches, (R_{Q21fe} and P_{Q12fe}) as shown schematically in Figure 9. Here the subscripts 1 and 2 refer to the F_1 and F_2 components, respectively, and e and f refer to the parities of the rotational levels. Because the two satellite branches are normally very weak, they may not even be observed in many cases. If the spin splitting of the spectral lines is not fully resolved, one would see only one R branch and one P branch, instead of two for each branch. As in the B–X system of $^{13}C^{18}O^+$, if the bands are degraded to longer wavelengths,

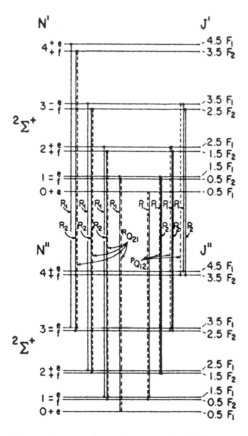

Figure 9   A schematic energy level diagram showing the first few rotational transitions in a band of a $^2\Sigma_u^+$–$^2\Sigma_g^+$ system.

Table 7 Equilibrium Molecular Constants[a] (in cm^{-1}, unless stated otherwise) of the X $^2\Sigma^+$, A $^2\Pi_i$, and B $^2\Sigma^+$ States of $^{13}C^{18}O^+$

Molecular constant	X $^2\Sigma^+$	A $^2\Pi_i$	B $^2\Sigma^+$
T_e	0.0	20731.04(1)	45877.04(1)
ω_e	2110.039(4)	1488.940(4)	1650.665(4)
$\omega_e x_e$	13.738(2)	12.411(2)	23.983(1)
$\omega_e y_e$	−0.0032(2)	0.0228(2)	
B_e	1.79468(2)	1.44258(2)	1.63374(2)
α_e	0.016440(3)	0.016791(4)	0.026003(7)
I_e(g cm^2)	15.5979(3) × 10^{-40}	19.4050(3) × 10^{-40}	17.1344(2) × 10^{-40}
r_e (Å)	1.11546(1)	1.24416(1)	1.16911(1)

[a] The number in parentheses indicates the uncertainty in the last digit and corresponds to one standard deviation.

R branch forms the head. In the present study the spin splitting of the rotational lines was observed in several bands but the satellite branches were not observed. The $|\gamma'-\gamma''|$ difference for the bands of this system is very small; the extent of the splitting in the spectral lines is also very small.

A method of "merging" procedure was used [11] to obtain various molecular parameters for the X $^2\Sigma^+$, A $^2\Pi_i$, and B $^2\Sigma^+$ states of $^{13}C^{18}O^+$. The reader is referred to that work [11] for further details of the "merging" procedure. The

Table 8 Isotope Shifts (in cm^{-1}) in the First Negative System of $^{13}C^{18}O^+$

Band	Observed	Calculated	Band	Observed	Calculated
1–0	+65.6	+65.0	3–5[a]	−272.7	−274.5
2–1	+36.1	+35.5	4–6[a]	−296.2	−298.6
0–0	−11.0	−11.6	0–3	−306.8	−307.2
1–1	−35.6	−36.3	1–4	−323.1	−323.6
2–2[a]	−58.9	−63.1	2–5	−341.5	−341.9
0–1	−112.7	−112.9	3–6[a]	−360.4	−361.9
1–2	−134.6	−134.9	4–7[a]	−382.4	−283.2
2–3	−158.5	−158.8	0–4[a]	−397.8	−400.1
0–2	−211.0	−211.4	1–5[a]	−412.8	−413.8
1–3	−230.2	−230.6	2–6[a]	−428.0	−429.3
2–4	−251.3	−251.8	3–7[a]	−445.8	−446.4

[a] The observed isotope shift is obtained as the difference in the vacuum wavenumbers of the band heads. For the remaining heads, it is the difference in the band origins.

equilibrium molecular constants of the X, A, and B states of $^{13}C^{18}O^+$ resulting from the analysis are listed in Table 7.

3. Isotope Shifts

The observed isotope shifts of all the 22 bands of the First Negative system of $^{13}C^{18}O^+$ relative to the corresponding bands of $^{12}C^{16}O^+$ (i.e., Δv ($^{12}C^{16}O^+ - ^{13}C^{18}O^+$)) are listed in Table 8. The isotope shifts calculated from Eq. (6) are also listed in the same table. In Eq. (6) the vibrational constants for the X and B states were taken from Haridass et al. [7,8]. In Eq. (6) the parameter $\rho = [\mu(^{12}C^{16}O)/\mu(^{13}C^{18}O)]^{1/2} = 0.9530$. The agreement between the observed and the calculated isotope shifts is good in the case of bands for which the band origin data were used and is satisfactory for those for which the band head data were used [11].

VII. $^{15}N_2^+$

A.   Earlier Work

Even though the First Negative (B $^2\Sigma_u^+$–X $^2\Sigma_g^+$) system of $^{14}N_2^+$ was first observed more than a century ago, Fassbender [80] was the first to perform a partial rotational analysis of a few bands and list the band head positions for 36 bands. The previous work done on this system has been reviewed by Lofthus [81], Tyte and Nicholls [82], and Lofthus and Krupenie [12]. Dick and colleagues [83], Gottscho and co-workers [84], Chevaleyre and Perott [85], and Klynning and Pages [86] reanalyzed several bands of this system. Gottscho and coresearchers [84] performed a complete deperturbation analysis of the bands of this system and presented precise constants for the X, A, and B states. They concluded that there is no evidence of $^4\Sigma_u^+$–B $^2\Sigma_u^+$ perturbations and that A $^2\Pi_{i,u}$ is the only state causing perturbations in the B $^2\Sigma_u^+$ state. Recently, Michaud et al. [87,88] did the rotational analysis of the B–X system of $^{14}N_2$ recorded under high-resolution Fourier spectrometry and observed new perturbations in the B $^2\Sigma^+$ ($v = 0$) state up to $N = 84$ by A $^2\Pi$ ($v = 10, 11, 12, 13$).

Wood and Dieke investigated the B–X system of $(^{14}N^{15}N)^+$ [89] and $^{15}N_2^+$ [90] and confirmed the vibrational numbering. In the latter paper, they have estimated the nuclear spin of ^{15}N from the intensity alternations in the rotational structure of the 1–0, 0–0, 1–2, 0–1, and 0–2 bands of $^{15}N_2^+$ but did not obtain any molecular constants from the analysis of the spectra.

Colbourn and Douglas [91] also reported the band origin of the 0–0 band of this system of $^{15}N_2^+$. In our laboratory, the B–X system of $^{15}N_2^+$ was excited in a hollow-cathode discharge tube of special design [22] with a very pure sample of $^{15}N_2$ by Reddy and Prasad [92].

B. Experimental Details

The design of the hollow-cathode discharge tube in which the B–X system of $^{15}N_2^+$ was excited is described in Sec. II and was recorded by photographing the cathode glow of the discharge. Nitrogen-15 gas rated at a purity of 99.9% of ^{15}N (Merck Sharpe and Dohme Canada Limited) was used. A DC voltage of 1100 V applied between the electrodes maintained the discharge at current of -65 mA. The emission spectrum was photographed under medium dispersion in the spectral region 3750–5170 Å on a 2 m Bausch and Lomb dual-grating spectrograph in the first order of a 600 grooves/mm grating blazed at 2.5 µm. The high-resolution spectra of the bands of this system were photographed on the Bausch and Lomb instrument in the third order of a 1200 grooves/mm grating blazed at 1.0 µm and also on a 3.4 m Jarrell-Ash spectrograph in the second and third orders of a 1200 grooves/mm grating blazed at 1.4 µm. The exposure times for the high-resolution spectra varied from 5 s (for the strong lines of the 0–0 band) to 5 hr (for the 0–3, 1–4, and 2–5 bands). Kodak Spectrum Analysis No. 1, 103 a-O, and 103-F plates were used to photograph the spectra. The slit width was maintained at 20 µm on the Bausch and Lomb spectrograph but varied from 15 µm to 30 µm on the Jarrell-Ash spectrograph, depending on the intensity of the band. The reciprocal dispersions of the spectra vary from 0.51 Å/mm at 4200 Å in third order to 0.95 Å in the second order. An Fe-Ne hollow-cathode lamp was used as the source for the reference spectra whose wavelengths were taken from work by Crosswhite [59]. The accuracy of the measurements was $\pm$ 0.002 Å.

C. Analysis of the Spectra

Fifteen bands of the First Negative system of $^{15}N_2^+$ degraded to shorter wavelengths have been identified. In addition, some bands of the Second Positive $(C\,^3\Pi_u–B\,^3\Pi_g)$ system of $^{15}N_2$ can also be seen. In most of the excitation conditions, the bands of this system are generally overlapped by those of the second positive system. Under favorable conditions, the intensity of the Second Positive system may be reduced but it is very difficult to suppress it completely. The weak 3–5 and 3–6 bands of the First Negative system of $^{15}N_2^+$ could not be photographed under high resolution. The vacuum wavenumbers of the band heads of all the remaining 13 bands measured on plates photographed under high resolution, their relative intensities, and the vibrational quantum numbers are given in Table 9.

The rotational structure of the 1–3 band of the First Negative system of $^{15}N_2^+$ photographed under high resolution on the Bausch and Lomb spectrograph is shown in Figure 10. It is clearly shown in this figure that the rotational lines with the odd N values are stronger than those with the even N values, giving

Table 9 Band Heads and Band Origins of the First Negative
$(B\ ^2\Sigma_u^+ - X\ ^2\Sigma_g^+)$ System of $^{15}N_2^+$

Band head (cm^{-1})	Band origin^a (cm^{-1})	Relative intensity^b	Assignment $v'-v''$
27968.4	27997.49^c	m	2–1
27827.6	27855.985(6)	m	1–0
25729.0	25753.656(6)	m	1–1
25538.4	25562.783(6)	vs	0–0
23862.4	23884.296(6)	w	2–3
23659.9	23681.709(7)	m	1–2
23438.7	23460.493(5)	s	0–1
21854.2	21873.709(6)	w	2–4
21620.9	21640.266(6)	m	1–3
21369.1	21388.491(6)	s	0–2
19876.8	19894.179(14)	vw	2–5
19612.5	19629.869(7)	w	1–4
19329.8	19347.246(6)	m	0–3

^{a} The number in the parentheses indicates the uncertainty in the last digit
and corresponds to one standard deviation.
^{b} Abbreviations for relative intensities vs, s, m, w, and vw represent very
strong, strong, medium, weak, and very weak, respectively.
^{c} Calculated from the term values of $v' = 2$ and $v'' = 1$ levels.

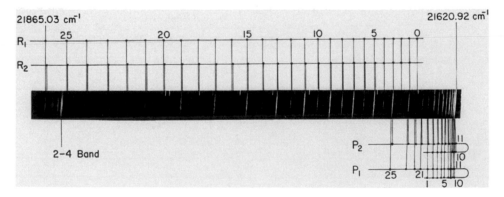

Figure 10 Rotational structure of the 1–3 band of the First Negative System $(B\ ^2\Sigma_u^+ - X\ ^2\Sigma_g^+)$ of $^{15}N_2^+$ photographed on the Bausch and Lomb spectrograph in the third order of a 1200 grooves/mm grating.

alternation of intensity in the band. Similar alternation of intensity is observed for all the bands of this system. The electron spin splitting of the rotational lines of the 1–3 band can also be clearly seen for the spectral lines with $N \geq 5$. Similar splitting of the rotational lines is observed in the other bands with $v' = 1$ for $N \geq 5$ and in those with $v' = 0$ for $N \geq 19$. For the bands with $v' = 2$, for which the rotational lines are identified only up to $N = 19$, the spin splitting is not observed. The satellite branches $^PQ_{12}$ and $^RQ_{21}$ are not observed in any of the bands of this system. The rotational quantum numbers and the vacuum wavenumbers of the spectral lines of all the twelve bands analyzed are listed by Albritton and colleagues [93]. The vacuum wavenumbers of the spectral lines of all the branches of a band were simultaneously fitted to the expressions of the R_1, R_2, P_1, and P_2 branches (see Fig. 9), which are given by

$$R_1(N) = \upsilon_0 + F_1'(N + 1) - F_1''(N) \tag{12}$$

$$R_2(N) = \upsilon_0 + F_2'(N + 1) - F_2''(N) \tag{13}$$

$$P_1(N) = \upsilon_0 + F_1'(N + 1) - F_1''(N) \tag{14}$$

$$P_2(N) = \upsilon_0 + F_2'(N + 1) - F_2''(N) \tag{15}$$

where υ_0 is the band origin

$$F_1(N) = B_\mathrm{v}N(N + 1) - D_\mathrm{v}N^2(N + 1)^2 + 1/2\gamma_\mathrm{v}N \tag{16}$$

and

$$F_2(N) = B_\mathrm{v}N(N + 1) - D_\mathrm{v}N^2(N + 1)^2 + 1/2\gamma_\mathrm{v}(N + 1) \tag{17}$$

and the molecular constants were estimated by the method of least-squares. The $(\upsilon_{obs} - \upsilon_{calc})$ values, obtained from the least-squares fits of the vacuum wavenumbers of individual bands, are given in parentheses in Table 2 of Ref. [7]. However, these are not given for the perturbed lines and also for a very few unperturbed lines that are excluded from the analysis. In general, the standard deviation of such a least-squares fit is -0.03 cm^{-1}.

Using the method of merging proposed by Albritton and colleagues [93] and Coxon [94] in which the molecular constants obtained from the analysis of the individual bands by least-squares fits are combined together, giving due consideration to their uncertainties and the correlations existing among them, a unique set of molecular constants for the B and X states of the First Negative system of $^{15}N_2^+$ was obtained. The band origins thus obtained from the merged least-squares fit are included in Table 9. The uncertainty in the origin of the 2–5 band appears to be high because the number of spectral lines used in the analysis is limited. The B_v, D_v, and γ_v values of states X and B, also obtained from the

Table 10 Rotational Constants[a] (in cm^{-1}) of the X $^2\Sigma_g^+$ and B $^2\Sigma_u^+$ States of $^{15}N_2^+$

Vibrational level	X $^2\Sigma_g^+$			B $^2\Sigma_u^+$		
	B_v	$D_v \times 10^6$	$-\gamma_v \times 10^3$	B_v	$D_v \times 10^6$	$\gamma_v \times 10^3$
0	1.79520(8)	5.5(1)	18.3(4)	1.93731(9)	6.6(2)	1.1(9)
1	1.77817(8)	5.6(1)	18.8(4)	1.91709(7)	6.4(1)	1.2(4)
2	1.76114(8)	5.8(1)	17.2(5)	1.89548(12)	16.0(3)	
3	1.74342(8)	5.5(1)	19.1(5)			
4	1.72639(10)	6.2(2)	19.1(6)			
5	1.70783(23)	4.0(6)				

[a] The number in the parentheses indicates the uncertainty in the last digit and corresponds to one standard deviation.

same merged least-squares fit, are listed in Table 10. The γ_0 value of the B state was estimated exclusively from the wavenumber data of $N = 33$–43 in the 0–0 band and was not included in the merged fit. The value of γ_2 of state B could not be estimated because the spin splitting was not observed in any of the bands with $v' = 2$. Similarly, γ_5 of state X could not be obtained because the $v'' = 5$ level was observed only in the 2–5 band. The B_e and α_e values of states X and B were obtained from the expression

$$B_v = B_e - \alpha_e(v + 1/2) \tag{7-a}$$

using the B_v values listed in Table 10. The B_e and α_e values thus obtained are listed in Table 11. The equilibrium internuclear distance r_e and the corresponding

Table 11 Equilibrium Molecular Constants[a] (in cm^{-1}, or otherwise specified) of the X $^2\Sigma_g^+$ and B $^2\Sigma_g^+$ states of $^{15}N_2^+$

Molecular constant	X $^2\Sigma_g^+$	B $^2\Sigma_g^+$
T_e	0.0	25460.258(9)
ω_e	2132.584(9)	2342.811(9)
$\omega_e x_e$	15.063(4)	24.656(3)
$\omega_e y_e$	−0.0324(4)	
B_e	1.80392(8)	1.94775(10)
α_e	0.01720(1)	0.02046(4)
r_e (Å)	1.1162	1.0742
I_g(g cm^2)	1.5518×10^{-39}	1.4372×10^{-39}

[a] The number in the parentheses indicates the uncertainty in the last digit and corresponds to one standard deviation.

moment of inertia I_e of states X and B, obtained from their respective B_e values, are also presented in the same table.

D. Vibrational Analysis and Isotope Shifts

All the origins of the bands of the B–X system of $^{15}N_2^+$, except that of the 2–1 band (see Table 9), were fitted to the expression

$$
\begin{aligned}
\upsilon_0 = \upsilon_e &+ \omega_e'(v' + 1/2) - \omega_e'x_e'(v' + 1/2)^2 + \omega_e'y_e'(v' + 1/2)^3 \\
&- \omega_e''(v'' + 1/2) + \omega_e''x_e''(v'' + 1/2) - \omega_e''y_e''(v'' + 1/2)^3
\end{aligned}
\tag{18}
$$

The system origin υ_e ($=T_e$) and the vibrational constants were obtained simultaneously from this fit and are listed in Table 11. Here the term value T_e of state B is the same as υ_e of the B–X system. Because only three vibrational levels ($v = 0$, 1, and 2) are observed for the B $^2\Sigma_u^+$ state, $\omega_e y_e$ could not be estimated for it.

The isotope shifts of the bands of the First Negative system of $^{15}N_2^+$ are obtained as the differences of the origins of the $^{14}N_2^+$ bands of this system and those of the corresponding bands of $^{15}N_2^+$ [i.e., $\nu_0(^{14}N_2^+) - \nu_0(^{15}N_2^+)$]. The values of $\nu_0(^{14}N_2^+)$ for the bands of this system were either directly taken from Jakubek et al. [6] or calculated from the vibrational constants reported by them. The values of $\nu_0(^{15}N_2^+)$ are the experimental values obtained by Albritton and colleagues [93]. The observed isotope shifts thus obtained are listed in Table 12. The isotope

Table 12 Isotope Shifts in the First Negative System of $^{15}N_2^+$

Band	Isotope shift $\Delta\nu(^{14}N^{2+}-^{15}N_2^+)$	
	Observed	Calculated
2–1	84.37	84.33
1–0	81.70	81.99
1–1	9.26	9.53
0–0	3.24	3.48
2–3	−54.16	−54.05
1–2	−61.01	−60.76
0–1	−69.21	−68.98
2–4	−120.06	−119.93
1–3	−129.03	−128.85
0–2	−139.44	−139.27
2–5	−183.79	−183.56
1–4	−195.14	−194.72
0–3	−207.68	−207.36

shifts were also calculated from Eq. (6) using the vibrational constants of the X and B states of $^{14}N_2{}^+$ taken from Chevaleyre et al. [85] and the value of $\rho = [\mu(^{14}N_2{}^+/\mu(^{15}N_2{}^+)]^{1/2} = 0.9662$ and are listed in the same table. The agreement between the observed and calculated isotope shifts is very good.

E. Perturbations in the B $^2\Sigma_u^+$ State of $^{15}N_2{}^+$

Among the 12 bands of the first negative system of $^{15}N_2{}^+$, for which the rotational structure is analyzed, some irregularities are observed only in the 0–0, 0–1, 0–2, and 0–3 bands. This indicates that the $v = 0$ level of the B $^2\Sigma_u^+$ state of $^{15}N_2{}^+$ is perturbed. In the 0–1 and 0–3 bands, the rotational lines are identified only up to $N = 21$ and it is difficult to make any conclusions from their structure regarding the perturbations. In the 0–0 and 0–2 bands the spectral lines are identified up to $N = 43$ and 26, respectively, and the perturbations in the $v = 0$ level are confirmed with the help of similar irregularities in the structure of these two bands. From the vacuum wavenumbers of the rotational lines in the 0–0 band it has been noticed that the spectral lines belonging to $R_1(N)$ and $R_2(N)$ branches are completely identified, whereas the lines of $P_1(N)$ and $P_2(N)$ branches with $N = 22$–32 are not. Hence, only the lines of the $R_1(N)$ and $R_2(N)$ branches were used in the analysis of the perturbation. Using the appropriate molecular constants, the wavenumbers of the $R_1(N)$ and $R_2(N)$ lines of the 0–0 band were calculated. The deviations ($v_{obs} - v_{calc}$) for these lines are plotted against N' in Figure 11. It can be seen in this figure that the maximum perturbations are observed at $N' = 24$ and 29 of the F_1 levels and at $N' = 21$ and 26 of the F_2 levels. Two extra lines are observed at the maximum perturbations for R_1 (23) ($N' = 24$) and $R_2(20)$ ($N' = 21$) corresponding to F_1 and F_2 levels respectively. The plots here indicate that this is a heterogeneous perturbation for which $\Delta\Lambda \neq 0$. In such a situation, a Σ_u^+ state can be perturbed only by a Π_u state. Since the maximum perturbations are occurring at four different N values and the perturbed state is the $^2\Sigma_u^+$ state, the perturbing Π_u state is obviously a $^2\Pi_u$ state, which is identified as A $^2\Pi_{i,u}$ on the basis of its proximity to B $^2\Sigma_u^+$.

The molecular constants provided by Jakubek et al. [6] for A $^2\Pi_{i,u}$ of $^{14}N_2{}^+$ and the value of $\rho(= 0.9662)$ were used to obtain the corresponding constants of $^{15}N_2{}^+$. Using these constants of state A of $^{15}N_2{}^+$, the $(T_e + G(v))$ and B_v values of the vibrational levels of state A were calculated. Because the B_v values of state A are smaller than the B_0 value of state B, only the vibrational levels of state A, which are slightly above the $v = 0$ level of state B, can perturb the latter. The $v = 0$ level of state B is at 26625.33 cm^{-1} with respect to the minimum of the potential energy curve of state X. Only the levels with $v \geq 10$ lie above the $v = 0$ level of state B. Hence, any vibrational level with $v \geq 10$ of state A might be perturbing the B $^2\Sigma_u^+$ state. The actual perturbing vibrational level is found by the method of trial and error. According to Kronig [95], at the points of maxi-

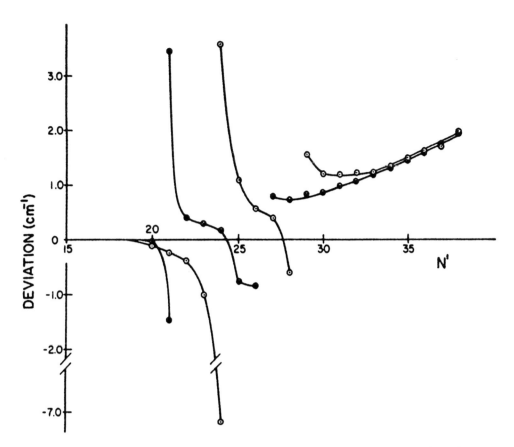

Figure 11 Plot of deviations ($\nu_{obs} - \nu_{calc}$) of the rotational lines of the R_1 and R_2 branches versus the rotational quantum number N' of the upper state in the 0–0 band. The lines joining the dots represent the R_1 branch, and those joining the crosses represent the R_2 branch.

mum perturbation, both the perturbed and the perturbing levels have equal energy and the same J value. Using this principle, the B_v value of the perturbing level was calculated from the corresponding F(J) value, which is the difference between the total energy of the perturbed rotational level ($T_e + G(v) + F(J)$ of state B) and the sum of the electronic and vibrational terms of the perturbing level ($T_e + G(v)$ of state A). The approximate value of B_v for state A, $v = 10$, thus obtained is 1.333 cm^{-1} and the corresponding value calculated from the known values of B_e and α_e is 1.445 cm^{-1}. Levels A, $v \geq 11$, are too high to perturb the observed rotational levels of the B, $v = 0$, state. The calculated position of the A, $v = 11$, is 28466.1 cm^{-1} above the X $^2\Sigma_g^+$ state. It is thus concluded that the level, $v = 10$ is perturbing the B, $v = 0$, level. No perturbations are observed in the $v = 1$ and 2 levels of state B within the observed rotational structure (up to $N = 34$ and 18, respectively).

VIII.   NO

A.   Introduction

Strutt (96) was the first to observe the γ (A $^2\Sigma^+$–X $^2\Pi_r$) system of NO. Over the years numerous investigations have been carried out on different types of spectra of NO: electronic spectra, vibration–rotation spectra, pure rotation spectra, rotational Raman spectra, magnetic rotation spectra, among others. For the previous work on the spectra of NO, the reader is referred to work by Miescher and Huber [97] and Huber and Herzberg [5]. Cisak and coresearchers [98] measured a part of the γ system of $^{14}N^{16}O$ and $^{15}N^{16}O$ recorded under medium resolution of a quartz prism spectrograph and obtained the vibrational constants and isotopic displacements. Engleman and colleagues [99] studied the 1–0 band of the γ system of $^{15}N^{16}O$ and $^{14}N^{18}O$ (in addition to some bands of $^{14}N^{16}O$). Engleman and Rouse [100] studied 17 bands of the γ system of $^{14}N^{16}O$ in the spectral region 2700–3100 Å under high resolution and determined the molecular constants of $v = 0$ to 16 of the X $^2\Pi$ state and $v = 0$ to 5 of the A $^2\Sigma^+$ state. Freedman and Nicholls [101] reanalyzed the high-resolution data of the 1–0 and 0–0 bands of the γ system of $^{14}N^{16}O$ and the 1–2 band of $^{14}N^{18}O$ and $^{15}N^{16}O$, obtained by Engleman and co-workers [99], using the perturbation technique described by Zare and coresearchers [102].

Griggs et al. [103] recorded the infrared 1–0 and 2–0 bands of $^{15}N^{18}O$ in the X $^2\Pi$ state under high resolution and obtained rotational and vibrational constants. Amiot and colleagues [104] recorded the high-resolution Fourier spectra of the infrared 1–0 band of $^{14}N^{16}O$, $^{14}N^{17}O$, $^{14}N^{18}O$, and $^{15}N^{16}O$ and the 2–1 band of $^{14}N^{16}O$, all in the X $^2\Pi$ state and obtained accurate molecular constants for $v = 0$, 1, and 2 of $^{14}N^{16}O$ and $v = 0$ and 1 of the other three isotopomers. Amiot and Guelachvili [105] investigated the infrared 1–0 and 2–1 bands of $^{14}N^{16}O$ and the infrared 1–0 band of $^{15}N^{17}O$ and $^{15}N^{18}O$ with a Fourier spectrometer. Molecular constants were determined by the use of the "direct approach" method. Valentin and coresearchers [106] and Henry et al. [107] reported the analysis of the infrared 1–0, 2–0, and 3–0 bands of $^{14}N^{16}O$ in the X$^2\Pi$ state. Teffo and colleagues [108] recorded the fundamental band (1–0), the high-energy satellite X $^2\Pi_{3/2}$–X $^2\Pi_{1/2}$ (1–0) subband, and the overtone bands 2–0 and 3–0, of $^{15}N^{16}O$ and $^{15}N^{18}O$, and obtained the vibrational constants and the vibrational dependence of the rotational constants. Precise Λ doubling and hyperfine structure constants of $^{14}N^{16}O$ were obtained by Meerts and Dymanus [109] using the molecular beam technique. Patel and Kerl [110] used optoacoustic spectroscopy for the study of $^{14}N^{16}O$ and obtained accurate values of the Λ doubling for the $X\,^2\Pi_{1/2}\ v = 0$ and 1 levels. Dale and coresearchers [111] studied the high-resolution laser magnetic resonance and infrared–radiofrequency double-resonance spectra of the 1–0 and 2–1 bands of the ground $X\,^2\Pi$ state of $^{14}N^{16}O$, $^{15}N^{16}O$, $^{14}N^{17}O$, and $^{14}N^{18}O$.

The γ (A $^2\Sigma^+$–X $^2\Pi_r$) system of the NO isotopomers was recently investigated by Wang and colleagues in our laboratory [112]. In the present section we report the band head measurements of the bands 1–0 and 0–v'' with $v'' = 0$–4 of the γ system of the nitric oxide isotopomers $^{14}N^{16}O$, $^{15}N^{16}O$, $^{14}N^{18}O$, and $^{15}N^{18}O$ in the spectral region 2140–2730 Å, excited in the anode glow of a two-column hollow-cathode discharge tube and photographed under medium resolution. The rotational structure of the 0–1, 0–2, and 0–3 bands of $^{15}N^{18}O$ was also recorded under high resolution. The experimental data of these three bands were combined with those of the infrared 1–0, 2–0, and 3–0 (9, 11, 14) absorption bands of the $X\,^2\Pi_r$ state of $^{15}N^{18}O$ in a global fit, using an effective Hamiltonian for the $^2\Pi$ state given by Brown and colleagues [61] and the matrix elements for the $^2\Pi$ and $^2\Sigma^+$ states given by Amiot et al. [113], and molecular constants were derived for the A $^2\Sigma^+$ and X $^2\Pi$ states.

B. Experimental Details

The γ system of the nitric oxide isotopomers was excited in the anode column of a hollow-cathode discharge tube [20]. Gases nitrogen-15 and oxygen-18 with specified purities of 98% and 99.9%, respectively, were used (Matheson Gas Products). A DC voltage of 1100 V applied between the electrodes of the discharge tube maintained the discharge at a current of ~75 mA when the pressure inside the discharge tube was ~ 0.8 torr. Under optimum experimental conditions the discharge was steady and bright. The spectrum of $^{14}N^{16}O$ was obtained with natural presence of nitrogen-14, oxygen-16 of low-pressure air in the discharge tube. When nitrogen-15 alone was admitted into the discharge tube, $^{15}N^{16}O$ was excited and when oxygen-18 alone was admitted, $^{14}N^{18}O$ was excited. When both nitrogen-15 and oxygen-18 were admitted into the discharge tube at an approximate ratio of 1:2, $^{15}N^{18}O$ was excited.

In our laboratory the γ system of the NO isotopomers was photographed in the spectral region 2140–2730 Å under medium dispersion on a 2.0 m Bausch and Lomb dual-grating spectrograph, both in the first order of a 600 grooves/mm grating blazed at 2.5 μm (with a reciprocal dispersion of 4.1 Å/mm at 2157Å) and in the first order of a 1200 grooves/mm grating blazed at 1.0 μm (with a dispersion of 2.0 Å/mm at 2367 Å). The spectra of the γ system of $^{15}N^{18}O$ in the same spectral region were also photographed under high resolution on a 3.4 m Jarrell-Ash spectrograph equipped with a 1200 grooves/mm grating blazed at 1.4 μm in the fifth order (with a dispersion of 0.34 Å at 2455 Å). The slit width was set at 20 μm for the Bausch and Lomb spectrograph and at 25 μm for the Jarrell-Ash instrument. Iron arc and copper arc, both of DC type, were used as the sources of the standard spectra [114,115] recorded on the former spectrograph and an iron–neon hollow-cathode lamp was used as the standard source [59] for the high-resolution spectra recorded on the latter spectrograph in the fourth order of the grating. Kodak SWR photographic plates were used to record the spectra. Overlapping orders of the spectra were eliminated with a Hoya U-340 optical filter and a chlorine gas filter. The exposure times varied from 3 s to 4.5 hr depending on the spectrograph, the intensity of the band, and the type of filter used. The accuracies of measurement were approximately ± 0.012 Å for the medium-resolution spectra and ± 0.003 Å for the high-resolution spectra.

C.   Analysis of the Spectra

1.   Vibrational Structure

The γ (A $^2\Sigma^+$–X $^2\Pi_r$) system of the nitric oxide isotopomers $^{14}N^{16}O$, $^{15}N^{16}O$, $^{14}N^{18}O$, and $^{15}N^{18}O$ were photographed under medium resolution as stated above. The resulting spectra of $^{14}N^{16}O$ and $^{15}N^{18}O$ are shown in Figure 12. In this figure the isotopic shifts of the bands can be seen very clearly. As the bands are degraded to shorter wavelengths, P_{12ee} and P_{22ff}/Q_{12ef} branches of the $^2\Sigma^+$–$^2\Pi_{3/2}$ subsystem and P_{11ee} and Q_{11ef}/P_{21ff} branches of the $^2\Sigma^+$–$^2\Pi_{1/2}$ subsystem form four distinct heads of a band in the order of the increasing wavenumber. The vacuum wavenumbers for all the 24 band heads of the six bands of each of the four isotopomers of NO, and their vibrational assignments and relative intensities are listed in Table 13. In this table, the data of the 0–1, 0–2, and 0–3 bands of $^{15}N^{18}O$ were obtained from the spectra photographed under high resolution and the rest of the data is from the medium-resolution spectra. The γ system of $^{15}N^{18}O$ and several bands of $^{14}N^{18}O$ of this system were observed for the first time in our laboratory [112].

2.   Rotational Analysis of the 0–1, 0–2, and 0–3 Bands of $^{15}N^{18}O$

The rotational structure of a band degraded to shorter wavelengths of the A $^2\Sigma^+$– X $^2\Pi_r$ system gives rise to 12 branches designated as $^2\Sigma^+$–$^2\Pi_{3/2}$: P_{12ee}, P_{22ff}, Q_{12fe}, Q_{22fe}, R_{12ee}, R_{22ff}, and $^2\Sigma^+$–$^2\Pi_{1/2}$: P_{11ee}, P_{21ff}, Q_{11ef}, Q_{21fe}, R_{11ee}, and R_{21ff} and is shown schematically in Figure 13. In this figure the parity of the rotational levels in the e/f notation described by Brown et al. [60] is used and the assignments of +

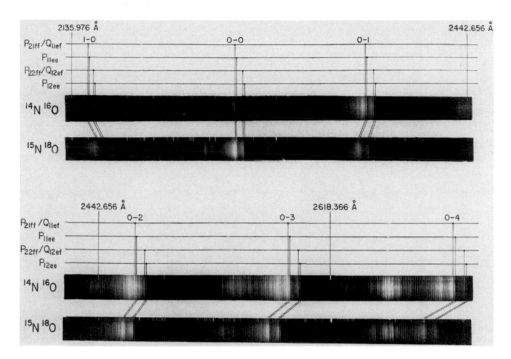

Figure 12 The gamma system of $^{14}N^{16}O$ and $^{15}N^{18}O$ excited in the anode column of a hollow-cathode discharge tube and photographed in the spectral region 2100–2730 Å under medium resolution.

and − signs proposed by Geuzebroek and coresearchers [116] on the basis of experimental evidence are adopted. In the rotational analysis carried out in the present work we used an effective Hamiltonian for the $^2\Pi$ state, which includes the terms representing the vibration and rotation of the nuclei, centrifugal distortion, fine structure, and the hyperfine structure discussed in detail by Brown and coresearchers [61], which is given by

$$H = H_0 + H_{rot} + H_{cd} + H_{fs} + H_{hfs} \tag{19}$$

See Amiot et al. [113] for details of the terms in Eq. (19). The corresponding matrix elements were also calculated by these authors. The matrix elements of the Hamiltonian for the upper $^2\Pi$ state and the lower $^2\Sigma^+$ state of the γ system of NO are listed in Ref. 113. Douay and co-workers [117] made some corrections for the matrix elements γ, γ_D, and γ_H given by Gatterer and Junkes [114]. The matrix elements of these Hamiltonians relevant to the present work are:

Table 13 Band Heads[a,b] (in cm^{-1}) of the A $^2\Sigma^+$–X $^2\Pi_r$ System of $^{14}N^{16}O$, $^{15}N^{16}O$, $^{14}N^{18}O$, and $^{15}N^{18}O$

$^{14}N^{16}O$	$^{15}N^{16}O$	$^{14}N^{18}O$	$^{15}N^{18}O$	Assignment $v'-v''$
46394.1 vw^2	46347.6 w	46329.3 vw	46280.6 w	1–0
46420.3	46370.9	46354.9	46305.5	
46520.3	46470.7	46453.9	46404.3	
46541.0	46488.4	46470.6	46426.6	
44051.7 w	44047.5 m	44049.6 m	44044.8 m	0–0
44076.4	44071.2	44072.8	44068.4	
44174.2	44172.1	44171.1	44171.2	
44194.4	44193.0	44187.9	44189.5	
42177.8 s	42207.8 s	42221.4 s	42252.31 s	0–1
42201.3	42230.0	42243.8	42273.92	
42300.8	42330.4	42344.6	42375.99	
42320.9	42350.9	42360.4	42394.27	
40332.5 s	40393.5 s	40422.3 s	40486.36 s	0–2
40354.3	40414.8	40442.9	40506.52	
40454.9	40517.7	40544.3	40609.25	
40473.1	40534.6	40560.8	40626.65	
38513.6 vs	38606.2 vs	38649.5 vs	38746.01 vs	0–3
38534.1	38625.9	38669.3	38764.86	
38636.2	38726.9	38770.9	38868.29	
38653.1	38745.2	38787.2	38884.54	
36723.8 s	36846.0 s	36903.3 s	37030.3 s	0–4
36742.8	36864.7	36920.4	37048.0	
36845.4	36964.9	37024.9	37152.0	
36864.1	36984.2	37039.3	37167.2	

[a] The four heads identified for each band are formed by the P_{12ee}, P_{22ff}/Q_{12ef}, P_{11ee} and P_{21ff}/Q_{11ef} branches in order of increasing wavenumber.
[b] The 1–0, 0–0, and 0–4 band heads are measured under the medium dispersion and the 0–1, 0–2, and 0–3 band heads of $^{15}N^{18}O$ are measured under the high dispersion.
[c] The abbreviations for the relative intensities vs, s, m, w and vw represent very strong, strong, medium, weak and very weak, respectively.

$$^2\Sigma^{e,f}: T_v + B_v(x+1)(x+1\mp 1) - D_v(x+1)^2(x+1\mp 1)^2$$
$$^2\Pi_{3/2}^{e,f}: T_v + 0.5A_v - 0.5A_{Dv}[(x+1)]^2 - 1] + B_v[(x+1)^2 + 1]$$
$$- D_v(x+1)^2[(x+1)^2 - 1]$$
$$^2\Pi_{1/2}^{e,f}: T_v - 0.5A_v - 0.5A_{Dv}[(x+1)^2 + 1] + B_v[(x+1)^2 + 1] \quad (20)$$
$$- D_v(x+1)^2[(x+1)^2 + 3] \mp 0.5p_v(x+1) \mp q_v(x+1)$$
$$\{^2\Pi_{3/2}, {}^2\Pi_{1/2}\}^{e,f}: -B_v[(x+1)^2 - 1]^{1/2} + 2.0D_v(x+1)^2[(x+1)^2 - 1]^{1/2}$$
$$\pm 0.5q_v(x+1)[(x+1)^2 - 1]^{1/2}$$

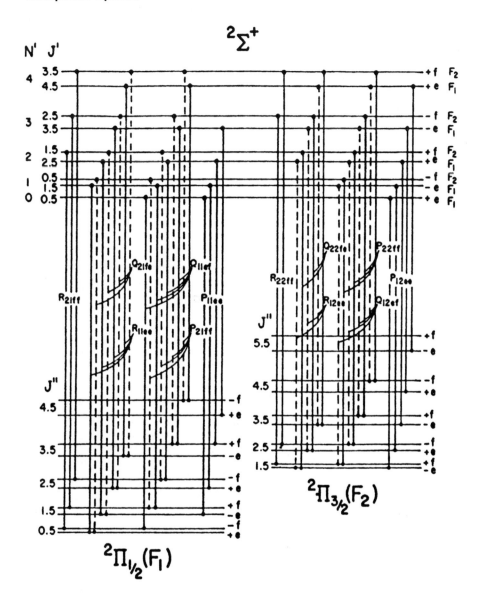

Figure 13 Schematic energy level diagram shows the first few rotational transitions of all 12 branches of a band of a $^2\Sigma^+ - ^2\Pi_r$ transition. Dashed lines are satellite branches.

where B_v and D_v are the rotational constants, A_v and A_{dv} are the spin-orbit constants, p and q are the Λ-doubling constants of the Π state, and $x = (J + 0.5) - 1$. The upper and lower state signs ($\pm$ or $\mp$) in the preceding matrix elements refer to the e/f levels, respectively.

Of the three bands of the γ system of $^{15}N^{18}O$ photographed under high resolution on the 3.4 Jarrell-Ash spectrograph, the rotational structure of the 0–2 band is shown in Figure 14. In this figure the four heads and the eight branches expected in a $^2\Sigma^+ - ^2\Pi_r$ transition are clearly identified. The pairs of branches P_{22ff} and Q_{12ef}, and R_{12ee} and Q_{22fe} of the $^2\Sigma^+ - ^2\Pi_{3/2}$ subsystem and Q_{11ef} and P_{21ff}, and Q_{21fe} and R_{11ee} of the $^2\Sigma^+ - ^2\Pi_{1/2}$ subsystem could not be resolved into two separate branches in the present work. The rotational quantum numbers and the vacuum wavenumbers of the spectral lines of the 0–1, 0–2 and 0–3 bands are listed by Brown and colleagues [61].

The wavenumber data of the lines of the 0–1, 0–2, and 0–3 bands of the γ system of $^{15}N^{18}O$ [112] can in principle be analyzed by one of the following methods. The first involves fixing the molecular constants of the X $^2\Pi$ state,

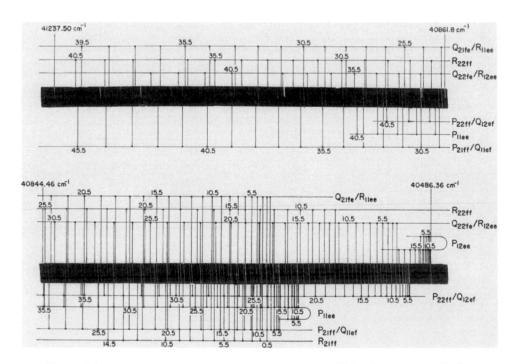

Figure 14   Rotational structure of the 0–2 band of the γ (A $^2\Sigma^+$ –X $^2\Pi$) system of $^{15}N^{18}O$, photographed on the 3.4 m Jarrell-Ash spectrograph in the fifth order of a 1200 grooves/mm grating.

which have been determined exclusively from the infrared vibration-rotation measurements [11,14], and then obtaining the molecular constants of the A $^2\Sigma^+$ state from our data of the γ system. The second entails analyzing the wavenumber data of the infrared vibration–rotation bands together with the γ system in a global fit. In our work [112] we have adopted the second method because the molecular constants of the ground $^2\Pi$ state of the γ system of $^{15}N^{18}O$ [103,105,109] were determined by different methods stated briefly here.

Griggs and coresearchers [103] used separate expressions for the $T(v,J)$ of the infrared 1–0 and 2–0 bands in the individual $^2\Pi_{1/2}$ and $^2\Pi_{3/2}$ components and did not take into account the interaction between them. Amiot and Guelachvili [105] adopted a "direct approach method" in the analysis of the 1–0 infrared band. In this method, the matrix elements of the Hamiltonian were derived considering that the main contributor to the Λ doubling is a $^2\Sigma^-$ state. Teffo and colleagues [108] analyzed the 2–0, 3–0 infrared bands and the $^2\Pi_{3/2}-^2\Pi_{1/2}$ (1–0) subband using the matrix elements given by Zare et al. [102].

The wavenumber data of the 0–1, 0–2, and 0–3 bands of the γ system of $^{15}N^{18}O$ and the infrared data of the vibration-rotation spectra of $^{15}N^{18}O$ arising from the ground $^2\Pi$ state [103,105,108] and estimated rotational constants of the A and X states are included. A simultaneous, weighted, nonlinear least-squares fit of all lines of these data was carried out. The weights are calculated from the expression $w = 1/\sigma^2$, where σ is the uncertainty for the spectral positions of the lines. The values of σ are 0.008 cm^{-1} for the data of Griggs et al. [103], 0.0002 cm^{-1} for those of Valentin and colleagues [106], 0.0005 cm^{-1} for those of Meerts and Dymanus [109], and in the range 0.01–0.1 cm^{-1} for our data. Residuals between the calculated and observed frequencies of the optical data for all three bands are given by Meerts and Dymanus [109] (Table 2). For a total of 1308 wavenumbers used in the fit, the variance was 1.46 and the root mean square deviation was only ~0.065 cm^{-1}. The molecular constants of the $v = 0$ level of the A state and those of the $v = 0, 1, 2,$ and 3 of the X state thus obtained are listed in Table 14. From this table it is clear that the molecular parameters for the $^2\Pi_r$ state show a smooth variation with $(v + \frac{1}{2})$. It should be mentioned that the Λ-doubling parameter q for $v = 2$ and 3 is fixed in the present analysis. The signs of the Λ-doubling parameters convey the information about the character of the perturber. Vivie and Peyerimhoff [118] concluded, based on theoretical considerations, that the $^2\Sigma^+$ and the G $^2\Sigma^-$ state are dominant perturbers and determine the magnitude of Λ splitting of the X $^2\Pi$ state. The sign of the parameter p depends not only on the symmetry and location of the perturber, but also on the sign of the effective spin-orbit coupling parameter between the Σ and the Π states. The sign of q is determined by the symmetry and location of the dominant perturber. Thus, if the value of the Λ-doubling parameter q is $+$ve, the dominant perturber is either Σ^+ above the Π state or Σ^- below the Π state, whereas if the value of q is $-$ve the dominant perturber is Σ^- above the Π state or Σ^+ below

Table 14   Rotational Constants[a] (in cm^{-1}) of the X $^2\Pi_r$ and A $^2\Sigma^-$ States of $^{15}N^{18}O$

Molecular constant	$v = 0$	$v = 1$	$v = 2$	$v = 3$
A $^2\Sigma^+$				
T	44130.244(2)			
B	1.812980(7)			
$D \times 10^6$	4.694(6)			
X $^2\Pi_r$				
$T(^2\Pi_{1/2})$	0	1793.30725(4)	3560.977(1)	5303.050(2)
B	1.548215(2)	1.532899(2)	1.517569(5)	1.50218(1)
$D \times 10^6$	4.551(5)	4.568(4)	4.595(6)	4.62(1)
A	123.1429(1)	122.9109(1)	122.665(2)	122.41(3)
$A_D \times 10^4$	1.518(7)	1.442(7)	1.37(4)	1.36(9)
$p \times 10^2$	1.067(1)	1.066(1)	1.08(1)	1.09(2)
$q \times 10^5$	7.6(2)	7.6(1)	1.0[b]	1.0[b]

[a] Number in parentheses is the uncertainty in the last digit and corresponds to one standard deviation.
[b] Fixed in the least-squares fit.

the Π state. The value of both p and q of the X $^2\Pi$ state are +ve in [112] (see also 119, 120). The B_v values of the $^2\Pi_r$ state thus obtained were fitted to the relation.

$$B_v = B_e - \alpha_e(v + 1/2) + \gamma_e(v + 1/2)^2 \tag{7-b}$$

and the equilibrium rotational constants B_e, α_e, and γ_e thus obtained are given in Table 15. The D_v values were fitted to the relation

$$D_v = D_e + \beta_e(v + 1/2) \tag{8-a}$$

and the resulting values of D_e and β_e are listed in Table 15. The equilibrium molecular constants A_e, α_A, and β_A are determined by fitting the values of the spin-orbit constant A_v given in Table 14 to the expression

$$A_v = A_e - \alpha_A(v + 1/2) + \beta_A(v + 1/2)^2 \tag{21}$$

and their values are given in Table 15. From the T values given in Table 14, we have determined the vibrational constants ω_e, $\omega_e x_e$, and $G(0)$ of $^2\Pi_r$; these are listed in Table 15. The $^2\Pi$ state neither strictly belongs to Hund's case (a) (large $|A|/B$) nor to Hund's case (b) (small $|A|/B$) in most of the actual cases, but belongs to mixed case [case (a) for the small rotation and case (b) for large rotation]. A Fortrat diagram for the 0–2 band of the γ system of $^{15}N^{18}O$ is shown in Figure 15. In this figure, with increasing J, the branches P_{11ee} and P_{22ff}, Q_{11ef} and Q_{22fe}, and R_{11ee} and R_{22ff}, draw closer together. This indicates the transition of X $^2\Pi$

Table 15 Equilibrium
molecular constants[a] (in cm^{-1})
of the X $^2\Pi_r$ State of $^{15}N^{18}O$

Constant	Value
$G(0)$	906.25(2)
ω_e	1818.92(2)
$\omega_e x_e$	12.809(5)
B_e	1.55585(2)
$\alpha_e \times 10^2$	1.527(2)
$\gamma_e \times 10^5$	$-1.8(5)$
$D_e \times 10^6$	4.537(3)
$\beta_e \times 10^8$	2.3(2)
A_e	123.258(3)
α_A	0.226(4)
$\beta_A \times 10^3$	$-4.5(9)$

[a] Number in parentheses is the uncertainty in the last digit and corresponds
to one standard deviation.

state from Hund's case (a) at lower J values to case (b) at higher J values [see,
for example, (21)]. The value of $A/B \approx 79$ for the ground $^2\Pi$ state.

D. Vibrational Analysis

The vibrational quanta $\Delta G(v + 1/2)$ between the levels v and $v + 1$ of a component of an electronic state are represented by [32]:

$$\Delta G(v + 1/2) = G(v + 1) - G(v) \tag{22}$$
$$= (\omega_e - \omega_e x_e) - 2\omega_e x_e(v + 1/2)$$

The observed vibrational quanta $\Delta G(v + \frac{1}{2})$ for the P_{21ff}/Q_{11ef} band heads in the
$^2\Pi_{1/2}$ component and for the P_{22ff}/Q_{12ef} band heads in the $^2\Pi_{3/2}$ component, of
$^{14}N^{16}O$, $^{15}N^{16}O$, and $^{14}N^{18}O$ (see Table 13) are separately fitted to a linear least-
squares program and the values of $(\omega_e - \omega_e x_e)$ and $2\omega_e x_e$ as per Eq. (22) were
obtained. The values of the vibrational constants ω_e and $\omega_e x_e$ obtained from the
observed band head data for the two components of the X $^2\Pi$ state for the three
isotopomers are listed in Table 16. In the same table we have listed the value of
$\Delta G'(\frac{1}{2})$ for $^{14}N^{16}O$, $^{15}N^{16}O$, and $^{14}N^{18}O$ obtained from the average value of $\Delta G'(\frac{1}{2})$
from the P_{21ff}/Q_{11ef} and P_{22ff}/Q_{12ef} band heads, since only two levels $v' = 0$ and
1 are observed for $A^2\Sigma^+$ in [112].

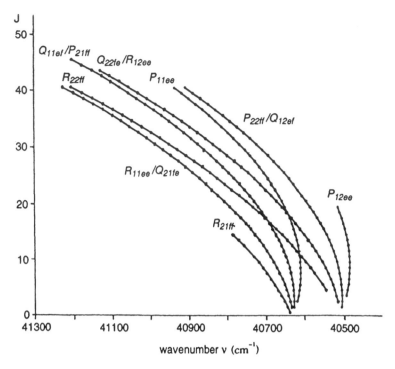

Figure 15   Fortrat diagram of the 0–2 band of the gamma system of $^{15}N^{18}O$.

Table 16   Vibrational Constants[a,b] (in cm^{-1}) of the $X\,^2\Pi_r$ and $A\,^2\Sigma^+$ states of $^{14}N^{16}O$, $^{15}N^{16}O$, and $^{14}N^{18}O$

		$^{14}N^{16}O$	$^{15}N^{16}O$		$^{14}N^{18}O$	
State	Constant	Observed value	Observed value	Calculated value	Observed value	Calculated value
$A\,^2\Sigma^+$	$\Delta G'(1/2)$	2345.3	2297.6		2282.4	
$X\,^2\Pi_{3/2}$	ω_e	1902.95(64)	1868.06(85)	1868.92(63)	1855.0(1.3)	1852.83(62)
	$\omega_e x_e$	13.91(27)	13.29(18)	13.42(26)	13.38(28)	13.19(26)
$X\,^2\Pi_{1/2}$	ω_e	1902.9(1.9)	1869.75(96)	1868.9(1.9)	1853.35(86)	1852.8(1.8)
	$\omega_e x_e$	14.07(42)	13.51(21)	13.57(41)	13.24(18)	13.34(40)

[a] Number in parentheses is the uncertainty in the last digit and corresponds to one standard deviation.
[b] Obtained from the bandhead data given in Table 13.

ACKNOWLEDGMENTS

This work was supported in part by Research Grant No. A-2440 awarded to S.P.R. by the Natural Sciences and Engineering Research Council of Canada. The assistance received from George Varghese during the preparation of the manuscript is also acknowledged.

REFERENCES

1. AB Pluvinel, F Baldet. Compt Rend 148: 759–762, 1909.
2. TA Dixon, RC Woods. Phys Rev Lett 34: 61–63, 1975.
3. NR Erickson, RL Snell, RB Loren, L Mundy, RL Plambeck. Astrophys J 245: L83, 1981.
4. PH Krupenie. Natl Bur Stand 5: 1966.
5. HP Huber, G Herzberg. Molecular Spectra and Molecular Structure. Vol. IV. Princeton, NJ: Van Nostrand-Reinhold, 1979.
6. Z Jakubek, R Kepa, A Para, M Rytel. Can J Phys 65: 94–100, 1987.
7. C Haridass, CVV Prasad, SP Reddy. Astrophys J 388: 669–677, 1992.
8. C Haridass, CVV Prasad, SP Reddy. J Mol Spectrosc 199:180–187, 2000.
9. CVV Prasad, SP Reddy. J Chem Phys 90: 3010–3014, 1989.
10. SP Reddy, CVV Prasad. J Chem Phys 91: 1972–1977, 1989.
11. CVV Prasad, SP Reddy. J Mol Spectrosc 144: 323–333, 1990.
12. A Lofthus, PH Krupenie. J Phys Chem Ref Data 6: 113–307, 1977.
13. RN Clayton. In: J Audouze, ed. CNO Isotopes in Astrophysics. Boston: Reidel, 1977.
14. SP Reddy, CVV Prasad. Astrophys J 331: 572–582, 1988.
15. IS McDermid, JB Laudenslager. J Quant Spectrosc Radiat Transfer 27: 483–492, 1982.
16. B Kivel, H Mayer, H Bethe. Ann Phys 2:57–80, 1957.
17. SP Langhoff, CW Bauschlicher Jr, H Patridge. J Chem Phys 89: 4909–4917, 1988.
18. MD Burrows, SL Banghcum, RC Oldernorg. Appl Phys Lett 46: 22–24, 1985.
19. G Herzberg, H Lew, JL Sloan, JKG Watson. Can J Phys 59:425–440, 1981.
20. SP Reddy, CVV Prasad. J Phys E Sci Instrum 22: 306–308, 1989.
21. G Herzberg. Molecular Spectra and Molecular Structure I. Spectra of Diatomic Molecules. Malabar, FL: Kreger, 1991, pp. 149, 261, 315.
22. A Fowler. Monthly Notices R Astron Soc 70: 176–182, 1909.
23. A Fowler. Monthly Notices R Astron Soc 70: 484–496, 1910.
24. F Baldet. Compt Rend 180: 271–273; 820–822, 1925.
25. RT Binge. Nature 116: 170–171L, 1925.
26. D Coster, HH Brons, H Bulthuis. Z Physik 79: 787–822, 1932.
27. R Schmid, L. Gerö. Z Physik 86: 297–313, 1933.
28. H. Bulthuis. Proc Acad Sci Amsterdam 38: 604–617, 1935.
29. KN Rao. Astrophys J 111: 306–313, 1950.
30. RK Asundi, RK Dhumwad, AB Patwardhan. J Mol Spectrosc 34: 528–532, 1970.

31. RK Dhumwad, AB Patwardhan, VT Kulkarni. J Mol Spectrosc 78: 341–343, 1979.
32. H Gagnaire, JP Goure. Can J Phys 54: 2111–2117, 1976.
33. JA Coxon, SC Foster. J Mol Spectrosc 93: 117–130, 1982.
34. DH Katayama, JA Welsh. J Chem Phys 75: 4224–4230, 1981.
35. RD Brown, RG Dittman, DC McGilvery. J Mol Spectrosc 104: 337–342, 1984.
36. BR Vujisić, DS Pesić. J Mol Spectrosc 128: 334–339, 1988.
37. CM Blackburn. Proc Natl Acad Sci 11: 28–34, 1925.
38. H Biskamp. Z Physik 86: 33–41, 1933.
39. RH Schmid. Phys Rev 42: 182–188, 1932.
40. LH Woods. Phys Rev 63: 431–432, 1943.
41. P Misra, DW Ferguson, KN Rao, E Williams Jr, CW Mathews. J Mol Spectrosc 125: 54–65, 1982.
42. F. Baldet. Compt Rend 178: 1525–1527, 1924.
43. RC Johnson. Proc R Soc (London) Ser A 108: 343–355, 1925.
44. H Bulthuis. Physica 1: 873–880, 1934.
45. KN Rao, KS Sarma. Mem Soc R Sci Liege Collect 13: 181–1986, 1953.
46. L Ćonkić, JD Janjić, DS Pesić, D Rakotoarijimy, BR Vujisić, S Weniger. Astrophys J 226: 1162–1170, 1978.
47. J Marchand, JD Incan, J Janin. Spectrochim Acta 25A: 605–609, 1969.
48. D Cossart, C Cossart-Magos. J Mol Spectrosc 141: 59–78, 1990.
49. KVLN Sastry, P Helminger, E Herbst, FC DeLucia. Astrophys J 250: L91–L92, 1981.
50. ND Piltch, PG Szanto, TG Anderson, CS Gudeman, TA Dixon, RC Woods. J Chem Phys 76: 3385–3388, 1982.
51. FC van den Heuvel, WL Meerts, A Dymanus. Chem Phys Lett 92: 215, 1982.
52. M Bogey, C Demuynck, JL Destombes. Mol Phys 46: 679–681, 1982.
53. M Bogey, C Demuynck, JL Destombes. J Chem Phys 79: 4704–4707, 1983.
54. PB Davies, WJ Rothwell. J Chem Phys 83: 5450–5452, 1985.
55. KN Rao. Astrophys J 111: 50–59, 1950.
56. Z Bembenek, U Domin, R Kepa, K Porada, M Rytel, M Zachwieja, Z Jakubek, JD Janjić. J Mol Spectrosc 165: 205–218, 1994.
57. TA Miller, VE Bondybey. Chem Phys Lett 50: 275–277, 1977.
58. VE Bondybey, TA Miller. J Chem Phys 69: 3597–3602, 1978.
59. HM Crosswhite. J Res Natl Bur Stand A Phys Chem A79: 17–68, 1975.
60. JM Brown, JT Hougen, KP Huber, JWC Johns, I Kopp, H Lefebvre-Brion, AJ Merer, DA Ramsay, J Rostas, RN Zare. J Mol Spectrosc 55: 500–503, 1975.
61. JM Brown, EA Colbourn, JKG Watson, FD Wayne. J Mol Spectrosc 74: 294–318, 1979.
62. C Amiot, JP Maillard, J Chauville. J Mol Spectrosc 87: 196–218, 1981.
63. M Douay, SA Rogers, PF Bernath. Mol Phys 64:425–436, 1988.
64. LT Earls. Phys Rev 48: 423–424, 1935.
65. BR Vujisić, DS Pešić. J Mol Spectrosc 128: 334, 1988.
66. RD Brown, RG Dittman, DC McGilvery, PD Godfrey. J Mol Spectrosc 101: 61, 1983.
67. Z Jakubek, R Kepa, JD Janjić. J Mol Spectrosc 125: 43, 1987.

68. BR Vujisić, DS Pešić, S Weniger, D Rakotoarijimy. Indian J Pure Appl Phys 18: 370, 1980.
69. CVV Prasad, GL Bhale, SP Reddy. J Mol Spectrosc 104: 165, 1984.
70. CVV Prasad, SP Reddy, M Sandys-Wunsch. J Mol Spectrosc 114: 436, 1985.
71. CVV Prasad, GL Bhale, SP Reddy. J Mol Spectrosc 121: 261, 1987.
72. CVV Prasad, SP Reddy. J Mol Spectrosc 121: 261, 1987.
73. L. Conkić, JD Janjić, DS Pešić, D Rakotoarijimy, VR Vujisić, S Weniger. Astrophys J 226: 1162, 1978.
74. RD Brown, RG Dittman, DG McGilvery, PD Godfrey. J Mol Spectrosc 101: 61, 1983.
75. Z Jakubek, R Kepa, Z. Rzeszut, M Rytel. J Mol Spectrosc 119: 280, 1986.
76. RR Shvangiradze, KA Oganezov, B Ya Chikhladze. Opt Spectrosc 8: 239–242, 1960.
77. D Pešić, J Janić, D Marković, M Rytel, T. Sewiec. Bull Chem Soc Belgrade 39: 249–255, 1974.
78. JD Janjić, DS Pešić. Astrophys J 209: 642–647, 1976.
79. DS Pešić, D Ž Marković. DS Janković. Fiz 7: 83–89, 1975.
80. M Fassbender. J Physik 30: 73, 1924.
81. A Lofthus. Spectroscopic Report No. 2. University of Oslo, Norway: Department of Physics, 1960.
82. DC Tyte, RW Nicholls. Identification Atlas of Molecular Spectra No. 3. London, Ontario, Canada: Department of Physics, The University of Western Ontario, 1965.
83. A Dick, W Benesch, HM Crosswhite, SG Tilford, RA Gottscho, RW Field. J Mol Spectrosc 69: 95–108, 1978.
84. RA Gottscho, RW Field, KA Dick, W Benesch. J Mol Spectrosc 74: 435–455, 1979.
85. J Chevaleyre, JP Perott. J Mol Spectrosc 85: 85–96, 1981.
86. L Klynning, P Pages. Phys Scripta 25: 543–560, 1982.
87. F Michaud, F Roux, S David, An-Dien Nguyen. Appl Opt 35: 2867–2873, 1996.
88. F Michaud, F Roux, SP Davis, An-Dien Nguyen, CO Laux. J Mol Spectrosc 203: 1–8, 2000.
89. RW Wood, GH Dieke. J Chem Phys 6: 734–739, 1938.
90. RW Wood, GH Dieke. J Chem Phys 8: 351–361, 1940.
91. EA Colbourn, AE Douglas. J Mol Spectrosc 65: 332–333, 1977.
92. SP Reddy, CVV Prasad. Astrophys J 331: 572–582, 1988; 337: 579, 1989.
93. DL Albritton, AL Schmeltekopf, RN Zare. J Mol Spectrosc 67: 132–156, 1977.
94. JA Coxon. J Mol Spectrosc 72: 252–263, 1978.
95. R de L Kroning. J Physik 50: 347, 1928.
96. RJ Strutt. Proc R Soy Ser A 93: 254, 1917.
97. E Miescher, KP Huber. Int Rev Sci, Phys Chem 3: 37–73, 1976.
98. H Cisak, J Danielak, M Rytel. Acta Phys Pol A 37: 67–70, 1970.
99. R Engleman, Jr PE Rouse, HM Peek, VD Baiamonte. Los Alamos Scientific Laboratory Report LA-4364, 1970.
100. R Engleman, Jr, PE Rouse. J Mol Spectrosc 37: 240–251, 1971.
101. R Freedman, RW Nicholls. J Mol Spectrosc 83: 223–227, 1979.

102. RN Zare, AL Schmeltekopf, WJ Harrop, DL Albritton. J Mol Spectrosc 46: 37–65, 1973.
103. JL Griggs, KN Rao, LH Jones, RM Potter. J Mol Spectrosc 22: 383–401, 1967.
104. C Amiot, R Bacis, G Guelachvili. Can J Phys 56: 251–265, 1978.
105. C Amiot, G Guelachvili. J Mol Spectrosc 76, 86–103, 1979.
106. A Valentin, A Henry, Ph. Cardinet, MF Le Maol, Da Wu Chen, KN Rao. J Mol Spectrosc 70, 9–17, 1978.
107. A Henry, MF Le Moal, Ph. Cardinet, A Valentin. J Mol Spectrosc 70: 18–26, 1978.
108. JL Teffo, A Henry, Ph. Cardinet, A Valentin. J Mol Spectrosc 82, 348–363, 1980.
109. WL Meerts, A Dymanus. J Mol Spectrosc 44: 320–346, 1972.
110. CKN Patel, RJ Kerl. Opt Commun 24: 294–296, 1978.
111. RM Dale, JWC Johns, ARW McKellar, M Riggin. J Mol Spectrosc 67: 440–458, 1977.
112. DX Wang, C Haridass, SP Reddy. J Mol Spectrosc 175: 73–84, 1996.
113. C Amiot, JP Maillard, J Chauville. J Mol Spectrosc 87: 196–218, 1981.
114. A Gatterer, J Junkes. Astrophysical Laboratory of the Vatican Observatory, 2nd ed. Cita Del Vaticano: Specola Vaticana, 1947.
115. AG Shenstone. J Opt Soc Am 45: 868, 1955.
116. FH Guezebroek, MG Tenenr, AW Kleyn, H Zacharias, S Stolte. Chem Phys Lett 187: 520–526, 1991.
117. M Douay, SA Rogers, PF Bernath. Mol Phys 64: 425–436, 1988.
118. R de Vivie, SD Peyerimhoff. J Chem Phys 90: 3660–3670, 1989.
119. RS Mulliken, A Christy. Phys Rev 38: 87–119, 1931.
120. JWC Johns, J Reid, DW Lepard. J Mol Spectrosc 65: 155–162, 1977.

4

Photoabsorption Cross-Section Measurements in the Ultraviolet and Vacuum Ultraviolet

Kouichi Yoshino
*Harvard–Smithsonian Center for Astrophysics,
Cambridge, Massachusetts*

I. INTRODUCTION

Although wavelength measurements in spectroscopy have been very well performed with photographic plates and films, photographic intensity measurements were limited by poor dynamic range, nonlinearity, and technical difficulties. However, after development of the photomultiplier there were many measurements of photoabsorption coefficients in the Ultraviolet (UV) and Vacuum Ultraviolet (VUV) region for many molecules during the period 1960 to 1980.

The ratio of the incident intensity $I_0(v)$ to the intensity transmitted $I(v)$ through a medium of column density N (cm^{-2}) is related to the absorption cross-section $\sigma(v)$ (cm^2 mol^{-1}) by the formula (Beer-Lambert law):

$$\ln \frac{I_0(v)}{I(v)} = \sigma N \tag{1}$$

The quantity σN is called *optical depth* (also opacity or optical thickness). Two other expressions also are used for absorption measurements: $I(v) = I_0(v)e^{-kpl}$ and $I(v) = I_0(v)10^{-\varepsilon cl}$. The absorption coefficients κ and ϵ are given in units of atm^{-1} cm^{-1} and dm^3 mol^{-1} cm^{-1}, respectively. In the above expressions l is the path length in units of cm, p is pressure in units of Torr or atm, and c is the concentration in units of mol dm^{-3}. Conversion factors between these absorption coefficients (cross sections) are given in Okabe's Table A-3 [1].

II. CROSS-SECTION MEASUREMENTS BEFORE 1980

Absorption intensity measurements are the observation of the optical density under known conditions such as the column density, path length, pressure of the media, and temperature. Some earlier measurements presented the absorption coefficients and the pathlength, but these data cannot be reduced to the absolute scale without knowledge of the other parameters. Problems with these measurements include estimation of $I_0(\nu)$, the resolution of the spectrometer, and the wavelength calibration. Most of instruments for the VUV region are single-path, and the intensity of the incident light $I_0(\nu)$ has to be assumed during the spectral scan. This reflects on the uncertainty of the absolute base for absorption. Absorption cross-section (coefficient) measurements were performed using relatively low resolution instruments (mostly 1 m spectrometer). As a result, the cross-section measurements for molecular bands with fine structures were severely distorted by the instrumental band widths, as pointed out by Hudson and Carter [2]. Wavelength calibration is essential to any spectroscopic measurements. Unfortunately, the early workers did not have accurate calibration of wavelengths. Therefore, we cannot simply compare the cross sections at the specific wavelength because they may be at the shifted wavelengths. The cross sections for small molecules up until 1977 were summarized by Okabe [1].

III. INSTRUMENTATION FOR CROSS-SECTION
MEASUREMENTS

It is important to have a clean and stable continuum background source to make absorption cross-section measurements. We have available a xenon lamp operated by DC discharge in the UV region, with molecular hydrogen (or deuterium) continuum from the bound to free molecular transitions in the wavelength region above 170 nm. Occasionally the hydrogen source was used as the background source below 170 nm, but the spectrum of hydrogen below 170 nm consists of many molecular rotational lines. For low-resolution measurements, the spectrum of molecular hydrogen looks like a continuum. The most convenient to use and strongest continuum source in the VUV region is synchrotron radiation source.

Temperature-dependent absorption cross sections have become important for applications in the atmospheric and astrophysical sciences. The absorbants in cells can be cooled by a bath of precooled liquid such as methanol down to their freezing point. Liquid nitrogen can used to cool down the gases to 78 K, but this application is limited to selected molecules, which have sufficient vapor pressure at this temperature. However, the supersonic expansion technique can provide very cold gases around 20 K. The temperature reached by the supersonic expansion technique depends on the expansion rate, and the observation is limited by small column density. Of course, the column density in the supersonic expanded gases cannot be estimated directly, but can be deduced from the known band oscillator strengths.

The absorbants can be separated from other medium using two windows. The shortest wavelength reached by available windows material (LiF) is 105 nm. The column density of gases can be obtained by

$$N = 9.65 \times 10^{18} \frac{Pl}{T} \tag{2}$$

where P is pressure in torr, path length l in cm, and temperature T in K. The absorption cell can be placed between the entrance slit and the background source, or beyond the exit slit. The later application sometimes has an advantage by limiting radiation to a monochromatic beam, and thus to gases that easily dissociate. However, high-resolution instruments have focal plane scanning, and therefore absorption cell must be placed at the entrance slit. Below 105 nm, the main body of the spectrometer can be used as an absorption cell by sealing the entrance slit to minimize leaking of gases through the slit. The relatively long pathlengths of the spectrometer mean that the pressure is low and therefore the accuracy of pressure measurements is also limited.

For the photoabsorption cross-section measurements of continuum and the broad bands, low-resolution instruments can be used. The resolution should be significantly better than the broad structures to observe. For line-by-line measurements of bands with fine structures, it is an essential to use high-resolution instruments. Most such measurements have been done using a 6 or 10 m grating spectrometer on photoelectric scan mode. Resolution of grating instruments is 0.3–0.4 cm^{-1}, and they are larger than the Doppler-limited spectrum of simple molecules. Only a few spectrometers remain available 6 m spectrometer at Center for Astrophysics (Cambridge, MA), the Advance Light Source (Berkeley, CA), the Photon Factory (KEK, Japan), and the 10 m spectrometer at Observatoire de Paris-Meudon (France). Two institutions have combined the synchrotron radiation and a high-resolution spectrometer into a facility for UV and VUV measurements, (*e.g.*, Photon Factory and the Advance Light Source).

The Fourier transform (FT) spectrometer is the most advanced instrument for high-resolution cross-section measurements. A few commercial FT spectrometers are available for measurements in UV region with resolution better than 0.1 cm^{-1}, similar to the Doppler widths of simple diatomic molecules. The only FT spectrometer capable of operating in the VUV region down to 140 nm is at Imperial College (London, UK), with resolution down to 0.025 cm^{-1} [3]. Some FT spectrometers have double detectors. One can be used for measurements of the background intensity and the other for absorption measurements.

IV. CROSS-SECTION MEASUREMENTS
OF THE PHOTOABSORPTION CONTINUUM

Cross-section measurements of photoabsorption continua can be performed using an instrument with relatively lower resolution. It is convenient to use such instrument to make a quick measurements to cover the entire wavelength region. Most of the uncertainty in the cross-section measurements with a single path spectrometer is estimating the background intensity level, $I_0(\nu)$, since it may vary with time and wavelength during the photoabsorption scan. It is also often difficult to keep constant pressures of gases in the absorption cell during the scan, especially with reactive or absorbent gases such as ozone and water. To overcome these problems, absolute cross sections of the gases can be measured at several fixed wavelength positions. At each position, the intensity can be recorded without gases in the absorption cell, and then with gases in it. Background intensity is estimated from the averages of two measurements (before and after) without gases. The full ranges of scanned cross sections is then calibrated using these absolute values at fixed wavelengths.

The wavelength scale of spectrometers, whether old or new (and now under computer control), must be calibrated. The method of calibration depends on the

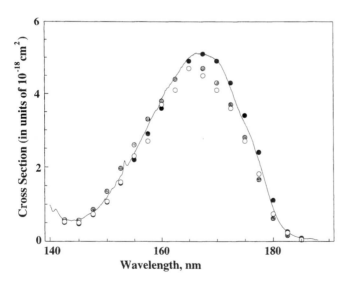

Figure 1 Absorption cross sections of H_2O at 295 K in the wavelength region 140 nm to 185 nm along with previous measurements [6]. The black circles show values from Lee and Suto [7], gray circles from Laufer and McNesby [8], and open circles from Watanabe and Zelikoff [9].

background source. Traditional sources such as the hydrogen (deuterium) lamp, Xe-lamp, and Ar-mini arc, often have impurity emission lines, which can be used as wavelength standards. However, a very pure continuum source such as synchrotron radiation has no lines. In this case, known absorption bands can be used for wavelength calibrations such as the forth positive bands of CO in the wavelength region 120–155 nm [4], and the Schumann-Runge bands of O_2 from 175 to 190 nm [5]. In the UV region, atomic and ionic lines from hollow-cathode lamps can also be used for the wavelength calibration.

Figure 1 illustrates the need for wavelength calibration and presents the results of recent absorption cross-section measurements of H_2O at 295 K [6] compared with previous measurements. Measurements by Lee and Suto [7] (black) agree completely with recent measurements [6] after a wavelength correction of -0.57 nm. The results of Laufer and McNesby [8] (gray circles) are shifted toward shorter wavelengths by 0.83 nm, and their cross sections around 170 nm are 5% lower than others. Early measurements of Watanabe and Zelikoff [9] (open circles) agree in wavelength but their cross sections are 10% lower than others. Some photoabsorption cross sections are pressure dependent, such as the Herzberg continuum of O_2. Pressure dependence usually suggests that the absorption involves two molecules as dimers and/or pressure induced absorption.

V. CROSS-SECTION MEASUREMENTS OF THE BANDS

Cross-section measurements for bands with many fine rotational structures are very limited, because of the lack of instrumental resolution. The best resolution with grating instruments is 0.3–0.4 cm^{-1}, compared with the Doppler widths of around 0.1 cm^{-1} for most of the simple molecules. Hudson and Carter[2] pointed out that plots of optical depth vs. column density are never linear, unless the spectral features are very much broader than the instrumental widths. All measured cross sections are weighted averages over the instrumental widths, so that the measured peak cross-section are lower limits to the true values; the measured wing cross sections are upper limits. A more detailed discussion on the effect of instrumental widths on cross-section measurements is found in the appendix of work by Stark et al. [10].

The first measurements of line-by-line cross sections of molecules were of the Schumann-Runge (S-R) bands of O_2 in the wavelength region 175–200 nm. The 6.6 m grating instrument used for these measurements was capable of a resolution of 0.4 cm^{-1}, but even this may not have been good enough to make such line-by-line measurements. However, the upper levels of the S-R bands in the VUV region are more or less predissociated, and so the rotational lines of the S-R bands are broadened. These predissociation linewidths vary with the vibrational and rotational quantum numbers, and are in the range 1–4 cm^{-1} [11].

The broadened line widths make it possible to obtain absolute cross sections with the grating instrument. The final results for cross sections of the S-R bands of oxygen at 300 K are presented in Figure 3 of Yoshino et al. [12].

Photoabsorption cross sections depend on temperature, based on distribution of the rotational population. For atmospheric application of the measurements of S-R bands, for example, cross sections at particular temperatures should

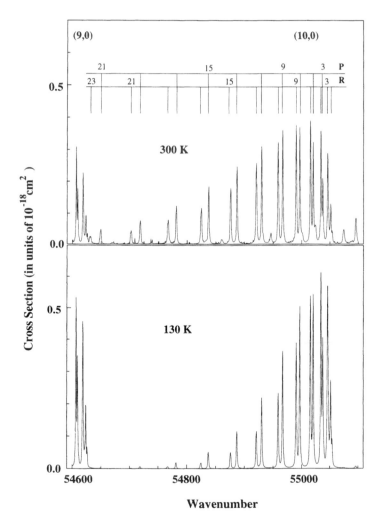

Figure 2 Temperature dependence of the photoabsorption cross sections. The (10,0) band of the S-R system is plotted at 300 K and 130 K. (From Ref. 13.)

be used. Minschwaner et al. [13] developed a model to calculate the cross sections for temperatures between 130 and 500 K at 0.5 cm^{-1} steps based on high-resolution experimental measurements at 300K and 79K [14]. As an example, the calculated cross sections of the (10,0) band at 130K and 300K are presented in Figure 2, demonstrating their temperature dependency.

For astrophysical applications, temperatures of the interstellar media (10–100 K) should be produced for laboratory measurement of spectra. By using the supersonic expansion technique, cross sections of CO bands at wavelength between 92.5 and 97.4 nm have been observed [15] at $\approx$ 20 K.

Even in transitions to highly excited levels, some bands (such as those of N_2, NO, and some O_2 bands) appear with fine absorption lines that are close to their Doppler widths (around 0.1 cm^{-1}). In this case, the traditional grating spectrometer cannot make reliable cross-section measurements because of limited resolution. However the FT spectrometers can operate with resolution better than 0.1 cm^{-1} to measure bands with Doppler-limited lines. Cross-section measurements of very weak Herzberg bands have been done in combination with FT spectrometer and the White cell [16,17]. To obtain good signal-to-noise ratios, FT spectrometry requires a stable and pure background source, and the ability to limit the band widths. This combination was realized by taking the IC VUV-FT spectrometer to the synchrotron radiation source at Photon Factory, thus making possible ultra-high-resolution cross-section measurements of NO and O_2 bands [18] in the VUV region.

Laser spectroscopy contributes to very high resolution measurements in the VUV region. However, the stability of the the laser sources makes it difficult to measure cross sections. So far, the only contribution to cross-section measurements is available from laser spectroscopy for A–X bands of CO [19].

VI. LINE AND BAND OSCILLATOR STRENGTHS

Integrated band cross sections are determined by summing $\sigma(v)$ over all of the rotational lines belonging to the (v', v'') band, and converting to the band oscillator strengths (f-values) by the relation

$$f_{v'v''} = \frac{mc^2}{\pi e^2} \frac{1}{\tilde{N}(v'')} \int \sigma(v)dv \tag{3}$$

in which $\tilde{N}(v'')$ is the fractional Boltzmann population of the vibrational level involved. The band oscillator strength is related to the line oscillator strength by

$$f_{v'v''J'J''} = \frac{S_{J''}^{J'}}{2J'' + 1} f_{v'v''} \tag{4}$$

where $S_{J'}^{J''}$ is the Hönl-London factor. The integrated cross-section is less affected by the instrumental function than are the peak cross-sections. The true intensity of the line comes from the the entire area of its absorption. Integrated cross sections of the line can be performed with the column density of the particular rotational level (J'') obtained using Boltzmann statistics. We could also obtain band oscillator strengths from integrated cross sections of the lines [19].

Band oscillator strengths are absolute values, and are independent of the temperature. The integrated area over the entire bands at two temperatures is the same. At the lower temperature, the rotational distribution is limited, and the peak intensity of the low J lines become higher, as demonstrated in Figure 2. This fact can be used when, for example, the column density cannot be measured, as with a supersonic jet. Total band absorption at the jet temperature can be combined with band oscillator strengths to deduce the column density [15].

Once band oscillator strengths are established, cross sections of any temperature can in principle be obtained with known Hönl-London factors. However, usually many of the excited levels are more or less perturbed by neighbor levels, and their intensities affected. In these cases, calculated cross sections from band oscillator strengths might not yield the correct values. Even after depertubation treatments using energy shifts from the perturbation, the calculated intensity effected by the perturbation is not reliable, because the intensities are very sensitive to interactions. Therefore, measurements of line oscillator strengths become important, especially in applications of astrophysical interest.

VII. RESULTS OF RECENT MEASUREMENTS

The results of UV cross sections are presented in Tables 2 and 3 of Chapter 14, for broad-cross sections and line-by-line data in HITRAN, respectively. Most of the VUV measurements are available from cfa-www.harvard.edu/amdata/ampdata/cfamols.html.

REFERENCES

1. H Okabe. Photochemistry of Small Molecules. New York: John Wiley & Sons, 1978.
2. RD Hudson, VL Carter. Bandwidth dependence of measured UV absorption cross sections of Argon. J Opt Soc Am 58:227–232, 1968.
3. AP Thorne, CJ Harris, I Wynne-Jones, RCM Learner, G Cox, A Fourier transform spectrometer for the vacuum ultraviolet: design and performance. J Phys E: Sci Instrum, 20:54–60, 1987.
4. SG Tilford, JD Simmons. Atlas of the observed absorption spectrum of Carbon Monoxide between 1060 and 1900 Å. Phys Chem Ref Data 1:147–188, 1972.

5. K Yoshino, DE Freeman, WH Parkinson. Atlas of the Schumann-Runge absorption bands of O_2 in the wavelength region 175–205 nm. Phys Chem Ref Data 13:207–227, 1984.

6. K Yoshino, JR Esmond, WH Parkinson, K Ito, T Matsui, Absorption cross section measurements of water vapor in the wavelength region 120 nm to 188 nm. Chem Phys 211:387–391, 1996.

7. LC Lee, M Suto. Quantitative photoabsorption and fluorescence study of H_2O and D_2O at 50–190 nm. Chem Phys 110:161–169, 1986.

8. AM Laufer, JR McNesby. Deutrium isotope effect in vacuum ultra-violet absorption coefficients of water and methane. Can J Chem 43:3487–3490, 1965.

9. K Watanabe, M Zelikoff. Absorption coefficient of water vapor in the vacuum ultra-violet. J Opt Soc Am 43:753–755, 1953.

10. G Stark, K Yoshino, PL Smith, K Ito, WH Parkinson. High-resolution absorption cross sections of carbon monoxide bands at 295 K between 91.7 and 100.4 nm. Astrophys J 369:574–580, 1991.

11. AS-C Cheung, K Yoshino, JR Esmond, SS-L Chiu, DE Freeman, WH Parkinson. Predissociation line widths of the (1,0)–(12,0) Schumann-Runge absorption bands of O_2 in the wavelength region 179–202 nm. J Chem Phys 92:842–849, 1990.

12. K Yoshino, JR Esmond, AS-C Cheung, DE Freeman, WH Parkinson. High resolution absorption cross sections in the transmission window region of the Schumann-Runge bands and Herzberg continuum of O_2. Planet Space Sci 40:185–192, 1992.

13. K Minschwaner, GP Anderson, LH Hall, K Yoshino. Polynomial coefficients for calculating O_2 Schumann-Runge cross sections at 0.5cm^{-1} resolution. J Geophys Res 97:10,103–10,108, 1992.

14. K Yoshino, DE Freeman, JR Esmond, WH Parkinson. High resolution absorption cross sections and band oscillator strengths of the Schumann-Runge bands of oxygen at 79K. Planet Space Sci 35:1067–1075, 1987.

15. K Yoshino, G Stark, JR Esmond, PL Smith, K Ito, M Matsui. High-resolution absorption cross section measurements of supersonic jet-cooled carbon monoxide between 92.5 and 97.4 nanometers. Astrophys J 438:1013–1016, 1995.

16. K Yoshino, JR Esmond, JE Murray, WH Parkinson, AP Thorne, RCM Learner, G Cox. Band oscillator strengths of the Herzberg I bands of O_2. J Chem Phys 103:1243–1249, 1995.

17. M-F Mérienne, A Jenouvrier, B Coquart, M Carleer, S Fally, R Colin, AC Vandaele, C Hermans. Fourier transform spectroscopy of the O_2 Herzberg bands. II. Band oscillator strengths and transition moments. J Molec Spectrosc 202:171–193, 2000.

18. T Imajo, K Yoshino, JR Esmond, WH Parkinson, AP Thorne, JE Murray, RCM Learner, G Cox. The application of a VUV Fourier transform spectrometer and synchrotron radiation source to measurements of: II. The $\delta(1,0)$ band of NO. J Chem Phys 112:2251–2257, 2000.

19. G Stark, BR Lewis, ST Gibson, JP England. High-resolution oscillator strength measurements for high-v' bands of the A $^1\Pi$(v')-X $^1\Sigma^+$(v" = 0) system of carbon monoxide. Astrophys J 505:452–458, 1998.

5

Ultraviolet and Vacuum Ultraviolet Laser Spectroscopy Using Fluorescence and Time-of-Flight Mass Detection

R. H. Lipson and Y. J. Shi
University of Western Ontario, London, Ontario, Canada

I. INTRODUCTION

The generation of tunable, monochromatic, and coherent short-wavelength light by nonlinear optics is a continuing and important research pursuit. Many different and novel chemical and physical applications using such sources have also been reported. These two subjects, however, are so vast that the review here by necessity will focus only on the principles and applications of ultraviolet (UV; 200 nm $\leq \lambda \leq$ 300 nm), vacuum ultraviolet (VUV; 100 nm $\leq \lambda \leq$ 200 nm), and extreme ultraviolet (XUV; $\lambda \leq$ 100 nm) so-called lasers generated by frequency doubling in anisotropic crystals, nonresonant tripling or third harmonic generation (THG), and two-photon resonant four-wave sum- and difference-mixing (FWSM and FWDM, respectively) in isotropic gases. The word laser is qualified to serve as a reminder that the production of coherent short-wavelength radiation is the result of parametric processes, and not population inversions. Furthermore, applications will be limited to those in which single-photon excitations of jet-cooled gas-phase molecules have been detected by either laser-induced fluorescence (LIF) or ion detection. This will unfortunately exclude a vast literature on lamp and synchrotron-based experiments, as well as experimental results on high-lying states in the UV, VUV, or XUV accessed by multiphoton absorption using longer-wavelength sources.

II. EXPERIMENTAL TECHNIQUES

A brief background review is provided here on a number of the different principles and experimental methods used in the UV and VUV laser spectroscopy of jet-cooled molecules.

A. Light Sources

Nonlinear optics began in 1961 with the pioneering work of Franken and coworkers, who doubled the red output of a ruby laser by second harmonic generation (SHG) in crystalline quartz [1]. Today, frequency doubling, using a battery of different crystals, is commercially available and so commonplace that the following discussion will focus primarily on third-order nonlinear phenomena. The theory of VUV generation is also well understood. A reiteration of some of the major points is nevertheless warranted, particularly for those whose interests lie not so much in laser physics but in potential applications. The interested reader is directed to several comprehensive reviews on this topic [2–4].

All nonlinear optical signals originate from an oscillating nonlinear polarization, $P(\omega)$ (dipole/unit volume), which can be induced in a material by the strong external electric fields, $E(\omega)$, of one or more incident laser beams. Under moderate laser powers, $P(\omega)$ can be written as a Taylor series in $E(\omega)$:

$$P(\omega) = N(\chi^{(1)} \cdot E(\omega) + \chi^{(2)} : E(\omega)E(\omega) + \chi^{(3)} : E(\omega)E(\omega)E(\omega) + \cdots) \quad (1)$$

where N, $\chi^{(1)}$, and $\chi^{(n)}$, $n \geq 2$, are the number density, linear susceptibility, and nth-order nonlinear susceptibilities tensors of the medium, respectively. Only the first two nonlinear terms, which can serve as sources for coherent light at new frequencies, will be considered.

The expression proportional to $\chi^{(2)}$ is responsible for second harmonic generation in anisotropic crystals. However, isotropic gases are most often used for VUV generation because nonlinear solid-state materials transparent to wavelengths shorter than the UV do not exist. The current limit is ≈ 190 nm for β-barium borate (BBO). Second harmonic generation in atomic media has also been demonstrated, but the mechanism is now generally accepted to be a third-order nonlinear effect involving two input laser photons and a third radial DC electric field produced by atom ionization [5]. In the absence of DC electric fields, and in isotropic media in which $\chi^{(2)}$ is identically zero, the leading term that can yield nonlinear effects, including THG, FWSM, and FWDM, involves the third-order nonlinear susceptibility, $\chi^{(3)}$.

1. Frequency Doubling

Wavelengths down to ≈ 210 nm with a frequency bandwidth ≤ 0.5 cm^{-1} are routinely generated in many spectroscopy labs by SHG of pulsed fundamentals in anisotropic crystals. Several groups, however, are using continuous wave (cw) lasers to obtain much higher resolution spectra. Two such systems are briefly described here.

Pratt and co-workers have described an ultra-high-resolution UV laser they have used to record fully rotationally resolved LIF excitation spectra of large polyatomic molecules [6]. In their experiments, a single-mode ring dye laser capable of stabilized scans of 90–120 GHz in the visible spectrum, and operating on Rhodamine 6G (595–630 nm), Kiton Red (620–640 nm) and/or DCM (630–700 nm) dye solutions, is pumped by ≈1 W of the 514.5 nm line of an argon–ion laser. UV light is generated by placing an LiIO$_3$ doubling crystal in the auxiliary waist of the ring laser to produce up to ≈1 mW output power in the UV. Phase matching the SHG crystals doubles the scan range to 180–240 GHz or 6–8 cm^{-1}. The spectrometer has a frequency bandwidth resolution of 2.5 MHz at an excitation wavelength of 300 nm, corresponding to an enviable resolving power of $v/\Delta v = 4 \times 10^8$.

A second intermediate resolution system reported by Bitto and Willmott [7] involves pulsed dye amplifying the cw output of a ring laser with a typical bandwidth of several MHz. Fourier transform limited pulses with up to a 100 mJ energy are obtained in this way using 550 mJ 532 nm light from a frequency-doubled Q-switched Nd:YAG laser. Approximately 150 MHz frequency bandwidth UV pulses with energies up to ≈300 μJ are subsequently generated by SHG in nonlinear crystals. Such a system is extremely versatile because it allows both precise high-resolution frequency-resolved *and* time-resolved experiments to be carried out.

2. Nonresonant Third Harmonic Generation

An energy level diagram for nonresonant tripling [8], where $\omega_{output} = 3\omega$, and ω is the input angular frequency of a single laser, is shown in Figure 1a. When a Gaussian beam of frequency ω and input power P_ω, is focused into an atomic medium, the resultant third harmonic output power $P_{3\omega}$ is given by [9]:

$$P(3\omega) = N^2[\chi^{(3)}(3\omega)]^2 P_\omega^3 F(b \, \Delta k) \tag{2}$$

The geometrical phase-matching function, F, describes the dispersive phase velocity relationship between the input and output waves, and depends on the product $b \, \Delta k$, where b is the confocal beam parameter of the focused radiation and Δk is the wave–vector mismatch between the input and output light beams:

$$\Delta k = k_{3\omega} - 3k_\omega = \frac{2\pi n(\lambda_{3\omega})}{\lambda_{3\omega}} - 3\frac{2\pi n(\lambda_\omega)}{\lambda_\omega} \tag{3}$$

The indices of refraction, $n(\lambda_{3\omega})$ and $n(\lambda_\omega)$, are those for the medium at the third harmonic and input laser beam wavelengths, $\lambda_{3\omega}$ and λ_ω, respectively. The product

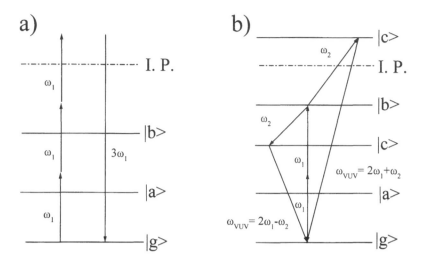

Figure 1 (a) Energy level scheme indicates a one-color nonresonant third harmonic generation process at the fundamental laser frequency ω (b) A four-level system indicates the relevant energy levels for both four-wave sum mixing at $2\omega_1 + \omega_2$, and four-wave difference mixing at $2\omega_1 - \omega_2$. IP, ionization potential.

$N^2 |\chi^{(3)}(3\omega)|$ is usually small for most media when the fundamental laser frequency does not match a resonance condition for an allowed transition. $P_{3\omega}$ in general cannot, however, be increased simply by using a larger gas pressure, because Δk is also a function of the number density N. Instead, the product $N^2 F(b\,\Delta k)$ [or $\Delta k^2 F(b\,\Delta k)$] must be optimized.

THG is typically, generated using a pulsed laser output that is tightly focused so that $b \ll$ the length of the medium L. This enhances the conversion efficiency by maximizing the P_ω^3 power dependence in Eq. 2. At the same time the phase matching function $F(b\,\Delta k)$ must also be considered for this focusing condition [9]:

$$F(b\,\Delta k) - \pi^2 (b\,\Delta k)^2\, e^{b\Delta k} \qquad \Delta k < 0$$
$$= 0 \qquad\qquad\qquad \Delta k \geq 0 \tag{4}$$

Equation (4) intimates that nonresonant THG is only possible in spectral regions in which the atomic medium is negatively dispersive. Gaps in the tuning curves are found because Δk is positive to the red of each allowed Rydberg state ← ground state transition due to anomalous dispersion. The function, $\Delta k^2 F(b\,\Delta k)$ is a maximum when $b\,\Delta k = -4$. THG can therefore be optimized by changing either the beam focus to vary b and/or changing N to alter Δk. Since the refractive index is a strong function of λ near a resonance line, coherent VUV generation at any allowed wavelength can be maximized by adding different amounts of a positively dispersive buffer gas to the nonlinear medium.

The rare gases (Xe [10,11], Kr [12,13], Ar [14], and Ne [15]), or metal vapors such as Hg [16], are most commonly employed for nonresonant THG (Fig. 2). The latter, however, must be produced in high-temperature heat pipe ovens, while the former can be housed in either a room-temperature gas cell with a MgF_2 or LiF output window for VUV generation, or expanded in a spatially localized supersonic jet (described below) making it a near-ideal "windowless" environment for XUV generation [17] (Fig. 3). Typical conversion efficiencies of $\approx 10^{-6}$–$10^{-3}\%$ leading to intense coherent VUV outputs are possible using commercial pulsed dye laser peak powers of 1–10 MW ($\approx$ 10 nsec pulse duration). It is also straightforward to triple the third harmonic output of a neodymium: yttrium aluminum garnet (Nd:YAG) laser at $\approx$355 nm to produce VUV light at $\approx$118 nm [18]. As described below, this source has been used extensively for analytical detection of molecules, despite its fixed wavelength, due to its technical simplicity.

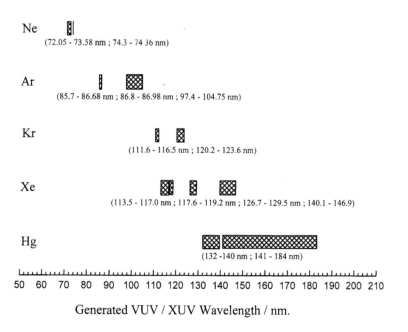

Ne   (72.05 - 73.58 nm ; 74.3 - 74.36 nm)

Ar   (85.7 - 86.68 nm ; 86.8 - 86.98 nm ; 97.4 - 104.75 nm)

Kr   (111.6 - 116.5 nm ; 120.2 - 123.6 nm)

Xe   (113.5 - 117.0 nm ; 117.6 - 119.2 nm ; 126.7 - 129.5 nm ; 140.1 - 146.9)

Hg   (132 -140 nm ; 141 - 184 nm)

50 60 70 80 90 100 110 120 130 140 150 160 170 180 190 200 210

Generated VUV / XUV Wavelength / nm.

Figure 2 Main spectral regions in the VUV and XUV where coherent light has been generated experimentally by nonresonant third harmonic generation, using Ne, Ar, Kr, Xe, or Hg as nonlinear media.

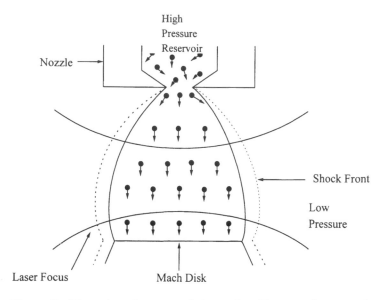

High
Pressure
Reservoir

Nozzle ——→

———— Shock Front

Low
Pressure

Laser Focus   Mach Disk

Figure 3 Illustration of a supersonic jet produced by expanding a gas from a region of high pressure into vacuum. The orientation of the arrows on the dots (indicating atoms and/or molecules) shows that the velocity distribution is narrower below the pinhole than in the high-pressure reservoir and, therefore, that the resultant translational temperature is lower. The spatially localized nature of the expansion located with the shock fronts make it a near-ideal windowless environment for XUV generation with focused laser beams.

3. Two-Photon Resonance-Enhanced Four-Wave Mixing

As shown in Figure 1b, two-photon resonance-enhanced four-wave mixing is a two-color variation of third-harmonic generation, where three input fundamental photons from two lasers, two at angular frequency ω_1, and one at angular frequency ω_2, are used to produce a fourth output photon at the sum $(2\omega_1 + \omega_2)$ and/or the difference $(2\omega_1 - \omega_2)$ frequency, both of which can lie in the VUV or XUV (Fig. 1b) [19,20]. The expression "four-wave mixing" is appropriate because three input waves are used to generate a fourth output signal.

In an expression similar to Eq. (2), the intensity of the generated VUV light, P_{VUV}, for sum-mixing is governed by a $P_1^2 P_2$ incident laser power dependence for beams at frequencies ω_1 and ω_2, respectively, and by a phase-matching function $F(b\,\Delta k)$ that is identical to Eq. (4) when a tight focusing geometry is used. However, the wave vector mismatch here is now defined as $\Delta k = k_{VUV} - 2k_1 - k_2$, where k_1, k_2, and k_{VUV} are the wave vectors for the input and output light beams, respectively. Although it may appear at first glance that resonant four-wave sum-mixing, like nonresonant tripling, is only possible in spectral regions where the nonlinear medium is negatively dispersive, most atomic elements used for this purpose are, to a good approximation, inherently negatively dispersive for virtually every VUV and XUV wavelength in their tuning ranges. Their first resonance lines, which carry most of the oscillator strength, come to the blue of the majority of the input laser frequencies used in the nonlinear optical process. However, the resultant broad tunability is often only achieved at the expense of the optimal phase-matching conditions.

The dispersion criterion for FWDM is less stringent in a tight focussing situation since Δk can be either positive or negative. The phase matching function

$$F(b \, \Delta k) = \pi^2 \, e^{-b|\Delta k|} \tag{5}$$

is a maximum when $\Delta k = k_{vuv} - 2k_1 + k_2 = 0$ [9] and, ordinarily, the FWDM conversion efficiency is about an order of magnitude better than that for FWSM.

One significant difference between resonance-enhanced four-wave mixing and nonresonant THG is that an appreciable VUV output conversion efficiency (on the order of 1 part in 10^7–10^5) for the former can be achieved using moderate incident pulsed laser powers by judiciously tuning frequencies ω_1 and ω_2 to be either one- or two-photon resonant with allowed transitions (Ω_{ag}, Ω_{bg}) in the medium. This approach minimizes one or more of the resonance denominators in the expression for the third-order nonlinear susceptibility [21]:

$$\chi^{(3)}(2\omega_1 \pm \omega_2; \omega_1, \omega_1, \pm \omega_2) = \tag{6}$$

$$\frac{1}{\hbar^3} \sum_{abc} \frac{\langle g|e_4\hat{\mu}|a\rangle \, \langle a|e_1\hat{\mu}|b\rangle \, \langle b|e_2\hat{\mu}|c\rangle \, \langle c|e_3\hat{\mu}|g\rangle}{(\Omega_{ag} - \omega_1) \, (\Omega_{bg} - 2\omega_1) \, (\Omega_{cg} - 2\omega_1 \mp \omega_2)}$$

Here, e_k is the unit vector identifying the polarization of wave vector k, $\hat{\mu}$ is the electric dipole moment operator, Ω_{jg} are the complex frequencies for the $j \leftarrow g$ transitions, and the summation is over all states.

Two-photon resonance enhancement is most commonly used experimentally since it can substantially increase $\chi^{(3)}$ with minimal two-photon absorption in the medium, provided that the pulse durations of the fundamental laser beams are reasonably short (≤ 10 ns). The one-photon resonance condition, on the other hand, can lead to substantial reduction in the coherent output due to strong linear absorption of the fundamental. Additional resonance enhancement is also possible if the VUV output frequencies correspond to allowed transitions in the medium, Ω_{cg}. This effect is usually small when the unstructured continuum of an atomic medium is used, due to the small magnitude of the bound-continuum $\langle c|e_3 \hat{\mu}|g\rangle$ matrix element [22]. A better conversion efficiency, however, is possible if the generated VUV wavelength is resonant with a transition to an autoionized level embedded in the continuum. Likewise, VUV light generated at wavelengths longer than the ionization potential of the nonlinear medium can often be in resonance with strongly allowed Rydberg state ← ground state transitions. While the conversion efficiency can then be very high (up to 5%) [23], the possibility of linear reabsorption is also large, and the generated light intensity can change dramatically with wavelength. While this latter effect is the anithesis of what is desired in a tunable spectroscopic light source, the resultant spectral fluctuations can handled by appropriate normalization techniques. At all times, input laser frequencies for atomic transitions leading to nonzero electric dipole matrix elements in the numerator of Eq. (6) must also be chosen.

In our laboratory, VUV light is routinely generated between ~174 and 66 nm by four-wave mixing in Mg and Hg vapor, as well as in Xe and Kr gas [24,25]. Energy level diagrams and tuning curves for these elements are shown in Figure 4.

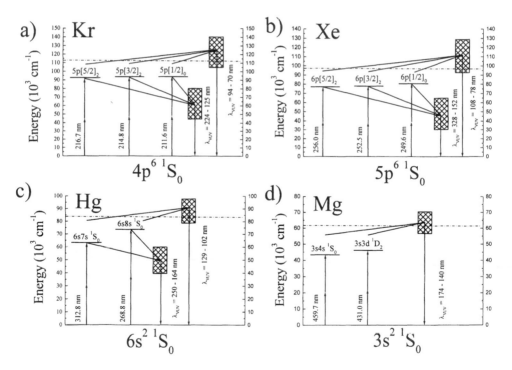

Figure 4 Energy level diagrams of the relevant atomic levels of the nonlinear media used at the University of Western Ontario for four-wave mixing. (a) Kr; (b) Xe; (c) Hg; (d) Mg. The wavelength regions generated by either four-wave sum- and/or difference mixing are depicted as hatched regions. The wavelengths (nm) of the fundamental laser beams indicated are those providing two-photon resonance enhancement of the third-order nonlinear susceptibility of the nonlinear media. IP, ionization potential of the nonlinear medium.

Table 1 Comparison of Second- and Third-Generation Synchrotrons and VUV Lasers

Properties	Second-generation synchrotron [84]	Third-generation synchrotron [85]	Four-wave mixing sources
E-range	10^{-1}–3×10^4 eV	Optimized for 10 eV to 1 keV	6–19 eV
Intensity	$<10^9$ Photons/pulse/ 0.1% band pass	Up to 10^{13} photons/sc/ 0.01% band pass	10^{12} photons/pulse
Resolution	10^{-3} eV (0.03 Å) at 5–40 eV; 10^{-2} eV (0.02 Å) at 40–100 eV	<10 meV at 100 eV; $E/\Delta E$ as high as 10^5	≤ 0.5 cm^{-1}
Repetition rate	~ 1–500 MHz	<500 MHz; > 1.5 MHz	Typically <50 Hz
Pulse duration	>100 ps	<70 ps	≤ 10 nsec

It is common to obtain VUV photon numbers exceeding 10^{10} per pulse by generating the fundamental beams using commercial dye lasers. The monochromaticity of the VUV is then determined essentially by these sources. Current commercial pulsed dye lasers have narrow output line widths, often <0.07 cm^{-1}, which yield VUV line widths <0.12 cm^{-1}. The corresponding frequency-resolving power, $\upsilon/\Delta\upsilon$, is better than 10^6 at the shortest wavelengths. This, combined with the resultant high spectral brightness due to the low divergence of the VUV output, represents a considerable improvement over the best monochromators available, and current synchrotron sources (Table 1). Gas-phase spectral resolution at the experimental level will be a convolution of the source line width, Doppler broadening, and other broadening mechanisms such as predissociation and/or autoionization. Individual rotational lines having a full-width at half maximum of $\leq \approx 0.5$ cm^{-1} in the VUV can usually be easily resolved.

B. Supersonic Jets

Supersonic jets have been established as a powerful tool for gas-phase spectroscopy since they can produce rovibrationally cold molecules, radicals, and clusters [26]. A free jet, shown schematically in Figure 3, arises when a gas is expanded adiabatically from a region of high pressure (stagnation pressure P_{stag}) through a pinhole of diameter D into vacuum. When D is larger than the mean free path between the gas molecules, the large number of resultant binary collisions dramatically narrows the velocity distribution of the sample; that is, all the atoms and/or molecules in the expansion travel in the same direction with the same speed. The local speed of sound drops rapidly away from the pinhole, as the square root of the translational temperature, T_{trans}, characterizing the width of the gaseous velocity distribution. At the same time only a finite amount of thermal energy in the high-pressure reservoir can be converted into a directed mass flow. Thus, the maximum velocity a molecule can attain in a supersonic jet is actually only slightly larger than the maximum velocity of a room-temperature Boltzmann distribution. The term supersonic is used, however, because the ratio of the approximately constant flow velocity over the local speed of sound, M (Mach number) can easily exceed unity.

The relationship among temperature (T), pressure (P), and density (ρ), in the expansion to the Mach number and the stagnation temperature, pressure, and density $(T_{stag}, P_{stag}, and \rho_{stag})$ behind the pinhole is given by [27]:

$$\frac{T}{T_{stag}} = \left(\frac{P}{P_{stag}}\right)^{(\gamma-1)/\gamma} = \left(\frac{\rho}{\rho_{stag}}\right)^{\gamma-1} = \left(1 + \frac{\gamma-1}{2} M^2\right)^{-1} \tag{7}$$

where γ is the heat capacity ratio, C_p/C_v. Since calculations predict M to vary as $(x/D)^{\gamma-1}$, where x is the distance between a point on the flow axis and the nozzle pinhole, it follows that the thermodynamic properties described by Eq. (7) are strongly position-dependent within the isentropic core.

Molecules seeded in a monoatomic gas (typically He or Ar) will attempt to equilibrate with T_{trans} by collisionally transferring their energy stored in vibrational and rotational degrees of freedom to the forward directional mass flow. This can often result in highly nonequilibrium internal energy distributions, because the relaxation rates are different for different motions of the molecule. As a rule of thumb, the translational, vibrational, (T_{vib}), and rotational (T_{rot}) temperatures of the molecules are ordered $T_{trans} \leq T_{rot} < T_{vib}$. At the same time, the gas density also drops precipitously away from the pinhole. Phase transitions that require three or more body collisions therefore become extremely improbable. Accordingly, the resultant internal *gas phase* rotational temperatures are almost always lower than the freezing point of a bulk sample, and often dramatically $(T_{rot} \leq 10\ K$ is not uncommon). The populations and, therefore, temperatures, of the various degrees of freedom become "frozen" at that point in the expansion where the number of collisions has effectively dropped to zero. Van der Waals clusters can also be produced in a jet, but usually at P_{stag} conditions higher than those needed simply to cool a stable monomer.

The position of the Mach disk, X_M, formed when the jet pressure equals the background gas pressure in the vacuum chamber, $P_{chamber}$, can be estimated from

$$X_M = 0.67D \sqrt{\frac{P_{stag}}{P_{chamber}}} \tag{8}$$

The gas beyond X_M is much hotter than that in the isentropic core and, therefore, the laser focus within the jet should be located between the nozzle pinhole and X_M. This condition, however, is very easy to achieve under typical but modest experimental conditions where, for example, $P_{stag} = 2$ atm, $P_{chamber} = 1.3 \times 10^{-7}$ atm ($\approx 10^{-4}$ torr), $D = 1.0$ mm, and X_M is ≈ 2.6 m.

Although the discussion above is valid for continuous free jet expansions, the same formalism is usually assumed to also hold for pulsed jets (often in the 200 μs range). These devices are in wide use since they provide lighter gas loads for the pumping system due to their reduced duty cycle.

C. Detection Methods

A jet environment is poorly suited to direct absorption measurements because the density of the absorbing molecules is generally very low and the absorption path length is short. Thus, only a small fraction of the incident light intensity can be absorbed, and that decrease must be detected over fluctuations in the UV and VUV intensity, which can be quite pronounced due to the nonlinear processes used to generate the light. It is fortunate that considerably more sensitive detection methods have been developed to overcome these problems. Here, the rudiments of two such approaches, namely laser-induced fluorescence (LIF) and ion detection by time-of-flight (TOF), mass spectrometry are presented briefly.

1. Laser-Induced Fluorescence

Laser-induced fluorescence involves the detection of optical emission from molecules that have been excited by the absorption of laser radiation. Such an approach for jet-cooled molecules requires excited states with a relatively short lifetimes ($\leq$μs). In this regard, unlike the UV and VUV, LIF in the infrared and microwave regions is extremely impractical because single electronic vibrational and rotational level lifetimes can be orders of magnitude longer.

UV and VUV photons usually have more than enough energy to excite electronic transitions and, therefore, the resultant spectra can provide detailed information about both electronic states involved. Two different experimental arrangements are possible. In the first, emission from a specific level excited at a fixed frequency is spatially separated in a monochromator and detected with a photomultiplier tube to yield a dispersed fluorescence or emission spectrum. Measurements of the resultant line spacings yield structural information about the lower state. As an alternative, excitation spectra can be recorded by monitoring the total undispersed emission as a function of the tunable exciting light frequency. As the analysis below shows, the result is equivalent to an absorption spectrum, and provides detailed excited state information [28].

Consider the experimental arrangement and energy level diagrams in Figure 5a and b, respectively. If collisions are neglected, each laser photon absorbed

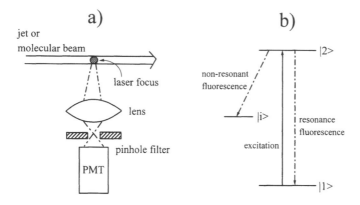

Figure 5 (a) Schematic diagram of a typical experimental arrangement required to record laser-induced fluorescence excitation spectra. (b) Three-level energy level diagram shows a single photon transition, and the possible resonant (=excitation frequency) and nonresonant fluorescence decay channels.

will be emitted with unit quantum efficiency. The relationship between the number of photons detected per second at incident laser frequency ω, $N_p(\omega)$, by an optical system/photomultiplier combination having collection efficiency (Φ) and quantum efficiency (QE), respectively, and the number of photons absorbed per second within the focal volume ΔV, $N_A (\omega)$, is given by:

$$N_P(\omega) = N_A(\omega) \cdot \Phi \cdot QE \tag{9}$$

N_A in turn can be expressed as

$$N_A(\omega) = \sigma(\omega) \cdot N \cdot \Delta V \cdot F(\omega) \tag{10}$$

where $\sigma(\omega)$ is the absorption cross section at incident laser frequency ω, N is the number of molecules per cm^3, and $F(\omega)$ is the incident laser flux in photons cm^{-2} s^{-1}. An excitation spectrum involves recording

$$\frac{N_P(\omega)}{F(\omega)} = \sigma(\omega) \cdot N \cdot \Delta V \cdot \Phi \cdot QE \tag{11}$$

If Φ and QE are independent of ω; (i.e., if they are insensitive to changes in fluorescence over the range of absorption frequencies) the normalized ratio in Eq. (11) represents the true absorption spectrum. As an example, consider a 1 mW continuous-wave UV laser operating at 303 nm (corresponding to $\approx 2 \times 10^{15}$ photons/s) with a beam cross-sectional area given by $\Delta V/\Delta x$. With a collection efficiency of $\Phi = 0.1$ and a quantum efficiency of QE $= 0.2$,

$$N_P = \sigma(\omega) \cdot N \cdot \Delta x \cdot (4 \times 10^{13}) \tag{12}$$

Assuming a photomultiplier dark count $N_D = 200$ counts s^{-1}, a signal-to-noise ratio $N_P/N_D \geq 1$ can be achieved for $\sigma(\omega) \cdot N \cdot \Delta x \geq 5 \times 10^{-12}$, which is several orders of magnitude more sensitive than conventional absorption spectroscopy. Although, in principle, excitation spectroscopy involves light detection against a zero background, specific care must be taken to eliminate scattered light that could be several orders of magnitude larger than the fluorescence. A spatial pinhole filter before the photomultiplier can help significantly in this regard.

2. Ion Formation and Detection by TOF Mass Spectrometry

Instead of monitoring absorptions through LIF, a second laser photon can be used to ionize the excited species. A $(1 + 1')$ resonance-enhanced multiphoton ionization (REMPI) scheme is shown in Figure 6. In general, the integers m and n in $(m + n)$ REMPI refer to the number of photons required to populate and subsequently ionize the state of interest, respectively [29–31]. A prime associated with the second integer indicates that the ionizing photon comes from a different laser than that used to excite the species of interest. The resultant ions can be collected with efficiency $\Phi = 1$ and detected with an electron multiplier, chan-

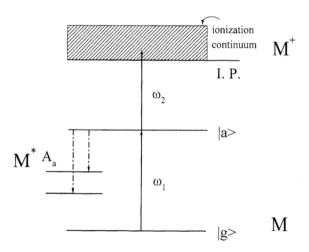

Figure 6   Energy level scheme shows a $(1 + 1')$ resonance-enhanced multiphoton ionization (REMPI) scheme leading to the production of a molecular ion, M$^+$, from ground state neutral M ($|g\rangle$), via the production of an excited-state M* ($|a\rangle$). Also indicated are possible competitive spontaneous emission relaxation pathways for the resonantly excited M* intermediate states with rate A_a.

neltron, or microchannel plate (MCP) having a quantum efficiency (QE) also close to unity. The ionizing laser need not be tunable as long as its energy $\hbar\omega_2$ > IP − E_a, where IP is the ionization potential of the particle and E_a is the term value of the excited state.

The pumping and relaxation scheme shown in Figure 6 can be readily analyzed using

$$\frac{dN_a}{dt} = (N_g - N_a)\sigma_{ag}n_1 - N_a\sigma_I n_2 - A_a N_a \tag{13}$$

where N_g and N_a are number densities in levels $|g\rangle$ and $|a\rangle$, respectively; σ_{ag} is the absorption cross section for the transition between levels $|g\rangle$ and $|a\rangle$ at frequency ω_1; σ_I is the ionization cross section out of level $|a\rangle$ at frequency ω_2; n_1 and n_2 are the photon fluxes (in $cm^{-2}s^{-1}$); and A_a is the spontaneous relaxation rate out of state $|a\rangle$. When Eq. (13) is solved for N_a in the steady-state approximation, the ion signal rate S_1 is then given by:

$$S_I = N_a \cdot \sigma_I \cdot n_2 \cdot \Phi \cdot QE \cdot \Delta V \tag{14}$$

where, as before, ΔV is the interaction volume. Using the solution for N_a from Eq. (13), and identifying $N_g \cdot \sigma_{ag} \cdot n_1 \cdot \Delta V$ as n_a, the number of photons absorbed per second on the first transition, Eq. (14) can be rewritten as:

$$S_I = n_a \frac{1}{1 + [(\sigma_{ag} \cdot n_1 + A_a)/(\sigma_I \cdot n_2)]} \tag{15}$$

which shows that if $\sigma_I \cdot n_2 \gg \sigma_{ag}n_1 + A_a$, nearly every absorbed photon can be detected. It should be kept in mind, however, that ionization cross sections are usually orders of magnitude smaller than transition cross sections. Therefore, the intensity of the ionizing laser should be much larger than that of the exciting laser, which can easily saturate the first step.

Recording the total ion yield as a function of ω_1 will generate an excitation spectrum, but mass selection prior to detection is more common. This additional step becomes particularly important if there is more than one absorbing species present in the jet or if the species has many naturally occurring isotopomers. In this regard, a TOF mass spectrometer [32,33] offers several advantages that make it the instrument of choice for jet spectroscopy.

A TOF mass spectrometer operates on the simple principle that an isoenergetic packet of different ions will spatially separate according to their mass/charge ratios (m/e) over a fixed distance. As a consequence, the mass of one ion can be differentiated from another by measuring the time it takes to traverse that fixed distance; that is, its time-of-flight. TOF instruments have almost unitary transmission that is approximately independent of ion mass and mass resolution, and an infinite mass range. Their main deficiency is related to their ultimate mass-resolving power. The two major limiting factors are the initial spread in the velocity of ions formed and the spatial spread of ions created in different volumes of the ionization region. The former can be alleviated by using a jet with its flow axis perpendicular to the flight tube axis. The velocity component along the flight tube direction can be reduced further by skimming the free jet to make a low divergent molecular beam.

The second problem can be minimized by operating the TOF mass spectrometer under the so-called Wiley-McLaren space focusing condition. Consider the geometry of the linear TOF mass spectrometer shown schematically in Figure 7. It consists of three grids: a repeller plate, a ground plate, and an acceleration plate. Space focusing can be understood by considering two ions of the same mass initially with zero component of velocity along the flight tube direction created at two different points $S \pm \delta$. Although both ions enter the flight tube, the ion at $S - \delta$ will actually gain more energy than the mass at $S + \delta$, which spends less time in the acceleration region. At some point down the flight tube, therefore, the ion starting from $S - \delta$ will overtake the one starting from $S + \delta$, even though the latter was initially ahead of the former. The optimum mass resolution will be achieved if the detector is placed at this takeover position. In practice, the flight tube distance is fixed and the acceleration and repeller plate voltages are adjusted to obtain the optimum resolution. In a 1 m TOF mass spectrometer, where the separations between the repeller and acceleration plates and

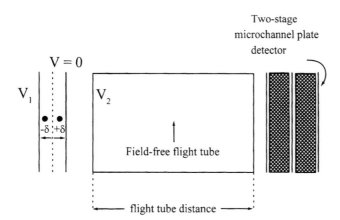

Figure 7   Schematic diagram shows the geometry of a linear time-of-flight mass spectrometer needed to understand space-focussing. V_1 and V_2, voltages on the repeller plate and acceleration grid, respectively.

the ground plate are 2.16 cm and 2.41 cm, respectively, space focusing will be achieved when $V_1 = +360$ V is applied to the repeller plate and $V_2 = -2800$ V is applied to the acceleration grid.

A low-duty cycle is also often cited as another limitation of TOF mass spectrometers, because a new ion bunch can only be injected into the instrument after the previous one has been completely detected. The use of relatively low repetition rate laser systems (characteristically 10–20 Hz) essentially eliminates this problem.

III.   APPLICATIONS OF UV AND VUV LASERS

A.   Spectroscopy

Although numerous examples exist, this chapter will use in part recent studies of ion-pair state ↔ valence state transitions to illustrate the practical use of UV and VUV LIF and TOF excitation spectra in jets. An ion-pair state is one that, instead of producing two neutral fragments upon dissociation, diabatically yields a pair of oppositely charged ions. Note that all molecules have ion-pair states in numbers comparable to or even larger than their conventional valence states [34], but specific examples will be drawn only from our studies of the group 12 (formerly IIB) metal monohalide radicals and the halogens.

1.   Group 12 Monohalide Radicals

Unstable free radicals are some of the most challenging species to produce and characterize spectroscopically. Yet their chemical significance becomes immediately apparent when one realizes that these highly reactive open shell molecules are often the transient intermediates produced and consumed in the elementary reaction steps of complex reaction mechanisms; for example, those associated with combustion and atmospheric chemistry. As reviewed by Miller and co-workers [35], a supersonic jet provides a near-requisite environment for cooling free radicals and for minimizing their reactions, since by simple dilution all that they will encounter are inert rare gas atoms.

The group 12 metal monohalide radical excimers, MX, M = Hg, Cd, Zn; X = I, Br, Cl were first studied by Wieland over 60 years ago [36]. Interest in these species became acute, however, when laser emission was found on their ion-pair B $^2\Sigma^+$ states → X $^2\Sigma^+$ ground state transitions [37]. The mercuric monohalide radicals are particularly noteworthy because their coherent outputs lie in the important blue–green spectral region. These laser systems are also attractive because excited MX radicals can be produced by photolytically or electrically dissociating an MX_2 precursor, which subsequently reforms, resulting in a cyclical production of the excited state [38].

MX radicals are analogues of the better-known and, in some instances, commercially important rare gas monohalide excimers such as XeCl or KrF. Replacing Xe with its large ionization potential, IP = 97834.4 cm^{-1} [39], by Hg (IP = 84184.1 cm^{-1} [39]), substantially decreases the separation between the ion-pair state minimum and the ground state dissociation limit. The resultant configurational mixing between B and X due to an avoided crossing between the Coulomb branch of the ion-pair state and the repulsive ground state therefore imparts a significant bound ionic character to the latter, which, for the most part, is absent for the rare gas species. This mixing likewise decreases the binding energies of the group 12 metal monohalide ion-pair states ($\approx$35000 cm^{-1}) relative to those for their rare gas monohalide counterparts ($\approx$44000 cm^{-1}).

Due to the very different shapes of the potential energy curves involved, the strongest Franck-Condon-allowed transitions originate from low v'-levels of the HgX B-state and terminate on high vibrational levels of X $^2\Sigma^+$. This expectation was quantitatively confirmed by Tellinghuisen and co-workers, who photographed emission spectra of single isotopomer samples excited in a mild Tesla discharge [40]. They found, for example, that the potential energy curve of the B $^2\Sigma^+$ ion-pair state of HgBr dissociating to Hg$^+$(^{2}S$_{1/2}$) + Br$^-$(^{1}S$_0$) is very deep, $D'_e \approx 39100$ cm^{-1}, due to the strong long-range Coulombic potential, $V(R) \propto R^{-1}$, binding the ion-pair together. Nevertheless, the long excited-state equilibrium bond length ($R'_e \approx 3.2$ Å) and small vibrational and rotational constants ($\omega'_e \approx 136$ cm^{-1}; $B'_e \approx 0.03$ cm^{-1}) observed are direct manifestations of its antibonding nature, as predicted by simple molecular orbital theory. In converse fashion, the X $^2\Sigma^+$ ground state is considerably shallower ($D''_e \approx 5527$ cm^{-1}), even though its equilibrium bond length is markedly shorter (by ≈ 0.57 Å).

Experiments in which the unstable radical excimers are probed in absorption are rarer, and exceedingly more challenging. Franck-Condon considerations dictate that transitions from lowest vibrational levels of the ground state will probe the inner wing of the B-state potential energy curve in the UV, exciting very high-lying closely spaced vibrational levels. Furthermore, many rovibronic levels of the radicals are expected to be populated in a discharge. These factors, combined with numerous naturally occurring isotopomers, conspire to yield near continuum-like spectra that at best are difficult to interpret. This was demonstrated by Eden and co-workers, who used transient absorption spectroscopy to pump optically the B-state of HgBr radical from its ground state [41].

Our group found that these problems could be alleviated by generating rotationally cold radicals in a corona-excited supersonic expansion (CESE) [42], and using a tunable UV laser to excite LIF. In this way, we could obtain the first precise "absorption" spectra involving the higher v'-levels of the HgX B-states.

Details about the experimental arrangement can be found elsewhere [43]. In brief, 1–2 mJ tunable UV radiation generated by frequency doubling the output of a Nd:YAG-pumped dye laser was focussed below the pinhole of a quartz bulb nozzle. The nozzle was sealed into a flange mounted on a chamber evacuated to <50 mtorr with a Roots blower/mechanical pump combination. HgX_2 precursor inside the nozzle was resistively heated to produce a vapor pressure above the pinhole. The vapor was entrained in ≈ 2 atm He and coexpanded in a continuous supersonic jet. To conserve samples, the gas delivery could be pulsed by connecting a solenoid valve outside the heater bulb nozzle inside the chamber.

HgX radicals were generated by exciting a discharge in the expanding gas. Positive voltage (≈ 400 V/10 mA) was applied to a stainless steel electrode running concentrically down the long axis of the bulb, while the chamber itself acted as the grounded cathode. Undispersed fluorescence was collected, filtered, and imaged onto the cathode of a photomultiplier. Care was taken not to image the corona-discharge from both inside the nozzle and the brightest region of the jet. Scattered light was reduced by temporally gating the data acquisition system $\approx$ 50 nsec after the peak of the scattered light.

Unlike the transient absorption spectra discussed above, the HgX B $^2\Sigma^+(v')$ $\leftarrow$ X$^2\Sigma^+(v'')$ excitation spectra are highly structured due to rotational cooling [43]. The (29,0) band of Hg^{35}Cl shown in Figure 8, for example, exhibits well-resolved peaks due to the naturally occurring isotopes of Hg. The intensity of

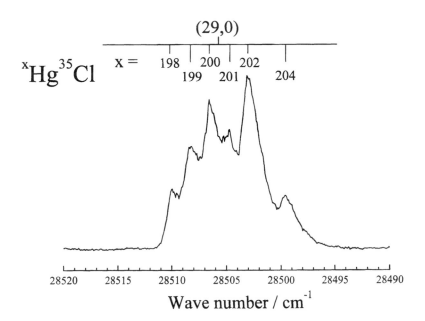

Figure 8   Close-up view of the (29,0) band of the Hg^{35}Cl B $^2\Sigma^+ \leftarrow$ X $^2\Sigma^+$ fluorescence excitation spectrum shows the isotopic structure due to the numerous isotopes of Hg.

each isotopomer peak within the band nicely reflects the normalized natural abundance of the particular Hg atom involved relative to ^{202}Hg (^{198}Hg: ^{199}Hg: ^{200}Hg: ^{201}Hg: ^{202}Hg: ^{204}Hg = 0.335: 0.565: 0.776: 0.444: 1.00: 0.230) [44]. Isotopomers containing ^{196}Hg, however, were not detected for a $(v',0)$ Hg^{35}Cl/Hg^{37}Cl doublet, since its normalized natural abundance is only 0.005.

The transition frequency of the Hg^{35}Cl component for any $(v' \gg 0,0)$ HgCl band will always be larger than that for Hg^{37}Cl due to simple mass scaling. Care must, however, be exercised in the analysis of high vibrational levels of an ion-pair state because vibrational isotope shifts can not only equal but often well exceed the local vibrational spacings, $\Delta G'_{v'+1/2}$. To a first approximation the vibrational level, v^o, where the isotope shift equals the vibrational spacings, can be calculated from [45]:

$$v^o = \frac{\rho_i}{1 - \rho_i} \tag{16}$$

where ρ_i is defined here as $\rho_i = \sqrt{\mu(^{200}Hg^{35}Cl/\mu_i)}$ and μ_i is the reduced mass of a particular HgCl isotopomer. The crossover vibrational level is calculated using Eq. (16) to be $v^o \approx 42$ for HgCl. As illustrated in Figure 9 the weaker Hg^{37}Cl doublet component will appear at a higher frequency than the nearest Hg^{35}Cl for $v' < 42$, and at a lower frequency for $v' > 42$. However, the vibrational quantum number of the Hg^{37}Cl band is actually one higher than its Hg^{35}Cl.

Such crossovers in HgCl are expected for integer multiples of v^o near $v' = 84$, 126, 168, and so on, with an additional unit increase in the relative numbering of the Hg^{x}Cl bands per v^o interval. The ^{x}HgX isotope shifts confirm the v^o value observed and, therefore, the absolute v'-numbering. Auxiliary information of this type or the location of the electronic origin is usually required to establish the crossover region being probed. This is because the standard deviation of "local" least-squares fit for all choices of nv^o; $n = 1, 2, \ldots$ will be of similar quality. Crossovers within a doublet component for adjacent HgCl isotopomers are only expected to become noticeable for $v' > 2600$, a spectral region far outside the observed Franck-Condon envelope.

Although the HgCl spectra are rotationally cold, the frequency bandwidth of the laser system was such that rotational structure could not be resolved. The radicals produced in the CESE, however, were found to be vibrationally warm. Thus, Franck-Condon minima in the hot-band vibrational progressions could be used to establish B-state equilibrium bond lengths. B ← X transitions probe the inner turning points of the excited state potential energy curve, where the high v' wave functions have their largest and $\approx$ constant amplitudes. The intensity minima therefore reflect the ground state wave function nodal structure superimposed on the electronic transition Franck-Condon factors.

Since a definite minimum in the $v'' = 2$ progression was found for HgCl at $(v', v'') = (34,2)$, the B-state equilibrium bond length was established by first

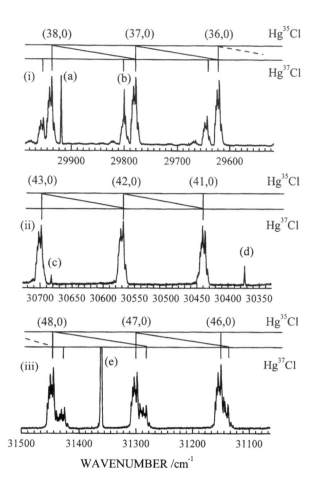

Figure 9 Overview of the HgCl B $^2\Sigma^+$ ← X $^2\Sigma^+$ fluorescence excitation spectrum in the region of the first "crossover" point at (42, 0). Bands having the same vibrational assignment but belonging to a radical with a different Cl isotope are connected by a diagonal line. The features labeled (a)–(e) are atomic transitions of He and Hg originating from metastable levels. (i) B–X vibrational bands just to longer wavelengths of (42,0). In this region the Hg^{35}Cl-Hg^{37}Cl isotope shifts are less than the local vibrational spacings. (ii) B–X vibrational bands near (42,0). In this region the Hg^{35}Cl–Hg^{37}Cl isotope shifts are approximately equal to the local vibrational spacings. (iii) B–X vibrational bands just to shorter wavelengths of (42,0). In this region the Hg^{35}Cl–Hg^{37}Cl isotope shifts are greater than the local vibrational spacings.

constructing RKR-Morse potential curves [46] for both X $^2\Sigma^+$ and B $^2\Sigma^+$. Here the derived vibrational constants were used in an RKR calculation [47] to determine the width of the potential as a function of v. The inner wings of the potentials were then calculated using Morse functions that incorporated R_e values from the literature. Morse parameters were ascertained using the vibrational frequency and dissociation energy for each electronic state. The outer turning points followed immediately.

Vibrational wave functions were evaluated numerically [48] in both potentials over the range of the observed levels and Franck-Condon factors derived. A series of RKR-Morse/Franck-Condon factor calculations were then carried out with $R_e(X)$ constrained, but $R_e(B)$ varied systematically until the predicted intensity minima agreed with the experimental observations. The calculated positions of the intensity minima were found to be very sensitive to the relative position of the two potentials energy curves because, in absorption, transitions from v'' = 2 probe an extremely steep portion of the B-state inner wing. For example, a decrease of $\approx$ 0.004 Å in $R_e(B)$ for HgCl dropped the minima location by $\approx$ one quantum in v'. The best $R_e(B)$ for HgCl was found to be 2.960 Å, with an estimated error of $\pm$ 0.002 Å.

2. Diatomic Halogens

The ion-pair states of the diatomic halogens are arguably the best characterized to date. A halogen ground state is found to be adequately described by the single molecular orbital configuration: $\sigma_{(g)}^2\pi_{(u)}^4\pi_{(g)}^4\sigma_{(u)}^0$, where the exponents indicate the number of electrons in each molecular orbital, and the g/u subscripts apply only to the homonuclear molecules. By convention, only the exponents are listed and, therefore, the ground state configuration is designated by 2440 [49]. There are a total of 20 ion-pair states of an AB halogen. Half of that number can be organized into four tiers that dissociate diabatically to the A$^-$($^1S_0$) + B$^+$(^3P, ^1D, ^1S), and, therefore, have separations that reflect the energy intervals of the positive ion, B$^+$. The remaining charge transfer states correlate with the much higher energy A$^+$ + B$^-$ limit, provided the ionization potential, IP, of B < IP (A). The 20 states that dissociate to a common set of asymptotes for the homonuclear species when A = B are delineated instead by inversion symmetry. Due to a $\Delta\Omega = 0$ propensity rule governing single-photon valence to ion-pair state excitations [50], the most intense VUV spectral feature is the transition from the $X0_{(g)}^+$ halogen ground state to the first tier $0_{(u)}^+$ ion-pair state having the strongly antibonding 1441 configuration. After a long and often confusing history, it is now widely accepted that the $0_{(u)}^+$ state should be labeled E and D for the heteronuclear and homonuclear halogens, respectively.

Like the group 12 monohalide radicals, the potential energy curve of the 1441 state of a given halogen (and all its other ion-pairs states) exhibits schizophrenic properties. It has a large dissociation energy relative to the ground state (on the order of 4 eV), but a longer equilibrium bond length, and much smaller vibrational and rotational energy level spacings. Spectroscopic difficulties arise because Franck-Condon-allowed single-photon transitions probe the inner wing of the 1441 ion-pair state potential energy curve, exciting very high, very closely spaced vibrational levels in the VUV. Extensive room-temperature rotational congestion and isotopic fine structure provide additional complications.

Many of these problems could be minimized by exciting a supersonically cooled halogen with tunable, coherent, and monochromatic VUV radiation. Detailed analyses of the single-photon VUV spectra of IX, $X = I$ [51], Br [52], Cl [45], BrY, $Y = Br$ [53], Cl [54], and Cl_2 [55] at vibrational and, sometimes, rotational precision have been completed by our group, in many instances, for the first time. Initially, VUV LIF proved to be a successful detection scheme. However, as our results for BrCl demonstrate, TOF mass detection provides an alternative and an often more powerful method for unravelling complex excitation spectra.

Until recently, the only quantitative information about the excited $E0^+$ charge transfer state of BrCl was provided by single-isotopomer emission spectra of the $E0^+-B0^+$ band system excited in a mild Tesla discharge. That analysis was based only on transitions involving $v_E \leq 2$ [56]. Low-resolution VUV synchrotron absorption and fluorescence excitation spectra of BrCl have also been reported [57]. Near 145 nm both Rydberg state $\leftarrow$ ground state and the E–X band systems are observed in absorption, while only the ion-pair state is seen in fluorescence excitation. Like the other interhalogens, its overall intensity was found to be strongly modulated as a function of frequency, a phenomenon attributed to interactions between $E0^+$ and isoenergetic Rydberg states.

Details about our apparatus shown schematically in Figure 10, which combines VUV LIF and TOF mass detection, have been published [25,54]. The system was built to allow XUV generation below 105 nm by four-wave mixing in a pulsed supersonic jet of Xe or Kr gas located in the first of two vacuum chambers. Differential pumping effectively eliminates all window material in the path of the XUV beam, allowing the generated light to enter the second chamber for experiments. The entire vacuum system is maintained at a base pressure $\leq 10^{-6}$ torr using diffusion pumps.

Since BrCl excitation wavelengths lie between 146.8 nm and 143.3 nm, those four-wave mixing experiments were done not in a free jet but in ≈ 5 torr Kr gas housed in a static cell mounted to the front end of the first vacuum chamber. BrCl molecules, formed by mixing Br_2 and excess Cl_2 prior to jet expansion in He at a total stagnation pressure of 34 psia, were excited by one photon of the VUV laser and subsequently ionized with a second photon from the fundamental beam in a $(1 + 1')$ REMPI process. The resultant ions were mass dispersed in a linear TOF mass spectrometer ($m/\Delta m \geq 500$). Valve opening was synchro-

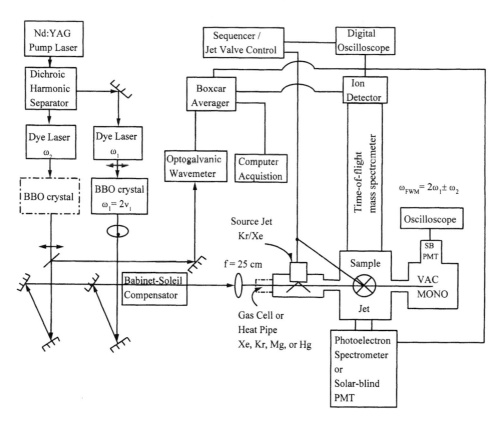

Figure 10 Schematic diagram of the experimental apparatus used to record VUV and XUV laser/time-of-flight mass spectra of jet-cooled molecules. Instruments indicated by dashed boxes are optional. Their use depends on the range of VUV or XUV wavelengths being generated.

nized to the arrival of the laser pulse inside the ionization volume with homemade electronics. The timing was chosen to provide as many rotational lines as possible without strong overlapping between vibrational bands.

Simultaneous VUV LIF detection with a solar-blind photomultiplier tube mounted 180 degrees to the axis of the flight tube was accomplished by constructing the repeller plate from a fine mesh with $\approx 90\%$ optical transmission. Scattered light into the photocathode was reduced using light baffles placed around the phototube and the path of the laser beams. In other experiments that will not be considered here, the photomultiplier was replaced with a dispersive photoelectron spectrometer to measure the kinetic energies of electrons formed by REMPI.

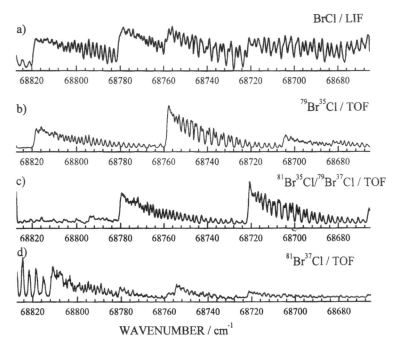

Figure 11 A small portion of the VUV laser excitation spectrum of BrCl near 145.4 nm. The strongest vibrational band of the E0$^+$ ← X0$^+$ transition for the ^{79}Br^{35}Cl isotopomer in this spectral region is assigned to (182,0). (a) VUV LIF excitation spectrum of a naturally abundant isotopic sample of BrCl. (b) VUV laser/time-of-flight (TOF) mass spectrum of the ^{79}Br^{35}Cl isotopomer. (c) VUV laser/TOF mass spectrum of the ^{81}Br^{35}Cl/^{79}Br^{37}Cl isotopomer pair (d) VUV laser/TOF mass spectrum of the ^{81}Br^{37}Cl isotopomer. Additional excitation spectra obtained by monitoring ^{79}Br and ^{81}Br atomic fragments produced by predissociation confirm that the bands in trace (c) are due to ^{81}Br^{35}Cl.

Small segments of the VUV BrCl E0$^+$–X0$^+$ LIF and TOF excitation spectra near 145.4 nm are presented in Figure 11 [54]. Four naturally occurring isotopomers contribute to the fluorescence excitation spectrum (Fig. 11a): ^{79}Br^{35}Cl (natural abundance = 38.41%), ^{81}Br^{35}Cl (37.36%), ^{79}Br^{35}Cl (12.28%), and ^{81}Br^{37}Cl (11.95%). Even though rotationally resolved vibronic bands are readily discerned, their isotopomer identities are difficult to establish based on natural abundances due to the strong perturbative intensity anomalies. The TOF spectra shown in Figure 11 b–d are considerably simpler, although large fluctuations in the band intensities are still evident in the TOF detection mode. Even though the mass spectrometer could not resolve ^{79}Br^{37}Cl and ^{81}Br^{35}Cl, all the bands in Figure 11c were subsequently assigned to ^{81}Br^{35}Cl by separately monitoring ^{79}Br$^+$ and ^{81}Br$^+$

daughter fragment ions produced by excited state predissociation followed by ionization, all within the VUV laser pulse duration.

The excited state vibrational numbering was initially estimated by fitting the isotopic transition term values, v_i, to the following mass-reduced mixed near-dissociation expansion (NDE) appropriate for a R^{-1} long-range potential [58]:

$$v_i = T'_e - D'_e - \frac{Ry}{X^2}\left(1 + \frac{a_3}{X}\right) \tag{17}$$

where X is defined as $[(v_D + \frac{1}{2}) - \rho_i(v' + \frac{1}{2})]$, $\rho_i = \sqrt{\mu(^{79}Br^{35}Cl)/\mu_i}$, where μ_i is the reduced mass of a particular BrCl isotopomer, T'_e and D'_e are the E-state electronic origin and dissociation energy, respectively, and Ry is the Rydberg constant for $^{79}Br^{35}Cl$. The second and third terms in Eq. (17) represent the correct limiting behavior for the term values in a Coulomb potential, while the fourth term proportional to a_3 is empirical and accounts for the fact that the vibrational levels observed in emission lie well below the dissociation limit. It is not coincidental that the limiting form of Eq. (17) is very similar to the Rydberg term expression for H-like systems:

$$T_n = IP - \frac{Ry}{(n - \delta)^2} \tag{18}$$

where T_n is the electronic state term value, Ry, is the Rydberg constant, and n is the principal quantum number. The quantum defect δ, which characterizes the deviation of the atom from pure hydrogen behavior, decreases with increasing orbital angular momentum ℓ. In NDE theory v_D is usually interpreted as the noninteger vibrational quantum number of the highest level supported by the potential. However, just as there are an infinite number of electronic states in the H-atom, an ion-pair state potential energy curve supports an infinite number of vibrational levels below its dissociation threshold. Therefore, $|v_D|$ can be viewed instead as a vibrational quantum defect. Its magnitude is equal to the quantum number n of a level in a fictional Coulomb potential having the same outer wing as the molecular curve and the same term value as $v' = 0$ of the ion-pair state. We found that the high vibrational ion-pair state term values of the iodine containing interhalogens could often be fitted to an NDE with fewer parameters than a Dunham series.

Calculations based on a fit to Eq. (17) established that approximately 120 $(v', 0)$ isotopic bands with $179 \leq v' \leq 213$ are located in the excitation wavelength region. Appreciable Franck-Condon factors for hot band transitions from $v'' = 1$ were also predicted at the short wavelength end of the experimental observations, and for $(v', v'' = 2)$ transitions at the longer wavelengths. The lowest-energy $^{79}Br^{35}Cl$ band from these calculations could be assigned to $(180,0)$. Ultimately, the best v'-numbering chosen was the one that minimized the standard deviation, σ, of a least-squares fit of the isotopic band origins and transition band head transition wave numbers to a mass-reduced Dunham expansion. Then the lowest $^{79}Br^{35}Cl$ energy band was assigned to (172,0), corresponding to a difference of -8 from the estimate provided by the NDE calculation. This difference, however, is not considered serious in light of the extensive frequency and intensity perturbations observed. Furthermore, precise excited-state rotational constants for most of the observed bands were also derived for the first time due to the monochromaticity of the VUV laser used.

VUV spectroscopy of the $2\ ^1\Sigma_u^+(v') \leftarrow X\ ^1\Sigma_g^+(v'' = 0)$ transition of elemental Cl_2 provides an excellent example of LIF and TOF detection together solving a difficult problem that can only be partially understood using one or the other technique. Until recently, the $2\ ^1\Sigma_u^+ - X\ ^1\Sigma_g^+$ band system defied quantitative analysis. That situation has now been significantly rectified by the judicious application of theory [59] and VUV laser experimentation [55]. In diabatical terms, the inner wing of the 2-state is formed by a strong avoided crossing between the inner wall of the third-tier ion-pair state that dissociates to $Cl^+(^1D_g) + Cl^-(^1S_g)$ and the outer wing of the $4p\pi$ Rydberg state. Double-well formation results from an additional perturbation at longer bond lengths, with the fourth tier ion-pair state dissociating to $Cl^+(^1S_g) + Cl^-(^1S_g)$. Due to the curvature of the potential near its predicted minimum at $R_e' \approx 2.1$ Å, the vibrational frequency of excited state, ω_e' was expected to be larger than that for a pure Rydberg state (≈ 628–665 cm^{-1}), valence state (≈ 260 cm^{-1}), or the ground state (≈ 560 cm^{-1}). VUV laser/ TOF mass spectra confirm this expectation (Fig. 12, $\omega_e' \approx 760$ cm^{-1}). Furthermore, it can be concluded from measured vibrational isotope shifts that the features previously attributed to the $(v', v'') = (0,0)$ band near 78125 cm^{-1} actually belong to the (1,0) band.

It is interesting that the intensity distributions in the isotopically resolved TOF spectra are anomalous due to what is believed to be an additional resonance enhancement in the second step of the $(1 + 1')$ REMPI process by a superexcited state embedded in the first ionization continuum of the molecule. As a consequence, rotational analyses were carried out in large part by computer simulating the LIF spectra. The results for $^{35}Cl_2$ 2-X (1,0) transition are shown in Figure 13. An intensity alternation of 5:3 evident in the P-branch for the odd and even J-levels, respectively, in Figure 13a, arises because the nuclear spin(I) of $^{35}Cl = 3/2$. However, the R-branch was unresolved and therefore the absolute J-numbering could not be established by ground state combination differences. Due to the limited number of J-levels populated at room temperature, the J-numbering could nevertheless be deduced almost by inspection to within ± 2. The best J-numbering used to determine constants was that producing a minimum least-squares fit standard deviation and the best simulation ($\delta J = 0$; Fig. 13b). Changing the J-numbering by $\delta J = +2$ (Fig. 13 c) or by $\delta J = -2$ (Fig. 13d), as required by the intensity alternation, produced simulations in much poorer agreement with the

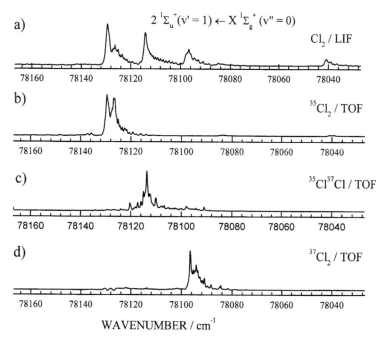

Figure 12 VUV laser excitation spectra of the $2\ ^1\Sigma_u^+(v' = 1) \leftarrow X\ ^1\Sigma_g^+(v'' = 0)$ transition of Cl_2 (a) VUV laser-induced fluorescence excitation spectrum of a naturally abundant isotopic sample of Cl_2. VUV laser/TOF mass spectra for the (b) $^{35}Cl_2$, (c) $^{35}Cl^{37}Cl$, and (d) $^{37}Cl_2$ isotopomers are shown.

experimental observations. Different simulations were carried out over exactly the same frequency range but using the new band origins and rotational B'-constants derived from refitting the observed transition frequencies to a Dunham expansion. As a consequence, a J-numbering was not considered acceptable if the resultant simulation could only match the observed spectrum by frequency shifting it wholesale to do so. The resultant $R_{v'=1} = 2.0991(46)$ Å is in very good agreement with theoretical expectations.

Yamanouchi and Tsuchiya have reviewed the steadily growing VUV laser spectroscopy literature up to 1994 [60]. Table 2 provides a revised catalogue of atomic and molecular systems studied with VUV lasers since 1990. Even a cursory inspection of our new list and the associated references shows that such short-wavelength sources are now not only being used to obtain new spectra and/or to rerecord known spectra with a higher-frequency resolution but also extensively in reaction dynamics, for coherent control, and for ion spectroscopy by pulsed field ionization methods.

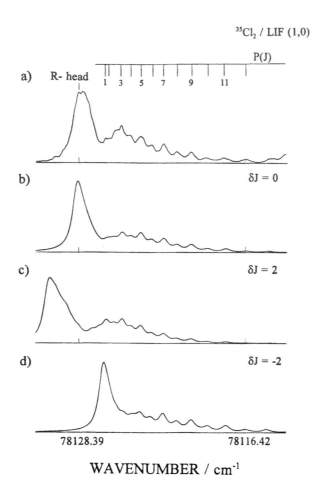

Figure 13 (a) A small portion of the VUV laser-induced fluorescence excitation spectrum assigned to the $2\ ^1\Sigma_u^+(v' = 1) \leftarrow X\ ^1\Sigma_g^+(v'' = 0)$ transition of $^{35}Cl_2$. (b) Best simulation, $\delta J = 0$. (c) Simulation obtained by changing the best set $\{J\}$ in the P-branch by $\delta J = +2$. (d) Simulation obtained by changing the best set $\{J\}$ in the P-branch by $\delta J = -2$.

B. Analytical Chemistry

Although lasers are becoming routine fixtures in many analytical chemistry labs [61], the specific combination of VUV lasers and mass spectrometry provides unique opportunities for organic molecule detection that are unavailable to other experimental methods based on either longer wavelength coherent sources or nonoptical strategies. For example, unambiguous molecular identification by con-

Table 2 VUV and XUV Laser Studies of Atomic and Molecular Systems

Molecular or atomic species	Electronic states	Spectroscopic method	References
H(D)	np Rydberg states, n = 3, 4, 7, 8	EIT	86
	np Rydberg states, n = 4 ~ 8	EIT	87
	3p (← 1s); 3p (← 2s)	EIT	88–93
	2p Rydberg states	EIT	94–98
	2p ^{2}P (← 1s ^{2}S)	LIF	99–123
	2p ^{2}P (← 1s ^{2}S)	PAS	124
	2p ^{2}P (← 1s ^{2}S)	PTS	125–142
	2p ^{2}P (← 1s ^{2}S)	HRTOF	143–152
	2p ^{2}P (← 1s ^{2}S)	DSTOF	153, 154
H$^-$	Near $n = 2$ threshold	DS	155–157
H$_2$, D$_2$	H $\bar{H}$ $^1\Sigma_g^+$	REMPI	158–160
	B $^1\Sigma_u^+$, B' $^1\Sigma_u^+$, C $^1\Pi_u$, D $^1\Pi_u$	REMPI	161–164
	I' $^1\Pi_g$, EF $^1\Sigma_g^+$, GH $^1\Sigma_g^+$	REMPI	165
	EF $^1\Sigma_g^+$, GK $^1\Sigma_g^+$, H $^1\Sigma_g^+$, I $^1\Pi_g$, J $^1\Delta_g$	OODR	166
	Singlet gerade s- and d-Rydberg states with J = 0, 1, 2	OODR	167
	B' $^1\Sigma_u^+$, B $^1\Sigma_u^+$, C $^1\Pi_u$	OODR	168–170
	B $^1\Sigma_u^+$, C $^1\Pi_u$	PTS	171
	H$^+$(1s^0) + H$^-$(1s^2)	TIPPS	172
	Rydberg states: Q(1) np π, n = 5, 7, 14 ~ 22	SS	173
	R(1)np1, n = 15 ~ 21		
	P(1)npσ, n = 8, 20 ~ 21		
	R(1)np3, n = 12 ~ 15		
	R(0)np2, n = 14, 15		
	R(0)np0, n = 14		
	Rydberg states, n = 30–49	PI	174
	B' $^1\Sigma_u^+$, B^1 Σ_u^+	LIF, PHOFEX	175–178

Table 2 Continued

Molecular or atomic species	Electronic states	Spectroscopic method	References
H_2^+	$X\ ^2\Sigma_g^+$	PFI-ZEKE	179,180
HD	$H\ \bar{H}\ ^1\Sigma_g^+$, $B''\ \bar{B}\ ^1\Sigma_g^+$	REMPI	158, 159, 181
	High gerade Rydberg states close to the first few rotational levels of the ground vibronic state of HD^+	REMPI	182
	$B'\ ^1\Sigma^+$, $B\ ^1\Sigma^+$, $C\ ^1\Pi$	OODR	170
	Rydberg states near second dissociation limit at ~84 nm	LIF	183
He	$2\ ^1P$	REMPI	159
	2^1S	TPDFS	184
C	$2p^2\ ^1D$, $2p^2\ ^3P$	LIF	185, 186
CH_3^+	$\tilde{X}\ ^1A_1'$	PFI-ZEKE	187
O	$3s\ ^1D°\ (\leftarrow 2p\ ^1D)$, $3s\ ^3S°\ (\leftarrow 2p\ ^3P_{2,1,0})$	LIF, PHOFEX	188–208
CH_4, CD_4	$\tilde{A}\ ^1T_2$	PEX	141
		PEX	108
CH_4^+	$\tilde{X}\ ^2B_1$	PFI-ZEKE	209
OH^+, OD^+	$X\ ^3\Sigma^-$	PFI-ZEKE	210
NH_3	$\tilde{B}\ ^1E''$	LIF	211
NH_3^+	$\tilde{X}\ ^2A_2''$	PFI-ZEKE	212
H_2O	$\tilde{C}\ ^1B_1$	OODR	213
H_2O, D_2O	$\tilde{B}^1A_1$	PEX, PHOFEX	142, 214
	$nd \leftarrow 1\ b_1$ Rydberg series, $n \geq 6$; $n = 6$–11, converging to H_2O^+ (000) and (100) vibrational states, respectively	REMPI	215
H_2O^+, D_2O^+	$\tilde{X}\ ^2B_1$	PFI-ZEKE	216
H_2O^+, H_2S^+	$\tilde{X}\ ^2B_1$	PFI-ZEKE	217–219
CD_2H_2	Lowest energy $^1T_2$ state	PEX	129
$CD_2H_2^+$	$\tilde{X}\ ^2B_2$	PFI-ZEKE	220

Ne	$3s[3/2]_1$, $3s'[3/2]_1$	REMPI	221
HF⁺	$X\ ^2\Pi_{1/2,\,3/2}$	PFI-ZEKE	222
ND₃	$\tilde{B}\ ^1E''$, $\tilde{D}\ ^1E'$	LIF, REMPI	223
ND₄⁺	$\tilde{X}\ ^1A_1$	PFI-ZEKE, MATI	224
C₂H₂, C₂D₂	$3R'2_0^1$, $3R''2_0^1$, $4R2_0^0$, $4R''2_0^0$, $4R''2_0^1$, $E2_0^1$, $E4_0^2$, $E5_0^2$	PEX	124, 143, 153, 154
HCN	$\tilde{A}\ ^1A''$	LIF	225
	$^1\Pi$ Rydberg state ($\ldots 1\pi^3 3s\sigma^1$ dominant configuration)	PEX	135, 138
HCN⁺	$\tilde{X}\ ^2\Pi$, $\tilde{A}\ ^2\Sigma^+$	PFI-ZEKE	226
CO	$X\ ^1\Sigma^+$	LIF	185, 227
	$B\ ^1\Sigma^+$	REMPI, LIF	163, 228–230
		CC	231, 232
	$a'\ ^3\Sigma^+$	EX	233
	$A\ ^1\Pi$	LIF	234–244
		AS	245
	$C\ ^1\Sigma^+$	REMPI	246
	$E\ ^1\Pi$	REMPI	247
	$W\ ^1\Pi$, $L\ ^1\Pi$, $K\ ^1\Sigma^+$, $L'\ ^1\Pi$	REMPI	248–253
	$A\ ^1\Pi$, $L\ ^1\Pi$	FWM	254–255
	$j\ ^3\Sigma^+$, $a\ ^3\Pi$, $k\ ^3\Pi$	LRF	256–258
	$np\sigma$ Rydberg states, $n = 5 \sim 8$		
	$nd\sigma$, δ Rydberg states, $n = 4 \sim 9$		
CO⁺	$X\ ^2\Sigma^+$	PFI-ZEKE	259
	$A\ ^2\Pi_{3/2,1/2}$	PFI-ZEKE	260
N₂	$b'\ ^1\Sigma_u^+$, $b\ ^1\Pi_u$, $o\ ^1\Pi u$	REMPI	261, 262
	$c_4'\ ^1\Sigma_u^+$, $c_3\ ^1\Pi u$	REMPI	263
N₂⁺	$X\ ^2\Sigma_g^+$, $A\ ^2\Pi_u$, $B\ ^2\Sigma_u^+$	PFI-ZEKE	259, 264–267
HCO, DCO	$\tilde{A}\ ^2A''$	PHOFEX, LIF	268–270
HCO, DCO	$\tilde{X}\ ^2A'$	LIF	237, 238

Table 2 Continued

Molecular or atomic species	Electronic states	Spectroscopic method	References
NO	$B\,^2\Pi_{3/2}$, $E\,^2\Sigma^+$, $H\,^2\Sigma^+$, $H'\,^2\Pi$, $O\,^2\Sigma^+$, $O'\,^2\Pi$	FWM	271–274
	ns Rydberg states, $n = 7, 9\sim12$	ARPES	275
	nf Rydberg states, $n = 5, 8\sim11, 13$		
NO⁺	$a\,^3\Sigma^+$	PFI-ZEKE	276
	$X\,^1\Sigma^+$	PFI-ZEKE	277, 278
S	^1D, ^3P	DS, LIF	279–283
	$3p^3ns\,^3S^o$, $n = 5\text{–}7$; $3p^3np\,^3P_{0,1,2}$, $n = 4,5$	LIF	284
	$3p^3nd\,^3D^o$, $n = 4,5$		
O₂	$O^+ + O^-(^2P_{3/2,\,1/2})$	TIPPS	285
	$B\,^3\Sigma^-_u$, $C'^3\Pi_u$	AS	286
	$f'\,^1\Sigma^+_u$, $f\,^1\Sigma^+_u$, $j\,^1\Sigma^+_u$	AS	287
O₂⁺	$X\,^2\Pi_{1/2g,3/2g}$, $a\,^4\Pi_u$	PFI-ZEKE	218, 288–290
SiH₄	$^1T_2$ states $(1a_1^2 2a_1^2 1t_2^6 3a_1^2 2t_2^5 4p,\ 3d,\ 5s\ \sigma^*\ (t_2)$ configurations)	PEX	291
H₂S		CC	292
	$HS^+ + H^-$, $HS^- + H^+$	TIPPS	293
H₂S, D₂S	$\tilde{B}\,^1A_1$	PEX	134, 138, 140
Cl	$4s\,^2P_{3/2,\,1/2} \leftarrow (3p\,^2P_{3/2,\,1/2})$	LIF	111, 294
HCl, DCl	Rydberg states in the energy range between their $^2\Pi_{3/2}$ and $^2\Pi_{1/2}$ ionic thresholds	AI	295
	$j\,^3\Sigma^-(0^+)$, $H\,^1\Sigma^+(0^+)$, $m\,^3\Pi(1)$	CC	232, 296
	$H^+ + Cl^-(^1S_0)$	TIPPS	297
HCl⁺	$\tilde{X}\,^2\Pi_{3/2,\,1/2}$	PFI-ZEKE	298
F₂	$C\,^1\Sigma^+_u$, $H\,^1\Pi_u$	REMPI	299

Species	State	Method	Ref.
Ar	$4s'[1/2]_1$	REMPI	300
	Rydberg states with $n \geq 27$ located below the $^2P_{3/2}$ ionic threshold of Ar^+.	PFI-ZEKE	301
	Rydberg states with $n = 60 \sim 200$	ODDR, PFI-ZEKE	302
Ar^+	$3p^5\ ^2P_{1/2,3/2}$	PFI-ZEKE	174, 303, 304
CO_2	$3p\pi_u\ ^3\Sigma_u^-$ Rydberg states	AS	305
CO_2^+	$\tilde{X}\ ^2\Pi_{3/2g,\,1/2g}$	PFI-ZEKE	174, 306, 307
N_2O^+	$\tilde{A}\ ^2\Sigma^+$	PFI-ZEKE	308
	$\tilde{X}\ ^2\Pi$	PFI-ZEKE	309
CH_3SH^+, $CH_3CH_2SH^+$ $CH_3SCH_3^+$		PFI-ZEKE	310, 311
ClO	$C\ ^2\Sigma^-$	LIF	199, 312–314
OCS	$2\ ^1\Sigma^+$	PHOFEX	315–317
Zn	$3d^94s^24p\ ^1P_1^0$, $3d^94s^24p\ ^3D_1^0$	EIT	318
Cl_2	$1\ ^1\Sigma_u^+$, $2\ ^1\Sigma_u^+$, $2\ ^3\Pi_u$	LIF, REMPI	319, 55
CS_2	Rydberg states $[^2\Pi_{1/2}]\,np\sigma_u$, $n = 15 \sim 31$	PI	320
	$[^2\Pi_{1/2}]\,np\pi_u$, $n = 17 \sim 22$		
	$[^2\Pi_{1/2}]\,nf_u$, $n = 14 \sim 25$		
CS_2^+	$\tilde{X}\ ^2\Pi_{1/2,3/2}$	PFI-ZEKE	320
CH_3PO_2		PI	321
Ar_2^+	$A^2\Sigma_{1/2u}^+$, $C\ ^2\Pi_{1/2u}$, $C\ ^2\Pi_{3/2u}$	PFI-ZEKE	322, 323
HBr^+	$A^2\Sigma^+$	PFI-ZEKE	324
Kr	$4p^55p[1/2]_0$, $4p^55s[3/2]_1^0$	EIT	325–327
Pb	$6p^2\ ^3P_0$, $6p^2\ ^3P_2$, $6p7s\ ^3P_1$, $6p9s\ ^3P_1$	EIT	328
CH_3OPO_2		PI	321
CH_2Cl_2O		PI	329
BrCl	$E\ 0^+$	LIF, REMPI	330

Table 2 Continued

Molecular or atomic species	Electronic states	Spectroscopic method	References
HI	Rydberg states converging to the $^2\Pi_{1/2}$ ionic threshold of HI$^+$	PI, ARPES	331, 332
	ns, np, nd, nf, ng Rydberg states, n = 6, 7	ARPES	333
	Q[$^3\Pi(1, 0^+)$], $^1\Pi(1)$]	DS	334
HI, DI	5sσ, 5dπ, 5dδ	CC	335–340
CH$_3$I		CC	341
XeRg (Rg = Ne, Ar, Kr)	Rg + Xe* (6s[3/2]$_1^o$, 6s'[1/2]$_0$)	LIF, REMPI	342–345
XeRg (Rg = Ar, Kr)	Rg + Xe*5d[3/2]$_1^o$	REMPI	346
Kr$_2$	Kr + Kr*(4d, 5p', 6s)	REMPI	347
Xe$_2$	Xe + Xe*(5d, 6p, 6s')	REMPI	348, 349
	Xe + Xe*(5d, 6p)	PES	350
	B′ I$_u$	LIF	351
	Xe + Xe*(5d[3/2]$_1$, 5d[5/2]$_3$, 7s[3/2]$_1$, 7s[3/2]$_2$)	LIF	352
Xe$_2^+$	A $^2\Sigma_{1/2u}^+$, B$^2\Pi_{3/2g}$, C $^2\Pi_{3/2u}$	PFI-ZEKE	353
Kr$_N$, N = 2 –2000	near Kr* 5s $^3P_2$	LIF	354

Studies are listed in approximate order of increasing mass; carried out from 1990 to date. AI, autoionization spectroscopy; AS, absorption spectroscopy; ARPES, angle-resolved photoelectron spectroscopy; CC, coherent control; DS, Doppler spectroscopy; DSTOF, Doppler-selected time-of-flight spectrometry; EIT, electromagnetically induced transparency; EX, excitation spectroscopy; FWM, four-wave mixing spectroscopy; HRTOFS, high-n Rydberg time-of-flight spectrometry; LIF, laser-induced fluorescence spectroscopy; LRF, laser-reduced fluorescence spectroscopy; MATI, mass-analyzed threshold ionization spectroscopy; OODR, Optical–optical double-resonance spectroscopy; PAS, photofragment action spectroscopy; PES, photoelectron spectroscopy; PEX, photoexcitation spectroscopy; PFI-ZEKE, pulsed field ionization–zero kinetic energy photoelectron spectroscopy; PHOFEX, photofragment excitation spectroscopy; PI, photoionization spectroscopy; PTS, photofragment transnational spectroscopy; REMPI, resonance-enhanced multiphoton ionization spectroscopy; SS, stark spectroscopy; TIPPS, threshold ion-pair production spectroscopy; TPDFS, two-photon Doppler-free spectroscopy.

ventional methods such as electron ionization (EI) [62], chemical ionization (CI) [63], and fast atom bombardment (FAB) [64] very often requires understanding the complex fragmentation patterns that depend on the thermal energy in the parent species, as well the excitation and ionization methods used. Furthermore, ion yields from surface-adsorbed molecules using secondary ion mass spectrometry (SIMS) [65] and matrix-assisted laser desorption and ionization (MALDI) [66] are strongly influenced by solid matrix effects. It should be appreciated, however, that the number of neutral molecules and fragments produced by SIMS and MALDI are expected to be orders of magnitude larger than the direct ion yields. A significant enhancement in detection sensitivity could consequently be achieved if these neutral species were postionized following thermal, particle-induced, or laser-induced desorption. The combination of postionization following surface desorption and TOF mass spectrometry has been termed ''surface analysis by laser ionization'' (SALI) [67,68]. Its inherent value is that it decouples surface desorption from postionization, thereby allowing the two steps to be independently optimized.

Both strengths and weaknesses emerge when REMPI postionization is compared to a nonresonant single-photon SALI process. REMPI has been shown to be particularly successful for elemental analysis, and different schemes have now been devised for almost every atom in the periodic table [69]. This approach is highly sensitive, selective, and as noted above, uniformly efficient since a focussed laser can ionize almost every species in its focal volume with relative ease. Nevertheless, surface composition measurements can be limited by large ion signal variations due to laser shot-to-shot fluctuations.

Unlike elemental analysis, however, few good things can be said about the application of REMPI to organic molecule detection. First, multiphoton absorption tends to result in facile and virtually unavoidable nonspecific fragmentation [70]. Furthermore, since there are no guarantees that every organic molecule will absorb a given UV/visible wavelength, electronic structures and oscillator strengths for each individual molecule must be known. A useful illustration of this problem is the facile detection of polycyclic aromatic hydrocarbons (PAHs) generated by ionizing cigarette smoke at 266 nm [71]. Due to resonance enhancement at that fundamental wavelength, PAH peaks dominate the resultant mass spectrum even though they make up only a small percentage of the total mass of the chemicals present. REMPI fundamentally yields the opposite of a structurally ideal mass spectrum that has a single peak at the parent ion mass.

Nonresonant ionization with VUV lasers minimizes many of these problems. Prior spectroscopic knowledge of the molecule is not required, and quantitative measurements become possible since single-photon ionization cross sections tend to be more uniform from molecule to molecule than those for REMPI. Most significant is that the first ionization potential of most organic molecules lies in the 7–13 eV range and, therefore, VUV lasers sources can be viewed as a ''uni-

Table 3 Molecular Systems Detected by Nonresonant VUV Ionization

Compounds	Methods	References
Automotive exhaust (oxygenates and aromatics) and gasoline additives (methyl t-butyl ether and t-amyl methyl ether)	SPI-ITMS	355
Methyl radicals (from gas phase reactions of CO^+ and N_2O^+ with CH_4)	SPI-TOFMS	356
C_{18} alkyl siloxane SAM on silicon; 1:1 mixture of polystyrene and p-fluoropolystyrene; polystrene; glutathione tripeptide	LD + SPI-ITMS	357
Biotin; biotinylated SAM on gold	LD + SPI-TOFMS	358
Thiolate, disulfide, and sulfide SAM on gold	LD + SPI-TOFMS	359
Straight-chain alkanethilates SAM on gold	LD + SPI-TOFMS	360
SiCl and $SiCl_2$ etch products from the thermal etching of Si(100) surface with Cl_2	SPI-TOFMS	361
Gramicidin S. (cyclic decapeptide); gramicidin D (linear peptide); fullerenes	SPI-TOFMS	362
Fullerenes	SPI-TOFMS; LD + SPI + FTMS	363, 364
Rubber vulcanizates	LD + SPI-TOFMS	364, 365
Di- and tripeptides containing glycine, alanine, leucine, and proline	SPI-TOFMS	366
Tripeptide glycine-alanine-leucine	SPI-TOFMS	367
Linear, branched, and cyclic hexanes	SPI-TOFMS	368
Aliphatic compounds: n-alkanes, alkenes, ketones, carboxylic acids, ethers, aldehyde, and amines	SPI-TOFMS	369
Poly-methyl methacrylate, poly-ethylene glycol, polystyrene	SALI	370
2,3,7,8-tetrachlorodibenzo-p-dioxin; 7-methylguanine	LD + SPI-TOFMS	371
Teflon, Nylon 66	SALI	372
1,1,1-Trideuterio-2,5-dithiahex-3-yne	IRMPD + SPI-TOFMS	373

Acetic anhydride	IRMPD +SPI-QMS	374
Disilane (Si_2H_6)	SPI-TOFMS	375
Methanol dimer ($CD_3OH)_2$	SPI-TOFMS	376
Propyne (C_3H_6), allene (C_3H_4) photodissociation products	SPI-TOFMS	377
Silicon	LD + SPI-QMS	378
(C_3–C_8) iodoalkanes	PD + SPI-TOFMS	379
(C_3–C_8) alkanethiols	PD + SPI-TOFMS	380
Copper hexafluoroacetylacetonate ($Cu(hfac)_2$)	PD + SPI-TOFMS	381
(η^5-Cyclopentadienyl)-$Fe(CO)_2R$ and (η^5-Indenyl)-$Fe(CO)_2R$ (R = alkyl, aryl)	PD + SPI-TOFMS	382
n-R-amines (R = heptyl, nonyl, undecyl, dipentyl)	MALD + SPI-TOFMS	383
(C_5–C_8) Branched alkenes and dienes	PD + SPI-TOFMS	384
Diacetylene (C_4H_2) and photochemical products	PD + SPI-TOFMS	385–387
n = 30, 38, 40, 50, 60 alkanes and pentaerythritol tetraester oils	SPI-TOFMS	388
Nitrobenzene ($C_6H_5NO_2$)	PD + SPI-TOFMS	389
As_2, As_4, Ga	SPI-TOFMS	390
(C_5–C_{14}) n-alkenes	PD + SPI-TOFMS	391
1,2,4-octene	PD + SPI-TOFMS	392
Ethylacetylene	pyrolysis + SPI-TOFMS	393
Tert-butyl nitrite	flash pyrolysis + SPI-TOFMS	394
Alkyl halides: XY, X = CH_3, C_2H_5; Y = Cl, Br	IPIS	395

FTMS, Fourier transform mass spectrometry; IPIS, ion photofragment imaging spectrometry; IRMPD, infrared multiphoton dissociation; LD, laser desorption; PD, photodissociation; SALI, surface analysis by laser ionization; SAM, self-assembled monolayers; SPI-ITMS, single-photon ionization–ion trap mass spectrometry; SPI-TOFMS, single-photon ionization–time-of-flight mass spectrometry; SPI-QMS, single-photon ionization–quadrupole mass spectrometry.

versal detector'' [72]. This burgeoning area of analytical chemistry has recently been reviewed [73].

Our preliminary examination of derivatized organic collectors used in mineral sulfide processing can demonstrate the virtues of soft nonresonant VUV laser ionization [74]. Table 3 lists other organic systems probed in a similar way (mainly but not exclusively with 118 nm light generated by nonresonant tripling of the 355 nm output of a Nd:YAG laser in Xe).

Valuable trace metals are most commonly separated and concentrated from crude ore by flotation [75]. Mineral grains (size $\leq 150 \, \mu m$), produced by crushing and grinding, are rendered hydrophobic in aqueous solution by the addition of organic molecules called collectors. The coated grains attach themselves to air bubbles that float to the solution surface where they are removed and analyzed, usually by mass spectrometry. As noted above, standard methods such as FAB, CI, EI, and SIMS almost always exhibit numerous fragmentation peaks. It therefore becomes notoriously difficult to correlate a particular fragment with a given parent molecule in mixtures of like compounds. This problem is not insignificant. Mining companies will often empirically add many collector molecules with different relative concentrations to their flotation circuits to enhance the collection efficiency. Given that billions of tons of ore are processed yearly by flotation, any technique that can help directly to optimize this process could result in a significant cost savings.

One of the most important classes of organic collectors are the alkali salts of O-alkyl dithiocarbonates, $ROC(S)S^-$, commonly known as xanthates [76]. Since the organic R-group is a short alkyl chain of 2–5 carbon atoms, xanthates will selectively adsorb to the metal sulfide because either there is no adsorption at the nonsulfide mineral/water interface or, if there is adsorption, the organic groups are too short to impart a hydrophobic character to the grain. Selectivity between different xanthate collectors can often only be achieved by the addition of separate chemicals, called depressants or activation modifiers [77], and/or by changes in solution conditions (pH, for example). For this reason, several collectors are often added to flotation circuit, and the conditions changed empirically to achieve the best results.

Xanthate salts have exceedingly low vapor pressures. Therefore, as proof-of-principle, volatile xanthate ester derivatives [78], $ROC(S)SR'$, were ionized using VUV wavelengths near 129.7 nm generated by four-wave difference mixing in Kr. Here R and R' are either identical or different alkyl chains. The esters were chosen for study even though their utility as collectors has not been established, to demonstrate the potential that the combination of VUV lasers and TOF mass spectrometry has for the analysis of organic molecules by parent mass.

Typical results for $CH_3SC(S)OC_2H_5$ under focused two-photon UV ($\lambda \approx$ 216 nm; power density was $\approx 10^8 \, W/cm^2$) and unfocused nonresonant VUV illumination are shown in Figure 14a and b, respectively [74]. Since the UV excita-

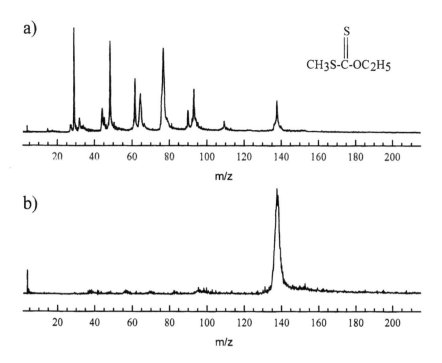

Figure 14 TOF mass spectrum of *O*-ethyl, *S*-methyl dithiocarbonate by (a) ultraviolet (UV) two-photon ionization and (b) VUV laser single-photon ionization.

tion wavelength chosen is one-photon resonant with the sulfur $n \rightarrow \sigma^*$ molecular transition [79], the parent mass peak intensity in Figure 14a is markedly reduced due to extensive fragmentation. The dominant feature in the VUV/TOF mass spectrum shown in Figure 14b is the parent ion, and signals due to low molecular mass ions are effectively absent. This result can also be compared with the mass spectra for $CH_3SC(S)OC_2H_5$ by obtained by EI and FAB (Fig. 15). Like the UV REMPI/TOF spectrum, a weak parent ion peak and extensive fragmentation are observed.

Individual VUV/TOF mass spectra of $CH_3SC(S)OC_2H_5$, $C_2H_5SC(S)OC_2H_5$, i-$C_3H_7SC(S)OC_2H_5$, and n-$C_3H_7SC(S)OC_3H_7$-i are shown in Figure 16. A similar spectrum for a mixture of these esters can be compared in Figure 17. Here the composition of the liquid mixture was established empirically to yield similar gas phase intensities for each parent ion. It can be concluded immediately that VUV single-photon ionization coupled with TOF mass spectrometry can readily identify components of gas-phase xanthate mixtures by parent mass. Furthermore, initial calculations suggest that the detection sensitivity of the experimental appa-

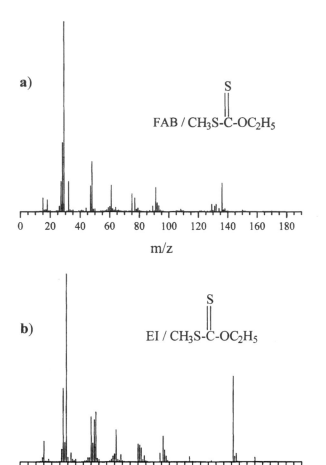

Figure 15 (a) Fast atom bombardment (FAB) mass spectrum of *O*-ethyl, *S*-methyl dithiocarbonate. (b) Electron impact (EI) mass spectrum of *O*-ethyl, *S*-methyl dithiocarbonate.
The electron energy was 70 eV.

ratus may well be in the femtomole range. Few other commercially available
analytical schemes match the performance hinted by our results. It may eventually
be possible to use VUV laser SALI methods to characterize mineral samples
from processing streams in situ. Our VUV laser/TOF mass spectrometer is currently being modified for true SALI experiments by the addition of thermal and
laser-induced desorption capabilities.

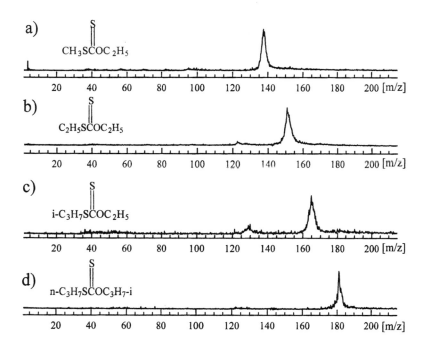

Figure 16 Individual VUV laser/TOF mass spectra for (a) $CH_3SC(S)OC_2H_5$; (b) $C_2H_5SC(S)OC_2H_5$; (c) $i\text{-}C_3H_7SC(S)OC_2H_5$; and (d) $n\text{-}C_3H_7SC(S)OC_3H_7\text{-}i$.

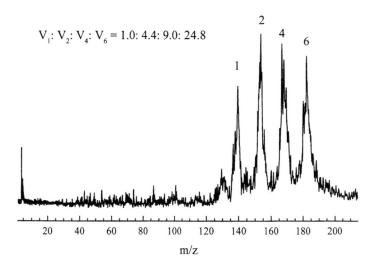

Figure 17 VUV laser/TOF mass spectrum for a mixture of (1) $CH_3SC(S)OC_2H_5$; (2) $C_2H_5SC(S)OC_2H_5$; (4) $i\text{-}C_3H_7SC(S)OC_2H_5$; and (6) $n\text{-}C_3H_7SC(S)OC_3H_7\text{-}i$, in a volume ratio of $V_1: V_2: V_4: V_6 = 1.0: 4.4: 9.0: 24.8$.

IV. CONCLUSIONS

A goal of the literature reviews presented here is to chronicle the diverse and often unexpected spectroscopic uses that have been found for VUV lasers. While crystal ball gazing is, at best, fraught with danger, it is not unreasonable to expect short-wavelength coherent sources to have a significant impact on emerging ''new'' research in the ''old'' field of classic chemical kinetics. Of course, state-selective reaction dynamics must be viewed as a twinned discipline to spectroscopy and, in this regard, VUV lasers are playing an important and growing role (Table 2). However, the highly nonequilibrium conditions encountered in supersonic free jets are not well-suited to so-called conventional chemical kinetics because critical parameters such as the temperature and concentration are very difficult to characterize and control.

Rowe and co-workers at the University of Rennes I have overcome these basic limitations by constructing a jet system known as a cinétique de réaction en ecoulement supersonique uniforme (CRESU; reaction kinetics in uniform supersonic flow) in which these properties can be well defined [80].

The heart of a CRESU is an axially symmetrical converging–diverging (Laval) nozzle that is shown and compared schematically with a free jet and a skimmed molecular beam in Figure 18 [81]. When such a nozzle is properly designed, the true thermodynamic temperature of the continuous isentropic flow obeys the relationship given by Eq. (7). Unlike the free jets described above, however, the Mach number is very close to constant everywhere along the expansion. Separate Laval nozzles are nevertheless needed for each buffer gas and each desired flow temperature and number density.

Designing a Laval nozzle that can produce a uniform flow downstream of the nozzle exit requires finding the inverse solution to a nonlinear second-order Navier-Stokes equation with viscosity terms. This type of calculation can be carried out perturbatively since uniform flow also implies that the nonlinear terms are small.

In hydrodynamic terms, operation with small viscosity effects at the walls of the nozzle means that the Reynolds number of the flow, Re, which is defined as the ratio of the inertial to viscous terms in the Navier-Stokes equation, is large. For a given reservoir temperature and Mach number, Re scales as $\sqrt{Q_M} \, P_{chamber}$ where Q_M is the mass flow.

A jet with minimal cluster formation and flow turbulence is produced at Rennes from relatively large-diameter nozzles at a chamber pressure of $\approx 10^{-1}$ torr with Re between 1000 and 2500. However, huge pumping capacities are required. The experimental system uses two large Roots pumps (12000 m^3/hr) in parallel backed by two additional Roots blowers with 8000 and 2000 m^3/hr pumping speeds, respectively. Pulsed Laval nozzle systems have been designed

a) Free Jet

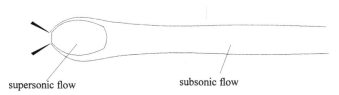

supersonic flow   subsonic flow

b) Molecular Beam

skimmer

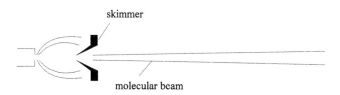

molecular beam

c) Uniform flow of a C.R.E.S.U.

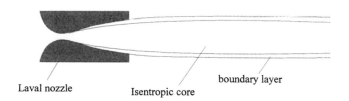

boundary layer

Laval nozzle   Isentropic core

Figure 18 Schematic diagrams of the nozzles used to generate (a) a very cold free jet with large variations in density and pressure along the flow axis produced by expanding a high-pressure gas through a small diameter pinhole; (b) a low divergent molecular beam produced by placing a skimmer in the free jet; and (c) a uniform supersonic flow of constant density and temperature produced by expanding a moderate pressure through a convergent–divergent Laval nozzle.

to reduce these enormous demands [82]. The gas reservoir behind the Rennes nozzle can also be cooled to liquid nitrogen temperatures (77 K), resulting in supersonic flow temperatures as low as 13 K, and translated forward and backward along the flow axis.

One of the most significant advantages of a CRESU system is that it provides an excellent environment in which to study reactions between neutral species by pulsed laser photolysis–laser-induced fluorescence (PLP-LIF). One tripled Nd:YAG laser beam (≈ 75 mJ per pulse) is typically used to generate an unstable neutral radical or atom by the photolysis of a stable radical precursor. The output of a second Nd:YAG pumped dye or optical parametric oscillator (OPO) is then used to probe the radical concentration at a given time delay (jitter ≤ 1 ns) by laser-induced fluorescence after reaction with another reactant in the expansion. A radical concentration–decay curve in the supersonic flow can be obtained by systematically changing the delay time between the two laser pulses. Due to the high velocities in the expansion, and because the supersonic flow is only uniform over tens of centimeters, the time scale probed for any reaction in a CRESU corresponds to ≈ 100–500 μs. Experiments are by necessity limited by these time and distance constraints to the study of processes having rate constants exceeding $\approx 10^{-12}$ cm^3 molecule^{-1} s^{-1}. A high-efficiency light collection system downstream of the nozzle exit and coupled to a photomultiplier, is mounted in such a way that the supersonic flow passes between them undisturbed.

Many important reactions in the intersteller medium involve H, O, C, and N atoms that have strong single-photon-allowed transitions lying in the VUV. Only one experiment to date has been reported that combines a CRESU and VUV laser excitation, but its results are very encouraging. Reactions between ground state C(^{3}P) atoms with O$_2$ and NO were studied by monitoring LIF at jet temperatures between 295 K and 15 K from C(^{3}P) atoms excited on their 3s ^{3}P $\leftarrow$ 2p ^{3}P transitions near 166 nm with a VUV laser generated by four-wave difference-mixing in Xe [83]. Not only were precise temperature-dependent rate coefficients deduced, but in the temperature range studied dramatic relaxation effects could also be observed for the populations of the individual ground state spin-orbit components. Results of this sort should significantly help to verify many of the kinetic assumptions currently being used to model interstellar and atmospheric chemistry.

ACKNOWLEDGMENTS

Research carried out at the University of Western Ontario (UWO) was supported by the Natural Sciences and Engineering Research Council of Canada (NSERC) and UWO. The authors thank Dr. S. S. Dimov, Dr. K. J. Jordan, Dr. P. Wang, Dr. X. K. Hu, Dr. D. M. Mao, Dr. N. A. McDonald, and Mr. J. Vanstone for their substantial contributions to the UWO research effort over the years. RHL thanks CNRS France, B. R. Rowe, J. B. A. Mitchell, and the experimental astrophysics group at the University of Rennes I for a visiting research fellowship, their hospitality, and use of Figure 18. YJS is grateful for an Ontario Graduate Scholarship.

REFERENCES

1. PA Franken, AE Hill, CW Peters, G Weinreich. Generation of optical harmonics. Phys Rev Lett 7: 118–119, 1961.
2. W Jamroz, BP Stoicheff. Generation of tunable coherent vacuum-ultraviolet radiation. Progr Optics 20: 326–380, 1983.
3. CR Vidal. Four-wave frequency mixing in gases. In: LF Mollenauer, JC White, eds. Tunable Lasers, Topics in Applied Physics 59. Berlin: Springer-Verlag, 1987, pp 57–113.
4. JW Hepburn. Generation of coherent vacuum ultraviolet radiation: applications to high-resolution photoionization and photoelectron spectroscopy. In: AB Meyers, TR Rizzo, eds. Laser Techniques of Chemistry, Vol. XXIII. New York: John-Wiley & Sons, 1995, pp 149–183.
5. W. Jamroz, PE LaRocque, BP Stoicheff. Resonantly enhanced second-harmonic generation in zinc vapor. Opt Lett 7: 148–150, 1982.
6. WA Majewski, JF Pfanstiel, DF Plusquellic, DW Pratt. High resolution optical spectroscopy in the ultraviolet. In: AB Meyers, TR Rizzo, eds. Laser Techniques of Chemistry, Vol. XXIII. New York: John-Wiley & Sons, 1995, pp 101–148.
7. H Bitto, PR Willmott. The dynamics and its power dependence of $S_1(^1B_{3u})$ pyrazine studied with transform-limited nanosecond pulses. Chem Phys 165: 113–121, 1992.
8. R Hilbig, G Hilber, A Lago, B Wolff, R Wallenstein. Tunable coherent VUV radiation generated by nonlinear optical frequency conversion in gases. Comments Atm Mol Phys 18: 157–180, 1986.
9. GC Bjorklund. Effects of focusing on third-order nonlinear processes in isotropic media, IEEE J Quantum Electron QE-11: 287–296, 1975.
10. R Hilbig, R Wallenstein. Enhanced production of tunable VUV radiation by phase-matched frequency tripling in krypton and xenon. IEEE J Quantum Electron QE-17: 1566–1573, 1981.
11. W Zapka, D Cotter, U Brackmann. Dye laser frequency tripling at 106 nm. Opt Commun 36: 79–81, 1981.
12. D Cotter. Conversion from 3371 to 1124 Å by nonresonant optical frequency tripling in compressed krypton gas. Opt Lett 4: 134–136, 1979; D Cotter. Tunable narrow-band coherent VUV source for the Lyman-alpha region. Opt Commun 31: 397–399, 1979; H Langer, H Puell, H Rohr. Lyman-alpha (1216 Å) generation in krypton. Opt Commun 34: 137–142, 1980.
13. R Mahon, TJ McIlrath, DW Koopman. Nonlinear generation of Lyman-alpha radiation. Appl Phys Lett 33: 305–307, 1978.
14. R Hilbig, R Wallenstein. Tunable XUV radiation generated by nonresonant frequency tripling in argon. Opt Commun 44: 283–288, 1983; EE Marinero, CT Rettner, RN Zare, AH Kung. Excitation of H_2 using continuously tunable coherent XUV radiation (97.3–102.3 nm). Chem Phys Lett 95: 486–491, 1983.
15. R Hilbig, A Lago, R Wallenstein. Tunable XUV radiation generated by nonresonant frequency tripling in neon. Opt Commun 49: 297–301, 1984; R Hilbig, G Hilber, A Timmermann, R Wallenstein. Broadly tunable VUV radiation generated by fre-

quency mixing in gases. In: SE Harris, TB Lucatorto, eds. Laser Techniques in the Extreme Ultraviolet. AIP Conference Proceedings 119. New York: American Institute of Physics, 1984, pp 1–9.

16. R Hilbig, G Hilber, R Wallenstein. Nonresonant tripling and sum-frequency mixing in Hg. Appl Phys B 41: 225–230, 1986.

17. CT Rettner, EE Marinero, RN Zare, AH Kung, Pulsed jets: novel nonlinear media for generation of vacuum ultraviolet and extreme ultraviolet radiation, J Phys Chem 88: 4459–4465, 1984.

18. LJ Zych, JF Young. Limitation of 3547 to 1182 Å conversion efficiency in Xe. IEEE J Quantum Electron QE-14: 147–149, 1978.

19. RT Hodgson, PP Sorokin, JJ Wynne. Tunable coherent vacuum-ultraviolet generation in atomic vapors. Phys Rev Lett 32: 343–346, 1973.

20. PR Herman, PE LaRocque, RH Lipson, W Jamroz, BP Stoicheff. Vacuum ultraviolet laser spectroscopy III: laboratory sources of coherent radiation tunable from 105 to 175 nm using Mg, Zn, and Hg vapors. Can J Phys 63: 1581–1588, 1985.

21. JA Armstrong, N Bloembergen, J Ducuing, PS Pershan. Interactions between light waves in a nonlinear dielectric. Phys Rev 127: 1918–1939, 1962; BJ Orr, JF Ward. Perturbation theory of the non-linear optical polarization of an isolated system. Mol Phys 20: 513–526, 1971.

22. EU Condon, GH Shortley. The Theory of Atomic Spectra. Cambridge (UK): Cambridge University Press, 1964.

23. CH Muller III, DD Lowenthal, MA DeFaccio, AV Smith. High-efficiency, energy-scalable, coherent 130 nm source by four-wave mixing in Hg vapor. Opt Lett 13: 651–653, 1988.

24. RH Lipson. Laser spectroscopies in the VUV and XUV. Physics Can 55: 223–232, 1999.

25. RH Lipson, SS Dimov, P Wang, YJ Shi, DM Mao, XK Hu, J Vanstone. Vacuum ultraviolet and extreme ultraviolet lasers: principles, instrumentation, and applications. Instrum Sci Tech 28: 85–118, 2000.

26. RE Smalley, L Wharton, DH Levy. Molecular optical spectroscopy with supersonic beams and jets. Accts Chem Res 10: 139–145, 1977.

27. DR Miller. Free jet sources. In: G Scoles, ed. Atomic and Molecular Beam Methods, volume I. New York: Oxford University Press, 1988, pp 14–53.

28. W Demtröder, Visible and ultraviolet spectroscopy: physical aspects. In: G Scoles, ed. Atomic and Molecular Beam Methods. volume 2. New York: Oxford University Press, 1992, pp 213–260.

29. PM Johnson, CE Otis. Molecular multiphoton spectroscopy with ionization detection. Annu Rev Phys Chem 32: 139–157, 1981.

30. SL Anderson. Multiphoton ionization state selection: vibrational mode and rotational-state control. Adv Chem Physics 82: 177–212, 1992.

31. MNR Ashfold, JD Howe. Multiphoton spectroscopy of molecular species. Annu Rev Phys Chem 45: 57–82, 1994.

32. WC Wiley, IH McLaren. Time-of-flight mass spectrometer with improved resolution. Rev Sci Instrum 26: 1150–1157, 1955.

33. DJ Auerbach. Velocity measurements by time-of-flight methods. In: G Scoles, ed. Atomic and Molecular Beam Methods. volume 1. New York: Oxford University Press, 1988, pp 362–379.

34. KP Lawley, RJ Donovan. Spectroscopy and electronic structure of ion-pair states. J Chem Soc Faraday Trans 89: 1885–1898, 1993.

35. XQ Tan, TG Wright, TA Miller. Electronic spectroscopy of free radicals in supersonic jets. In: JM Hollas, D Phillips, eds. Jet Spectroscopy and Molecular Dynamics. London: Blackie Academic & Professional, 1995, pp 74–117.

36. K Wieland. Bandenspektren der quecksilber-, cadmium- und zinkhalogenide. Helv Phys Acta 2: 47–94, 1929; K Wieland. Absorptions- und Fluoreszenzspektren dampfförmiger quecksilberhalogenide. I HgJ$_2$. Z Phys 76: 801–813, 1932; K Wieland. Absorptions- und Fluoreszenzsperktren dampfförmiger quecksilberhalogenide. II. HgBr$_2$ and HgCl$_2$. Z Phys 77: 157–1655, 1932; K Wieland. Das langwellige emissions- ünd flüoreszenzspektrum (5700-3000 angstrom) von natürlichem HgCl ünd von künstlich angereichertem HgCl37. Helv Phys Acta 14: 420–464, 1941.

37. R Burnham. Discharge pumped mercuric halide dissociation lasers. Appl Phys Lett 33: 156–159, 1978; YE Gavrilova, VS Zrodnikov, AO Klementov, AS Podsosonny. Excimer HgI* laser excited by an electric discharge. Sov J Quantum Electron 10: 1457–1459, 1980; WT Whitney. Sustained discharge excitation of HgCl and HgBr B $^2\Sigma_{1/2}^+ \to$ X $^2\Sigma_{1/2}$ lasers. Appl Phys Lett 32: 239–241, 1978; JG Eden. VUV-pumped HgCl laser. Appl Phys Lett 33: 495–497, 1978; JH Parks. Laser action on the B $^2\Sigma_{1/2}^+ \to$ X $^2\Sigma_{1/2}^+$ band of HgCl at 5576 Å. Appl Phys Lett 31: 192–194, 1977; JG Eden. Green HgCl (B $^2\Sigma_{1/2}^+ \leftarrow$ X $^2\Sigma_{1/2}^+$) laser. Appl Phys Lett 31: 448–450, 1977.

38. AW McCown, MN Ediger, JG Eden. Quenching kinetics and small signal gain spectrum of the ZnI photodissociation laser. Opt Common 40: 190–194, 1982.

39. CE Moore. Atomic energy levels. Vol. III. Natl Bur Stand Circ No. 467. Washington, DC: USGPO, 1971.

40. J Tellinghuisen, JG Ashmore. The B $\to$ X transition in ^{200}Hg ^{79}Br Appl Phys Lett 40: 867–869, 1982; J Tellinghuisen, JG Ashmore. Mixed representations for diatomic spectroscopic data: application to HgBr. Chem Phys Lett 102: 10–16, 1983; J Tellinghuisen, PC Davies, P Berwanger, KS Viswanathan. B $\to$ X transitions in HgCl and HgI. Appl Phys Lett 41: 789–791, 1982; KS Viswanathan, J Tellinghuisen. The B ($^2\Sigma^+$) $\to$ X ($^2\Sigma^+$) transition (4050–4500Å) in HgI. J Mol Spectrosc 98: 185–198 (1983).

41. DP Greene, KP Killeen, JG Eden. X $^2\Sigma \to$ B $^2\Sigma$ absorption band of HgBr: optically pumped 502 nm laser. Appl Phys Lett 48: 1175–1177, 1986; DP Greene, KP Killeen, JG Eden. Excitation of the HgBr B $^2\Sigma_{1/2}^+ \to$ X $^2\Sigma_{1/2}^+$ band in the ultraviolet. J Opt Soc Am B3: 1282–1287, 1986.

42. PC Engelking. Spectroscopy of jet-cooled ions and radicals. Chem Rev 91: 399–414, 1991.

43. RH Lipson, KJ Jordan, HA Bascal. Fluorescence excitation spectra of jet-cooled HgBr radicals. J Chem Phys 98: 959–967, 1993; KJ Jordan, HA Bascal, RH Lipson, M Melchior. The B $^2\Sigma^+ \leftarrow$ X $^2\Sigma^+$ transition of HgI. J Mol Spectrosc 159: 144–

155, 1993; HA Bascal, KJ Jordan, RH Lipson. Ion-pair spectroscopy of HgCl. Can J Chem 71: 1615–1621, 1993.

44. P DeBièvre, M Gallent, NE Holden, IL Barnes. Isotopic abundances and atomic weights of the elements. J Phys Chem Ref Data 13: 809–891, 1984.

45. RH Lipson, AR Hoy. Vacuum Ultraviolet Laser Spectra of ICl. J Chem Phys 90, 6821–6826, 1989.

46. J Tellinghuisen, SD Henderson. The use of Morse-RKR curves in diatomic calculations. Chem Phys Lett 91: 447–451, 1982.

47. RJ LeRoy. Program manual for RKR1: a computer program implementing the first-order RKR method for determining diatom potential energy curves from spectroscopic constants. University of Waterloo, Chemical Physics Research Report CP-425, 1992.

48. RJ LeRoy. Program manual for LEVEL: a computer program for solving the radial Schrodinger equation for bound and quasibound levels and calculating (if desired) expectation values and Franck-Condon intensity factors. University of Waterloo, Chemical Research Report, CP-300, 1991.

49. RS Mulliken. Iodine revisited. J Chem Phys 55, 288–309, 1971; JCD Brand, AR Hoy. Multiphoton spectra and states of halogens. Appl Spectrosc Rev 23: 285–327, 1987.

50. RN Zare, DR Herschbach. Charge transfer model for alkali halide electronic transition strengths. J Mol Spectrosc 15: 462–472, 1965.

51. AR Hoy, RH Lipson. Reinvestigation of the Cordes band system of I_2 using a vacuum ultraviolet laser. Chem Phys 140: 187–193, 1990.

52. RH Lipson, AR Hoy. A vacuum ultraviolet laser study of IBr. Mol Phys 68: 1311–1319, 1989.

53. RH Lipson, AR Hoy, MJ Flood. Direct vibrational numbering of the $D0_u^+$ ion-pair state of Br_2. Chem Phys Lett 149, 155–160, 1988; RH Lipson, AR Hoy. VUV laser spectroscopy of ion-pair states of Br_2. J Mol Spectrosc 143: 183–198, 1989.

54. SS Dimov, RH Lipson, T Turgeon, JA Vanstone, P Wang, DS Yang. Vacuum ultraviolet/time-of-flight mass spectroscopy: ion-pair spectra of $^{79}Br^{35}Cl$. J Chem Phys 100, 8666–8672, 1994; P Wang, SS Dimov, RH Lipson. Vibronic analysis of the ion-pair $(E0^+)$ ground state $(X0^+)$ transition of BrCl. J Chem Phys 107: 3345–3351, 1997.

55. P Wang, IV Okuda, SS Dimov, RH Lipson. Mass-resolved VUV laser spectra in the vicinity of the Rydberg minimum of the $1\ ^1\Sigma_u^+$ state of Cl_2. J Phys Chem 229: 370–376, 1994; P Wang, SS Dimov, G Rosenblood, RH Lipson. Vibrational reassignment of the $2\ ^1\Sigma_u^+ \leftarrow X\ ^1\Sigma_g^+$ transition of Cl_2. J Phys Chem 99: 3984–3989, 1995; P Wang, SS Dimov, RH Lipson. Rotational analyses for selected bands of the $2\ ^1\Sigma_u^+ \leftarrow X\ ^1\Sigma_g^+$ transition of Cl_2. J Phys Chem A 101, 4555–4559, 1997; P Wang, IV Okuda, SS Dimov, RH Lipson. Rotationally resolved vacuum ultraviolet laser spectra of the $^{37}Cl_2\ 1\ ^1\Sigma_u^+ \leftarrow X\ ^1\Sigma_g^+$ transition. J Mol Spectrosc 190, 213–225, 1998.

56. DK Chakraborty, PC Tellinghuisen, J Tellinghuisen. The emission of BrCl: analysis of the $D' \rightarrow A'$ and $E \rightarrow B$ transitions. Chem Phys Lett 141: 36–40, 1987; SW Brown, CJ Dowd, Jr., J Tellinghuisen. The $E \rightarrow B$ transition in $^{81}Br^{37}Cl$. J Mol Spectrosc 132: 178–192, 1988.

57. A Hopkirk, D Shaw, RJ Donovan, KP Lawley, AJ Yencha. Vacuum-ultraviolet absorption, fluorescence excitation, and dispersed fluoresence spectra of BrCl. J Phys Chem 93: 7338–7342, 1989; A Kvaran, AJ Yencha, KP Lawley, RJ Donovan. Analyses of the 290–400 nm oscillatory continua due to transitions from the $E(O^+)$ ion pair state of BrCl. Mol Phys 75: 197–207, 1992.

58. RJ LeRoy, RB Bernstein. Dissociation energy and long-range potential of diatomic molecules from vibrational spacings of higher levels. J Chem Phys 52: 3869–3879, 1970; RJ LeRoy, W-H Lam. Near dissociation expansions in the spectroscopic determination of diatom dissociation energies in the spectroscopic determination of diatom dissociation energies: method, and application to $BeAr^+$. Chem Phys Lett 71, 544–548, 1980; KJ Jordan, RH Lipson, NA McDonald, RJ LeRoy. Jet emission spectra of CdI and HgI and near dissociation theory analyses for CdI and ZnI. J Phys Chem 96, 4778–4787, 1992; SH Pan, FH Mies. Rydberg-like properties of rotational-vibrational levels and dissociation continuum associated with alkali-halide charge-transfer states. J Chem Phys 89: 3096–3103, 1988.

59. SD Peyerimhoff, RJ Buenker. Electronically excited and ionized states of the chlorine molecule. Chem Phys 57: 279–296, 1981.

60. K Yamanouchi, S Tsuchiya. Tunable vacuum ultraviolet laser spectroscopy: excited state dynamics of jet-cooled molecules and van der Waals complexes. J Phys B: Atm Mol Opt Phys 28: 133–165 (1995).

61. RN Zare. Laser chemical analysis. Science 226: 298–303, 1984; R Snook. Laser techniques for chemical analysis. Chem Soc Rev 26: 319–326, 1997.

62. JH Futrell. In: TD Märk, GH Dunn, Eds. Electron Impact Ionization. Wien New York: Springer-Verlag, 1985, pp 364–374.

63. HM Fales, Y Nagai, GWA Milne, HB Brewer, Jr., TJ Bronzert, LJ Pisano. Use of chemical ionization mass spectrometry in analysis of amino acid phenylthiohydantoin derivatives formed during Edman degradation of proteins. Anal Biochem 43: 288–299, 1971; MA Baldwin, FW McLafferty. Direct chemical ionization of relatively involatile samples. Application to underivatized oligopeptides. Org Mass Spectrom 7: 1353–1356, 1973.

64. M Barber, RS Bordoli, GJ Elliot, RD Sedgwick, AN Taylor. Fast atom bombardment mass spectrometry. Anal Chem 54: 645A–656A, 1982; RM Caprioli. Continuous-flow fast atom bombardment mass spectrometry. Anal Chem 62: 477A–485A, 1990.

65. A Benninghoven, WK Sichtermann. Detection, identification and structural investigation of biologically important compounds by secondary ion mass spectrometry. Anal Chem 50: 1180–1184, 1978.

66. M Karas, U Bahr. Matrix-assisted laser desorption-ionization (MALDI) mass spectrometry of biological molecules. In: RM Caprioli, A Malorni, G Sindona, eds. Mass Spectrometry in Biomolecular Sciences, Nato ASI Series Vol. 475. Dordrecht, The Netherlands: Kluwer Academic Publishers, 1996, pp 33–49.

67. CH Becker, KT Gillen. Surface analysis by nonresonant multiphoton ionization of desorbed or sputtered species. Anal Chem 56: 1671–1674, 1984.

68. CH Becker, KT Gillen. Surface Analysis of contaminated GaAs: comparison of new laser-based techniques with SIMS. J Vac Sci Technol A 3: 1347–1349, 1985.

69. GS Hurst, MG Payne, SD Kramer, JP Young. Resonance ionization spectroscopy and one-atom detection. Rev Mod Phys 51: 767–819, 1979.

70. EW Schlag, HJ Neusser. Multiphoton mass spectrometry. Acc Chem Res 16: 355–360, 1983.

71. BD Morrical, DP Fergenson, KA Prather. Coupling two-step laser desorption/ionization with aerosol time-of-flight mass spectrometry for the analysis of individual organic particles. J Am Soc Mass Spectrom 9: 1068–1073, 1998.

72. U Schühle, JB Pallix, CH Becker. Sensitive mass spectrometry of molecular adsorbates by stimulated desorption and single-photon ionization. J Am Chem Soc 110: 2323–2324, 1988.

73. DJ Butcher. Vacuum ultraviolet radiation for single-photon mass spectrometry: a review. Microchem J 62: 354–362, 1999.

74. YJ Shi, XK Hu, DM Mao, SS Dimov, RH Lipson. Analysis of xanthate derivatives by vacuum ultraviolet laser-time-of-flight mass spectrometry. Anal Chem 70: 4534–4539, 1998.

75. I Persson. Absorption of ions and molecules to solid surfaces in connection with flotation of sulphide minerals. J Coord Chem 32: 261–342, 1994; HL Shergold. Flotation in mineral processing. In: JI Ives, ed. The Scientific Basis of Flotation. The Hague: Martinus Nijhoff Publishers, 1984, pp 229–287; VI Klassen, VA Mpokrousov. An Introduction to the Theory of Flotation. London: Butterworth, 1963; AM Gaudin. Flotation. New York: McGraw-Hill, 1957.

76. PK Ackerman, GH Harris, RR Klimpel, FF Aplan. Evaluation of flotation collectors for copper sulfide and pyrite I. Common sulfhydryl collectors. Int J Min Process 21: 105–127, 1987.

77. JY Kim, SL Chryssoulis. Influence of lead ions in sulfide flotation—the application of laser-ionization mass spectrometry. Min Metall Process 69–76, 1996.

78. I Degani, R Fochi. The phase-transfer synthesis of O,S-dialkyl dithiocarbonates from alkyl halides and alkyl mthanesulfonates. Synthesis 5: 365–368, 1978.

79. KJ Rosengren. Electronic absorption spectra of unconjugated alkyl thials and thiones. Acta Chem Scand 16: 2284–2292, 1962.

80. IR Sims, J-L Queffelec, A Defrance, C Rebrion-Rowe, D Travers, P Bocherel, BR Rowe, IWM Smith. Ultralow temperature kinetics of neutral-neutral reactions. The technique and results for the reactions $CN + O_2$ down to 13 K and $CN + NH_3$ down to 25 K. J Chem Phys 100: 4229–4241, 1994.

81. IWM Smith, BR Rowe. Reaction kinetics at very low temperatures: laboratory studies and interstellar chemistry. Accts Chem Res 33: 261–268, 2000.

82. DB Atkinson, MA Smith. Design and characterization of pulsed supersonic expansions for chemical applications. Rev Sci Instrum 66: 4434–4446, 1995.

83. D Chastaing, SD Le Picard, IR Sims. Direct kinetic measurements on reactions of atomic carbon, $C(^3P)$, with O_2 and NO at temperatures down to 15 K. J Chem Phys 112: 8466–8469, 2000.

84. I Munro, AP Sabersky. In: H Winich, S Doniach, eds. Synchrotron Radiation Research. New York: Plenum Press, 1980.

85. Characteristics of the Advanced Light Source: A Brief Summary, Lawrence Berkeley Labs. http://www-als.lbl.gov/als/workshops/alscharacter.html.

86. DW Tokaryk, GZ Zhang, BP Stoicheff, Phys Rev A 59: 3116–3119, 1999.

87. GZ Zhang, DW Tokaryk, BP Stoicheff. Phys Rev A 56: 813–819, 1997.
88. M Katsuragawa, GZ Zhang, K Kakuta. Opt Commun 129: 212–216, 1996.
89. RSD Sihombing, M Katsuragawa, GZ Zhang, K Hakuta. Phys Rev A 54: 1551–1555, 1996.
90. RI Thompson, BP Stoicheff, GZ Zhang, K Hakuta. Appl Phys B 60: S129–S139, 1995.
91. GZ Zhang, M Katsuragawa, K Hakuta, RI Thompson, BP Stoicheff. Phys Rev A 52: 1584–1593, 1995.
92. RI Thompson, BP Stoicheff, GZ Zhang, K Hakuta. Quantum Opt 6: 349–358, 1994.
93. GZ Zhang, K Hakuta, BP Stoicheff. Phys Rev Lett 71: 3099–3102, 1993.
94. K Hakuta, L Marmet, BP Stoicheff. Phys Rev A 45: 5152–5159, 1992.
95. L Marmet, K Hakuta, BP Stoicheff. J Opt Soc Am B9, 1038–1046, 1992.
96. K Hakuta, L Marmet, BP Stoicheff. Phys Rev Lett 66: 1042–1045, 1991.
97. K Hakuta, L Marmet, BP Stoicheff. In: M Ducloy, E Giacobino, G Camy, eds. Laser Spectroscopy X. Singapore: World Scientific, 1991, pp 301–306.
98. L Marmet, K Hakuta, BP Stoicheff. Opt Lett 16: 261–263, 1991.
99. RA Brownsword, M Hillenkamp, T Laurent, RK Vatsa, H-R Volpp, J Wolfrum. J Phys Chem A, 101: 5222–5227, 1997.
100. RA Brownsword, M Hillenkamp, T Laurent, RK Vatsa, H-R Volpp. J Chem Phys 106: 4436–4447, 1997.
101. RA Brownsword, M Hillenkamp, T Laurent, RK Vatsa, H-R Volpp, J Wolfrum. J Phys Chem A 101: 995–999, 1997.
102. RA Brownsword, M Hillenkamp, T Laurent, RK Vatsa, H-R Volpp, J Wolfrum. J Chem Phys 106: 1359–1366, 1997.
103. RA Brownsword, M Hillenkamp, T Laurent, RK Vatsa, H-R Volpp. J Chem Phys 106: 9563–9569, 1997.
104. RA Brownsword, T Laurent, RK Vatsa, H-R Volpp, J Wolfrum. Chem Phys Lett 249: 162–166, 1996.
105. N Tada, K Tonokura, K Matsumoto, M Koshi, A Miyoshi, H Matsui. J Phys Chem A 103: 322–329, 1999.
106. H-R Volpp, J Wolfrum. In: J Wolfrum, H-R Volpp, R Rannacher, J Warnatz, eds. Gas Phase Chemical Reaction Systems: Experiments and Models 100 years after Max Bodenstein. Springer Series in Chemical Physics 61. Heidelberg: Springer, 1996, pp 14–31.
107. RA Brownsword, T Laurent, RK Vatsa, H-R Volpp, J Wolfrum. Chem Phys Lett 258: 164–170, 1996.
108. RA Brownsword, M Hillenkamp, T Laurent, RK Vatsa, H-R Volpp, J Wolfrum. Chem Phys Lett 266: 259–266, 1997.
109. S Koppe, T Laurent, PD Naik, H-R Volpp, J Wolfrum. Can J Chem 72: 615–624, 1994.
110. W Yi, R Bersohn. Chem Phys Lett 206: 365–368, 1993.
111. K Tonokura, Y Matsumi, M Kawasaki, S Tasaki, R Bersohn. J Chem Phys 97: 8210–8215, 1992.
112. S Koppe, T Laurent, PD Naik, H-R Volpp, J Wolfrum, T Arusi-Parpar, I Bar, S Rosenwaks. Chem Phys Lett 214: 546–552, 1993.

113. T Laurent, PD Naik, H-R Volpp, J Wolfrum, T Arusi-Parpar, I Bar, S Rosenwaks. Chem Phys Lett 236: 343–349, 1995.

114. W Yi, S Satyapal, N Shafer, R Bersohn. J Chem Phys 99: 4548–4553, 1993.

115. A Tezaki, S Okada, H Matsui. J Chem Phys 98: 3876–3883, 1993.

116. Y Matsumi, K Tonokura, M Kawasaki, HL Kim. J Phys Chem 96: 10622–10626, 1992.

117. S Satyapal, R Bersohn. J Phys Chem 95: 8004–8006, 1991.

118. W Yi, A Chattopadhyay, R Bersohn. J Chem Phys 94: 5994–5998, 1991.

119. K-H Gericke, M Lock, FJ Comes. Chem Phys Lett 186: 427–430, 1991.

120. J Park, R Bersohn. J Chem Phys 93: 5700–5708, 1990.

121. S Satyapal, GW Johnston, R Bersohn, I Oref. J Chem Phys 93: 6398–6402, 1990.

122. GW Johnston, S Satyapal, R Bersohn, B Katz. J Chem Phys 92: 206–212, 1990.

123. M Koshi, F Tamura, H Matsui. Chem Phys Lett 173: 235–240, 1990.

124. P Löffler, D Lacombe, A Ross, E Wrede, L Schnieder, KH Welge. Chem Phys Lett 252, 304–310, 1996.

125. DH Mordaunt, MNR Ashfold, RN Dixon, P Löffler, L Schnieder, KH Welge. J Chem Phys 108: 519–526, 1998.

126. MNR Ashfold, DH Mordaunt, SHS Wilson. Comm Atm Mol Phys 32: 187–196, 1996.

127. DH Mordaunt, MNR Ashfold, RN Dixon. J Chem Phys 104: 6460–6471, 1996.

128. SHS Wilson, CL Reed, DH Mordaunt, MNR Ashfold, M Kawasaki. Bull Chem Soc Jpn 69: 71–76, 1996.

129. AJR Heck, RN Zare, DW Chandler. J Chem Phys 104: 3399–3402, 1996.

130. SHS Wilson, JD Howe, KNR Rosser, MNR Ashfold, RN Dixon. Chem Phys Lett 227: 456–460, 1994.

131. SHS Wilson, MNR Ashfold, RN Dixon. J Chem Phys 101: 7538–7547, 1994.

132. DH Mordaunt, MNR Ashfold. J Chem Phys 101: 2630–2631, 1994.

133. IR Lambert, GP Morley, DH Mordaunt, MNR Ashfold, RN Dixon. Can J Chem 72: 977–984, 1994.

134. GP Morley, IR Lambert, DH Mordaunt, SHS Wilson, MNR Ashfold, RN Dixon, CM Western. J Chem Soc Faraday Trans 89: 3865–3875, 1993.

135. GP Morley, IR Lambert, MNR Ashfold, KN Rosser, CM Western. J Chem Phys 97: 3157–3165, 1992.

136. BA Balko, J Zhang, YT Lee. J Chem Phys 94: 7958–7966, 1991.

137. RE Continetti, BA Balko, YT Lee. Chem Phys Lett 182: 400–405, 1991.

138. MNR Ashfold, IR Lambert, DH Mordaunt, GP Morley, CM Western. J Phys Chem 96: 2938–2949, 1992.

139. X Xie, L Schnieder, H Wallmeier, R Boettner, KH Weige, MNR Ashfold. J Chem Phys 92: 1608–1616, 1990.

140. L Schnieder, W Meier, KH Weige. J Chem Phys 92: 7027–7037, 1990.

141. DH Mordaunt, IR Lambert, GP Morley, MNR Ashfold, RN Dixon, CW Western, L Schnieder, KH Welge. J Chem Phys 98: 2054–2065, 1993.

142. DH Mordaunt, MNR Ashfold, RN Dixon. J Chem Phys 100: 7360–7375, 1994.

143. P Löffler, E Wrede, L Schnieder, JB Halpern, WM Jackson, KH Welge. J Chem Phys 109: 5231–5246, 1998.

144. L Schnieder, K Seekamp-Rahn, E Wrede, KH Welge. J Chem Phys 107: 6175–6195, 1997.
145. J Zhang, CW Riehn, M Dulligan, C Wittig. J Chem Phys 103: 6815–6818, 1995.
146. L Schnieder, K Seekamp-Rahn, J Borkowski, E Wrede, KH Welge, FJ Aoiz, L Bañares, MJ D'Mello, VJ Herrero, V Sáez Rábanos, RE Wyatt. Science 269: 207–210, 1995.
147. J Zhang, M Dulligan, C Wittig. J Phys Chem 99: 7446–7452, 1995.
148. C Jaques, L Valachovic, S Ionov, E Böhmer, Y Wen, J Segall, C Wittig. J Chem Soc Faraday Trans 89: 1419–1425, 1993.
149. J Segall, Y Wen, R Singer, C Wittig, A García-Vela, RB Gerber. Chem Phys Lett 207: 504–509, 1993.
150. J Segall, Y Wen, R Singer, M Dulligan, C Wittig. J Chem Phys 99: 6600–6606, 1993.
151. J Segall, Y Wen, R Lavi, R Singer, C Wittig. J Phys Chem 95: 8078–8081, 1991.
152. L Schnieder, K Seekamp-Rahn, F Liedeker, H Steuwe, KH Welge. Faraday Discuss. Chem Soc 91: 259–269, 1991.
153. J-H Wang, Y-T Hsu, K Liu. J Phys Chem A 101: 6593–6602, 1997.
154. L-H Lai, D-C Che, K Liu. J Phys Chem 100: 6376–6380, 1996.
155. P Balling, HH Andersen, CA Brodie, UV Pedersen, VV Petrunin, MK Raarup, P Steiner, T Andersen. Phys Rev A 61: 022702-1–022702-11, 2000.
156. P Balling, P Kristensen, HH Andersen, UV Pedersen, VV Petrunin, L Praestegaard, HK Haugen, T Andersen. Phys Rev Lett 77: 2905–2908, 1996.
157. HH Andersen, P Balling, P Kristensen, UV Pedersen, SA Aseyev, VV Petrunin, T Andersen. Phys Rev Lett 79: 4770–4773, 1997.
158. E Reinhold, W Hogervorst, W Ubachs, L Wolneiwicz. Phys Rev A 60: 1258–1270, 1999.
159. W Hogervorst, KSE Eikema, E Reinhold, W Ubachs. Nucl Phys A 63: 353c–362c, 1998.
160. E Reinhold, W Hogervorst, W Ubachs. Phys Rev Lett 78: 2543–2546, 1997.
161. E Reinhold, W Hogervorst, W Ubachs. J Mol Spectrosc 180: 156–163, 1996.
162. PC Hinnen, W Hogervorst, S Stolte, W Ubachs. Can J Phys 72: 1032–1042, 1994.
163. RM Rao, J Dvorak, RJ Beuhler, MG White. J Phys Chem B 102: 9050–9060, 1998.
164. PC Hinnen, W Hogervorst, S Stolte, W Ubachs. Appl Phys B 59: 307–310, 1994.
165. E Reinhold, A de Lange, W Hogervorst, W Ubachs. J Chem Phys 109: 9772–9782, 1998.
166. K Tsuchiyama, J Ishii, T Kasuya. J Chem Phys 97: 875–882, 1992.
167. H Rottke, KH Welge. J Chem Phys 97: 908–926, 1992.
168. EF McCormack, EE Eyler. Phys Rev Lett 66: 1042–1045, 1991.
169. EE Eyler. Comments. Atm Mol Phys 24: 299–310, 1990.
170. EE Eyler, N Melikechi. Phys Rev A 48: R18–R21, 1993.
171. A Stolow, BA Balko, EF Cromwell, J Zhang, YT Lee. J Photochem Photobiol A.: Chem 62: 285–300, 1992.
172. R Shiell, XK Hu, Q Hu, JW Hepburn. Faraday Discuss 115: 331–343, 2000.
173. HH Fielding, TP Softley. Chem Phys Lett 185: 199–205, 1991.
174. HH Fielding, TP Softley, F Merkt. Chem Phys 155: 257–265, 1991.

175. A Balakrishnan, V Smith, BP Stoicheff. Phys Rev A 49: 2460–2469, 1994.
176. A Balakrishnan, V Smith, BP Stoicheff. Phys Rev Lett 68: 2149–2152, 1992.
177. A Balakrishnan, BP Stoicheff. J Mol Spectrosc 156: 517–518, 1992.
178. LM Dobeck, HM Lambert, W Kong, PJ Pisano, PL Houston. J Phys Chem A 103: 10312–10323, 1999.
179. F Merkt, TP Softley. J Chem Phys 96: 4149–4156, 1992.
180. F Merkt, SR Mackenzie, TP Softley. J Chem Phys 99: 4213–4214, 1993.
181. E Reinhold, W Hogervorst, W Ubachs. Chem Phys Lett 296: 411–416, 1998.
182. F Merkt, H Xu, RN Zare. J Chem Phys 104: 950–961, 1996.
183. A Balakrishnan, M Vallet, BP Stoicheff. J Mol Spectrosc 162, 168–171, 1993.
184. SD Bergeson, A Balakrishnan, KGH Baldwin, TB Lucatorto, JP Marangos, TJ McIlrath, TR O'Brian, SL Rolston, CJ Sansonetti, J Wen, N Westbrook. Phys Rev Lett 80: 3475–3478, 1998.
185. CEM Strauss, SH Kable, GK Chawla, PL Houston, IR Burak. J Chem Phys 94: 1837–1849, 1991.
186. D Chastaing, SD Le Picard, IR Sims. J Chem Phys 112: 8466–8469, 2000.
187. JA Blush, P Chen, RT Wiedmann, MG White. J Chem Phys 98: 3557–3559, 1993.
188. N Taniguchi, K Takahashi, Y Matsumi, SM Dylewski, JD Geiser, PL Houston. J Chem Phys 111: 6350–6355, 1999.
189. LL Springsteen, S Satyapal, Y Matsumi, LM Dobeck, PL Houston. J Phys Chem 97: 7239–7241, 1993.
190. M Abe, Y Sato, Y Inagaki, Y Matsumi, M Kawasaki. J Chem Phys 101: 5647–5651, 1994.
191. K Takahashi, Y Matsumi, M Kawasaki. J Phys Chem 100: 4084–4089, 1996.
192. K Takahashi, M Kishigami, Y Matsumi, M Kawasaki, AJ Orr-Ewing. J Chem Phys 105: 5290–5293, 1996.
193. K Takahashi, M Kishigami, N Taniguchi, Y Matsumi, M Kawasaki. J Chem Phys 106: 6390–6397, 1997.
194. K Takahashi, N Taniguchi, Y Matsumi, M Kawasaki, MNR Ashfold. J Chem Phys 108: 7161–7172, 1998.
195. J Miyawaki, K Yamanouchi, S Tsuchiya. Chem Phys Lett 180: 287–292, 1991.
196. J Miyawaki, T Tsuchizawa, K Yamanouchi, S Tsuchiya. Chem Phys Lett 165: 168–170, 1990.
197. AMS Chowdhury, Y Matsumi, M Kawasaki. Chem Lett 77–78, 1997.
198. Y Matsumi, AMS Chowdhury. J Chem Phys 104: 7036–7044, 1996.
199. K Takahashi, R Wada, Y Matsumi, M Kawasaki. J Phys Chem 100: 10145–10149, 1996.
200. AMS Chowdhury, M Kawasaki. Laser Phys 6: 1175–1179, 1996.
201. AMS Chowdhury. Laser Phys 7: 1058–1062, 1997.
202. AMS Chowdhury, Y Matsumi, M Kawasaki. Laser Phys 7: 946–951, 1997.
203. Y Matsumi, SM Shamsuddin, Y Sato, M Kawasaki. J Chem Phys 101: 9610–9618, 1994.
204. A Miyoshi, K Tsuchiya, N Yamauchi, H Matsui. J Phys Chem 98: 11452–11458, 1994.
205. Y Matsumi, Y Inagaki, GP Morley, M Kawasaki. J Chem Phys 100: 315–324, 1994.

206. M Abe, Y Sato, Y Inagaki, Y Matsumi, M Kawasaki. J Chem Phys 101: 5647–5651, 1994.
207. SM Shamsuddin, Y Inagaki, Y Matsumi, M Kawasaki. Can J Chem 72: 637–642, 1994.
208. AMS Chowdhury. Laser Chem 17: 191–203, 1997.
209. R Signorell, F Merkt. J Chem Phys 110: 2309–2311, 1999.
210. RT Wiedmann, RG Tonkyn, MG White, K Wang, V McKoy. J Chem Phys 97: 768–772, 1992.
211. X Li, CR Vidal. J Chem Phys 101: 5523–5528, 1994.
212. B Niu, MG White. J Chem Phys 104: 2136–2145, 1996.
213. WL Glab. J Chem Phys 107: 5979–5982, 1997.
214. AH Zanganeh, JH Fillon, J Ruiz, M Castillejo, JL Lemaire, N Shafizadeh, F Rostas. J Chem Phys 112: 5660–5671, 2000.
215. MJJ Vrakking, YT Lee, RD Gilbert, MS Child. J Chem Phys 98: 1902–1915, 1993.
216. RG Tonkyn, R Wiedmann, ER Grant, MG White. J Chem Phys 95: 7033–7040, 1991.
217. K Wang, MT Lee, V McKoy, RT Wiedman, MG White. Chem Phys Lett 219: 397–404, 1994.
218. F Merkt, R Signorell, H Palm, A Osterwalder, M Sommavilla. Mol Phys 95: 1045–1054, 1998.
219. RT Wiedmann, MG White. Proc SPIE: Opt Methods State Time-Resolved Chem 1638: 273, 1992.
220. R Signorell, M Sommavilla, F Merkt. Chem Phys Lett 312: 139–148, 1999.
221. KSE Eikema, W Ubachs, W Hogervorst. Phys Rev A 49: 803–808, 1994.
222. A Mank, D Rodgers, JW Hepburn. Chem Phys Lett 219: 169–173, 1994.
223. X Li, CR Vidal. J Chem Phys 102: 9167–9173, 1995.
224. R Signorell, H Palm, F Merkt. J Chem Phys 106: 6523–6533, 1997.
225. DM Jonas, X Zhao, K Yamanouchi, PG Green, GW Adamson, RW Field. J Chem Phys 92: 3988–3989, 1990.
226. RT Wiedmann, MG White. J Chem Phys 102: 5141–5151, 1995.
227. RL Miller, SH Kable, PL Houston, I Burak. J Chem Phys 96: 332–338, 1992.
228. L Fleck, RJ Beuhler, MG White. J Chem Phys 106: 3813–3816, 1997.
229. SS Dimov, CR Vidal. Chem Phys Lett 221: 307–310, 1994.
230. V Kleiman, K Trentelman, Y Huang, RJ Gordon. Chem Phys Lett 222: 161–166, 1994.
231. SP Lu, SM Park, Y Xie, RJ Gordon. J Chem Phys 96: 6613–6620, 1992.
232. RJ Gordon, S-P Lu, SM Park, K Trentelman, Y Xie, L Zhu, A Kumar, WJ Meath. J Chem Phys 98: 9481–9486, 1993.
233. T Sykora, CR Vidal. J Chem Phys 108: 6320–6330, 1998.
234. GC McBane, SH Kable, PL Houston, GC Schatz. J Chem Phys 94: 1141–1149, 1991.
235. R Jimenez, SH Kable, JC Loison, CJSM Simpson, W Adam, PL Houston. J Phys Chem 96: 4188–4195, 1992.
236. DW Neyer, SH Kable, J-C Loison, PL Houston, I Burak, EM Goldfield. J Chem Phys 97: 9036–9045, 1992.
237. DW Neyer, X Luo, PL Houston. J Chem Phys 98: 5095–5098, 1993.

238. DW Neyer, X Luo, I Burak, PL Houston. J Chem Phys 102: 1645–1657, 1995.
239. Y Inagaki, M Abe, Y Matsumi, M Kawasaki, H Tachikawa. J Phys Chem 99: 12822–12828, 1995.
240. YS Choi, CB Moore. J Chem Phys 103: 9981–9988, 1995.
241. M Abe, Y Inagaki, LL Springsteen, Y Matsumi, M Kawasaki, H Tachikawa. J Phys Chem 98: 12641–12645, 1994.
242. SA Buntin, RR Cavanagh, LJ Richter, DS King. J Chem Phys 94: 7937–7950, 1991.
243. FJ Schlenker, F Bouchard, IM Waller, JW Hepburn. J Chem Phys 93: 7110–7118, 1990.
244. YS Choi, P Teal, CB Moore. J Opt Soc Am B 7: 1829–1834, 1990.
245. A Jolly, JL Lemaire, D Belle-Oudry, S Edwards, D Malmasson, A Vient, F Rostas. J Phys B: Atm Mol Opt Phys 30: 4315–4337, 1997.
246. W Ubachs, PC Hinnen, P Hansen, S Stolte, W Hogervorst, P Cacciani. J Mol Spectrosc 174: 388–396, 1995.
247. P Cacciani, W Hogervorst, W Ubachs. J Chem Phys 102: 8308–8320, 1995.
248. W Ubachs, KSE Eikema, W Hogervorst, PC Cacciani. J Opt Soc Am B 14: 2469–2476, 1997.
249. KSE Eikema, W Hogervorst, W Ubachs. Chem Phys 181: 217–245, 1994.
250. KSE Eikema, W Hogervorst, W Ubachs. J Mol Spectrosc 163: 19–26, 1994.
251. PF Levelt, W Ubachs, W Hogervorst. J Chem Phys 97: 7160–7166, 1992.
252. PC Hinnen, S Stolte, W Hogervorst, W Ubachs. J Opt Soc Am B 15: 2620–2625, 1998.
253. W Ubachs, KSE Eikema, PF Levelt, W Hogervorst, M Drabbles, WL Meerts, JT Ter Meulen. Astrophys J 427: L55–L58, 1994.
254. K Tsukiyama, M Tsukakoshi, T Kasuya. Appl Phys B 50: 23–28, 1990.
255. K Tsukiyama, M Momse, M Tsukakoshi, T Kasuya. Opt Commun 79: 88–92, 1990.
256. A Mellinger, CR Vidal. J Chem Phys 101: 104–110, 1994.
257. A Mellinger, CR Vidal. Chem Phys Lett 238: 31–36, 1995.
258. A Mellinger, CR Vidal, Ch Jungen. J Chem Phys 104: 8913–8921, 1996.
259. W Kong, D Rodgers, JW Hepburn, K Wang, V Mckoy. J Chem Phys 99: 3159–3165, 1993.
260. W Kong, JW Hepburn. J Phys Chem 99: 1637–1642, 1995.
261. W Ubachs, KSE Eikema, W Hogervorst. Appl Phys B 57: 411–416, 1993.
262. W Ubachs, I Velchev, A de Lange. J Chem Phys 112: 5711–5716, 2000.
263. PF Levelt, W Ubachs. Chem Phys 163: 263–275, 1992.
264. H Palm, F Merkt. Chem Phys Lett 284: 419–422, 1998.
265. F Merkt, TP Softley. Phys Rev A 46: 302–314, 1992.
266. JW Hepburn. J Chem Phys 107: 7106–7113, 1997.
267. ER Grant, MG White. Nature 354: 249–250, 1991.
268. J-C Loison, SH Kable, PL Houston, I Burak. J Chem Phys 94: 1796–1802, 1991.
269. SH Kable, J-C Loison, DW Neyer, PL Houston, I Burak, RN Dixon. J Phys Chem 95: 8013–8018, 1991.
270. SH Kable, J-C Loison, PL Houston, I Burak. J Chem Phys 92: 6332–6333, 1990.
271. K Tsukiyama, M Tsukakoshi, T Kasuya. J Chem Phys 94: 883–888, 1991.
272. K Tsukiyama, M Tsukakoshi, T Kasuya. J Chem Phys 92: 6426–6431, 1990.

273. A Sugita, M Ikeda, K Tsukiyama. Appl Phys B 67: 253–256, 1998.
274. S Hayashi, T Suzuki, T Ichimura, K Tsukiyama. Appl Phys B 65: 555–561, 1997.
275. J Guo, A Mank, JW Hepburn. Phys Rev Lett 74: 3584–3587, 1995.
276. W Kong, D Rodgers, JW Hepburn. J Chem Phys 99: 8571–8576, 1993.
277. H Palm, F Merkt. Chem Phys Lett 270: 1–8, 1997.
278. RT Wiedmann, MG White, K Wang, V McKoy. J Chem Phys 98: 7673–7679, 1993.
279. G Nan, I Burak, PL Houston. Chem Phys Lett 209: 383–389, 1993.
280. G Nan, DW Neyer, PL Houston, I Burak. J Chem Phys 98: 4603–4609, 1993.
281. G Nan, PL Houston. J Chem Phys 97. 7865–7872, 1992.
282. GC McBane, I Burak, GE Hall, PL Houston. J Phys Chem 96: 753–755, 1992.
283. A Mank, C Starrs, MN Jego, JW Hepburn. J Chem Phys 104: 3609–3619, 1996.
284. U Berzinsh, L Caiyan, R Zerne, S Svanberg, E Biemont. Phys Rev A 55: 1836–1841, 1997.
285. JDD Martin, JW Hepburn. Phys Rev Lett 79: 3154–3157, 1997.
286. BR Lewis, PM Dooley, JP England, K Waring, ST Gibson, KGH Baldwin, H Partridge. Phys Rev A 54: 3923–3938, 1996.
287. BR Lewis, JP England, RJ Winkel, Jr., SS Banerjee, PM Dooley, ST Gibson, KGH Baldwin. Phys Rev A 52: 2717–2733, 1995.
288. W Kong, JW Hepburn. Can J Phys 72: 1284–1293, 1994.
289. W Kong, D Rodgers, JW Hepburn. Chem Phys Lett 203: 497–502, 1993.
290. M Braunstein, V McKoy, SN Dixit, RG Tonkyn, MG White. J Chem Phys 93: 5345–5346, 1990.
291. Th Glenewinkel-Meyer, JA Bartz, GM Thorson, FF Crim. J Chem Phys 99: 5944–5950, 1993.
292. VD Kleiman, L Zhu, X Li, RJ Gordon. J Chem Phys 102: 5863–5866, 1995.
293. R Shiell, XK Hu, Q Hu, JW Hepburn. J Phys Chem A 104: 4339–4342, 2000.
294. V Skorokhodov, Y Sato, K Suto, Y Matsumi, M Kawasaki. J Phys Chem 100: 12321–12328, 1996.
295. M Drescher, A Brockhinke, N Bowering, U Heinzmann, H Lefebvre-Brion. J Chem Phys 99: 2300–2306, 1993.
296. SM Park, S-P Lu, RJ Gordon. J Chem Phys 94: 8622–8644, 1991.
297. JDD Martin, JW Hepburn. J Chem Phys 109: 8139–8142, 1998.
298. RG Tonkyn, RT Wiedmann, MG White. J Chem Phys 96: 3696–3701, 1992.
299. PF Levelt, KSE Eikema, S Stolte, W Hogervorst, W Ubachs. Chem Phys Lett 210: 307–314, 1993.
300. I Velchev, W Hogervorst, W Ubach. J Phys B: At Mol Opt Phys 32: L511–L516, 1999.
301. F Merkt, A Osterwalder, R Seiler, R Signorell, H Palm, H Schmutz, R Gunzinger. J Phys B: Atm Mol Opt Phys 31: 1705–1724, 1998.
302. F Merkt, H Schmutz. J Chem Phys 108: 10033–10045, 1998.
303. JDD Martin, JW Hepburn, C Alcaraz. J Phys Chem A 101: 6728–6735, 1997.
304. F Merkt. J Chem Phys 100, 2623–2628, 1994.
305. XF Yang, J-L Lemaire, F Rostas, J Rostas. Chem Phys 164: 115–122, 1992.
306. RT Wiedmann, MG White, H Lefebvre-Brion, C Cassart-Magos. J Chem Phys 103: 10417–10423, 1995.

307. F Merkt, SR Mackenzie, RJ Rednall, TP Softley. J Chem Phys 99: 8430–8439, 1993.

308. W Kong, D Rodgers, JW Hepburn. Chem Phys Lett 221: 301–306, 1994.

309. RT Wiedmann, ER Grant, RG Tonkyn, MG White. J Chem Phys 95: 746–753, 1991.

310. Y-S Cheung, J-C Huang, CY Ng. J Chem Phys 109: 1781–1786, 1998.

311. Y-S Cheung, CY Ng. Int J Mass Spectrom 187: 533–543, 1999.

312. Y Matsumi, S Nomura, M Kawasaki, T Imamura. J Phys Chem 100: 176–179, 1996.

313. Y Matsumi, SM Shamsuddin, M Kawasaki. J Chem Phys 101: 8262–8263, 1994.

314. Y Matsumi, SM Shamsuddin. J Chem Phys 103: 4490–4495, 1995.

315. CD Pibel, K Ohde, K Yamanouchi. J Chem Phys 101: 836–839, 1994.

316. A Hishikawa, K Ohde, R Itakura, S Liu, K Yamanouchi, K Yamashita. J Phys Chem A 101: 694–704, 1997.

317. K Yamanouchi, K Ohde, A Hishikawa, CD Pibel. Bull Chem Soc Jpn 68: 2459–2464, 1995.

318. KH Hahn, DA King, SE Harris. Phys Rev Lett 65: 2777–2779, 1990.

319. T Tsuchizawa, K Yamanouchi, S Tsuchiya. J Chem Phys 93: 111–120, 1990.

320. JC Huang, YS Cheung, M Evans, CX Liao, CY Ng, CW Hsu, P Heimann, H Lefebvre-Brion, C Cossart-Magos. J Chem Phys 106: 864–877, 1997.

321. JH Werner, TA Cool. Chem Phys Lett 275: 278–282, 1997.

322. R Signorell, F Merkt. J Chem Phys 109: 9762–9771, 1998.

323. R Signorell, A Wuest, F Merkt. J Chem Phys 107: 10819–10822, 1997.

324. A Mank, T Nguyen, JDD Martin, JW Hepburn. Phys Rev A 51: R1–R4, 1995.

325. C Dorman, I Kucukkara, JP Marangos. Phys Rev A 61: 013802, 1999.

326. C Dorman, JP Marangos, JC Petch. J Mod Opt 45: 1123–1135, 1998.

327. C Dorman, JP Marangos. Phys Rev A 58: 4121–4132, 1998.

328. AJ Merriam, SJ Sharpe, H Xia, D Manuszak, GY Yin, SE Harris. Opt Lett 24: 625–627, 1999.

329. JH Werner, TA Cool. Chem Phys Lett 290: 81–87, 1998.

330. SS Dimov, RH Lipson, T Turgeon, JA Vanstone, P Wang, DS Yang. J Chem Phys 100, 8666–8672, 1994; P Wang, SS Dimov, RH Lipson. J Chem Phys 107: 3345–3351, 1997.

331. A Mank, M Drescher, T Huth-Fehre, N Bowering, U Heinzmann, H Lefebvre-Brion. J Chem Phys 95: 1676–1687, 1991.

332. A Mank, M Drescher, T Huth-Fehre, G Schonhense, N Bowering, U Heinzmann. J Electron Spectrosc 52: 661–670, 1990.

333. A Mank, M Drescher, A Brockhinke, N Bowering, U Heinzmann. Z Phys D 29: 275–289, 1994.

334. DJ Gendron, JW Hepburn, J Chem Phys 109, 7205–7213, 1998.

335. L Zhu, V Kleiman, X Li, SP Lu, K Trentelman, RJ Gordon. Science 270: 77–80, 1995.

336. JA Fiss, L Zhu, K Suto, G He, RJ Gordon. Chem Phys 233: 335–341, 1998.

337. L Zhu, K Suto, JA Fiss, R Wada, T Seideman, RJ Gordon. Phys Rev Lett 79: 4108–4111, 1997.

338. JA Fiss, L Zhu, RJ Gordon. Phys Rev Lett 82: 65–68, 1999.

339. RJ Gordon, L Zhu, T Seideman. Acc Chem Res 32: 1007–1016, 1999.
340. JA Fiss, A Khachatrian, L Zhu, RJ Gordon, T Siedeman. Faraday Discuss. 113: 61–76, 1999.
341. VD Kleiman, L Zhu, JA Fiss, RJ Gordon. J Chem Phys 103: 10800–10803, 1995.
342. CD Pibel, K Yamanouchi, J Miyawaki, S Tsuchiya, B Rajaram, RW Field. J Chem Phys 101: 10242–10251, 1994.
343. CD Pibel, K Ohde, K Yamanouchi. J Chem Phys 105: 1825–1832, 1996.
344. T Tsuchizawa, K Yamanouchi, S Tsuchiya. J Chem Phys 92: 1560–1567, 1990.
345. S Liu, A Hishikawa, K Yamanouchi. J Chem Phys 108: 5330–5337, 1998.
346. DM Mao, XK Hu, YJ Shi, RH Lipson. J Chem Phys 111: 2985–2990, 1999.
347. DM Mao, XK Hu, JH Leech, YJ Shi, RH Lipson. J Chem Phys 114: 4025–4035, 2001.
348. DM Mao, XK Hu, YJ Shi, J Ma, RH Lipson. Can J Phys 78: 433–447, 2000.
349. DM Mao, XK Hu, SS Dimov, RH Lipson. J Mol Spectrosc 181: 435–445, 1997.
350. DM Mao, XK Hu, YJ Shi, RH Lipson. Chem Phys 257: 253–261, 2000.
351. CD Pibel, K Yamanouchi, S Tsuchiya. J Chem Phys 100: 6153–6159, 1994.
352. K Tsukiyama, T Kasuya. J Mol Spectrosc 151: 312–321, 1992.
353. RG Tonkyn, MG White. J Chem Phys 95: 5582–5589, 1991.
354. AV Kanaev, L Museur, MC Castex. J Chem Phys 107: 4006–4014, 1997.
355. DJ Butcher, DE Goeringer, GB Hurst. Anal Chem 71: 489–496, 1999.
356. J Li, VM Bierbaum, SR Leone. Chem Phys Lett 313: 76–84, 1999.
357. O Komienko, ET Ada, J Tinka, MBJ Wijesundara, L Hanley. Anal Chem 70: 1208–1213, 1998.
358. JL Trevor, DE Mencer, KR Lykke, MJ Pellin, L Hanley. Anal Chem 69: 4331–4338, 1997.
359. JL Trevor, KR Lykke, MJ Pellin, L Hanley. Langmuir 14: 1664–1673, 1998.
360. JL Trevor, L Hanley, KR Lykke. Rapid Commun. Mass Spectrom 11: 587–589, 1997.
361. N Materer, RS Goodman, SR Leone. J Vac Sci Technol A 15: 2134–2142, 1997.
362. CH Becker, KJ Wu. J Am Soc Mass Spectrom 6: 883–888, 1995.
363. P Wurz, KR Lykke. J Chem Phys 95: 7008–7010, 1991.
364. KR Lykke, P Wurz, DH Parker, MJ Pellin. Appl Opt 32: 857–866, 1993.
365. KR Lykke, DH Parker, P Wurz, JE Hunt, MJ Pellin, DM Gruen, JC Hemminger, RP Lattimer. Anal Chem 64: 2797–2803, 1992.
366. C Koster, J Grotemeyer. Org Mass Spectrom 27: 463–471, 1992.
367. CH Becker, LE Jusinski, L Moro. Int J Mass Spectrom Ion Processes 95: R1–R4, 1990.
368. RJJM Steenvoorden, PG Kistemaker, AE De Vries, L Michalak, NMM Nibbering. Int J Mass Spectrom Ion Processes 107: 475–489, 1991.
369. SE Van Bramer, MV Johnston. J Am Soc Mass Spectrom 1: 419–426, 1990.
370. JB Pallix, U Schühle, CH Becker, DL Huestis. Anal Chem 61: 805–811, 1989.
371. U Schühle, JB Pallix, CH Becker. J Am Chem Soc 110: 2323–2324, 1988.
372. U Schühle, JB Pallix, CH Becker, J Vac Sci Technol A6: 936–940, 1988.
373. S Ruhman, Y Hass, J Laukemper, M Preuss, H Stein, D Feldmann, KH Welge. J Phys Chem 88: 5162–5167, 1984.
374. D Feldmann, J Laukemper, KH Welge. J Chem Phys 79: 278–282, 1983.

375. K Tonokura, T Murasaki, M Koshi. Chem Phys Lett 319: 507–511, 2000.
376. S-T Tsai, J-C Jiang, YT Lee, AH Kung, SH Lin, C-K Ni. J Chem Phys 111: 3434–3440, 1999.
377. C-K Ni, JD Huang, YT Chen, AH Kung, WM Jackson. J Chem Phys 110: 3320–3325, 1999.
378. T Gonthiez, T Gibert, P Brault, C Boulmer-Leborgne, C Olivero, MC Castex. Appl Phys A 69 [Suppl.]: S171–S173, 1999.
379. PL Ross, MV Johnston. J Phys Chem 99: 4078–4085, 1995.
380. PL Ross, MV Johnston. J Phys Chem 97: 10725–10731, 1993.
381. JA Bartz, DB Galloway, LG Huey, T Glenewinkel-Meyer, FF Crim. J Phys Chem 97: 11249–11252, 1993.
382. JA Bartz, TM Barnhart, DB Galloway, LG Huey, T Glenewinkel-Meyer, RJ McMahon, FF Crim. J Am Chem Soc 115: 8389–8395, 1993.
383. CD Mowry, MV Johnston. J Phys Chem 98: 1904–1909, 1994.
384. SE Van Bramer, MV Johnston. Org Mass Spectrom 27: 949–954, 1992.
385. RE Bandy, C Lakshminarayan, RK Frost, TS Zwier. Science, 258: 1630–1633, 1992.
386. RK Frost, GS Zavarian, TS Zwier. J Phys Chem 99: 9408–9415, 1995.
387. RE Bandy, C Lakshminarayan, RK Frost, TS Zwier. J Chem Phys 98: 5362–5374, 1993.
388. E Nil, HE Hunziker, MS de Vries. Anal Chem 71: 1674–1678, 1999.
389. DB Galloway, JA Bartz, LG Huey, FF Crim. J Chem Phys 98: 2107–2114, 1993.
390. PG Strupp, AL Alstrin, RV Smilgys, SR Leone. Appl Opt 32: 842–846, 1993.
391. SE Van Bramer, PL Ross, MV Johnston. J Am Soc Mass Spectrom 4: 65–72, 1993.
392. SE Van Bramer, MV Johnston. Anal Chem 62: 2639–2643, 1990.
393. J Boyle, L Pfefferle. J Phys Chem 94: 3336–3340, 1990.
394. P Chen, SD Colson, WA Chupka, JA Berson. J Phys Chem 90: 2319–2321, 1986.
395. K Suto, Y Sato, CL Reed, V Skorokhodov, Y Matsumi, M Kawasaki. J Phys Chem A 101: 1222–1226, 1997.

6

Spectroscopy and Applications of Diatomic and Triatomic Molecules Assisted by Laser Light at 157.6 nm

Alciviadis-Constantinos Cefalas and Evangelia Sarantopoulou
National Hellenic Research Foundation, Theoretical and Physical Chemistry Institute, Athens, Greece

I. THE MOLECULAR FLUORINE LASER

Laser radiation from molecular fluorine transitions at 156.71, 157.48, and 157.59 nm was observed for the first time in 1978 by J. R. Woodworth and J. K. Rice [1,2]. They used e-beam excitation to achieve population inversion in a mixture of He (1500 mbar) and F_2 (5.3 mbar). The observed stimulated emission was assigned to the $^3\Pi_g \rightarrow {}^3\Pi_u$ transitions of the F_2 molecule. The intensity of the laser pulse was 7 MW/cm^2 and the efficiency was 3.5%. The experimental setup used by Woodworth et al. was complicated and the system had to be pumped and filled back after each shot. Plummer et al. [3] in 1979 used a different more convenient technique by pumping mixtures of F_2 and He in a fast ultraviolet (UV)–preionized discharge. This technique proved successful for pumping of the excimer lasers and it was more flexible. The cathode electrode whose position was fixed consisted of a stainless steel mesh. The discharge volume could be varied by moving one electrode in order to find the optimum electrode spacing for higher laser output. In 1985 Cefalas et al. [4] developed a simple UV preionized F_2 laser of the fast-discharge type to measure the small signal gain by the passive cell absorption method. The gain with this equipment at optimum working conditions of 2 atm total gas pressure and 2 cm electrode spacing was found to be 3.2% cm^{-1}. The output energy was measured as 12 mJ per pulse and the pulse duration was 10 ns despite the use of gases with high concentration of impurities (0.1% impurities). With this experimental apparatus it was proved that the molecular fluorine laser had the potential to be a powerful VUV laser similar to the excimer lasers. Following its development, higher output energy of 15 mJ per pulse was achieved by Ishchenko et al. [5]. The above values of output energies were half those predicted by the theoretical calculations of Ohwa and Obara [6], if one considers only the dissociative collision of F_2 molecules by ion–ion recombination of energy transfer reaction and neglecting the direct excitation of F_2 molecules by electron impact or energy transfer from He*-, He**-, and He$_2^*$-excited atoms or clusters.

The predictions of Ohwa and Obara were confirmed by Yamada et al. [7], who developed an F_2 laser that delivered 112 mJ per pulse at 8 atm total gas pressure. The electric circuit in this apparatus was similar to that developed by Cefalas et al. (1985) of the fast charge transfer type with UV preionization. The innovating point in the geometry of this cavity was the small distance between the electrodes of 10 mm, providing stable discharge even at higher gas pressure and hence improving the laser's performance. In an effort to investigate the physical parameters and limits of the F_2 molecular laser, Cefalas et al. used two discharge laser heads driven by a spark-gap switch to measure the small-signal gain and the saturation intensity in the oscillator–amplifier configuration [8,9]. The small-signal gain coefficient was measured to be 5.2 ± 0.4% cm^{-1} at 3 atm total pressure and 1.5 cm electrode spacing. It was 4.1 ± 0.4% cm^{-1} at 2 atm total pressure and 2 cm electrode spacing. The values of saturation intensities were found to be 5MW/cm^2 and 4.6 MW/cm^2, respectively. The dependences of the energy output and efficiency of the F_2 laser on the pump power (up to 40 MW/cm^3) have been studied by Kuznetsor and Sulakshin [10]. The maximum lasing efficiency was 0.05% and the laser radiation energy was 120 mJ at pressure of the He–F_2 mixture of 3 atm. A theoretical kinetics model successfully described the characteristics of the F_2 laser output. The authors concluded that at high gas pressure operation, where dissociation of the F_2 (A′) state is accelerated, higher discharge pumping power is achieved [11]. The small-signal gain and the saturation intensity of the discharge pumped F_2 laser, operated at higher pressures (<10 atm) and higher excitation rates (7–39 MW/cm^3), were measured [12] with a similar oscillator–amplifier configuration to that used by Cefalas et al. [8,9]. The small-signal net gain with this apparatus now reaches 37 ± 4% cm^{-1} at an excitation rate of approximately 26 MW/cm^3 for a 6-atm gas pressure. The saturation intensity, which depends on a nonsaturation absorption coefficient, was estimated. The output energy and the temporal behavior of a molecular F_2^* laser pumped by a coaxial electron beam have been measured in gas mixtures of He/F_2 and He/Ne/F_2 [13]. The highest output energy of 172 mJ has been obtained in a mixture of He/Ne/F_2 (19.9%/80%/0.1%) at a pressure of 12 bar, corresponding to a specific output energy of 10.8 Joule/lt and an intrinsic efficiency of 2.6%. An electron beam pumped molecular F_2^* laser with pulse width up to 160 ns, and output energy of 1.7 J (optical flux of 4.6 MW/cm^2) has been realized [14] by the same group. The widths of the laser pulses seem to be limited by the duration of the excitation pulse (160 ns). For specific output powers up to 100 kW/cm^3, no signs of self-terminating laser pulses, due to bottlenecking in the lower laser level, have been observed. The application of a prepulse–main pulse excitation scheme utilizing a saturable magnetic switch in combination with x-ray preionization has resulted in the generation of long optical pulses from a molecular fluorine laser [15]. Optimum laser pulse durations of 70 ns (full width at half maximum) have been obtained in a gas mixture of helium and 3 mbar fluorine at a total pressure of 2 bar. Laser pulse duration was limited by instabilities in the electric discharge. The laser pulse duration was found to decrease with increasing fluorine pressure and to saturate with increasing current density.

II. PHOTOLITHOGRAPHY AT 157 nm

Despite the fact that the molecular fluorine laser emits strong monochromatic radiation at 157.6, its nontunability in the VUV is a serious drawback for a wide range of applications. However, this disadvantage is balanced by the high energy emitted per photon. The molecular fluorine laser has the potential to be used in the near future in many areas of science and technology. Indeed, next-generation microelectronics circuits will have minimum dimensions below 100 nm. It is envisioned that 157 nm laser lithography will be the next step in optical lithography. At 157 nm, under VUV illumination of the mask target, lithographic features with dimensions of 0.10 µm on the photoresist have be achieved by an F_2 laser lithographic stepper developed by Exitech UK [16]. The high-repetition-rate molecular fluorine laser developed by Lambda Physik in Germany was likewise the driving force for developing 157 nm photolithography. The first to propose the F_2 laser in photolithography was White et al. in 1984 [17]. A 2 kHz repetition rate discharge-pumped molecular fluorine laser oscillating at 157 nm was developed. It has achieved an average power of 22 W at the repetition rate of 2 kHz with a newly developed solid-state pulse power module with 6 J/pulse input energy [18]. Single-line operation at 157.6 nm was achieved by means of a prism assembly. Laser operation at repetition rates up to 1 kHz without signs of power saturation results in an average power of 15 W. The energy stability with these devices was comparable to the stability of the ArF laser [19].

Research in new photoresists for 157 nm photolithography [20–26] requires the investigation of basic optical and photochemical studies at these wavelengths. The high value of the absorption coefficient of the polymeric materials in the VUV, from 10^4 to 10^6 cm^{-1}, imposes restrictions on selection of the photoresists for 157 nm photolithography, demanding high-purity materials and de-fect-free thin films. There is a complete bond breaking of all the organic mole-cules at these wavelengths. The parent molecule disintegrates into small fragments, atomic, diatomic or triatomic, which fly apart with supersonic speeds. This occurs because, for all the organic molecules, the dissociative excited states of the small radicals occupy the energy range above 6.2 eV (200 nm). Therefore the photodissociation process could impose serious problems on the optics of the projection system by contaminating it.

To overcome this serious problem, complete microscopic description of the photodissociation dynamics involved in the scission of the polymer chains, and the dynamics of the extraction of the photoproducts, is required at 157 nm. Until now various theories have been developed that analyze the role of the ex-cited species in ultraviolet laser materials ablation. The complexity and diversity of the processes involved in laser ablation dynamics, such as laser excitation of the absorbing molecules and energy transfer from the excited parent molecules into the internal and the translational degrees of freedom of the photofragments, require for their description the analytical expression of the potential energy sur-faces of the parent molecules and the photofragments, for both ground and the excited states [27].

As the first contribution in this subject, and in order to investigate basic photochemical mechanisms of the photodissociation dynamics of various poly-meric materials at 157 nm, we have used potential photoresistive materials with aliphatic and aromatic chains in their molecular structure [21–22]. Nylon 6.6 is a good example. The nylon 6.6 monomer has the potential to dissociate into small photofragments in which most of the photoresists used for 193 and 157 nm photolithography are likely to be dissociated.

Mass spectroscopy of nylon 6.6 at 157 nm [24] reveals that even at moder-ate laser energy there was a complete breaking of the polymeric chain bonds. The following fragments were observed for m/e larger than 30 atomic mass units (amu). The molecular photofragments from the photodissociation of the parent monomer were observed mainly between 20 and 30 amu (Fig. 1). Photofragments with two carbon atoms have a relatively higher probability of being dissociated from the parent monomer than heavier photofragments with four carbon atoms. The polymeric material dissociates into fragments with the predominant mass at 28 amu for both laser wavelengths. Therefore the amide group is mainly involved in the photodissociation process of nylon 6.6 in the VUV at 157 nm, which is the case for wavelengths shorter than 248 nm [21].

Experimental findings suggest that the bound potential energy surfaces of the excited states of the parent molecule correlate with the dissociative potential energy surfaces of the excited states of the molecular photofragments over a wide energy range above 5 eV. Photochemical dissociation of the aliphatic chain is

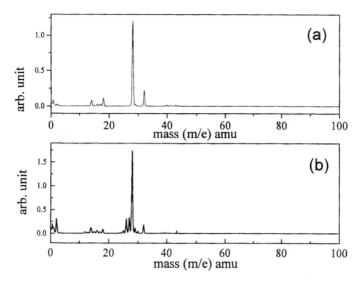

Figure 1 a. Background mass spectrum of nylon 6.6 at 10^{-7} mbar b. Mass spectrum of nylon 6.6 following photofragmentation at 157.6 nm.

the main ablative process at 157 nm. The same response to 157 nm photons has been found in molecules with aromatic structure [22].

The experimental apparatus for obtaining the mass spectrum consists mainly of the molecular fluorine laser source and the vacuum chamber, into which the quadrupole mass spectrometer and sample under investigation were placed (Figs. 2, 3). The sample was a membrane of pure amorphous nylon 6.6, 150 μm thick. The density of the sample was 0.25 gr/cm^3 and it was placed 1 mm apart from the quadrupole mass spectrometer (Baltzers QMG 311) at right angles to its axis (Fig. 2). The all stainless steel 316 vacuum chamber was evacuated to 10^{-6} mbar using a turbomolecular pump.

Laser sources used in these experiments were of the fast-discharge type, which have been described previously [8,9]. The laser head delivers 10 ± 1 mJ per pulse, and the pulse duration was ∼15 ns at FWHM.

The laser beam was focused on the nylon 6.6 sample, using a quartz lens of 40 cm focal length. After photodissociation of the parent molecule, the molecular photofragments were ionized using an electron gun inside an isopotential chamber (Welnet). The molecular ions were focused with an Eizen lens, and eventually directed alongside the quadrupole mass filter. The upper detection limit of this filter was 300 amu. After entering the mass filter, ions were deflected at right angles, and detected using a high-gain (∼10^8) secondary electron multiplier

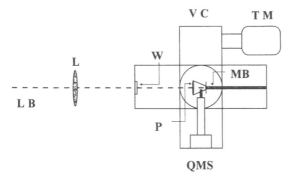

Figure 2 Experimental setup for obtaining mass spectrum of nylon 6.6. LB, laser beam; L, lens; VC, vacuum chamber; TM, turbomolecular pump; W, LiF, window; S, sample; P, photofragments; QMS, quadrupole mass spectrometer.

Figure 3 157 nm F_2 laser and mass spectrometer at the NHRF.

(SEM). The signal was then amplified and registered using a boxcar integrator and a computer. Using the same experimental techniques, several polymeric materials have been studied for both VUV absorption below 200 nm and photodissociation dynamic processes under 157 nm illumination. Si-based polymers seem to have the right value of absorbance of 4 μm^{-1} at 157 nm and therefore they can be imaged at the usual bilayer thickness of 0.1 μm. However, excessive outgassing at 157 nm at the moment is a serious drawback for their use in 157 nm photolithography [20]. Mass spectroscopic and outgassing studies at 157 nm suggest that a complete bond breaking of the aromatic and linear carbon chains is taking place even at low values of laser energy [20]. However, fluorocarbon polymeric materials seem to have the right absorption coefficient in the VUV and limited outgassing at 157 nm.

III. 157 nm SPECTROSCOPY OF DIATOMIC AND TRIATOMIC MOLECULES

A. OH

Photodissociation of the OH radical was studied at 157 nm via detection of the product H atoms with the resonance fluorescence technique [28]. OH radicals were produced in a fast-flow cell from the reaction between H and NO_2 and subsequently photodissociated by an excimer laser operating on the F_2 emission. The quantum yield for photodissociation of OH was measured and the photodissociation cross section calculated.

B. O_2

Direct photodissociation of molecular oxygen from the B $^3\Sigma_u^{-1}$ state have been studied by photoexcitation at 157 nm. The fine structure branching ratios and Doppler profiles of $O(^3P_j)$ photofragments were measured using a resonance-enhanced multiphoton ionization technique. Branching ratios of the $O(^3P_j)$, $j = $ 2, 1, 0, photofragments were measured [29]. The fine structure population of $O(P_3)$ produced in photodissociation of O_2 at 157 nm was measured in a pump and probe experiment [30]. The population of the ground state ($J = 2$) was found to be 93%, with only 6% in the $J = 1$ level and 1% in $J = 0$. The predominance of $J = 2$ is consistent with direct dissociation on the B $^3\Sigma_u$ potential energy surface. Based on available information in the literature, excitation of the $^3\Pi_u$ state and curve crossing to the $^5\Pi_u$ state are ruled out as possible origins of the population with $J < 2$. The populations of $J < 2$ are due to nonadiabatic transitions at large distances. Mechanisms consistent with the data include a Demkov-type interaction, which is caused by coupling by the radial kinetic energy operator. Photodissociation of O_2 at 157 nm has been studied using the photofragmentation translational spectroscopic technique [31]. Two product channels $O_2 + h\nu \rightarrow O(^1D) + O(^3P)$, $O_2 + h\nu \rightarrow O(^3P) + O(^3P)$ have been observed. The relative yields and anisotropy parameters of both channels were determined. Anisotropy mixing of dissociation resulting from a perpendicular excitation and a parallel-type excitation has been observed in the dissociation channel $O_2 + h\nu \rightarrow O(^3P) + O(^3P)$. The observed results can be used to look at the detailed dynamical processes of O_2 dissociation through the Schumann-Runge band.

C. HCl, DCl, CH₃Cl

Hydrogen chloride and methyl chloride were photodissociated at 157 nm [32]. Branching ratios for production of the resulting chlorine atoms in the $^2P_{1/2}$ state relative to the $^2P_{3/2}$ state were determined at 157 nm for HCl and CH_3Cl. Doppler profiles of the chlorine fragments have been measured and are interpreted by assuming their anisotropy parameters and translational energies. HCl undergoes a perpendicular optical transition. For CH_3Cl, the transition is a mixture of parallel and perpendicular types at 157 nm.

The H, D, and Cl atoms from the photodissociation of HCl and DCl at 157 were detected by laser-induced fluorescence (LIF) in the vacuum ultraviolet region [33]. Doppler profiles of H and D resonance lines at 121.6 nm in the LIF spectra indicate that the absorption of HCl and DCl at 157 nm is a mixture of perpendicular and parallel transitions. Fine-structure branching ratios were measured for the chlorine atom by LIF. Results suggest that nonadiabatic couplings during the break-up of HCl (DCl) in the excited states play an important role in determining the fine-structure branching ratios.

D. H₂

The photochemical desorption of molecular hydrogen was investigated by F_2 laser irradiation of a Si(111)-(1*1):H surface [34]. The photon energy of 7.9 eV used was in the region of the broad sigma–sigma* optical transition of SiH centered around 8.5 eV. Molecular dynamics calculations, based on a Tersoff-type interaction potential between silicon and hydrogen, describe the reaction of a hydrogen atom created in a direct bond-breaking process with a neighboring hydrogen atom to form molecular hydrogen. This secondary surface reaction preserves the nonthermal character of the desorption process.

E. OCS

The production of significant concentrations of electrons in high-intensity pulsed-laser photolysis of carbonyl sulfide (OCS) to $S(^1S)$ at 157 nm has been observed [35]. These results imply a photoionization cross section for $S(^1S)$ of 2.6×10^{-19}

cm^2. The authors have measured the quenching of $S(^1S)$ for a variety of conditions, including incident laser fluences of 100 mJ/cm^2 and initial $S(^1S)$ densities of 3×10^{16} cm^{-3}. Quenching by electrons can be minimized by adding diluents such as SF_6 or CF_4. They interpret their observed $S(^1S)$ decay rates in terms of quenching by $S(^3P)$ with a rate constant of 3×10^{-11} cm^{3}s^{-1} and by CS [and/or $O(^3P)$] with a rate constant near 2×10^{-10} cm^3 s^{-1}. The measured $S(^1S)$ lifetime decreases with increasing initial $S(^1S)$ density, reaching a value of about 1 μs at a $S(^1S)$ density of 3×10^{16} cm^{-3}. The quantum yield for $S(^1S)$ production was found to be high (near 0.8) and independent of incident laser flux or added diluent pressure. These storage times and production efficiencies are consistent with possible use of the $S(^1S)$ to $S(^1D)$ transition as an efficient high-energy-storage laser.

A F_2 laser at 157 nm was also used in a time of flight experiment to measure the translational energy distribution of the fragments of OCS from which the vibrational energy distribution was inferred [36]. It is highly inverted in spite of the fact that the CO bond distances are similar in CO and OCS. Vibrational population inversion, the authors believe, is related to the vibrational structure seen in absorption. A symmetrical type of vibration is excited during the dissociation,which tends to pull both the O and S atoms away from the C atoms. The quasivibrational trajectory ends with the sulfur atom separating, but in the process the CO vibration is strongly excited.

F. H$_2$O

Polarized photofluorescence excitation spectroscopy of H_2O was described by Andresen et al. [37]. The H_2O^* state $(A^1B_1)^3$ dissociates directly and rapidly with 2.7 eV excess energy. Water molecules were photolyzed by an F_2 laser at 157 nm [38,39]. The nascent OH $^2\Pi_{3/2}$, rotational state distribution, probed via laser-induced fluorescence, reveals a strong preference for populating the upper lambda-doublet component. The population inversion is found to be a function of both the initial temperature of the H_2O and the final rotational state in which the OH radical is formed. These results may provide a simple mechanism for the astronomical OH maser observed by others.

H_2O was excited to the lowest electronic excited state, $^1B_1$, at 157.6 nm. The OH product state distribution was completely analyzed by special LIF experiments probing the fragment distribution [40]. The OH rotational excitation is relatively low and can be described by a temperature parameter of approximately 500 K independently of the OH vibrational excitation. This is in accordance with former measurements as well as with theoretical calculations. No selective population of the electronic fine-structure levels was observed, which is in agreement with the expectations for a room-temperature experiment. The observed magnitude of the vibrational excitation [$P(\upsilon'' = 0):P(\upsilon'' = 1):P(\upsilon'' = 2):P(\upsilon'' = 3):P(\upsilon'' = 4) = 59.2:33.1:6.1:1.4:0.2$] is smaller than the one calculated. Calculations predicted the $\upsilon'' = 6$ level to be populated, whereas in the experiment, transitions probing the $\upsilon'' = 5$ state were not observed.

In similar experiments vibrationally mediated photodissociation, in excitation of an overtone of an O–H stretching vibration, has been used. The vibrationally excited molecule is dissociated with a second photon and initial vibrational excitation alters the decomposition dynamics. For example, the amount of vibrationally excited OH produced in vibrationally mediated photodissociation of H_2O depends strongly on the eigenstate prepared in the vibrational overtone excitation step. This is in agreement with theoretical calculations [41]. The authors describe experiments that exploit this sensitivity to control breaking of the O-H bond in HOD.

Laser-induced fluorescence of microwave-stimulated OH molecules from H_2O photodissociation was investigated as a first step in a series of laboratory experiments to understand features of astronomical OH masers [42]. The inversion between the $\Pi_{3/2} J = 7/2$ lambda-doublet states of OH generated via photodissociation of cold H_2O at 157 nm is shown to be 1.8:1. Within a microwave, Fabry–Perot cavity tuned to the resonance of one of the main hyperfine transitions in this lambda doublet, it is possible to stimulate all inverted OH molecules. The linewidth of the two main microwave transitions is measured as a function of microwave power and interaction time.

G. ICN

Dispersed CN B $^2\Sigma^+$ – $X^2\Sigma^+$ photofragment fluorescence polarization anisotropies measured [43] following iodo cyanide (ICN) dissociation at 157.6 nm vary widely and apparently erratically with emission wavelength. They cannot be converted directly to CN B $^2\Sigma^+$ rotational alignments because of spectral congestion. A novel linear regression technique is used to extract CN B $^2\Sigma^+$ populations and rotational alignments from fluorescence emission and polarization anisotropy measurements. The authors present a flexible procedure that allows one to consider many models for the population and alignment distributions. Criteria are established to identify the best models. The CN B $^2\Sigma^+$ vibrational branching ratios for $\upsilon' = 0:1:2:3:4$ are determined by linear regression to be $0.46:0.25:0.13:0.09:0.07$, with a distinct rotational population dependence within each vibrational level. Extracted CN B^2 Σ^+ alignments for $\upsilon' = 0,1,2$, and 3 are presented, and these range from -0.31 to nearly 0.2. The alignments vary smoothly with nuclear rotation N' for each υ', demonstrating that the scatter in the measured polarization anisotropies results from vibrational band overlap at different wavelengths. These results show the largest photofragment alignment variation with vibration and rotation that has been measured following a single-photon dissociation process. A model is presented to estimate partial channel CN B($\upsilon' = 0$) product populations, and a discontinuity in the experimental $\upsilon' = 0$ alignment is considered.

H. HDO

Experimental values for the absorption coefficient and the branching ratio of HDO at 157 nm were compared with theoretical values [44].

I. AsF$_3$

Visible fluorescence from one-photon excitation of AsF$_3$ at 157 nm was dispersed and the fluorescing species were attributed to the excited photofragment of AsF$_2$ [45]. The radiative lifetime of AsF$_2^*$ was measured to be 25.5 $\pm$ 1.8 μs. The quenching rate constant of AsF$_2^*$ by AsF$_3$ was $(1.51 \pm 0.05) \times 10^{-10}$ cm^3 s^{-1}. Ultraviolet AsF (A-X) emission is observed from two-photon excitation of AsF$_3$.

J. O$_3$

Ozone was generated in pure oxygen (p $\sim$ 5 kPa), synthetic air (p $\sim$ 7 kPa), and oxygen–argon mixtures (p $\sim$ 3 kPa) by irradiation of these gases with the VUV light of a repetitively pulsed (15 Hz) F$_2$ laser at 157.6 nm with maximum energy of about 4 mJ/pulse [46]. An absorption photometer measurement operating at 253.7 nm (Hg line) determines the ozone concentration as a function of oxygen and/or additive gas pressure, the repetition frequency of the laser, and the wall temperature of the reaction chamber. The temporal development of ozone concentration as a function of these parameters is calculated by means of rate equations for the species O(^{3}P), O$_2$(X $^3\Sigma_g^-$), O$_3$ (^{1}A$_1$), O(^{1}D), O$_2$(a$^1\Delta_g$), O$_2$(b^1 Σ_g^+), vibrationally excited O$_3^*$ (^{1}A$_1$), and the photon distribution. The maximum concentration of O$_3$ in the sealed-off chamber reaches 1.6% in pure O$_2$, 4.1% in air, and 1.2% in a 1:5 O$_2$/-Ar mixture at 3 kPa. The annihilation of O$_3$ by the wall and temperature-dependent volume processes (300 K $<$ T $<$ 395 K) was studied and experimental and theoretical results were compared.

K. CO$_2$

The branching ratio was measured for the production of O(^{3}P) in the photodissociation of CO$_2$ at 157 nm [47]. A gas mixture consisting of CO$_2$, H$_2$, and Ar was irradiated with an F$_2$ excimer laser, while the relative concentration of O(^{3}P) was monitored continuously using atomic resonance fluorescence. The O(^{1}D) product was removed by either reacting with H$_2$ or being quenched by CO$_2$. At high H$_2$ CO$_2$ ratio, a residual O(^{3}P) signal persisted due to the nascent photofragments of CO$_2$. Stern–Volmer analysis indicated that the fraction of O(^{3}P) produced is 5.9%. Control experiments using O$_2$ and N$_2$ and O as precursor molecules confirmed this interpretation of the data. A mechanism is proposed based on curve crossing from ^{2}B$_2$ to ^{3}B$_2$ potential energy surfaces of CO$_2$. Since the ^{1}B$_2$ state is bent, a substantial fraction of absorbed energy is initially in bending motion, resulting in a long-lived chaotic trajectory that has many opportunities to cross over to the triplet surface.

$O(2p \ ^3P_j)$ (j = 2, 1, and 0) fragments produced in the 157 nm photodissociation of CO_2 were detected by resonance-enhanced multiphoton ionization in a molecular beam [48]. Doppler profiles and fine-structure branching ratios were measured for the oxygen-atom photofragment in the ^3P_j states. Doppler profiles were analyzed to give an anisotropy parameter of β = 2.0 and an internal energy equivalent to 3.9 vibrational quanta of CO. The fine-structure populations were found to be 0.70, 0.16, and 0.14 for j = 2, 1, and 0, respectively. A mechanism is proposed in which complex on the $^1B_2$ surface undergoes intersystem crossing to the $^3B_2$ surface. A phase-space model with a constraint on the impact parameter is shown to be consistent with the observed energy release. The nonstatistical fine-structure population could be caused by long-range interactions on the triplet surface. In a bulb experiment, $O(^3P)$ was produced by quenching of $O(^1D)$. The fine-structure populations of the resulting $O(^3P_j)$ were 0.64, 0.25, and 0.11. This state distribution is consistent with a long-lived complex that decays to give statistical products.

Vibrational and rotational distributions of $CO(^1\Sigma_{g+})$ produced in the 157 nm photodissociation of CO_2 have been determined by measuring vacuum-ultraviolet laser-induced fluorescence spectra of the CO photoproduct [49]. The photodissociation of CO_2 is known to occur via two pathways; one yielding $O(^1D)$ and the other yielding $O(^3P)$. Spin conservation and previous experimental studies confirm that dissociation via the $O(^1D)$ channel is the dominant process. The available energy for this channel is sufficient to populate only the ground and first excited vibrational levels of CO. The authors measured the rotational distributions for CO in υ = 0 and υ = 1 and found them to be non-Boltzmann. In fact, a highly structured distribution with distinct peaks at J = 10, 24, 32, and 39 is observed for CO in υ = 0. A less structured population is displayed by molecules in υ = 1. The relative vibrational population (υ = 0/υ = 1) was determined to be 3.7. Doppler spectra of individual rovibronic transitions were also recorded. The profiles have widths in accord with the available translational energy, display the expected v perpendicular to J correlation, and are best described by an isotropic distribution of the velocity vectors with respect to the polarization direction of the dissociation light.

Photodissociation of CO_2 at 157 nm was studied by the photofragment-translational spectroscopy technique. Product time-of-flight spectra were recorded and center-of-mass translational energy distributions were determined [50]. Two electronic channels were observed; one forming $O(^1D)$ and the other $O(^3P)$. With previously determined anisotropy parameters of β = 2 for the $O(^3P)$ channel and β = 0 for the $O(^1D)$ channel, an electronic branching ratio of 6% $O(^3P)$ was obtained, consistent with previous results. The translational energy distribution for the $CO(\upsilon) + O(^3P)$ channel was very broad (over 30 kcal/mol) and appeared to peak near $CO(\upsilon = 0)$. The value of $\beta = 2$ for the $O(^3P)$ channel was confirmed by comparing Doppler profiles, derived from the authors' measured translational energy distribution, with previously measured Doppler profiles. This suggests that the $O(^3P)$ channel arises from a direct transition to an excited triplet state. The $O(^1D)$ channel had a structured time-of-flight that related to rovibrational distributions of the CO product. The influence of the excitation of the CO_2 (V_2) bending mode was investigated and shown to have a small but not negligible contribution. Based on a comparison of the authors' data with a previous VUV laser-induced fluorescence study, they obtain as their best estimate of the vibrational branching ratio $CO(\upsilon = 0)/CO(\upsilon = 1) = 1.9$, for the $CO(\upsilon) + O(^1D)$ channel.

L. SiH

SiH bonds on a hydrogenated Si(111) surface were directly broken by electronic excitation with the 7.9 eV photons of a F_2 laser [51]. Independence of the kinetic energy of the desorbing hydrogen and the linear dependence of the desorption yield on fluence prove the photochemical character of the photoprocess. Solution of the heat diffusion equation indicates a negligible surface heating for the fluences applied.

IV. VUV SPECTROSCOPY OF TRIATOMIC CLUSTERS

Investigation of the properties of small charged or neutral clusters is a subject of growing interest because they bridge the gap between a solid and molecular state of matter [52–59]. Experimental techniques developed to study the clusters and their ions can be classified into two main categories: the optical and the mass spectroscopic. The use of laser spectral techniques such as laser-induced fluorescence or two-photon ionization has been proved unsuccessful for large clusters. This is because both techniques require that the laser-induced transition populate an excited electronic state that survives long enough to be detected by either its fluorescence or its lowered ionization threshold. The large metal clusters have hundred of electronic states within the first few electronvolts. Those states, even if they interact weakly with each other, produce a very dense magnifold of vibronic levels. As a result, very complicated spectra arise or radiationless transitions between those levels that quickly lead to a degradation of electronic excitation. On the other hand the detection of cluster ion using mass spectroscopy is free of such problems for both large and small clusters. Mass spectroscopic studies provide basic information about the dissociation energy and kinetics of the clusters. This can be expressed by the appearance of strong cluster peaks in the mass spectrum denoting the stability of the particularly cluster structures.

It has been observed experimentally that stable clusters can exist in various sizes containing a definite number of atoms for a specific charge. This number is called "magic number" [59–62]. A key problem is to find a sufficiently sensitive detection scheme to detect them. Several models have been proposed to predict or explain the experimentally observed "magic number" [63–68].

Investigation of the electronic structure and dynamics of clusters using vacuum ultraviolet spectroscopy is very important area in the study of clusters. VUV experimental techniques have two advantages: systems with a large band gap can be studied and excitation of inner shell electrons allow us to obtain element-specific information.

The application of a tunable, intense, and coherent VUV light source to molecular spectroscopy and photodissociation chemistry is proposed by K. Yamanouchi [69] to study the high-lying electronically excited states of small jet-cooled molecules and van der Waals clusters. Combination of the VUV laser, generated by two-photon resonance, four-wave sum, or difference frequency mixing scheme, and the free jet expansion techniques is promising to derive precise and substantial information about the level structure and the dynamic processes characteristic of highly excited molecular species excited in the VUV wavelength region.

Moller et al. [70] provide an overview of recent developments in the field of cluster research using VUV radiation. Electronic excitation and relaxation dynamics in helium clusters, and the inner shell photoionization of NaCl clusters produced with a new pick-up cluster source, are also discussed. The near edge absorption structure at the Cl 2p edge contains information on the geometric structure of clusters.

Clusters can be formed in a molecular beam apparatus [71–75], in an oven using the gas aggregation technique [76–80], and by desorption from a surface by ion beam bombardment [81–83], but they can also be the result of photodissociation of a parent molecule [84–87].

Ionization of clusters can be achieved by electron impact [88–90], by one-photon ionization with VUV [91–93] light, by synchrotron radiation or multiphoton ionization with UV or visible laser light [94–96], by liquid metal ion source LMIS [97–99], and by laser-induced field vaporization (LIFV) [68,100–101]. An interesting experimental setup for photoluminescence spectroscopy on van der Waals clusters has been described by R. Karnbach et al. [102]. It consists of a molecular beam apparatus with a cluster beam installed behind a high-intensity VUV synchrotron radiation beamline. Special emphasis was given to the design of a very intense cluster source that can also be used for preparation of quantum clusters (He, H_2). To determine the cluster size, a time-of-flight mass spectrometer can be attached to the setup. In addition, an atomic cross jet is installed in the experimental chamber that can be used for mass separation or for doping of the clusters. Luminescence light can be recorded with several different detectors or spectrally analyzed with a secondary monochromator equipped with a position-sensitive detector. The pulsed nature of synchrotron radiation provides the basis for time-resolved measurements in the regime 100 ps–3 μs.

In the work of Wucher et al., neutral atoms and clusters desorbed from a solid germanium surface by 5 keV Ar^+ ion bombardment [74,103]. Charged clusters were detected directly by time-of-flight mass spectrometry (TOF-MS). The corresponding neutral species were postionized prior to mass and energy analysis. These authors employed two different photoionization schemes. The first was a single-photon ionization process using an F_2 molecular laser as an intense VUV source with photon energy in excess of all relevant ionization potentials. The second was a nonresonant multiphoton ionization process using a high-intensity laser delivering pulses of 250 femtosecond duration at a wavelength of 267 nm. The two processes were compared, and in both cases the available laser pulse energy was sufficient to saturate the ionization of Ge atoms and all detected Ge_2, Ge_3, and up to Ge_7 clusters. This is an indication that the results obtained with both photoionization techniques closely reflect the true cluster sputtering yields and, in particular, are not dominated by photon-induced fragmentation.

Relative yields of sputtered Ge_n clusters are found to obey a power law dependent on the cluster size n with an exponent around −6.5 [74]. Kinetic energy distributions exhibit a shift of the maximum toward lower energies with increasing cluster size. Asymptotic decay towards high energies is found to be virtually identical for all measured atoms and clusters. The experiments were repeated on two different single crystalline Ge surfaces <100>and<111> that could be reproducibly amorphized by ion bombardment and reannealed. As a result, the crystalline structure of the ion-bombarded surface was not found to play a significant role in the formation of sputtered clusters.

Neutral clusters of Ag, Al, Nb, and Ta were also formed using the technique of sputtering of polycrystalline surfaces. They were studied experimentally by nonresonant single-photon postionization using a UV or VUV laser beam and time-of-flight mass spectrometry [104]. The mass spectra were recorded at different laser intensities. As a result, photoionization cross sections and the yields of sputtered clusters were determined as a function of the cluster size. The cluster yields roughly exhibit a power law dependence on the cluster size, the exponent of which is found to be inversely correlated with the sputtering yield of the sample. This finding is of particular importance, since it rules out simple statistical combination models to describe the formation of large sputtered clusters. From the yield distributions it is inferred that, depending on the sputtering conditions, up to 46% of the sputtered atoms may be emitted in a bound state. Experimental results were compared with theoretical model descriptions of the cluster formation process.

Neutral silver atoms and small clusters Ag_n ($n = 1 \ldots 4$) were generated by bombarding a polycrystalline silver surface with Ar^+ ions of 5 keV [105,106]. The sputtered particles were ionized by a crossed electron beam or by the F_2 single photon at 7.9 eV from a pulsed VUV laser that permits nonresonant single-photon ionization (SPI) of all investigated species and is subsequently detected by a quadrupole mass spectrometer. Photoionization cross sections were evaluated from the laser intensity dependence of the measured mass spectrum. SPI ionization cross sections do not vary dramatically between silver atoms and different clusters. As a consequence, fragmentation influences encountered in previous studies with longer-wavelength lasers are practically eliminated from the determination of yields and kinetic energy distributions of the sputtered clusters. The resulting relative cluster sputtering yields (normalized to the yield of silver atoms) exhibit a power law dependence on the cluster size n according to n-delta with exponents of delta ranging from 4.3 to 7.4 depending on the nature and the bombarding energy of the primary ions. Kinetic energy distributions of the sputtered neutral atoms and clusters are evaluated up to clusters containing seven atoms. Asymptotic decay of the energy distribution towards high-emission energies becomes steeper from Ag to Ag_3 and remains practically constant for larger clusters. By in situ combination of both ionization mechanisms, absolute values of the ratio $\sigma(e)$ $(Ag_{(n)})/\sigma(e)(Ag)$ between the electron impact ionization cross sections of silver clusters and atoms could be determined for a fixed electron energy of 46 eV. These values can then be used to calibrate previously measured relative ionization functions. By calibrating the results using literature data measured for silver atoms, we present absolute cross sections for electron impact ionization of neutral Ag_2, Ag_3, and Ag_4 as a function of the electron energy between threshold and 125 eV.

V. LASER-INDUCED FLUORESCENCE AND MASS SPECTROSCOPY OF MERCURY DIATOMIC AND TRIATOMIC CLUSTERS ASSISTED BY LASER LIGHT AT 157.6 nm

Laser-induced fluorescence of the mercury clusters Hg_2 and Hg_3 in the spectra range from 300 to 510 nm has been obtained from the dissociation of $HgBr_2$ at 157.6 nm with an F_2 molecular laser. The excitation process involves two-photon absorption, which dissociates the molecule at 15.76 eV total photon energy with the subsequent formation of metallic clusters [85].

Photodissociation of $HgBr_2$ at 157.5 nm using the F_2 molecular laser was proven to be an efficient method for producing small mercury clusters Hg_2 and

Hg_3 in their excited states [107]. Although photodissociation of mercury halides by the ArF excimer laser at 193 nm (and at other excimer and dye lasers wavelengths) has been extensively studied [108–113], there are no reports of metallic cluster formation at this wavelength. In addition, a detailed study of the spectroscopy of the triatomic mercury clusters in electric discharges has been investigated well before VUV optical excitation [114,115].

The electronic structure of $HgBr_2$ and the HgBr molecules indicates that three-dipole allowed transitions can be excited between the low-lying electronic states of the $HgBr_2$ molecule, $1\ {}^1\Pi u$, $1\ {}^1\Sigma_u^+$, $2\ {}^1\Sigma_u^+$, and its ground state $1\ {}^1\Sigma_g^+$ [116–119]. VUV excitation at 157.5 nm stimulates the $1\ {}^1\Sigma_g^+ \rightarrow 2\ {}^1\Sigma_u^+$ transition, (Fig. 4).

$$HgBr_2(1\ {}^1\Sigma_g^+) + h\nu(7.88\ eV) \rightarrow HgBr_2(2\ {}^1\Sigma_u^+). \tag{1}$$

The $2\ {}^1\Sigma_u^+$ state of $HgBr_2$ at 7.88 eV correlates with the the $D({}^2\Pi_{3/2})$ and $C({}^2\Pi_{1/2})$ states of HgBr and the ${}^2P$ state of Br.

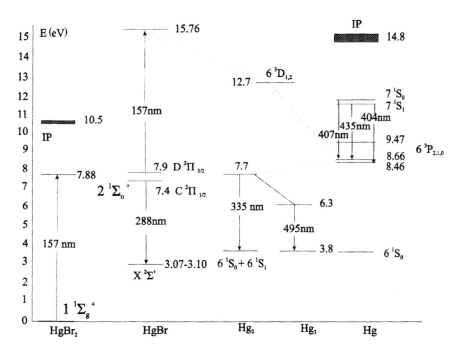

Figure 4 Energy pathway of formation of mercury atoms from the dissociation of $HgBr_2$ at 157.6 nm.

$$HgBr_2(2\ {}^1\Sigma_u^+) \rightarrow HgBr_2({}^2\Pi_{1/2}) + Br({}^2P)$$
$$HgBr_2(2\ {}^1\Sigma_u^+) \rightarrow HgBr_2({}^2\Pi_{3/2}) + Br({}^2P)$$
(2)

The $2\ {}^1\Sigma_u^+$ electronic state is strongly correlated with $D({}^2\Pi_{3/2})$ and $C({}^2\Pi_{1/2})$ electronic excited states of the HgBr diatomic following dissociation to HgBr and Br.

The small energy gap between the D state and the available energy from one photon at 157.5 nm can be bridged thermally at high temperatures. Therefore a fraction of the population is transferred to the D state [120] and the remaining part to the C state of HgBr. From the C and D states the molecules can either radiate to the X state or can absorb a second 157.5 nm photon and photoionize the HgBr molecule. The C $\rightarrow$ X transition at 285 nm is the strongest transition in the spectrum, while the D $\rightarrow$ X transition at 265 nm is ~10 times weaker. The overall absorption cross section at 157.5 nm has been found to be $(4.5 \pm 2.6) \times 10^{-16}$ cm^2 [107] in the temperature range between 40 and 90°C. This is considerably higher than the value of the absorption cross section at 193 nm, which has been found to be 3.8×10^{-17} cm^2 [121]. However, the theoretical calculations of Wadt [116] predict that the absorption cross section between 1 ${}^1\Sigma_g^+$ and 2 ${}^1\Sigma_u^+$ states at 157 nm should be smaller than the absorption cross section between 1 ${}^1\Sigma_g^+$ and 1 ${}^1\Sigma_u^+$ states at 193 nm by a factor of 35. The above discrepancy between the experimental and theoretical values indicates that other loss mechanisms than the absorption between the 1 ${}^1\Sigma_g^+$ and 2 ${}^1\Sigma_u^+$ states should play a major part in the annihilation of VUV photons by the HgBr$_2$ molecules. These absorption mechanisms include one-step ionization of HgBr from the C and D states and two-step ionization of HgBr$_2$ from its ground state. The C and D states of HgBr at 7.88 eV correlate with the $6^3P_{0,1}$ metastable states of Hg either via one-photon absorption and subsequent deexcitation to the $6^3P_{0,1}$ states:

$$HgBr({}^2\Pi_{1/2}, {}^2\Pi_{3/2}) + h\nu(7.88\ eV) \rightarrow HgBr^+(15.76\ eV)$$
$$HgBr^+(15.76\ eV) \rightarrow Hg(6D, 7S) + Br({}^2P)$$
$$Hg(6D, 7S) \rightarrow Hg(6P) + h\nu$$
(3)

or through thermal dissociation according to

$$HgBr({}^2\Pi_{1/2}, 6^3P_{0,1}) + kT \rightarrow Hg(6^3P_0, 6^3P_1) + Br({}^2P)$$
(4)

The energy gap between the ${}^2\Pi_{3/2}$ state of HgBr and the Hg(6^3P_0) state of mercury is between 0.58 and 0.53 eV [122,123]. Therefore, at 300°C less than ~0.1% of the population of the diatomic is thermally dissociating to the Hg(6^3P_0) + Br(^2P) atoms. The observed emissions at 404.6, 435.8, 407.8, 312, 313, 296, 365, 366, and 254 nm are from the allowed dipole transitions from the 7S, 6D, and 6P states of mercury

$$Hg(7\,^1S_0) \rightarrow Hg(6\,^3P_1) + h\nu(407.8 \text{ nm})$$
$$Hg(7\,^3S_0) \rightarrow Hg(6\,^3P_1) + h\nu(4358 \text{ nm})$$
$$Hg(7\,^1S_0) \rightarrow Hg(6\,^3P_0) + h\nu(404.6 \text{ nm})$$
$$Hg(6\,^1D_2) \rightarrow Hg(6\,^3P_2) + h\nu(365.5 \text{ nm}) \tag{5}$$
$$Hg(6\,^3D_2) \rightarrow Hg(6\,^3P_1) + h\nu(312.6 \text{ nm})$$
$$Hg(6\,^3D_1) \rightarrow Hg(6\,^3P_1) + h\nu(313.1 \text{ nm})$$
$$Hg(6\,^3P_1) \rightarrow Hg(6\,^1S_0) + h\nu(253.6 \text{ nm})$$
$$Hg(6\,^3D_1) \rightarrow Hg(6\,^3P_0) + h\nu(296.0 \text{ nm})$$

The kinetic processes associated with the formation of $Hg_2(AO_g^{\pm})$ from $6\,^3P_{0,\,1}$ and $6\,^1S_0$ states have been studied in detail [124]. Spectroscopy of the Hg_2 molecule has been given by the work of Niefer et al. [125], Callear et al. [126], and Drullinger et al. [127]. A collision between a metastable mercury atom $6\,^3P_0$ and a ground state atom $6\,^1S_0$ can create the mercury dimer Hg_2 in its excited states $AO_g^{\pm}$. Collision between a $6\,^3P_0$ mercury atom and a $6\,^1S_0$ ground state atom can form a mercury dimer either in the AO_g^+ state or in the DI_u state, depending on the thermal energy available. The $AO_g^{\pm}$ states 2800 cm^{-1} below the optically active DI_u state either can act as an energy reservoir for the formation of the Hg_3 cluster or can thermally populate the DI_u state of Hg_2, which gives the 335 nm band through de-excitation to the dissociative ground state.

$$Hg(6\,^3P_0) + Hg(6\,^1S_0) \rightarrow Hg_2(AO_g^-)$$
$$Hg(6\,^3P_0) + Hg(6\,^1S_0) \rightarrow Hg_2(AO_g^+)$$
$$\rightarrow Hg_2(DI_u)$$
$$Hg(AO_g^{\pm}) + kT(0.34 \text{ eV}) \rightarrow Hg_2(DI_u)$$
$$Hg(DI_u) \rightarrow 2Hg(6\,^1S_0) + h\nu(335 \text{ nm})$$

The spectral lines detected in the region between 410 and 426 nm originate from the de-excitation of the $Hg_2(PI_u)$ Rydberg state to the AO_g^+ state.

The $Hg_2(PI_u)$ state is created through a collision between an $Hg(7\,^3S_1)$ atom with a ground state atom $Hg(6\,^1S_0)$

$$Hg(7\,^3S_1) + Hg(6\,^1S_0) \rightarrow Hg_2(PI_u)$$
$$Hg_2(PI_u) \rightarrow Hg_2(AO_g^+) + h\nu(410\text{--}426 \text{ nm})$$

This de-excitation channel is another pathway of populating the reservoir of the $AO_g^{\pm}$ states.

The 335 nm band is much narrower than the one previously reported from electric discharges [128]. Transition from the $6\,^3D_{1,2}$ manifold to the $6\,^3P_1$ state is located at 313 nm. Transitions from the same $6\,^3D_{1,2,3}$ manifold to the $6\,^3P_2$ state were recorded as well. Two new progressions of lines at 349 nm, 353 nm, and 357 nm, and at 371 nm, 375 nm, and 380 nm had been observed. The first manifold has been observed previously in HgBr discharges and was attributed to the presence of N_2. However, the spectra have been recorded alongside of and sideways to the laser beam and the presence of N_2 should be excluded from our laser system.

The intensity of these spectral lines varies with temperature (they are very intense above 250°C). Their temporal behavior suggests that they are transitions from excited states of the Hg_2 cluster formed by the highly excited states to Hg without attempting any assignments. The dependence of the intensity of mercury lines at 404.6 nm and 366 nm on temperature, together with rate of formation arguments, suggest that these lines might originate from emissions from the highly excited states of Hg_2: for example, $(Hg(7^1D_2) + Hg(6^1S_0))$ to lower ones. As far as the progression of lines around 380 nm is concerned, there is a dependence between the intensity of these lines and the intensity of mercury lines at 365 nm that originate from the 6^3D manifold. The remaining part of the spectrum between 385 and 400 nm consists of a series of lines. Temporal evolution of these spectral lines suggests that they originate from transitions from the excited states of the mercury triatomic Hg_3. Similar long-lived states have been observed previously around 446 nm and have been attributed to the complex potential surfaces of Hg_3. The lifetime of excited states of the Hg_2 cluster is believed to be short [129], and the long decay rates observed are attributed instead to clusters' rates of formation in their excited states. This argument is also supported by the temporal evolution of the 335 nm band. Therefore, considering that the rates of formation are small, the long lifetimes could be attributed to Hg_3 clusters because of the complicated structure of their potential surfaces. Strong perturbations present in the Hg_3 molecule, alongside tunneling effects, could be the source of long decay rates. The broad continuum in the blue-green region of the spectrum taken in a mercury vapour column with maximum at 495 nm has been assigned to emission from the Hg_3 molecules [127]. The same band has been observed previously by Cefalas et al. [107] by laser-induced fluorescence. Niefer et al. [125], using selective excitation in mercury vapor, had assigned the observed absorption between 415 and 510 nm to the presence of the mercury cluster Hg_3. The sideways fluorescence of the 495 nm band between 480 nm and 510 nm indicates some vibrational structure with a maximum at 494 nm. The formation of Hg_3 originates from the collision of an Hg_2 molecule in its first excited state with an Hg atom in its ground state, since the band is observed even without the use of a buffer gas.

$$Hg_2(AO_g^\pm) + Hg(6^1S_0) \rightarrow Hg_3^*(AO_g^\pm)$$
$$\rightarrow Hg_3(XO_g^\pm) \rightarrow 3Hg + h\nu(495 \text{ nm band}) \quad (6)$$

The emitted radiation at 495 nm is highly directional and concentrated. Its superfluorescence nature [107] was manifested by the appearance of blue–green spots on a screen even 50 cm away from the fluorescence cell. Similar kinds of blue–green spots have been observed by Shingenari et al. [130] at 404 and 407 nm during the photodissociation of $HgBr_2$ at 193 nm. The appearance of these spots suggests that at 495 nm the gain exceeds the absorption for this excitation scheme. The blue–green spots do not appear on a regular basis, indicating that thermal gradients inside the heat pipe oven might cause strong refractive index variations. For excitation schemes that involve electrical discharges or selective excitation of mercury atomic levels at high temperatures, extension of the band towards the red is due to the formation of mercury clusters $Hg_n (n > 3)$.

VI. BISTATE QUANTUM SWITCHING IN MERCURY TRIATOMIC CLUSTERS

Quantum beats with a period from 25 to 100 ns have been observed in mercury triatomic clusters following de-excitation from excited electronic states of volcanic shape to the ground electronic state (Fig. 5). Observation of the quantum beats can be explained if we assume that the dynamic surface of the Hg_3 cluster has a symmetry of a double well. This is true for a large number of X_3 systems, including Cu_3 [131], Na_3 [132], P_3, As_3, Sb_3 Bi_3 [133] Ca_3, and Sc_3 [134]. Formation of the Hg_3 clusters was also observed by optical spectroscopy by J. Koperski et al. [135].

Because the vibrational and the electronic parts of the total wavefunctions are not any longer separable for the Hg_3 molecular system, the excited electronic states have a complex volcanic structure. The presence of potential energy barriers is the source of tunneling effects of the optical electron. In the simplest case, potential energy barriers exhibit saddle points with double minima. A double barrier can exhibit resonances where total transmission of electrons is taking place even for energies below the height of the potential barrier. This accounts for zero dispersion of the particle wave packet. It is responsible for the observed long duration of the quantum beats because the molecule oscillates, trapped between a symmetrical and an antisymmetrical wavefunction without any reflection losses. The presence of the double minima causes splitting of vibrational eigenfunctions in a symmetrical and an antisymmetrical component. The reason is the introduction of an additional translational symmetry in the total Hamiltonian of the system [136]. The amount of the energy split between the two parts depends on the molecular parameters and energy separation of the optical electron from the top of the potential barrier (Fig. 6). Energy splitting is related to the beat period. Hence the period of the quantum beat depends on the electron energy relative to the top of the barrier. It increases with increased energy separation from the top of the double barrier because total transmission probability of the optical electron through the energy potential barrier depends on its energy separation from the top of the potential barrier as well. The potential barrier over which

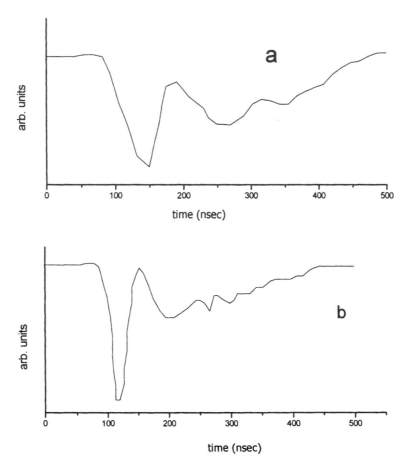

Figure 5   a. Mode competition at 445 nm. b. Mode competition at 446.9 nm.

quantum beats were observed was found to be 0.05 eV deep [137]. As a result, selective excitation in Hg_3 clusters could modulate the period of the quantum beat and the electron transmission probability through the potential barrier of the excited electronic states. This process simulates the switching function of a quantum transistor with its gate driven by light.

Double-charged Hg_3 clusters following the photodissociation of the $HgBr_2$ were observed using the mass spectroscopic technique. The parent $HgBr_2$ was expanded from a molecular beam apparatus, after being heated in the molecular beam chamber. Parent molecules were crossed with the F_2 molecular laser focused in the molecular beam with a MgF_2 lens with focal length of 20 mm (Fig. 7). The Hg_3 clusters were formed according to the mechanism described previ-

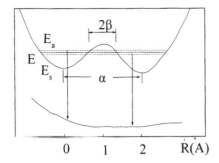

Figure 6 Potential energy surface for the 445 nm band. Symmetry splitting is indicated. Optical transitions are taking place between the excited and ground electronic states of the mercury trimer.

ously, in their excited electronic states and were observed with the quadruple mass spectrometer (Fig. 8). Double-charged triatomic mercury clusters were the smallest ever observed (Fig. 7). Berchignac at al. [77], using a different experimental technique, observed double-charged mercury clusters containing up to five atoms of mercury. The theoretical calculations performed by Tomanek [67] give for the number n in Hg_n^{2+} the value of 300; Jentsh et al. found $n = 8$ [68], and Sattler et al. found $n = 9$ [80].

To interpret the observed quantum beats and make the proper assignments,

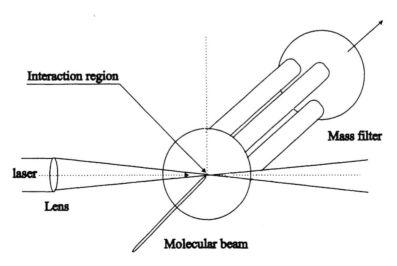

Figure 7 Geometry of detection of trimer mercury clusters. Direction of the laser light is perpendicular to the direction of the molecular beam and quadrupole mass filter.

a simple formula was derived that relates the experimentally measured splitting of the vibrational levels of the excited triatomic molecule to the shape of the potential surfaces showing double minima. The presence of the double minima imposes translational symmetry in the potential energy surfaces. We will consider the simplest case: the potential energy surface is symmetrical with respect to an axis that passes from the point equidistant from the two minima (Fig. 6). For the sake of simplicity, we will consider only one dimension x. The symmetry of the potential surface implies the relation

$$U(x + a) = U(x) \tag{7}$$

where a is the distance between the two minima of the potential surface $U(x)$.

Therefore we can introduce the translation operator $\hat{T}(a)$, which translates the space coordinate of a point at x by a distance a, and for a wave function $\Psi(x)$ we have

$$\hat{T}(a)\psi(x) = \psi(x + a) \tag{8}$$

Expanding $\psi(x + a)$ in a Taylor series around the point x we have

$$\psi(x + a) = \sum_0^\infty \frac{1}{m!} \frac{d^m\psi(x)}{dx^m} a^m = \left(\sum_0^\infty \frac{1}{m!} a^m \nabla^m \right) \psi(x) = \hat{T}(a)\psi(x) \tag{9}$$

and the translation operator $\hat{T}(a)$ is of the form

$$\hat{T}(a) = \exp(a\nabla) \tag{10}$$

The action of the translational operator on the potential function is of the form

$$\hat{T}(a)U(x) = U(x + a) = U(x) \tag{11}$$

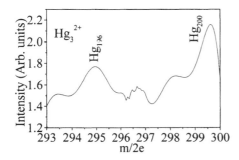

Figure 8   Mass spectrum of mercury doubly ionized trimer clusters.

Using the relationship above it is easy to see that the translation operator commutes with the potential energy of the system and the kinetic part of the Hamiltonian $\hat{H}$, where

$$\hat{H} = -\frac{\hbar^2}{2m}\frac{d^2}{dx^2} + U(x) \tag{12}$$

Therefore the translational operator commutes with the Hamiltonian of the system and the operators $\hat{H}$ and $\hat{T}(a)$ have the same set of wave functions, $\psi(x)$.

Suppose now that the eigenfunction $\psi(x)$ of the translation operator $\hat{T}(a)$ corresponds to the eigenvalue $T(a)$

$$\hat{T}(a)\psi(x) = T(a)\psi(x) \tag{13}$$

We can prove easily that

$$\hat{T}^m(a) = T(ma) \tag{14}$$

and if we write $T(a)$ in the form

$$T(a) = \exp(i\phi(a)) \tag{15}$$

it can be proved that

$$\phi(a) = ka \tag{16}$$

$$\hat{T}(a)\psi(x) = \exp(ika)\psi(x) = \psi(x + a) \tag{17}$$
$$\psi(x + a) = \exp(ika)\psi(x)$$

Therefore the eigenfunctions $\psi(x)$ of the Hamiltonian and the translation operator obey the so-called translation property of Eq. (17) (Bloch equations). The zero-order approximation now of the wave function $\psi(x)$, $\psi_0(x)$ enables the calculation of the energy eigenvalues of Eq. (12) to the first order.

$$\hat{H} = \left[-\frac{\hbar^2}{2m}\frac{d^2}{dx^2} + U(x)\right]\psi_0(x) = E_1\psi_0(x) \tag{18}$$

where the subscripts 0 and 1 refer to the zero and first-order approximations for the wave function and the energy. From Eq. (18) we find

$$E_1 = \frac{\int \psi_0^*(x)[-(\hbar^2/2m)(d^2/dx^2) + U(x)]\psi_0(x)dx}{\int \psi_0^*(x)\psi_0(x)dx} \tag{19}$$

The potential function can be expanded now around x in a Taylor series

$$U(x) = \sum_a U_a(x)\exp(ika) \tag{20}$$

Therefore Eq. (19) becomes

$$E_1 = I \int \psi_0^*(x) \left[-\frac{\hbar^2}{2m}\frac{d^2}{dx^2} + \sum_a U_a(x)\exp(ika) \right] \psi_0(x)dx \qquad (21)$$

where

$$I = \{ \int \psi_0^*(x)\psi_0(x)dx \}^{-1}$$

Now we have

$$E_1 = I \left\{ \int \psi_0^*(x) \left[-\frac{\hbar^2}{2m}\frac{d^2}{dx^2} + \sum_{a' \neq a} U_{a'}(x)\exp(ika') + U_a(x)\exp(ika) \right] \psi_0(x)dx \right\} \qquad (22)$$

taking into account the translational properties of the potential function $U(x)$ and the wave function $\psi(x)$ Eq. (17), we have

$$E_1 = I \left\{ \int \psi_0^*(x) \left[-\frac{\hbar^2}{2m}\frac{d^2}{dx^2} + \sum_{a' \neq a} U_{a'}(x - a') \right] \psi_0(x)dx \right.$$
$$\left. + U_a(x - a) \int \psi_0^*(x)\psi_0(x)dx \right\}$$

$$E_1 = I \left\{ \int \psi_0^*(x) \left[-\frac{\hbar^2}{2m}\frac{d^2}{dx^2} + \sum_{a' \neq a} U_{a'}(x - a') \right] \psi_0(x - a')\exp(ika)dx \right.$$
$$\left. + U_a(x - a) \int \psi_0^*(x)\psi_0(x)dx \right\} \qquad (23)$$

$$= E_a + F(a, k)\exp(ika)$$

where we have denoted

$$E_a = U_a(x - a)$$

and

$$F(a, k) = I \left\{ \int \psi_0^*(x) \left[-\frac{\hbar^2}{2m}\frac{d^2}{dx^2} + \sum_{a' \neq a} U_{a'}(x - a' - a) \right] \psi_0(x - a)dx \right\} \qquad (24)$$

The last integral describes the exchange of energy between the two parts of the barrier: right of the symmetry axis where the wave function is $\psi_0(x)$ and left of the symmetry axis where the wave function is $\psi_0(x - a)$.

The energy E_1 therefore depends quasicontinuously on the wave vector k and its real part changes from $E_{1\,min}$ to $E_{1\,max}$ as k changes:

$$E_1 = E_a + F(a, k)\cos(ka)$$
$$E_{1\ min} = E_a + F(a, k) \tag{25}$$
$$E_{1\ max} = E_a + F(a, k)$$

Hence the amount of the energy splitting as the particle moves through the double barrier is $\Delta E = 2F(ak)$. The mass of the cluster m is related to the energy E_1 through the equation

$$m^{-1} = \left| \frac{1}{\hbar^2} \frac{d^2 E}{dk^2} \right| \tag{26}$$

and therefore

$$m^{-1} = \left| \frac{1}{\hbar^2} \frac{dF}{dk^2}\cos(ka) - 2\frac{dF}{\hbar^2 dk}a\ \sin(ka) - \frac{Fa^2}{\hbar^2}\cos(ka) \right| \tag{27}$$

For a heavy particle like the triatomic mercury trimer, we do not expect the exchange integral to change a great deal with the wave vector k. Therefore, Eq. (27) becomes

$$m = \left| \frac{2\hbar^2}{\Delta E a^2 \cos(ka)} \right| \tag{28}$$

Equation 28 is our final result and relates the energy-splitting ΔE of the vibrational levels to the mass (m) of the triatomic and the distance (a) between the two minima of the potential surface. The observed period of the quantum beats (Fig. 9) is related to the energy-splitting ΔE of the vibrational levels by

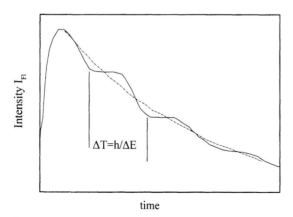

Figure 9 Time evolution of transitions between excited and ground electronic states of mercury trimer.

Table 1 Period T of Quantum Beats for the Three Emission
Bands of Mercury Triatomic Cluster

Band λ (nm)	Emission λ (nm)	Q B Period T (ns)
385	385, 387, 389	75, 50, 25
445	445, 447, 449, 451, 453	100, 75, 50, 25
485	482, 485, 488	300, 100, 75, 50, 25

The second column shows the wavelength of emission from different vibronic
levels.

the uncertainty relation $\Delta E \, \Delta T = h$. For a typical beat distance of 75 ns, $\Delta E = 5 \times 10^{-6}$ eV. From Eq. 28 we find that this value corresponds to a distance a between the minima of the potential surface of 0.2 nm for $k = 53 \times 10^{10}$ m^{-1}.

A formula that relates the molecular parameters to the amount of splitting ΔE is given by

$$\Delta E = \frac{\Delta E_o}{\pi} \exp\left(-\int_{-\beta}^{\beta} |k| \, dx \right) \tag{29}$$

where β is the distance between two turning points (Fig. 6), ΔE_o the energy difference between two neighboring vibronic levels near the top of the barrier, and k is the wave vector related to the reduced cluster mass and it is given by $k = (2m/\hbar^2 \, (E - U))^{1/2}$. $(E - U)$ is the distance of a vibronic level from the top of the barrier. Eq. (29) can be used to test the validity of our approximations, which have been derived from Eq. (28).

From spectroscopic data the energy difference between two neighboring vibrational levels is 0.04 eV. Hence, the first vibronic level below the top of the potential barrier will be situated below the top of the barrier at the most by the same amount of energy of 0.04 eV. Therefore, the wave vector k associated with the vibronic level near the top of the potential barrier is ~3.6 $\times$ 10^{11} m^{-1}. For a splitting ΔE of 5 $\times$ 10^{-6} eV, the distance 2β is estimated to be 0.02 nm and the distance α between the two minima is 0.2 nm. The period of the observed quantum beats for the three emission bands is indicated in Table 1. Fourier transform analysis of the signals indicates multiple frequencies over the same vibrational level. This is due to the fact that the potential energy surfaces are not symmetrical in space. Quantum beats are observed only for the vibrational levels indicated in Table 1 because symmetry splitting and tunneling is taking place only for those levels close to the top of the potential barrier, in agreement with the theoretical model. The transmission coefficient T of a particle for a double barrier described by the $U_0\cosh^{-2}(x/2\beta)$ dependence in a region near its top is given by the equation

$$T = \frac{\sinh^2(2\pi k\beta)}{\sinh^2(2\pi k\beta) + \cosh^2[1/2\pi\sqrt{32mU_0\eta^{-2}\beta^2 - 1}]} \tag{30}$$

From Eq. (30) if $1 - 32mU_0\eta^{-2}\beta^2 = (2n + 1)^2$, then the transmission coefficient is equal to 1. In this case the particle is transmitted through the barrier with zero reflection losses. For $n = 1$, $U_0 = 10^{-3}$ eV, in agreement with our experimental results regarding the high of the potential barrier over which beat oscillations were observed. The process simulates the switching properties of a molecular transistor with its gate driven by light.

REFERENCES

1. JR Woodworth, JK Rice. An efficient, high-power F_2 laser near 157 nm. J Chem Phys 69:2500–2504, 1978.
2. JK Rice, AK Hays, JR Woodworth VUV emissions from mixtures of F_2 and the noble gases-A molecular F_2 laser at 1575 A. Appl Physics Letts 31:31–33, 1977.
3. H Pummer, K Hohla, AM Diegelmann, JP Reilly. Discharge pumped F_2 laser at 1580 A. Optics Communics 28:104–106, 1979.
4. AC Cefalas, C Skordoulis, M Kompitsas, CA Nicolaides. Gain measurements at 157 nm in an F_2 pulsed discharge molecular laser. Optics Communics 55:423–426, 1985.
5. VN Ishchenko, SA Kochubei, AM Razhev. High-power efficient vacuum ultraviolet F_2 laser excited by an electric discharge. Sov J Quantum Electron. 16:707–709; translated from Kvantov Elektron Moskva 13:1072–1075, 1986.
6. M Ohwa, M Obara. Theoretical evaluation of high-efficiency operation of discharge-pumped vacuum-ultraviolet F_2 lasers. Appl Physics Letts. 51:958–960, 1987.
7. K Yamada, K Miyazaki, T Hasama, T Sato. High-power discharge-pumped F_2 molecular laser. Appl Physics Letts 54:597–599, 1989.
8. C Skordoulis, S Spyrou, AC Cefalas. Gain and saturation measurements in a discharge excited F_2 laser using an oscillator amplifier configuration. Appl Physics B 51:141–145, 1990.
9. C Skordoulis, E Sarantopoulou, S Spyrou, AC Cefalas. Amplification characteristics of a discharge excited F_2 laser. J Modern Optics 37:501–509, 1990.
10. AA Kuznetsov, SS Sulakshin. An F_2 laser-excited by a high-power proton-beam. Kvantov Elektron 18:22–22, 1991.
11. M Kakehata, T. Uematsu, F. Kannari, M. Obara. Efficiency characterization of vacuum ultraviolet molecular fluorine (F_2) laser (157 nm) excited by an intense electric-discharge. IEEE J Quantum Electron 27:2456–2464, 1991.

12. M Kakehata, CH Yang, Y Ueno, F Kannari. Gain and saturation intensity measurements of a discharge pumped F_2 laser at high-excitation rates. Appl Physics Letts 61:3089–3091, 1992.

13. FTJL Lankhorst, HMJ Bastiaens, PJM Peters, WJ Witteman. High specific laser output energy at 157-nm from an electron-beam-pumped He/Ne/F_2 gas-mixture. Appl Physics Letts 63:2869–2871, 1993.

14. FTJL Lankhorst, HMJ Bastiaens, PJM Peters, WJ Witteman. Long pulse electron beam pumped molecular F_2 laser. J Appl Physics 77:399–401, 1995.

15. HMJ Bastiaens, SJM Peeters, X Renard, PJM Peters, WJ Witteman. Long pulse operation of an x-ray preionized molecular fluorine laser excited by a pre pulse-main pulse system with a magnetic switch. Appl Physics Letts 72:2791–2793, 1998.

16. N Rizvi, D Milne, P Rumsby, M Gower. Laser micromachining—new developments and applications. Proc SPIE 3933:261–270, 2000.

17. JC White, HG Craighead, RE Howard, LD Jackel, OR Wood. VUV laser photolithography. AIP Conf Proc 119:324–329, 1984.

18. S Nagai, T Enami, T Nishisaka, J Fujimoto, O Wakabayashi, H Mizoguchi. Development of kHz F_2 laser for 157 nm lithography. In: Digest of Papers, Microprocesses and Nanotechnology '99. International Microprocesses and Nanotechnology Conference. Tokyo, Japan: Japan Society of Applied Physics, 234:64–65 1999.

19. T Hofmann, JM Hueber, P Das, S Scholler. Revisiting F_2 laser for DUV microlithography. Proc SPIE 3679:541–546, 1999.

20. AC Cefalas, E Sarantopoulou, E Gogolides, P Argitis. Absorbance and outgassing of photoresist polymeric materials for UV lithography below 193 nm including 157 nm lithography. Microelectron Engin 53:123–126, 2000.

21. AC Cefalas, N Vassilopoulos, Z Kollia, E Sarantopoulou, C Skordoulis. Mass spectroscopic studies and ablation characteristics of nylon 6.6 at 248 nm. Appl Phys A 70:21–28, 2000.

22. AC Cefalas, E Sarantopoulou, P Argitis, E Gogolides. Mass spectroscopic and degassing characteristics of polymeric materials for 157 nm photolithography. Appl Physics A 69: 929–933, 1999.

23. AC Cefalas, P Argitis, Z Kollia, E Sarantopoulou, TW Ford, AD Stead, A Maranka, CN Danson, J Knott, D Neely. Laser plasma x-ray contact microscopy of living specimens using a chemically amplified epoxy resist. Appl Phys Letts 72:25–27, 1998.

24. AC Cefalas, E Sarantopoulou. 157 nm photodissociation of polyamides, Microelect. Engin 53:464–468, 2000.

25. TM Bloomstein, MW Horn, M Rothschild, RR Kunz, ST Palmacci, RB Goodman, Lithography with 157 nm lasers. J Vac Sci Technol B 15:2112–2116 1997.

26. RR Kunz, TM Bloomstein, DE Hardy, RB Goodman, DK Downs, JE Curtin, Outlook for 157 nm resist design. SPIE 3678:13–23, 1999.

27. I. Petsalakis, G Theodorakopoulos, Electronic states of CF^+. Chem Phys 254:181–186, 2000.

28. RJ Cody, C Moralejo, JE Allen Jr. Photodissociation of the hydroxyl radical (OH) at 157 nm. J Chem Phys 95: 2491–2496, 1991.

29. Y Matsumi, M Kawasaki. Fine-structure branching ratios of the O(3Pj) atomic frag-

ments from photodissociation of oxygen molecules at 157 and 193 nm. J Chem Phys 93:2481–2486, 1990.

30. YL Huang, RJ Gordon. The multiplet state distribution of O(3Pj) produced in the photodissociation of O_2 at 157 nm. J Chem Phys 94:2640–2647, 1991.

31. JJ Lin, DW Hwang, YT Lee, XM Yang. Photodissociation of O_2 at 157 nm— experimental-observation of anisotropy mixing in the O_2 + hv−)O(P − 3) + O(P − 3) channel. J Chem Phys 109:1758–1762, 1998.

32. Y Matsumi, PK Das, M Kawasaki. Doppler spectroscopy of chlorine atoms generated from photodissociation of hydrogen-chloride and methyl-chloride at 157-nm and 193-nm. J Chem Phys 92:1696–1701, 1990.

33. K Tonokura, Y Matsumi, M Kawasaki, S Tasaki, R Bersohn. Photodissociation of hydrogen-chloride at 157 and 193 nm—angular-distributions of hydrogen-atoms and fine-structure branching ratios of chlorine atoms in the 2Pj levels. J Chem Phys 97:8210–8215, 1992.

34. U Wetterauer, A Pusel, P Hess. VUV photodesorption of molecular-hydrogen from the hydrogenated silicon(III) surface. Chem Phys Lett 300:397–402, 1999.

35. JK Rice, JR Woodworth. High-intensity laser photolysis, of OCS at 157 nm: $S^{1/s}$ production photoionization, and loss. J Appl Phys 50:4415–4421, 1979.

36. GS Ondrey, S Kanfer, R Bersohn. The collinear photodissociation of OCS at 157 nm. J Chem Phys 79:179–184, 1983.

37. P Andresen, GS Ondrey, EW Rothe, B Titze. The photodissociation of H_2O at 157 nm: full internal state distributions and alignment of nascent $OH(X^2\Pi)$ radicals. In: J Eichler, W Fritsch, IV Hertel, N Stolterfoht, U Wille, eds. 13th International Conference on the Physics of Electronic and Atomic Collisions. Berlin, West Germany: 1983.

38. P Andresen, GS Ondrey, B Titze. Creation of population inversions in the Lambda doublets of OH by the photodissociation of H_2O at 157 nm: a possible mechanism for the astronomical maser. Phys Rev Lett 50:486–488, 1983.

39. P Andresen, EW Rothe. Polarized LIF spectroscopy of OH formed by the photodissociation of H_2O by polarized 157 nm light. J Chem Phys 78:989–990, 1983.

40. K Mikulecky, K Heinz-Gericke, FJ Comes. Decay dynamics of $H_2O(^1B_1)$: full characterization of OH product state distribution. Chem Phys Lett 182:290–296, 1991.

41. RL Vander-Wal, JL Scott, FF Crim. Selectively breaking the O-H bond in HOD. J Chem Phys 92:803–805, 1990.

42. H Wurps, H Spiecker, JJ ter-Meulen, P Andresen. Laser induced fluorescence measurements of microwave stimulated OH molecules from H_2O photodissociation. J Chem Phys 105:2654–2659, 1996.

43. JA Guest, F Webster. Strong fragment alignment variation with internal state from ICN dissociation at 157.6 nm: linear regression modelling of CN B^2 sigma$^+$ products J Chem Physics 86:5479–5490, 1987.

44. N Shafer, S Satyapal, R Bersohn. Isotope effect in the photodissociation of HDO at 157.5 nm. J Chem Phys 90:6807–6808, 1989.

45. C Ye, M Suto, LC Lee, TJ Chuang. Radiative lifetime and quenching rate constant of AsF_2^* fluorescence. J Phys B 22:2527–2529, 1989.

46. B Freisinger, U Kogelschatz, JH Schafer, J Uhlenbush, W Viol. Ozone production in oxygen by means of F_2 laser at X = 157.6 nm. Appl Phys B 49:121–129, 1989.

47. Zhu Yi-Fei, RJ Gordon, The production of O(^{3}P) in the 157 nm photodissociation of CO_2. J Chem Physics 92:2897–2901, 1990.
48. Y Matsumi, N Shafer, K Tonokura, M Kawasaki, H Yu Lin, RJ Gordon. Doppler profiles and fine structure branching ratios of O (^3P_j) from photodissociation of carbon oxide at 157 nm. J Chem Phys 95:7311–7316, 1991.
49. RL Miller, SH Kable, PL Houston, I Burak. Product distributions in the 157 nm photodissociation of CO_2. J Chem Phys 96:332–338, 1992.
50. A Stolow, YT Lee. Photodissociation dynamics of CO_2 at 157.6 nm by photofragment translational spectroscopy. J Chem Phys 98:2066–2076, 1993.
51. A Pusel, U Wetterauer, P Hess. Direct breaking of SiH-surface bonds with 157 nm laser radiation. Laser Optom 31:48–50, 1999.
52. JM Soler, MR Beltran, K Michaelian, IL Garzon, P Ordejon, D Sanchez-Portal, E Artacho. Metallic bonding and cluster structure. Phys Rev B 61:5771–5780, 2000.
53. K Michaelian, N Rendon, IL Garzon. Structure and energetics of Ni, Ag, and Au nanoclusters. Phys Rev B 60:2000–2010, 1999.
54. U Meier, SD Peyerimhoff, F Grein. Ab initio MRD-CI study of neutral and charged Ga_2 Ga_3 and Ga_4 clusters and comparison with corresponding boron and aluminum clusters. Z Phys D 17:209–224, 1990.
55. Surface Science 156:1986; Z Phys D 3:1986.
56. A Dixon, C Colliex, R. Ohana, P Sudraud, J Van-de-Walle. Field-ion emission from liquid tin. Phys Rev Letts 46:865–868, 1981.
57. K Gamo, Y Inomoto, Y Ochiai, S Namba. B, Sb and Si liquid metal alloy ion sources for submicron fabrication. Jpn J Appl Phys 21:415–420, 1982.
58. C Brechignac, M Broyer, P Cahuzac, G Delacretaz, P Labastie, L Woste. Stability and ionization threshold of doubly charged mercury clusters. Chem Phys Lett 118:174–178, 1985.
59. AR Waugh. Current-dependent energy spectra of cluster ions from a liquid gold ion source. J Phys D 13: L203–208, 1980.
60. S Mukherjee, KH Bennmann. On the stability of charged metallic clusters. Surface Sci 156:580–583, 1985.
61. G Durand, JP Daudey, JP Malrieu. Singly and doubly charged clusters of (S^2) melals. Ab initio and model calculations on Mg_n^+ and Mgn^{++} ($n \leq 5$). J Phys 47:1335–1346, 1986.
62. S Martrenchard, C Jouvet, C Lardeux-Dedonder, D Solgadi, Observation of very small doubly charged homogeneous para difluorobenzene clusters. J Chem Phys 94:3274–3275, 1991.
63. K Sattler. The physics of microclusters. Adv Solid State Phys 23:1–12, 1983.
64. O Echt, K Sattler, E Recknagel. Magic numbers for sphere packings: experimental verification in free xenon clusters. Phys Rev Lett 47:1121–1124, 1981.
65. JG Gay, BJ Berne. Computer simulation of Coulomb explosions in doubly charged Xe microclusters. Phys Rev Lett 49:194–198, 1982.
66. P Pfau, K Sattler, R Pflaum, E Recknagel. Observation of doubly charged lead clusters below the critical size limit for Coulomb explosion of sphere configurations. Phys Lett A 104A:262–263, 1984.

67. D Tomanek, S Mukherjee, KH Bennemann. Simple theory for the electronic and atomic structure of small clusters. Phys Rev B 28:665–673, 1983.

68. Th Jentsch, W Drachsel, JH Block. Stability of doubly charged homonuclear trimeric metal clusters. Chem Phys Lett 93:144–147, 1982.

69. K Yamanouchi. Tuneable vacuum ultraviolet laser ideal light for spectroscopy of atoms, molecules, and clusters. Electron Spectrosc Rel Phenom 80:267–270, 1996.

70. T Moller, ARB Decastro, K Vonhaeften, A Kolmakov, T Laarmann, Lofken, C Nowak, F Picucci, M Riedler, C Rienecker, A Wark, M Wolff. Electronic-structure and excited-state dynamics of clusters—what can we learn from experiments with synchrotron radiation. Electron Spectrosc Rel Phenom 103:185–191, 1999.

71. OF Hagena. Formation of silver clusters in nozzle expansions. Z Phys D 20:425–8, 1991.

72. C Brechignac, M Broyer, P Cahuzac, G Delacretaz, P Labastie, L Woste. Size dependence of linear-shell autoionization lines in mercury clusters. Chem Physics Letts 120:559–563, 1985.

73. C Brechgnac, M Broyer, P Cahuzac, G Dalacretaz, P Labastie, L Woste, Stability and ionization thresholod of double charged mercury clusters. Chem Phys Lett 118:174–278, 1985.

74. R Heinrich, A Wucher. Yields and energy distributions of sputtered semiconductor clusters. Nuclear Instrum Methods Physics Res 140:27–38, 1998.

75. MA Ratner. Kinetics of cluster growth in expanding rare-gas jet. Low Temp Physics 25:266–273, 1999.

76. W Schulze, B Winter, J Urban, I Goldenfeld. Stability of multiply charged silver clusters. Z Phys D 4:379–381, 1987.

77. W Schulze, B Winter, I Goldenfeld. Generation of germanium clusters using the gas aggregation technique: stability of small charged clusters. J Chem Physics 87:2402–2403, 1987.

78. W Schulze, B Winter, I Goldenfeld. Stability of multiply positively charged homonuclear clusters. Phys Rev B 38:12937–12941, 1988.

79. A Hoareau, P Melinon, B Cabaud, D Rayane, B Tribollet, M Broyer. Fragmentation of singly and doubly charged lead clusters. Chem Phys Lett 143:602–608, 1988.

80. K Sattler, J Muhlbach, O Echt, P Pfau, E Rechnagel. Evidence for Coulomb explosion of doubly charged microclusters. Phys Rev Lett 47:160–163, 1981.

81. VI Matveev, PK Khabibullaev. Sizes and charges of clusters produced in the process of ion sputtering of a metal. Doklady Physics 44:789–792, 1999.

82. A Wucher, Z Ma, WF Calaway, MJ Pellin. Yields of sputtered metal clusters: the influence of surface structure. Surface Sci 304:L439–444, 1994.

83. A Wucher, M Wahl, H Oechsner. The mass distribution of sputtered metal clusters. Exp Nuclear Instrum Methods Physics Res 83:73–78, 1993.

84. S Nishigaki, W Drachsel, JH Block. Photon-induced field ionization mass spectrometry of ethylene on silver. Surface Sci 37:389–409, 1979.

85. C Skordoulis, E Sarantopoulou, SM Spyrou, C Kosmidis, AC Cefalas. Laser induced fluorescence (LIF) of Hg_2 and Hg_3 via dissociation of $HgBr_2$ at 157 nm. Z Phys D 18:175–180, 1991.

86. L Wiedeman, H Helvajian. Laser photodecomposition of sintered Yba_2Cu_3/O_{6+x}: ejected species population distributions and initial kinetic energies for the laser ablation wavelengths 351, 248, and 193 nm. J Appl Phys 70:4513–4523, 1991.

87. Proceedings Laser Induced Chemistry Conference, 18–22 Sept. 1989, Bechyne, Czechoslovakia. Spectrochim Acta 46A(4):1990.

88. M Ovchinnikov, BL Grigorenko, KC Janda, VA Apkarian. Charge localization and fragmentation dynamics of ionized helium clusters. J Chem Phys 108:9351–9361, 1998.

89. C Brechgnac, M Broyer, P Cahuzac, G Dalacretaz, P Labastie, L Woste. Stability and ionization threshold of double charged mercury clusters. Chem Phys Lett 118: 174–278, 1985.

90. D Rayane, P Melinon, B Cabaus, A Hoareau, B Tribollet, M Broyer. Electronic properties and fragmentation processes for singly and doubly charged indium clusters. J Chem Phys 90:3295–3299, 1989.

91. E Ruhl, PG F Bisling, B Brutschy, H Baumgartel. Photoionization of aromatic van der Waals complexes in a supersonic jet. Chem Phys Lett 126:232–237, 1986.

92. JR Grover, EA Walters, ET Hui. Dissociation energies of the benzene dimer and dimer cation. J Phys Chem 91:3233–3237 1987.

93. W Kamke, B Kamke HU Kiefl, IV Hertel. Photoionization and fragmentation of $(N_2O)_N$ clusters. J Chem Phys 84:1325–1334, 1986.

94. B Bescos, B Lang, J Weiner, V Weiss, E Wiedenmann, G Gerber. Real-time observation of ultrafast ionization and fragmentation of mercury clusters. Eur Phys J D 9:399–403, 1999.

95. T Baumert, R Thalweiser, V Weiss, G Gerber. Time-resolved studies of neutral and ionized Na_n clusters with femtosecond light pulses. Z Phys D 26:131–134, 1993.

96. JA Booze, T Baer. On the determination of cluster properties by ionization techniques. J Chem Physics 96:5541–5543, 1992.

97. A Dixon, C Colliex, R Ohana, P Sudraud, J Van de Walle. Field-ion emission from liquid tin. Phys Rev Lett 46:865–868, 1981.

98. J Van de Walle, P Joyes. Study of Bi_n^{p+} ions formed in liquid-metal ion sources. Phys Rev B 35:5509–5513, 1987.

99. Y Gotoh, T Kashiwagi, H Tsuji, J Ishikawa. Empirical relation between electric field at the ionization point and emission current of liquid copper, gold, germanium, and tin ion sources. Appl Phys A 64:527–532, 1997.

100. W Drachsel, T Jentsch, KA Gingerich, JH Block. Observation of doubly charged triatomic cluster ions in field evaporation. Surface Sci 156:173–182, 1985.

101. K Hata, M Ariff, K Tohji, Y Saito. Selective formation of C_{20} cluster ions by field evaporation from carbon nanotubes. Chem Phys Lett 308:343–346, 1999.

102. R Karnbach, M Joppien, J Stapelfeldt, J Wormer, T Moller. CLULU—an experimental setup for luminescence measurements on van der Waals clusters with synchrotron radiation. Rev Sci Instrum 64:2838–2849, 1993.

103. A Wucher, R Heinrich, RM Braun, KF Willey, N Winograd. Vacuum ultraviolet single-photon versus femtosecond multiphoton ionization of sputtered germanium clusters. Rapid Commun Mass Spectrom 12:1241–1245, 1998.

104. A Wucher, M Wahl. The formation of clusters during ion-induced sputtering of metals. Nucl Instrum Methods Phys Res B 115:581–589, 1996.

105. D Koch, M Wahl, A Wucher. Electron-impact and single-photon ionization cross-sections of neutral silver clusters. Z Phys D 32:137–144, 1994.

106. M Wahl, A Wucher. VUV Photoionization of sputtered neutral silver clusters. Nucl Instrum Methods Physics Res B 94:36–46, 1994.

107. AC Cefalas, C Skordoulis, C Nicolaides. Superfluorescent laser action around 495nm in the blue–green band of the mercury trimer Hg_3. Optics Commun 60:49–54, 1986.

108. TA Cool, JA McGarvey Jr, AC Eriandson. Two-photon excitation of mercury atoms by photodissociation of mercury halides Chem Phys Lett 58:108–113, 1978.

109. W Luthy, R Schiimiele, T Gerber. Spectral properties of a $HgBr_2$ photodissociation laser. Phys Lett 88:450–452, 1982.

110. Nai-Ho Cheung, JA McGarvey Jr, AC Eriandson, TA Cool. The vibrational distribution of HgBr (X^2 Sigma) molecules formed by photodissociation of $HgBr_2$ at 193 nm J Chem Phys 77:5467–5474, 1982.

111. T Efthimiopoulos, D Zevgolis, J Katsenos, D Zigos. Laser action from (B2E-X2E)HgBr induced by UV laser multiphoton dissociation of $HgBr_2$ measurements and experimental results. Proc SPIE 3423:68–71, 1998.

112. EJ Schimitschek, JE Celto, JA Trias. Mercuric bromide photodissociation laser Appl Phys Lett 31:608–610, 1977.

113. JH Schloss, JG Eden. Excited-state absorption in HgBr Appl Phys Lett 55:1282–1284, 1989; see also C Whitehurst, TA King. Multiphoton excitation of mercury atoms by photodissociation of HgX_2(X = Cl, Br, I). J Phys B 20:4053–4051, 1987.

114. RE Drullinger, MM Hessel, EW Smith. Experimental study of mercury molecules. J Chem Phys 66:5656–66, 1977.

115. AB Callear, KL Lai. Electronic absorption spectra of metastable dimers and trimers in optically excited mercury vapour. The carrier of the green emission. Chem Phys Lett 64:100–104, 1979.

116. WR Wadt. The electronic structure of $HgCl_2$ and $HgBr_2$ and its relationship to photodissociation J Chem Phys 72:2469–2478, 1980.

117. C Duzy, HA Hyman. Radiative lifetimes for the B to X transition in HgCl, HgBr, and HgI. Chem Phys Lett 52:345–348, 1977.

118. J Tellinghuisen, JG Ashmore. The B to X transition in ^{200}Hg ^{79}Br. Appl Phys Lett 40:867–869, 1982.

119. J Husain, JR Wiesenfeld, RN Zare. Photofragment fluorescence polarization following photolysis of $HgBr_2$ at 193 nmJ. Chem Phys 72:2479–2483, 1980.

120. AK Rai, SB Rai, OK Rai. Ultraviolet bands in diatomic mercury bromide. J Phys B 16:1907–1913, 1983.

121. BE Wilcomb, R Burnham, N Djew. UV absorption cross section and fluorescence efficiency of $HgBr_2$. Chem Phys Lett 75:239–242, 1980.

122. SH Linn, WB Tzeng, JM Jr Brom, CY Ng. Molecular beam photoionization study of $HgBr_2$ and HgI_2. J Chem Phys 78:50–61, 1983.

123. J Maya. Fluorescence yields of metal halide vapors excited by photodissociation. Appl Phys Lett 32:484–486, 1978.

124. M Stock, EW Smith, RE Drullinger, MM Hessel. Relaxation of the mercury 6^2P_0 and 6^3P states J Chem Phys 67:2463–2469, 1977.

125. RJ Niefer, J Supronowicz, JB Atkinson, L Krause. Laser-induced fluorescence of Hg_2 $H1_u$ excimer molecules. Phys Rev A 34:1137–1142, 1986; J Supronowicz, RJ Niefer, JB Atkinson, L Krause. Laser-induced fluorescence spectrum arising from $G0_u^+$ $X0_g^+$ bound-free transitions in Hg_2 L. J Phys B19:1153–1164, 1986; RJ Niefer, J Supronowicz, JB Atkinson, L Krause. Laser-induced fluorescence from the H, I, and J states of Hg_2. Phys Rev A 35:4629–4238, 1987; RJ Niefer, J Supronowicz, JB Atkinson, L Krause. Laser-induced fluorescence spectroscopy of the Hg_3 excimer. Phys Rev A 34:2483–2485, 1986; RJ Niefer, JB Aktinson, L Krause. Vibrational structure of some Hg_2 fluorescent bands. J Phys B16:3767–3773, 1983.

126. AB Callear, KL Lai. Electronic absorption spectra of metastable dimers and trimers in optically excited mercury vapour: the carrier of the green emission. Chem Phys Lett 64, 100, 1979; AB Callear, KL Lai. Vibrational structure of Rydberg transitions of metastable mecury dimers. Chem Phys Lett 75:234, 1980; AB Callear, KL Lai. Transient immediates in the flash photolysis of mercury vapour. Dimers, trimers and the enigmatic 2480 AA bands. Chem Phys 69:1, 1982; AB Callear, DR Kendall. A violet emission from optically excited mercury vapour: kinetics of the formation and decay of excited dimers and trimers. Chem Phys Lett 64:401, 1979; AB Callear, DR Kendall. Combination of metastable and ground-state mercury atoms into ungerade dimer states. Chem Phys Lett 70:215, 1980; AB Callear, DR Kendall. Luminescence from mixtures of mercury vapour and nitrogen following pulsed excitation with resonance radiation. Chem Phys 57:65, 1981.

127. RE Drullinger, MM Hessel, EW Smith. A theoretical analysis of mercury molecules. J Chem Phys 66:5656–5666, 1977; EW Smith, RE Drullinger, MM Hessel, J Cooper. A theoretical analysis of mercury molecules. J Chem Phys 66:5667–5681, 1977.

128. ER Mosburg, MD Wilke. Excimer densities and destruction mechanisms in a high pressure pure mercury positive column. J Chem Phys 66:5682–5693, 1977.

129. PJ Hay, TH Dunning, RC Raffeneti. Electronic states of Zn_2 ab initio calculations of a prototype for Hg_2. J Chem Phys 65:2679–2689, 1976.

130. T Shigenari, F Uesugi, H Takuma. Superfluorescent laser action at 404.7 nm from the transition of a mercury atom produced by photodissociation of $HgBr_2$ by an ArF laser. Opt Lett 7:362–364, 1982.

131. SR Langhoff, CW Bauschlicher Jr, SP Walch, BC Laskowski. Ab initio study of the ground state surface of Cu_3. J Chem Phys 85:7211–7215, 1986.

132. M Broyer, G Delacretaz, P Labastie, JP Wolf, L Woste. Spectroscopy of vibrational ground-state levels of Na_3. J Phys Chem 91:2626–2630, 1987.

133. K Balasubramanian, K Sumathi, D Dai. Group V trimers and their positive ions: the electronic structure and potential energy surfaces. J Chem Phys 95:3494–34505, 1991.

134. SP Walch, CW Bauschlicher Jr. On 3d bonding in the transition metal trimers: the electronic structure of equilateral triangle Ca_3, Sc_3, Sc_3^+, and Ti_3^+. J Chem Phys 83:5735–5742, 1985.

135. J Koperski, JB Atkinson, L Krause. First observations of laser-excited Hg_3 and Hg_2 Rg spectra in a supersonic expansion beam. J Mol spectrosc 187:181–192, 1998.

136. LD Landau, EM Lifshitz. Quantum mechanics. Oxford, England: Pergamon Press, 1981.
137. AC Cefalas, SM Spyrou, E Sarantopoulou. Observation of vibronic–electronic interaction in mercury triatomic clusters via real time quantum beats. Laser Ultrafast Proc 4:175–181, 1991.

7

Tunable Solid-State Ultraviolet and Vacuum Ultraviolet Lasers

Ilko K. Ilev and Ronald W. Waynant
U.S. Food and Drug Administration, Rockville, Maryland

I. INTRODUCTION

The ultraviolet (UV) and vacuum ultraviolet (VUV) spectral range is of great interest for many attractive industrial and biomedical applications. The research efforts in this range are focused on development of alternative laser methods and materials providing widely tunable short-wavelength laser generation. The all-solid-state laser concept for UV/VUV tunable generation is a superior alternative to the conventional excimer-gas-based UV/VUV lasers. This concept includes the development of all-solid-state nonlinear optical techniques as well as tunable UV/VUV solid-state active materials such as rare-earth-doped laser crystals. Several important factors have recently significantly contributed to the all-solid-state nonlinear techniques being recognized as one of the most effective and easy ways of getting UV- and VUV-tunable generation.

These factors can be summarized in the following main groups. First is the development of all-solid-state ultrashort-pulse nonlinear techniques based on high-intensity picosecond and femtosecond lasers that allow, for instance, use of only two sequential cascade processes to obtain UV/VUV tunable laser generation. Second is the development of new nonlinear upconversion crystals possessing high conversion efficiency and broad spectral transparency in the UV and VUV range down to 155 nm, which is below the shortest wavelength of 157 nm achieved by F_2-excimer laser. Third is the development of all-solid-state diode-pumped lasers that allow extremely compact and reliable laser systems to be commercially produced. The final group is the development of new fiber-optic materials for UV laser delivery and generation. Most of the fundamental nonlinear effects contribute directly to the process of generating UV- and VUV-tunable laser radiation that is the subject of this chapter. Most of it will therefore address the physical principles of nonlinear frequency conversion effects as well as commonly used nonlinear conversion techniques, including frequency upconversion and fiber Raman lasers. The final section briefly presents the basic features of other known solid-state sources of tunable UV/VUV radiation, including rare-earth-doped wide-band-gap crystals for UV/VUV fluorescence.

II. ALL-SOLID-STATE NONLINEAR OPTICAL TECHNIQUES FOR UV AND VUV TUNABLE LASER GENERATION

Among modern photonics technologies, nonlinear optics is one of the most intensively studied, especially since the invention of ultrashort high-intensity lasers. Short-pulse lasers provide high-intensity optical interactions and the nonlinear conversion effects acquire an important influence on the mechanism of these interactions. The first part of this section provides an analytical overview of the general principles and features of basic nonlinear effects in light of their application for development of all-solid-state UV/VUV-tunable lasers. It reviews the fundamental properties of nonlinear susceptibilities and main second- and third-order nonlinear frequency conversion processes. The second and third parts consider the main all-solid-state nonlinear techniques and optical materials used for UV/VUV-tunable generation. Significant results achieved by the nonlinear frequency upconversion techniques are also considered.

A. Nonlinear Frequency Upconversion Effects

1. Nonlinear Susceptibilities

In principle, the intensity of a laser source is significantly higher (orders of magnitude) than could be produced by any conventional light source. Laser intensity is comparable with the typical field strength of optical materials (10^9–10^{10}V/m). Hence, the laser intensity affects the optical medium and creates conditions for nonlinear optical processes. This effect has a bidirectional character. First, the laser emission affects the optical medium. As a consequence, the material response ceases to be a linear function of the applied electromagnetic field. In fact, optical constants that characterize the properties of optical material (for instance, refractive indexes and dielectric constants) are no longer constants and become nonlinear functions of the light intensity. Second, at the same time, there is observed an opposite effect: of the optical medium on the incidence light field. As a result, some important parameters of the light field are changed for instance, frequency, intensity, phase, propagation direction, among others.

In an analytical representation, the appearance of nonlinear optical effects is related to a conversion of the material equation that describes the dependence between the macroscopic polarization (**P**) induced by light propagating in the medium and the optical electric field (**E**) [1–8]. This equation is converted from a linear form:

$$\mathbf{P} = \chi^{(1)} \mathbf{E} \tag{1}$$

to a nonlinear form:

$$\mathbf{P} = \mathbf{P}^L + \mathbf{P}^{NL} = \chi^{(1)} \mathbf{E} + \chi^{(2)} \mathbf{EE} + \chi^{(3)} \mathbf{EEE} + \cdots \tag{2}$$

Here,

$$\mathbf{P}^L = \chi^{(1)} \mathbf{E} \tag{3}$$

is the linear polarization;

$$\mathbf{P}^{NL} = \chi^{(2)} \mathbf{EE} + \chi^{(3)} \mathbf{EEE} + \cdots \tag{4}$$

is the nonlinear polarization; $\chi^{(1)}$ is the linear susceptibility, a second-order tensor with typical values $\chi^{(1)} \approx 1$; $\chi^{(2)}$ is the quadratic (second-order) nonlinear susceptibility, a third-order tensor with typical values $\chi^{(2)} \approx 10^{-11}$–$10^{-12}$ m/V; and $\chi^{(3)}$ is the cubic (third-order) nonlinear susceptibility, a fourth-order tensor with typical values $\chi^{(3)} \approx 10^{-21}$–$10^{-22}$ m^2/V^2.

With increasing laser intensity, the high-order nonlinear polarization terms $\mathbf{P}^{(n>1)}$ become more and more important and will lead to a large variety of nonlinear optical effects. The study of the nonlinear susceptibilities $\chi^{(2)}$, $\chi^{(3)}$, $\chi^{(4)}$ . . . of the excited optical medium gives rich information about the energetic spectra of the materials. It also provides alternative techniques for development of effective laser wavelength converters including UV and VUV ranges. The nonlinear optical effects can be classified by the specific properties of the nonlinear susceptibilities as follows:

The linear susceptibility $\chi^{(1)}$ defines the complex refractive index of the optical medium and is connected with the dielectric constant of the free space ϵ_0 by the expression:

$$\varepsilon_1 = \varepsilon_0 (1 + \chi^{(1)}) \tag{5}$$

The second-order nonlinear susceptibility $\chi^{(2)}$ describes second-order nonlinear effects and it is related to the asymmetrical-type crystals, which have no center of inversion symmetry. Nonlinear effects included in this category are:

Second-harmonic generation associated with $|\chi^{(2)}(2\omega = \omega + \omega)|^2$
Sum or difference frequency generation—$\chi^{(2)}(\omega_3 = \omega_1 + \omega_2)$
Parametric fluorescence
Optical rectification

The third-order nonlinear susceptibility $\chi^{(3)}$ describes third-order nonlinear effects and is related to the symmetrical-type crystals, liquids, and gases for which $\chi^{(2)} = 0$. These nonlinear optical effects are:

Third-harmonic generation—$|\chi^{(3)}(3\omega = \omega + \omega + \omega)|^2$
Four-photon parametric mixing—$\chi^{(3)}(\omega_4 = \omega_1 + \omega_2 - \omega_3)$
Two-photon absorption—Im $\chi^{(3)}(\omega_2 = \omega_1 + \omega_2 - \omega_2)$
Stimulated Raman scattering—Im $\chi^{(3)}(\omega_{AS,S} = \omega_p \pm \omega_R)$
Optical Kerr effect—Re $\chi^{(3)}(\omega = 0 + 0 + \omega)$
Self-phase modulation

A simple illustration of the above-listed nonlinear optical effects is shown in Figure 1. We consider two waves ω_1 and ω_2 interacting in a quadratic nonlinear medium characterized by the second-order nonlinear susceptibility $\chi^{(2)}$ (Fig. 1a) or in a cubic nonlinear medium characterized by the third-order nonlinear suscep-tibility $\chi^{(3)}$ (Fig. 1b).

Since most of the mentioned nonlinear effects contribute directly to the process of generation of tunable UV and VUV laser emission [9–11] that is the subject of this chapter, the basic nonlinear effects will be discussed in some detail below. At the beginning, harmonic generation and sum frequency mixing will be considered. Next, nonlinear effects such as stimulated Raman scattering, four-photon mixing, and self-phase modulation will be discussed in view of their appli-cation for development of UV-tunable fiber Raman lasers.

2. Harmonic Generation

Second-Harmonic Generation. Second-harmonic generation (SHG) oc-curs in optical materials without an inversion center of symmetry, that is, in quadratic nonlinear media when $\chi^{(3)} = 0$ [1,2,12]. It is a nonlinear process whereby two pumping optical fields of the same frequency interact in the nonlin-ear material to produce a third field with a double higher frequency. In Figure 1a the SHG process is associated with the case when the frequencies of the inter-acting waves ω_1 and ω_2 are equal: $\omega_1 = \omega_2 = \omega$. Analytically, the SHG process can be described by a simple consideration. Let us assume to have a pumping electromagnetic field in the standard form for harmonic electromagnetic fields:

$$\mathbf{E} = \mathbf{A} \sin\omega t \qquad (6)$$

Figure 1   Basic nonlinear optical effects observed at nonlinear interaction of two elec-tromagnetic waves with frequencies ω_1 and ω_2 in an optical medium possessing quad-ratic- (a) or cubic- (b) nonlinear susceptibility. SHG, second harmonic generation; SFG, sum-frequency generation; DFG, difference-frequency generation; THG, third-harmonic generation; FPM, four-photon mixing.

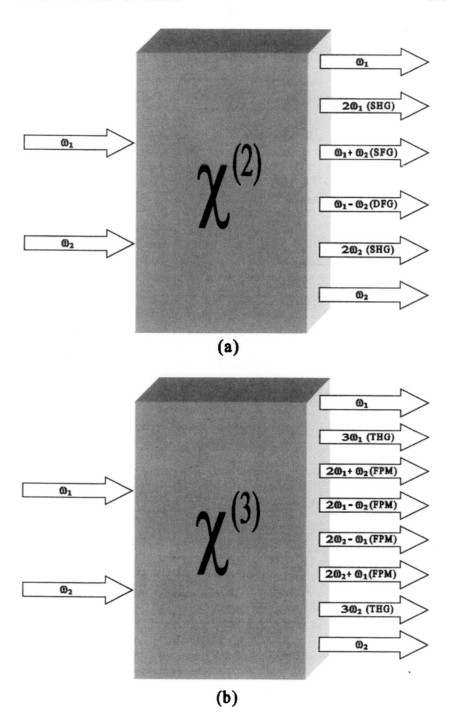

(a)

(b)

If we put this field in the nonlinear material Eq. (2), we get:

$$P = \chi^{(1)} A \sin\omega t + 1/2\chi^{(2)} A^2 - 1/2\chi^{(2)} A^2 \cos(2\omega t) \tag{7}$$

Hence, when we act with a harmonic electromagnetic field [Eq. (6)] on a quadratic nonlinear optical medium, a polarization wave P arises that contains three components:

$$P = P_{const} + P_\omega + P_{2\omega} \tag{8}$$

where $P_{const} = 1/2\chi^{(2)}A^2$ is the constant polarization component; $P_\omega = \chi^{(1)}A \sin \omega t$ is the polarization component with a frequency equal to the pumping wave frequency; and $P_{2\omega} = 1/2\chi^{(2)} A^2 \cos(2\omega t)$ is the polarization component with a frequency twice that of the pumping wave frequency.

The SHG process is most efficient when the fundamental and second-harmonic phase velocities are matched. Collinear phase matching occurs when the interacting waves have a common wave vector direction. In the case of SHG, the phase matching requires that the propagation constant of the source polarization $2k_\omega$ be equal to the propagation constant $k_{2\omega}$ of the second harmonic or:

$$\Delta k = 2k_\omega - k_{2\omega} = 0 \tag{9}$$

If we present the propagation constants by the refractive indexes (n_ω, $n_{2\omega}$) and wavelengths (λ_ω, $\lambda_{2\omega}$) of the fundamental and second-harmonic waves, respectively, the condition [Eq. (9)] can be rewritten in the form:

$$\Delta k = 2\pi[(n_{2\omega}/\lambda_{2\omega}) - (2n_\omega/\lambda_\omega) = 0 \tag{10}$$

Thus, the phase-matching condition reduces to:

$$n_{2\omega} = n_\omega \tag{11}$$

In practice, the most effective and commonly used phase-matching technique is to utilize anisotropic crystals possessing natural birefringence that can compensate for material dispersion. For example, in a negative ($n_e < n_0$) uniaxial crystal (see Fig. 2), the fundamental wave is directed into the crystal as an ordinary wave (o-wave) and the second-harmonic wave is generated as an extraordinary wave (e-wave). As can be seen from Figure 2, in the direction OA (so-called phase-matching direction), which forms angle θ_m (so-called phase-matching angle) with the crystal axis z, the phase-matching condition [11] is satisfied in the form:

$$n_{2\omega}^0 = n_\omega^e \tag{12}$$

The dependence [Eq. (12)] corresponds to the so-called type I phase-matching condition (ooe-phase-matching) that is one of the possible varieties. Table 1 shows entire possibilities for phase-matching at SHG in optical crystals. It is essential that while, at Type I matching, two fundamental waves have identical linear polarizations and the second-harmonic wave has perpendicular polariza-

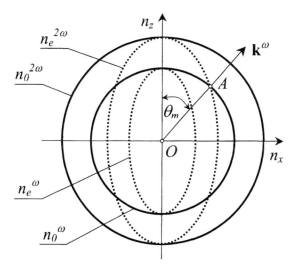

Figure 2   Phase matching at SHG in a negative ($n_e < n_o$) uniaxial crystal. The condition $n_{2\omega}^o = n_\omega^e$ [see Eq. (12)] is satisfied at $\theta = \theta_m$.

tion, at Type II matching, two fundamental waves should have reciprocally perpendicular polarizations.

To maximize the conversion efficiency in SHG, various techniques usually based on focusing of the fundamental gaussian laser beams or intracavity SHG are applied [2,4,5,7]. Figure 3 shows a typical optical arrangement of a continuous-wave intracavity SHG. The SHG crystal is placed at the beam waist of the intracavity laser mode. This technique provides a high fundamental intensity and SHG conversion efficiency. However, special measures should be taken to prevent SHG instability caused by the laser mode impurity.

In addition, the material equation for second-order nonlinear processes in the cartesian coordinate system can be presented in the form:

Table 1   Possible Phase-Matching Conditions at SHG in Nonlinear Crystals

Type of phase matching	Nonlinear crystal	
	Negative	Positive
Type I	ooe	eeo
Type II	oee	eoo

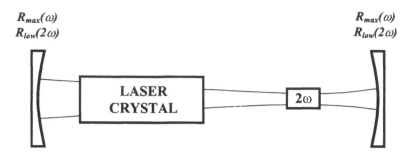

Figure 3 Typical optical arrangement for intracavity SHG of a solid-state laser. R_{max}, R_{low}, maximum and low reflectance, respectively, of the cavity mirrors.

$$P_i^{(2)} = \sum_{jk} \varepsilon_0 \chi_{ijk}^{(2)} E_j E_k \qquad (13)$$

In general case, $\chi_{ijk}^{(2)}$ is a three-dimensional tensor having 27 independent coefficients. However, when one applies a perturbation symmetry condition between the electromagnetic fields Ej and Ek, the number of independent coefficients reduces to 18. Hence, the three-dimensional tensor $\chi_{ijk}^{(2)}$ can be replaced by a more simple and convenient two-dimensional 3×6 tensor, well known as the Kleinman **d**-tensor. The use of the **d**-tensor coefficients is a commonly applied method for characterizing the properties of nonlinear optical media.

Third- and Higher-Order Harmonic Generation. Third-harmonic generation (THG) is a third-order nonlinear effect defined by the third-order nonlinear susceptibility $\chi^{(3)}$ [1,3]. It can occur in both acentric and centric material, including liquids and gasses. The THG is described by the four-rank tensor $|\chi^{(3)}(3\omega = \omega + \omega + \omega)|$. In the terms of Figure 1b, the THG process is achieved when one of the interacting frequencies ω_1 or ω_2 is equal to the double value of the other frequency: $\omega_2 = 2\omega_1$ or $\omega_1 = 2\omega_2$. Thus, the interaction is $\omega_1 + \omega_1 + \omega_1 = 3\omega_1$ or $\omega_2 + \omega_2 + \omega_2 = 3\omega_2$ (see Fig. 1b). This is the case of so-called direct THG when the THG process is realized entirely in a cubic nonlinear medium described by $\chi^{(3)}$. Figure 4 illustrates the THG mechanism by an energy-level diagram. However, the THG process is also possible in a quadratic nonlinear medium characterized by the second-order nonlinear susceptibility $\chi^{(2)}$ that is not completely zero. In this case THG is realized by frequency mixing of the output frequency and SHG frequency. This kind of THG is called the quasicubic or cascade cubic nonlinear process.

In an identical manner, higher-order harmonic generation processes can be interpreted. The energy diagrams in Figure 4 show also the mechanism for direct realization of fourth-harmonic generation (FOHG) and fifth-harmonic generation (FIHG). These processes are characterized by high-order nonlinear susceptibilit-

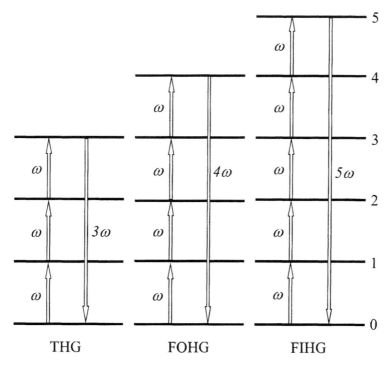

THG   FOHG   FIHG

Figure 4 Energy-level diagrams illustrate mechanism of direct third-(THG), fourth-(FOHG), and fifth- (FIGH) harmonic generation.

ies, $\chi^{(4)}$ for FOHG and $\chi^{(5)}$ for FIHG. They can be described by the expressions $(\omega + \omega + \omega + \omega) = 4\omega$ (FOHG) and $(\omega + \omega + \omega + \omega + \omega) = 5\omega$ (FIHG). However, the FOHG, for example, can be realized as a result of various cascade processes related to the lower-order susceptibilities $\chi^{(2)}$ and $\chi^{(3)}$. In this case, the FOHG can be considered as a cascade process defined by the dependence

$$\omega + \omega + \omega \rightarrow 3\omega \rightarrow 3\omega + \omega \rightarrow = 4\,\omega \qquad (14)$$

This means that the first stage of the cascade FOHG process is the third harmonic to be generated and the THG frequency to be mixed with the fundamental frequency in order for the FOHG frequency to be obtained. In the next sections, we will consider concrete examples of practical application of these high-order harmonic generation processes For effective frequency conversion, the third- and higher-order harmonic generation processes require fulfilling phase-matching conditions similar to these for the SHG [see Eqs. (9) to (12)].

3. Frequency Upconversion

According to the consideration in Sec. II.A.1. (see Fig. 1), the upconversion or sum-frequency generation (SFG) can be described as a second-order nonlinear effect at which two lower-energy photons, ω_1, ω_2, combine in a nonlinear medium to generate a higher-energy photon ω_3 where:

$$\omega_1 + \omega_2 = \omega_3 \tag{15}$$

The SFG process can be analyzed in a similar way to SHG (see Sec. II.A.1.a) [2–5,7]. Let us assume that two electromagnetic waves propagate in a quadratic nonlinear medium ($\chi^{(3)} = 0$) with different frequencies ω_1 and ω_2:

$$\mathbf{E}_1(z,t) = \mathbf{E}_{01} \cos(\omega_1 t - k_1 z) \tag{16}$$

$$\mathbf{E}_2(z,t) = \mathbf{E}_{02} \cos(\omega_2 t - k_2 z) \tag{17}$$

If we replace expression (16) and (17) in the nonlinear material Eq. (4), we obtain

$$\begin{aligned}\mathbf{P}^{NL} = \chi^{(2)}[\mathbf{E}_{01}^2 \cos^2(\omega_1 t - k_1 z) + \mathbf{E}_{02}^2 \cos^2(\omega_2 t - k_2 z) \\ + 2\mathbf{E}_{01}\mathbf{E}_{02} \cos(\omega_1 t - k_1 z) \cos(\omega_2 t - k_2 z)\end{aligned} \tag{18}$$

Eq. (18) includes the following components:

$$\mathbf{P}_0 = 1/2\chi^{(2)}(\mathbf{E}_{01}^2 + \mathbf{E}_{02}^2) \tag{19}$$

$$\mathbf{P}_{2\omega1} = 1/2\chi^{(2)} \cos[2(\omega_1 t - k_1 z)] \tag{20}$$

$$\mathbf{P}_{2\omega2} = 1/2\chi^{(2)} \cos[2(\omega_2 t - k_2 z)] \tag{21}$$

$$\mathbf{P}_{\omega1+\omega2} = \chi^{(2)}\mathbf{E}_{01}\mathbf{E}_{02} \cos[(\omega_1 + \omega_2)t - (k_1 + k_2)z] \tag{22}$$

$$\mathbf{P}_{\omega1-\omega2} = \chi^{(2)}\mathbf{E}_{01}\mathbf{E}_{02} \cos[(\omega_1 - \omega_2)t - (k_1 + k_2)z] \tag{23}$$

The availability of the polarization component [Eq. (22)] confirms the fact that an SFG with a higher frequency $\omega_3 = \omega_1 + \omega_2$ and a propagation constant $\mathbf{k}_3 = \mathbf{k}_1 + \mathbf{k}_2$ are generated in the medium. Similarly to the case with SHG (see Table 1 and Fig. 2), the SFG process also requires satisfaction of phase-matching conditions. In this case, these conditions have the following forms:

$$\mathbf{k}_3 = \mathbf{k}_{\omega1+\omega2} = \mathbf{k}_1 + \mathbf{k}_2 \tag{24}$$

$$n_2\omega_3 = n_1\omega_1 + n_2\omega_2 \tag{25}$$

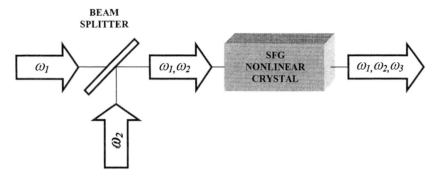

Figure 5 A basic SFG configuration in which a signal laser beam at ω_1 and a strong pumping laser beam at ω_2 are combined by a beam splitter in a nonlinear crystal to generate a laser beam at the sum frequency $\omega_3 = \omega_1 + \omega_2$.

A typical experimental SFG arrangement is demonstrated by Figure 5. A signal laser beam at ω_1 and a strong pumping laser beam at ω_2 are combined by a beam splitter in a nonlinear crystal to generate a laser beam at the sum frequency $\omega_3 = \omega_1 + \omega_2$.

Of a significant importance for the development of widely tunable (including in the UV and VUV) all-solid-state lasers is an SFG variation called optical parametric amplification (OPA) [2,4,5,7]. It can be considered as a reverse SFG process in which low-energy photons at frequency ω_s (the signal photons) are amplified by a nonlinear interaction with high-energy photons at frequency ω_p (the pump photons). To be valid according to the law of energy conservation, additional photons at a frequency ω_i (the idler photons) should be simultaneously generated. Thus, the SFG process can be represented by the dependence:

$$\omega_p = \omega_s + \omega_i \tag{26}$$

The basic practical applications of the SFG nonlinear processes are associated with development of tunable optical parametric oscillators (OPO). A principal OPO configuration is shown in Figure 6. The nonlinear crystal is commonly placed within an optical Fabry–Perot resonator. It forms by two dichroic dielectric mirrors that have either maximum transmittance at the pump wavelength and maximum reflectance at the signal wavelength or both at the signal and idler wavelengths. At a given threshold intensity of the pumping laser, the OPO cavity provides resonances for the signal or idler (or both) waves.

An important advanced OPO property is its potential for wide spectral tuning of the output laser emission. A conventional and simple method of OPO tuning is to rotate the crystal or to change its orientation angle θ_p (see Fig. 6) relative to the axis of the Fabry–Perot cavity. Concrete examples of widely tun-

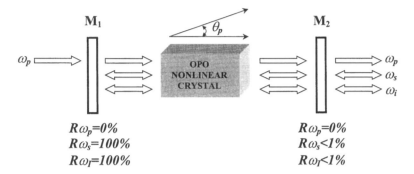

Figure 6 Principal OPO arrangement in which a nonlinear crystal, placed in a Fabry-Perot cavity with resonance frequencies ω_s and ω_i satisfying the OPO condition $\omega_p = \omega_s + \omega_i$ [see Eq. (26)], is pumped by a laser beam at ω_p. As a result, simultaneous amplification of the signal (at ω_s) and idler (ar ω_i) wave is obtained. The angle θ_p relates to the OPO crystal orientation.

able spectra obtained using SFG and OPO techniques are presented in the next section.

B. UV and VUV Solid-State Laser Sources Based on Harmonic Generation

In the last decade when high-intensity ultrashort and diode-pumped solid-state laser systems have become commercially available, development of tunable all-solid-state UV and VUV lasers represents an attractive alternative to the well-established excimer UV/VUV laser [9]. The reason for the dramatically increasing interest in the solid-state concept for short-wavelength lasers can be attributed to some of its advanced features: functional simplicity and compact construction; turnkey reliability; high average laser power and repetition rate; low operating cost; and removal of toxic gases. Among the known solid-state laser systems, the nonlinear frequency conversion, based on fundamental nonlinear effects described in Sec. II.A., are the most common short-wavelength laser techniques. Usually these techniques include generation of tunable UV and VUV laser radiation using harmonic generation or frequency mixing in nonlinear crystals. In this section, we will summarize the main results achieved by nonlinear frequency conversion techniques.

1. UV/VUV Material Limitations

With respect to the basic material characteristics of the nonlinear crystals as well as supporting optical elements, the UV and VUV ranges impose some specific limitations that can be summarized in the following main groups [4,6]: relatively short spectral transmittance; relatively low lifetime of the laser optics; availability of a temporal walk-off effect caused by the different group velocities in the case of ultrashort laser pulses; and small available sizes of some nonlinear crystals. An additional limitation exists for laser radiation with wavelength shorter than 185 nm: air absorption caused mainly by absorption in the Schumann–Runge bands of oxygen. Due to this absorption, laser radiation in this spectral range should be propagated in vacuum. In fact, the wavelength of 185 nm defines the boundary between the UV (185–400 nm) and VUV (105–185 nm) ranges. As an illustration of one of the basic UV/VUV material limitations, Figure 7 shows the temperature dependence of the short-wavelength transmission limit of various conventional UV/VUV window materials [11,13]. The cutoff wavelength of all materials drops with temperature, but no material has a cutoff below those of LiF at 105 nm. Despite the mentioned UV/VUV limitations, nonlinear crystals and optical materials are now widely commercially available in both the UV and VUV. Research efforts are directed to the development of new nonlinear crystals and optical materials with improved parameters.

2. Nonlinear Crystals for Harmonic Generation

The generation of optical harmonics is one of the simplest ways to obtain shortwave length laser emission. By using various kinds of nonlinear crystals for direct SHG or cascade harmonic generation processes, we easily can convert laser radiation from the visible and near-infrared (IR) range to the UV and VUV. At present, the most commonly used SHG nonlinear crystals transparent in the UV and partially in the VUV are KDP, ADP, KTP, LBO, and BBO. These crystals have been studies and fully characterized both theoretically and experimentally. Basic properties of the SHG crystals are summarized in Table 2. Some of the newly developed crystals such as CLBO, CBO, and LB4, which can be also used in harmonic generation processes, are presented in Table 3 (see Sec. II.C.2.) with regard to their application in nonlinear mixing processes. In accordance with the analysis described in Sec. II.A., the most important condition for a nonlinear process to be maximally effective is fulfilling the phase-matching conditions [see Eqs. (9), (11), (24), (25)]. In fact, these are the most important requirement characterizing the quality of the nonlinear crystal. For this reason, one of the most conventional methods for testing the nonlinear crystal quality is to study phase-matching-angle and temperature dependencies of the output laser power produced by nonlinear optical interactions. Each of the nonlinear crystals described in Table 2 possesses its own advanced features that may have significant importance in a given situation. For example, KDP is the crystal that can be grown to large sizes. However, it has relatively low nonlinear coefficients and small hygroscopicity. KTP, on the other hand, has about an order of magnitude higher nonlinear

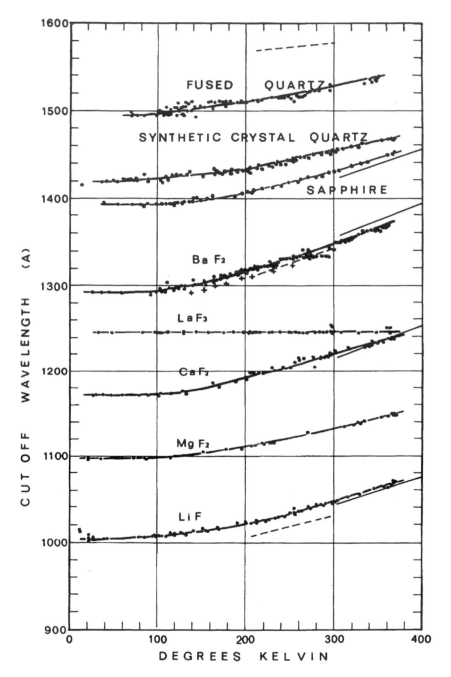

Figure 7   Short-wavelength transmission cutoff vs. temperature for various widely used UV optical materials. (From Refs. 11 and 13.)

Table 2 Basic Properties of Commonly Used Crystals for Harmonic Generation in the UV and VUV Range

Nonlinear crystal	Type of interaction	Cutoff of UV transmission (nm)	SHG cutoff (nm)	Nonlinearity d_{eff} (pm/V)	Damage threshold (GW/cm^2)	Max aperture diameter (mm)	References
KDP (KH$_2$PO$_4$)	ooe oee	177	487	$d_{36} = 0.39$	23	~400	4, 12, 22, 29, 55
ADP (NH$_4$H$_2$PO$_4$)	ooe oee	180	525	$d_{36} = 0.47$	25	~100	4, 12, 29, 30, 55
KTP (KTiO$_2$PO$_3$)	eoe	350	990	$d_{32} = 5.0$ $d_{31} = 6.5$	1	~10	4, 16, 22, 29, 30
LBO (LiB$_3$O$_5$)	ooe eoo	155	555	$d_{32} = 1.16$	25	~15	4, 19, 22, 30, 55, 56
BBO (β-BaB$_2$O$_4$)	ooe eoe	189	411	$d_{32} = 16.0$ $d_{31} = 0.08$	10	~15	4, 20, 22, 39, 55

Table 3 Properties of Nonlinear Optical Crystals Used for SFG Processes in UV and VUV Ranges

Nonlinear crystal	Crystal type	Transparency [µm]	Nonlinearity d_{eff} [pm/V]	Minimum wavelength (nm) achievable at:			References
				SHG	FOHG	SFG	
LBO (LiB_3O_5)	Negative Biaxial	0.160–3.2	$d_{32} = 1.16$	277	242.5	160	4, 19, 22, 30, 55, 56
BBO (β-BaB_2O_4)	Negative Uniaxial	0.189–3.5	$d_{32} = 16.0$ $d_{31} = 0.08$	204.8	189	189	4, 20, 22, 39, 55
CBO (CsB_3O_5)	Positive Biaxial	0.167–3.0	$d_{14} = 1.08$	272.8	236.3	167	53, 55, 56, 63
CLBO ($CsLiB_6O_{10}$)	Negative Uniaxial	0.180–2.75	$d_{36} = 0.86$	236.7	211.7	180	23, 24, 54, 55, 56
KB5 ($KB_5O_84H_2O$)	Positive Biaxial	0.182–1.5	$d_{31} = 0.04$ $d_{32} = 0.003$	217	194.8	162	18, 55, 56, 60, 64, 65
LB4 ($Li_2B_4O_7$)	Negative Uniaxial	0.160–3.3	$d_{31} = 0.15$	243.8	218.3	160	55, 56
KBBF ($Kbe_2BO_3F_2$)	Negative Uniaxial	0.155–3.66	$d_{11} = 0.76$	164	155	155	55, 56, 57

coefficient and is nonhygroscopic. Among the other crystals presented, it has the longest cutoff wavelength in the UV, but only can be grown to relatively small sizes. The situation with both BBO and LBO is quite similar. They have high nonlinearity, but relatively small aperture sizes. In the case of ultrashort pumping pulses, the short length of the nonlinear crystal has some advantages related to neglecting the influence of such unwanted nonlinear effects such as SFM and two-photon absorption, as well as the ultimate use of the crystal transparency window. In the next section, we will present some of the basic results on the UV/VUV frequency conversion obtained by harmonic generation.

3. UV/VUV Frequency Conversion Using Harmonic Generation

UV and VUV laser radiation has been generated by means of either direct SHG or cascade high-order harmonic generation processes. The range of pumping laser sources used includes both solid-state lasers with fixed wavelength (such as nanosecond [14–30], picosecond [31–34] and cw [35–41] Nd:doped lasers) and tunable lasers (such as dye [42–46], Ti:Sapphire [47–52], etc.).

Harmonic Generations of Nd:Doped Lasers. Since the invention of laser four decades ago, and especially recently with the extensive introduction of laser-diode pumped systems, neodymium-doped lasers have played a basic role in solid-state laser technology. In combination with nonlinear frequency conversion techniques, these lasers provide one of the most effective and low-cost methods of short-wavelength laser generation. An enormous number of papers have been written on the generation of harmonics using Nd:YAG, Nd:YLF, Nd:YAB, Nd: glass, and other Nd:doped lasers. These works cover the range of laser parameters including pulsed [14–34] and continuous wave (cw) [35–41] regimens, nanosecond [14–30] and picosecond [31–34] pulses, and external [35,36] and intracavity [37–41] techniques for harmonic generation.

In the nanosecond range, a highly efficient SHG of a Q-switched Nd:YAG laser ($\lambda = 1.064$ µm, 9-ns pulse duration, 10-Hz repetition rate, and 200-MW/cm^2 pump-power density) was reported by Xie et al. [14]. The authors used a 14 mm LBO crystal and achieved more than 70% energy conversion efficiency. Up to 59% energy conversion efficiency of SHG has been obtained at frequency doubling of a Q-switched Nd:YAG laser ($\lambda = 1.064$ µm, 45-ns pulse duration, 10-Hz repetition rate, and 380-MW/cm^2 pump-power density) using a 3.9 mm KTP crystal [15]. Chen et al. [16] demonstrated a cascade generation of the second ($\lambda = 532.1$ nm), third ($\lambda = 354.7$ nm), fourth ($\lambda = 266.1$ nm) and fifth ($\lambda = 212.8$ nm) harmonics of either nanosecond Q-switched Nd:YAG laser ($\lambda = 1.064.2$ µm, 8 ns pulse duration, 10-Hz repetition rate, and 250-MW/cm^2 pump-power density) or picosecond mode-locked Nd:YAG laser ($\lambda = 1.064.2$ µm, 1 ns pulse duration, and 2-GW/cm^2 pump-power density). The authors used different kinds of BBO crystals and obtained various energy conversion efficiencies as follows: 58% (68%) for SHG; 23% (no measurement) for THG; 12% (29%) for FOHG; and 4% (15%) for FIHG when the nanosecond (picosecond) Nd:YAG laser is used, respectively. Komatsu and colleagues [17] reported the growth and UV applications of a large-sized (over 50 mm diameter) LBO crystal. The crystal was tested at generation of the second (achieved 20% conversion efficiency), fourth, and fifth harmonics of a Q-switched Nd:YAG laser. Umenura and Kato [18] presented the first experimental study on the sixth-harmonic generation (λ = 177.4 nm) of a Q-switched Nd:YAG laser (λ = 1.0642 μm, 12 ns pulse duration, and 80-MW/cm^2 pump-power density). They used a 10 mm KBO crystal (see Table 3 for its basic parameters) and achieved an approximate 1 kW output peak power.

In the picosecond range, many authors, starting from one of the earliest works of Shapiro [31] and Glenn [32], have studied SHG process both theoretically and experimentally. A high-efficiency SHG of a picosecond Nd:YAG laser (λ = 1.064 μm, 35-ps pulse duration, and 1.5–4 GW/cm^2 pump-power density) was demonstrated by Hung et al. [33]. In this experiment, an energy conversion efficiency of 60–65% is achieved when a 15 mm LBO crystal is used. Ebrahimzadeh and colleagues [34] used a frequency-doubled, mode-locked, Q-switched, diode-laser-pumped Nd:YLF laser (λ_ω = 1.047 μm, $\lambda_{2\omega}$ = 523.5 nm, 182-MHz-mode-locked and 500-Hz-Q-switched repetition rate, 55 ps pulse duration of a single pulse in a train of 105 ns) to pump an LBO-based parametric oscillator.

In the cw regime of Nd:YAG lasers, harmonic generation processes have been studied using both external [35,36] and intracavity [37–41] techniques for harmonic generation. An external-cavity SHG of a cw Nd:YAG laser (λ = 1.064 μm, 18-W pump power) in a 6 mm LBO crystal has reported by Yang et al. [35]. In this case, the conversion efficiency was 36%. Ou et al. [36] obtained a much higher conversion efficiency (85%) for frequency doubling of a cw TEM$_{00}$-mode Nd:YAIO$_3$ laser (λ_ω = 1.08 μm, $\lambda_{2\omega}$ = 540 nm, 700-mW pump power). As a nonlinear crystal, a 10 mm KTP crystal inside an external ring cavity is used Hemmati [37] demonstrated a cw diode-pumped self-frequency-doubled Nd:YAB laser (λ_ω = 1.063 μm, $\lambda_{2\omega}$ = 531 nm, 1-W pump power). The conversion efficiency obtained from a self-frequency-doubling Nd:YAB crystal is 4%. The first all-solid-state cw UV laser source with an output power more than 1 W was presented by Oka et al. [38]. The authors used an intracavity-frequency-doubled cw Nd:YAG laser (λ_ω = 1.064 μm, $\lambda_{2\omega}$ = 532 nm, 2.9-W SHG output power) as a pumping source for the fourth-harmonic generation ($\lambda_{4\omega}$ = 266 nm, 1.5-W output pump) in a 5 mm long BBO crystal placed inside an external ring cavity.

Harmonic Generations of Tunable Lasers.   Using a tunable dye laser (λ = 410–430 nm, 8 ns pulse duration, and 120 MW/cm^2 pump-power density) pumped by a Q-switched Nd:YAG laser (λ_p = 532.1 nm), Kato [42] obtained a SHG-based UV laser radiation tunable in the range 204.8–215 nm. The SHG process is realized in a 9 mm BBO crystal whose phase-matching angle is changed from 90 to 72 degrees to obtain a continuously tunable output emission. Shorter-wavelength UV radiation (up to 201.1 nm) is generated by an SFG process when the dye laser's is SHG mixed with the Nd:YAG pumping wavelength. Kato [43] mixed the SHG of a tunable dye laser (λ_1 = 613.1–595 nm) and the fundamental wavelength (λ_2 = 1.0642 μm) of a Q-switched Nd:YAG laser by an SFG process in an 8-mm LBO crystal. Tunable UV radiation in the range 232.5–238 nm was thus obtained. Clab and Hessler [44] used an 8.2 mm BBO crystal to generate continuously tunable UV laser radiation near 200 nm (λ = 197.4–203.2 nm) by SFG mixing of tunable dye laser radiation (λ = 592.2–608 nm, 5 ns pulse duration, and 50 MW/cm^2 pump-power density) with its SHG (λ = 296.1–304.8 nm, 5 ns pulse duration, and 50 MW/cm^2 pump-power density).

Harmonic Generations of Ti:Sapphire Lasers. Among modern laser technology systems, Ti:sapphire lasers are one of the most suitable laser sources for the development of all-solid-state nonlinear-conversion-based laser systems in the UV and VUV. Typically, Ti-sapphire lasers are tunable near 800 nm. However, using frequency quadruplers based on two sequential SHG stages or other cascade nonlinear processes, it is easy to move to the 200 nm spectral region, providing high peak powers and high conversion efficiencies.

Generation of optical harmonics of Ti:sapphire lasers has been realized for both pulsed (including nanosecond, picosecond, and femtosecond pulses) and cw Ti:sapphire lasers. The first SHG of a nanosecond Ti:sapphire laser (λ = 700–900 nm tuning range; 12 ns and 25 ns pulse duration at λ = 780 nm and λ = 715 nm, respectively; 25 Hz repetition rate; and 35 mJ output energy at λ = 780 nm) was demonstrated by Skripko et al. [47]. The authors used a 5 mm LBO crystal and obtained an SHG tuning range in the interval 350–450 nm with a conversion efficiency of 30%. In the picosecond range, Nebel and Beigang [48] reported on the generation of second, third, and fourth harmonic of a mode-locked Ti:sapphire laser (λ = 720–850 nm tuning range; 1.5 ps pulse duration; 82-MHz repetition rate; and 1.75 W output power at λ = 790 nm). They used three types of nonlinear crystals (LBO, BBO and LiJO$_3$). By tuning over the whole range of the Ti:sapphire laser, they obtained a broad tunable range from 205 to 525 nm. In the femtosecond range, an intracavity UV SHG in a 0.057 mm BBO crystal was realized by Edelstein and co-workers [49]. The authors used a femtosecond, passively mode-locked, dye laser (λ_{max} = 620 nm, 49 fs pulse duration) as pumping source and obtained UV SHG femtosecond radiation with the following parameters: 43 fs pulse duration, 100 MHz repetition rate, 20 mW power. Rotermund and Petrov [50] described a high-repetition femtosecond laser system operating in the 200 nm spectral range based on the FOHG of a femtosecond Ti:sapphire laser. They used a Tsunami Ti:sapphire laser (Spectra-Physics) with an 85 fs pulse duration that provides a tunable range from 760 to 845 nm with an average power of 2 W at 82-MHz repetition rate. The shortest wavelength achieved with this system is 193.7 nm [51] Bourzeix et al. reported UV generation at 205 nm by two frequency-doubling steps of a cw Ti:sapphire laser ($\lambda = 820$ nm, 2.2 W power). The first SHG ($\lambda = 410$ nm) was realized by a 10.7-mm-long LBO crystal and the second ($\lambda = 205$ nm) by a 13.8 mm BBO crystal.

C. UV and VUV Tunable Solid-State Laser Sources Based on Nonlinear Frequency Upconversion

The all-solid-state concept for generation of UV/VUV tunable laser radiation is maximally comparable with the nonlinear frequency upconversion techniques based on the second-order nonlinear optical effects. Basic physical principles of these nonlinear effects are presented in Sec. II.A.1. and II.A.3. In practice, the SFG and OPO frequency upconversions are the most extensively applied frequency upconversion techniques [53–75]. In this section, we summarize the basic results achieved by these techniques on UV/VUV tunable laser generation.

1. Nonlinear Crystals for Frequency Upconversion

Basic parameters of both conventionally used (BBO, LBO, and KB5) and some newly developed (CBO, CLBO, and LB4) nonlinear crystals applicable for UV/VUV SFG and OPO are shown in Table 3, similarly to Table 2 (see Sec. II.B.2). In fact, the well-known nonlinear crystals KDP and ADP presented in Table 2 have been also widely used for UV SFG since they have a cutoff wavelength of UV transmission below 200 nm. However, because their near-IR transmission is limited to about 1.55 μm, they can not be used for SFG processes below 190 nm. BBO and LBO, both discovered by Chen and co-workers in 1984 and 1988, respectively [20,19], are among the most commonly used frequency upconversion crystals. BBO possesses a large spectral transmission range with a small absorption, a large temperature phase-matching bandwidth and relatively high breakdown threshold. It has the largest birefringence among all-VUV transparent nonlinear crystals. LBO combines a very high damage threshold and a high nonlinearity with a relatively large acceptance angle bandwidth and low birefringence. It has a small walk-off between fundamental and generated radiation that allows the use of long crystals and the generation of laser beams with good spatial mode quality.

LBO and BBO are especially suitable for frequency conversion of high-intensity laser radiation. Recently developed borate nonlinear crystals CBO (introduced by Wu et al. in 1993 [53]) and CLBO (introduced by Mori et al. [54]) offer advanced features in nonlinear upconversion processes. The CBO crystal is transparent far into the UV (up to 167 nm) and has relatively high nonlinearity and damage threshold. However, the tuning range of this crystal is limited to about 184 nm due to the small birefringence and the vanishing nonlinear optical constant. Because of its adequate birefringence and larger effective nonlinear constant, CLBO has greater potential than LBO for effective frequency conversion below 190 nm. Unfortunately, it has a cutoff wavelength at 180 nm. In addition, CLBO is hydroscopic. That creates serious troubles also during crystal fabrication. KB5 has been used as a nonlinear crystal for frequency conversion in the UV and VUV for more than three decades. It possesses one of the shortest cutoff wavelengths of UV transmission, which allows it to be used for SFG processes below 190 nm. The nonlinearity of KB5 is smaller than that of BBO and the crystal is hydroscopic. LB4 has relatively lower nonlinear coefficients. However, its effective nonlinearity is maximum at an angle of 90 degrees. Moreover, this crystal has a higher damage threshold than LBO and its hygroscopicity is smaller than for other nonlinear borate crystals. LB4 has a significant potential in the VUV region below 180 nm. The minimum estimated SFG wavelengths that can be achieved by LB4 and KB5 are 170.3 nm and 164.6, respectively [55,56]. In addition, one of the most recently developed nonlinear crystals is also included in Table 3. It is the KBBF crystal introduced by Chen et al. in 1996 [57]. Among all the crystals presented, KBBF has the shortest cutoff wavelength UV transmission. It transmits down to 155 nm, which is below 157 nm, the shortest wavelength achieved by the F_2-excimer laser. Although KBBF is not yet commercially available, the initial experiments with this crystal, when VUV SHG in the range 184.7–200 nm is observed, as well as some estimated minimum SFG wavelengths [55] confirm its promising SFG features.

2. UV/VUV Laser Generation Using Nonlinear Frequency Upconversion

In accordance with the previous analyses on short-wavelength frequency conversion processes and techniques (see Sec. II.A.1, II.A.3., II.C.2.), among the most suitable and practical techniques leading to UV/VUV-tunable generation are SFG and OPO frequency upconversions. Here, we will consider the main results achieved by these techniques in both nanosecond and ultrashort pulse ranges.

In the earlier frequency upconversion experiments conducted in the nanosecond range [59–67], Kato [59] has demonstrated a UV-tunable SFG at 207.3–217.4 nm by a type I frequency mixing of the fundamental ($\lambda = 622$ nm, 10 ns pulse duration, 10 Hz repetition rate, and 4.2 MW pump-power) and second harmonic ($\lambda = 311$ nm and 1.4 MW pump power) of a visible dye laser in a KB5 crystal. Kato [60], using the same type KB5 crystal, obtained a tunable UV SFG up to 196.6 nm when the fourth harmonic of a Q-switched Nd:YAG laser ($\lambda = 266$ nm, 12 ns pulse duration, 10 Hz repetition rate and 20 MW power) is mixed with a near-IR dye laser ($\lambda = 721–770$ nm and 1.2 MW power). Muckenheim et al. [61] used an 8 mm BBO crystal to generate tunable (188.9–197 nm) UV SFG radiation down to the end of the transparency range below 190 nm. They realized type I SFG mixing between a tuned dye laser (λ = 780–950 nm, 15 ns pulse duration and 4.5 mJ energy) and a frequency-doubled dye laser at fixed wavelength λ = 497 nm. A sum-frequency mixing [$\omega_3 = \omega_1 + \omega_2$, see Eq. (26)] of the fourth (λ_1 = 266 nm) or the fifth (λ_1 = 213 nm) harmonic of a pulsed Nd:YAG laser and the tunable infrared radiation from an OPO using a BBO crystal (BBO-OPO, λ_2 = 1.22–1.6 µm) has been presented by Borsutsky et al. [62]. The SFG is realized in a LBO crystal and produces tunable laser radiation in the range λ_3 = 188–242 nm. A similar SFG arrangement, but using a CBO crystal, has been presented by Kato [63]. The authors obtained a tunable SFG up to 185 nm by sum-frequency mixing of the fifth harmonic (λ = 213 nm) of a Nd:YAG laser and the output of a KTP-OPO tuned in the mid-IR range (2.8–3.2 µm).

In the range of short (picosecond [68,69] and femtosecond [55,56,70–74]) laser pulses, Nebel and Beigang [68] have employed a cw mode-locked tunable Ti:sapphire laser (λ = 720–850 nm, 1.4 ps pulse duration, 82 MHz repetition rate, and 1.5 W power at λ = 790 nm) for a sum-frequency mixing of the fundamental and third harmonic (λ = 240–285 nm, 150 mW power) of this laser. Figure 8 shows the experimental arrangement of the SFG system. The authors used an 8-mm-long LBO crystal for the SHG (40% conversion efficiency), a 9 mm BBO crystal for the THG (30% conversion efficiency), and an 8.5 mm BBO crystal for the SFG (0.7% conversion efficiency from the fundamental radiation). The shortest wavelength they obtained is 192.5 nm. An all-solid-state UV pico-second optical parametric generator using a type II phase-matched LBO crystal is described by Izawa et al. [69]. As a pumping laser, the fourth harmonic of an actively mode-locked Q-switched Nd:YAG laser ($\lambda_{4\omega}$ = 266 nm, 70 ps pulse duration, 3 GW/cm^2 pumping intensity) was used. The optical parametric generator produced a narrow linewidth (0.1 nm FWHM) UV radiation at λ = 326.7 nm. The first experimental realization of a tunable femtosecond fourth harmonic generation in the near-VUV range (189–200 nm) by sum-frequency mixing between the fundamental and third harmonic was demonstrated by Ringling et al. [70]. The nonlinear SFG scheme is shown in Figure 9. The authors used an amplified femtosecond Ti:sapphire laser (λ = 756–800 nm, 150 fs pulse duration, 300 µm single-pulse energy at 1 kHz repetition rate) and obtained tunable UV pulses with energies as much as 4 µJ at λ = 200 nm. Utilizing the same type of femtosecond amplified Ti:sapphire laser, Seifert et al. demonstrated a SFG of 200 fs VUV (172.7–187 nm) pulses in a 1 mm LBO crystal [71]. The sum-frequency mixing is realized between the Ti:sapphire's fourth harmonic and parametrically generated tunable IR pulse (λ = 1.65–2.15 µm). Petrov et al. [56] have presented a tunable VUV SFG of 100 fs pulses down to 170 nm employing a 107 µm LB4 crystal. They used a sum-frequency mixing of the fourth harmonic ($\lambda_{4\omega}$ = 189–210 nm)

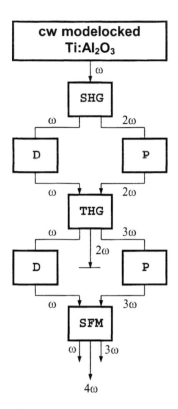

Figure 8 A typical experimental configuration of frequency upconversion system based on sum-frequency mixing (SFM) of the second and third harmonic of a cw mode-locked Ti:sapphire laser. D, compensation of the group-velocity-difference-caused temporal delay; P, polarization adjustment. (From Ref. 68.)

of a commercial femtosecond Ti:sapphire laser (λ_ω = 756–840 nm, 100 fs pulse duration, 1 kHz repetition rate, 600 µJ energy) and tunable laser radiation of a near-IR optical parametric generator (λ = 1.6–2.5 µm). In this experiment, a VUV SFG in the range 170–185 nm with conversion efficiency of 4% is achieved.

D. UV Tunable Fiber Raman Lasers

Optical fibers (OFs) are an effective solid-state medium for broadband optical frequency conversions in the spectral range covering the UV, VIS, and near-IR. When powerful laser emission is propagated in OFs, conditions for effective development of such basic nonlinear processes as stimulated Raman scattering

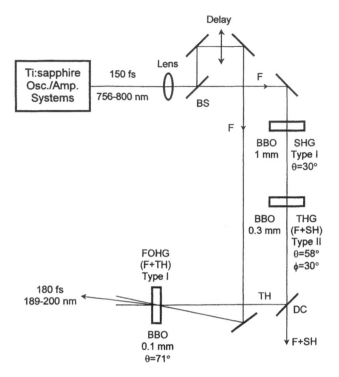

Figure 9 An experimental noncollinear SFG set up. BS, beam splitter; F, fundamental; SH, second harmonic; TH, third harmonic; DC, dichroic mirror; θ, φ, crystal angles. (From Ref. 70.)

(SRS), self-phase modulation (SFM), and four-photon mixing (FPM) are created [76–78]. The high efficiency of these processes in OFs can be attributed to the simultaneously achievement of a long effective length of interaction (10^2–10^4 m) and a high density of the exciting power (higher than 10^8 W/cm^2) because of the microscopic fiber core diameters. As a result of the mutual operation of the nonlinear effects, broadband continuum spectra are generated in OFs. In principle, multicascade SRS and, to a lesser degree FPM, form the discrete structure of the separate spectral components. SPM and FPM lead to broadening of these components and thus the spectrum is filled and a broadband continuum is obtained. The continuum spectra provides attractive tunable (discrete or continuum) laser sources for practical applications.

Nonlinear effects in OFs, in particular, SRS and its application for the development of fiber Raman lasers, have been investigated considerably more intensively in the near-IR and visible spectral regions, in comparison with the UV, mainly due to very low attenuation losses and low material dispersion of OFs in these regions. Nevertheless, investigations of SRS and its applications in the UV region have been motivated by continuously increasing interest in applications in various research areas associated with shorter wavelength range, as well as the availability of conventional UV fibers with low losses reduced to Rayleigh scattering limit.

Nonlinear properties of the silicon-based OFs (on the base of SiO_2) are determined by the third-order nonlinear susceptibility $\chi^{(3)}$. These media are centrosymmetrical and the quadratic nonlinear susceptibility $\chi^{(2)}$ is zero (see Sec. II.A.). The various nonlinear effects are related to the two $\chi_S^{(3)}$ components according to the expression:

$$\chi^{(3)} = \chi_R^{(3)} - i\chi_I^{(3)} \tag{27}$$

where the real (nonresonance) part $\chi_R^{(3)}$ determines SFM, FPM, and optical Kerr effect, and the imaginary part $\chi_I^{(3)}$ determines SRS and stimulated Brillouin scattering. In comparison with typical conventional nonlinear media having non-waveguide structure, the value of the nonlinear susceptibility $\chi^{(3)}$ for SiO_2 is significantly lower. For instance, while $\chi^{(3)}$ for SiO_2 is 1.4×10^{-22} m^2/V^2, $\chi^{(3)}$ for the nitrobenzene is 39×10^{-22} m^2/V^2, and for sulfur-carbonate it is 159×10^{-22} m^2/V^2 [76]. This obvious disadvantage of OFs is compensated multiply by their unique property to maintain high levels of exciting power density at long distances.

1. Stimulated Raman Scattering in Optical Fibers

Stokes and Anti-Stokes Raman Scattering. Raman scattering is an inelastic photon–phonon interaction of exciting light emission with vibrations (optical phonons) of the crystal lattice. The incidence electromagnetic field with optical frequency ω_p causes changes in the electron configuration as well as displacements of the molecular nuclei. As a result, molecules of the optical medium start to vibrate with their own frequencies ω_R (so-called Raman frequencies), which are specific unique features of each optical medium. An opposite effect is incurred too. The molecular vibrations with frequencies ω_R modulate the incidence light field with frequency ω_p. In this way, light emission with combined frequencies $\omega_S = \omega_p - \omega_R$ and $\omega_{AS} = \omega_p + \omega_R$ occurs. Here, the emission with lower-frequency ω_S corresponds to Stokes scattering (red satellite) and the emission with higher-frequency ω_S to anti-Stokes scattering (blue satellite).

From a quantum-mechanical point of view, we can consider each molecule as a quantum system that contains discrete electron levels and vibration sublevels. Consider two vibration levels of an electron level (Fig. 10). The energy level k is assumed to be a basic (nonexcited) level, the level n is a higher (excited) level, and the energy difference between k and n is $\hbar\omega_R$, where $\hbar$ is Planck's constant

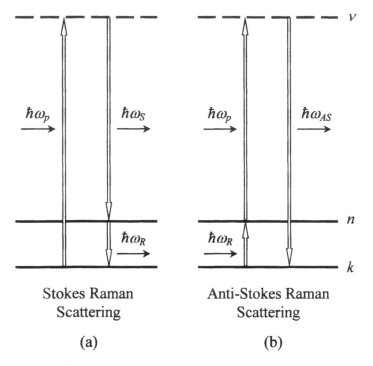

Stokes Raman   Anti-Stokes Raman
Scattering   Scattering

(a)   (b)

Figure 10 Energy-level diagrams of (a) Stokes ($\omega_S = \omega_p - \omega_R$) and (b) anti-Stokes ($\omega_{AS} = \omega_p + \omega_R$) Raman scattering.

and ω_R is the Raman frequency. In this case, the Stokes Raman component with frequency $\omega_S = \omega_p - \omega_R$ is obtained (Fig. 10a) when a light energy quantum $\hbar\omega_p$ excites the molecule from basic level k to a virtual higher level v possessing a relatively short relaxing time. Then the molecule relaxes to the other level n and a Stokes light quantum with smaller energy $\hbar\omega_S = \hbar\omega_p - \hbar\omega_R$ is emitted. As can be seen, the Stokes component is obtained from molecules initially only on the basic level. The anti-Stokes Raman component with frequency $\omega_{AS} = \omega_p + \omega_R$ is formed (Fig. 10b) only from molecules that are already excited to the upper level n by a preliminary Stokes process. An energy quantum $\hbar\omega_p$ falls on a molecule excited to the level n. After its fast relaxation from a higher virtual level v, the molecule returns to the basic level k, emitting an anti-Stokes light quantum with higher energy $\hbar\omega_{AS} = \hbar\omega_p + \hbar\omega_R$. The intensity of the Stokes Raman components is significantly stronger than that of the anti-Stokes components, because the molecule population of the basic level k is much higher than of the excited level n.

Spontaneous Raman Scattering. Spontaneous Raman scattering in optical media is observed at weak exciting light fields. In this case, the molecule vibrations have a fluctuation (incoherent) character. Hence, the scattering light waves are also incoherent and have 10^6–10^8 times smaller intensity than that of the exciting light.

The most frequently used OFs are produced on a base of oxide glasses (SiO_2). These fibers possess an amorphous structure. In such optical media, frequencies of the molecule vibrations form nonhomogeneous broadened bands that are mutually covered and create a continuum frequency distribution. Because of this, the spontaneous Raman spectrum of silicon glasses has a continuum character and covers a frequency shift range from 0 to 1000 cm^{-1}. Figure 11 shows

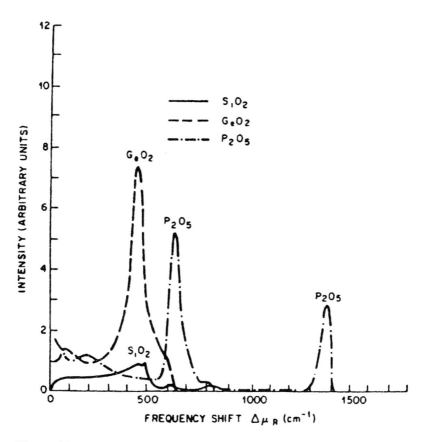

Figure 11 Typical spontaneous Raman scattering spectra of some oxide glasses such as SiO_2, GeO_2, and P_2O_5, which are used in OF production. (From Refs. 77 and 79.)

typical spontaneous Raman spectra of oxide glasses used for fiber production [77,79]. The intensity maxima of the spontaneous Raman scattering for both SiO_2- and GeO_2- doped fibers are in the range 440–460 cm^{-1}. The Raman spectrum of P_2O_5-doped fibers has two maxima, at 650 cm^{-1} and 1330 cm^{-1}, respectively.

Stimulated Raman Scattering. SRS is observed at strongly (laser) excited light fields. It can be characterized by the following features. First, the incidence light wave excites phase-matched stimulated vibrations and the scattered light is coherent (see Fig. 12a). Second, stimulated vibrations are accumulated, the vibration level population is changed, and conditions for anti-Stokes Raman scattering are created. Third, besides the Stokes and anti-Stokes first-order components, the scattered light contains components of higher orders ($\omega_p \pm n\,\omega_R$), that is, a cascade (or multi-Stokes) conversion of optical frequencies is developed (Fig. 12b). Fourth, with regards to the level of the exciting light power, the SRS is a threshold process. Finally, the spectral width of the SRS generation is narrower and the intensity of the SRS components is much stronger (about e^{10}–e^{20} times) than at spontaneous Raman scattering.

The process of SRS in OFs has specific behavior due mainly to the fact that the OF, as a waveguide medium with long effective length and small transverse dimensions, provides maintenance of high pump power density over a long distance. Spatial evolution of the SRS along the fiber length can be described in

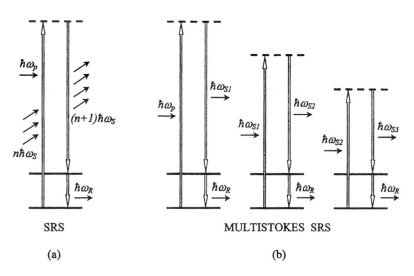

Figure 12 Energy-level diagrams of (a) stimulated and (b) cascade (or multi-Stokes) Raman scattering.

the following manner. In the initial part of the OF length, a spontaneous Raman scattering signal is obtained. This signal is weak; however, it covers the whole frequency region corresponding to the spontaneous Raman scattering. When the pumping power reaches the SRS threshold, the weak signal from the spontaneous scattering is amplified according to an exponential law that can be expressed [76] as;

$$S(\omega_S) \sim \exp[g_R(\omega_S)] \tag{28}$$

Here $S(\omega_S)$ is the spectrum observed at Stokes frequencies and $g_R(\omega_S)$ is the Raman gain. The exponential amplification is strongly nonuniform and it is maximally for those lines at the maximum of spontaneous spectra, that is, with maximum Raman gain. This process of nonuniform amplification of SRS lines leads to a significant reduction of their spectral linewidth. On the other hand, because of the exponential increasing power of the first Stokes order (S_1), an effect of pump power depletion appears. Then, the S_1 power reaches the threshold level needed for excitation of the second Stokes order (S_2). At this moment, saturation of the Raman gain is achieved and the S_2 power increases exponentially. Hence, here S_1 is in the role of pumping emission. The processes leading to the S_2 generation (and to the generation of higher-order Stokes components S_3, S_4, S_5, etc., respectively) are identical with those for the S_1 generation. Thus, a multi-Stokes generation is realized (see Fig. 12b) and the Stokes part of the SRS spectrum is formed. Simultaneously, each Stokes component prepares the optical medium for generation of the corresponding anti-Stokes component from the SRS spectrum.

For analytical description of the SRS in OFs, two basic theoretical models developed by Smith [80] and AuYeung and Yariv [81,82] can be applied. Smith's model is used at relatively weak nonlinear interactions when the pump power threshold for the SRS generation should be estimated. AuYeung and Yariv's model, improved by Yijiang and Snyder [83], is applied at an effective Stokes (especially at multi-Stokes) conversion and includes the effects of pump depletion and Raman gain saturation. According to Smith's model, the evolution of a Stokes power signal $P_S(L)$ along the fiber length can be presented in the form:

$$P_S(L) = P_S(0) \exp[(-\alpha L) + (g_R P_p(0)/A_{eff})L_{eff}] \tag{29}$$

Here, $P_p(0)$ is the pump power at the fiber beginning, g_R is the Raman gain of the SRS in the medium, α is the linear attenuation coefficient. L_{eff} is the effective length of the OF and can be characterized by the expression [76]:

$$L_{eff} = [1 - \exp(-\alpha L)]/\alpha. \tag{30}$$

A_{eff} is the effective transverse area of the OF that, in the case of multimode OF, coincides with the transverse area of the OF core. However, in the case of single-mode or graded-index, OF is given by the formula $A_{eff} = 2\pi(\omega_{eff})^2$, where the effective radius of the OF core ω_{eff} can be estimated by the dependence:

$$\omega_{\text{eff}} = a[0.65 + 1.619/(V_p)^{3/2} + 2.879/(V_p)^6] \tag{31}$$

Here, a is the OF core radius and V_p is the normalized frequency at pumping laser wavelength λ_p. The Raman threshold power P_{th}^R is defined as the input pump power $P_p(0)$ at which the power of the Stokes component $P_S(L)$ and the pump power $P_p(L)$ at the OF end are equal. Using the above dependencies, we can obtain a practical formula that allows the Raman threshold power P_{th}^R to be estimated:

$$P_{\text{th}}^R = 16(bA_{\text{eff}})/g^R L_{\text{eff}} \tag{32}$$

where b is a coefficient depending on the polarizing properties of both the pumping laser emission and OF.

The efficiency of the SRS in OFs depends on the properties of both optical medium and exciting laser emission. These properties can be summarized in three main groups: (1) material parameters, (2) waveguide parameters, and (3) parameters of the pump laser emission. The first group includes such material parameters of the fiber-optic medium as spontaneous Raman spectra (see Fig. 11), Raman gain, material absorption among others. From Figure 11's Raman spectra of glasses doped with SiO_2, GeO_2, and P_2O_5, it can be seen that the maximum-frequency Raman shift (1330 cm^{-1}) has P_2O_5-doped OF. Larger Raman shifts of 3000 cm^{-1} and 4200 cm^{-1} are achieved when D_2- and H_2-doped OFs are used, respectively [84,85]. Moreover, according to the same Raman spectra, the Raman gain of GeO_2-doped OFs is about 10 times higher than that of SiO_2-doped OFs. However, the SiO_2-doped OFs have low attenuation losses (below 1 dB/km), while the losses in GeO_2-doped OFs exceed 50 dB/km. P_2O_5-doped OFs have similar behavior. Hence, depending on the concrete requirements of the output Raman spectra, optimum material OF parameters should be found. The second group of OF parameters includes such waveguide OF parameters as attenuation losses (α), effective OF length (L_{eff}), and transverse area (A_{eff}), optical birefringence, among others. In accordance with the dependencies [29–32], a decreasing of attenuation losses leads to increasing effective OF length. In combination with a decreasing of the OF transverse dimensions this leads to decreasing SRS threshold power and increasing of SRS output power. Here, as with the material OF parameters, we have to optimize waveguide properties of the OFs depending on our purposes. The third group of parameters includes properties of the pumping laser emission: pump power, pulse duration, and spectral linewidth. For an effective SRS process in OF, the pump laser power should be within the interval between two limitations. It should be higher than the SRS threshold power and lower than the critical laser power required to damage the OF material. Typical experimental data for pump power densities needed for SRS in OF are power densities in the interval $10^8–10^{11}$ W/cm^{-1} [76,77]. Depending on the laser pulse duration, two basic regimens of the SRS in OFs are observed [76]. If the pulse duration is longer than a typical time for transverse relaxation of excited states, a quasistationary (or quasi-cw) SRS regime in the OF medium is settled. It is a situation typical for pump laser pulses with nanosecond (1–100 ns) duration generated by lasers with Q-switched modulation. When the pulse duration is shorter than the specific relaxation time, we have a nonstationary SRS regimen in the OF. This regimen is typical for ultrashort laser pulses (<100 ps). In this case, an effect of increasing of the group-velocity dispersion occurs and special techniques for compensation of this effect need to be applied.

2. Four-Photon Mixing in Optical Fibers

Four-photon mixing (FPM) is a nonlinear optical process in which two exciting photons with frequencies ω_{p1} and ω_{p2} are converted into a Stokes ω_S and an anti-Stokes ω_{AS} photon by analogy with SRS, because $\omega_{AS} > \omega_{p1} > \omega_S$ (see Sec. II.A.1). In basic FPM variant, the exciting photons possess equal frequencies ($\omega_{p1} = \omega_{p2} = \omega_p$) that, in combination with the signal frequencies ω_{AS} and ω_S, satisfy the law of energy conservation

$$2\omega_p = \omega_{AS} + \omega_S \tag{33}$$

The FPM frequencies ω_{AS} and ω_S are symmetrically disposed towards ω_p and are shifted by the interval $\Delta\omega_F$ that can be expressed by the dependence

$$\Delta\omega_F = \omega_p - \omega_S = \omega_{AS} - \omega_p \tag{34}$$

The FPM frequency shifts of $\Delta\omega_F$ in OFs cover a broad spectral range between 100 cm^{-1} and 5500 cm^{-1} [76,77,86,87].

The principal difference between the SRS and FPM is that while the SRS is a self-matched process, the FPM needs a phase-matching condition to be satisfied. This condition is necessary because of differences existing between the propagation constants of the interacting waves and, in the case of fiber-optic media, it can be described by the expression (see Fig. 13):

$$\Delta k = 2k_p - k_{AS} - K_S = 0 \tag{35}$$

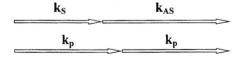

Figure 13 Collinear phase-matching condition [$\Delta k = 2k_p - k_{AS} - k_S = 0$, Eq. (35)] in an OF required at four-photon mixing (FPM) between the pumping, anti-Stokes, and Stokes waves having wave vectors k_p, k_{AS}, and k_S, respectively.

Here, k_p, k_{AS}, and k_S are wave vectors of exciting, anti-Stokes, and Stokes waves, respectively. In nonwaveguide media with normal material dispersion, collinear phase matching is impossible because always $(k_{AS} + k_S) > 2k_p$. However, in the case of waveguide media (including OFs), it is possible for the material dispersion to be compensated and, therefore, collinear phase matching conditions over long interaction lengths can be created. In multimode OFs, phase-matching conditions are realized when the interacting waves (exciting, Stokes, and anti-Stokes) are propagated in waveguide modes for which the material dispersion is compensated by the mode dispersion. In single-mode OFs, the material dispersion is compensated by variations of the waveguide dispersion.

3. Self-Phase Modulation in Optical Fibers

SPM is a nonlinear effect (similar to the optical Kerr effect in the time scale) defined by the intensity dependence of the refractive index of the optical medium. If a light pulse with intensity $I(t)$ is propagated in an OF, this dependence can be written in the form [8]:

$$n(t) = n_0 + n_2 I(t) \tag{36}$$

where n_0 and n_2 are linear and nonlinear parts of the refractive index $n(t)$, respectively. SPM is a threshold process and the minimum exciting power P_{th}^S necessary for obtaining it is given by an expression similar to the Eq. (32) related to the SRS threshold power P_{th}^R [8,76]:

$$P_{th}^S = (\lambda_p A_{eff})/(2 \pi n_2 L_{eff}) \tag{37}$$

Using dependencies from Eqs. (29–32) and (37), a comparable estimation of the threshold powers P_{th}^S and P_{th}^R can be made. If we assume the following typical values of the input parameters: $\lambda_p = 1.064$ μm; $2a = 50$ μm; $L = 1$ km; $\alpha = 4.6 \times 10^{-6}$ cm^{-1} (2 dB/km); $g_R = 9.20 \times 10^{-12}$ cm/W; and $n_2 = 3.2 \times 10^{-16}$ cm^2/W [8,76], we obtain $P_{th}^R = 850$ W and $P_{th}^S = 12$ W, and for the relation $P_{th}^R/P_{th}^S = 70$.

Spectral broadening at SPM is observed predominantly at short picosecond pump pulses when the combined action of SPM and group-velocity dispersion occurs. Then, an abnormal dispersion occurs in OFs, conditions for propagation of optical solitons are created, or, at normal dispersion, pulse shortening occurs. SPM is observed also at nanosecond pump pulses. In this case, the spectral broadening is attributed to the existence of a fine structure in the pump pulse containing subnanosecond pulses.

4. UV Raman Scattering in Optical Fibers

The broadband continuum spectra generated in OFs as result of mutual action of such nonlinear effects described in the previous section as SRS, FPM, and

SPM provide an attractive basis for development of tunable fiber Raman lasers (FRLs). These laser sources offer a highly efficient and low-threshold means of nonlinear frequency conversion and widely tunable generation covering the UV, visible, and near-IR spectral ranges. Two basic operating FRL regimens have been established. *First* is the single-pass FRL (Fig. 14a) in which there is no cavity feedback and multimode cascade Stokes emission is generated. *Second* is the fiber Raman oscillator (Fig. 14b), in which an OF is placed inside a Fabry-Perot cavity. The laser emission can be widely tuned if a frequency-selective element is used, such as prism–mirror combination, grating, or intracavity interferometer [76,77]. As a nonlinear medium for the FRLs, small-core single-mode fibers are usually used. Nevertheless, multimode graded-index fibers possess several essential advantages, from the nonlinear frequency conversion point of view, such as more efficient launching and maintenance of high pump power, higher overall Stokes power generation, and broadband frequency conversion. They can therefore, be applied in FRL systems for the generation of widely tunable emission, especially in the nanosecond range in which effects such as SPM and group-velocity dispersion are negligible [76,77]. In the visible range, broadband nonlinear frequency conversion in both single-mode and multimode OFs is obtained by pumping with various laser sources such as the second harmonic ($\lambda_p = 532$ nm) of Q-switched or mode-locked Nd : YAG laser or the dye and argon ($\lambda_p = 514$) laser [76,77,88]. The study of nonlinear frequency conversion in OFs is especially intensive in the near-IR around the fiberoptic windows at wavelengths 0.85 μm, 1.3 μm, and 1.55 μm, where the material dispersion and attenuation losses have minimal values. In this range, broadband spectra by both single-pass

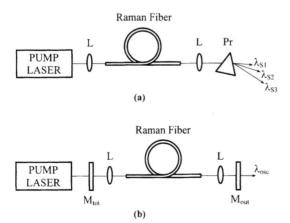

Figure 14   Two basic optical arrangements of conventional fiber Raman lasers (FRLs): (a) single-pass FRL and (b) fiber Raman oscillator.

and oscillation generation in single-mode or multimode OFs are demonstrated [76,77]. As a conventional pumping source, Nd:YAG laser at $\lambda_p = 1.064$ μm or $\lambda_p = 1.32$ μm is used.

The development of FRLs in the UV range is more complicated, due mainly to the relatively higher attenuation losses and material dispersion of OFs in this range. However, the recent progress in manufacturing UV OFs with low attenuation losses comparable to those of near-IR OFs, as well as the increasing interest in various laser applications associated with shorter wavelength range, have stimulated research efforts directed toward development of UV FRLs.

The first experimental results with FRLs on the short wavelength side have been demonstrated by Lin and Stolen [88]. They used a 20 kW 10 ns blue dye laser to pump a 7 μm core-diameter 19.5 m silica-core OF. A combination of SRS and FPM produced a broad continuum spectrum in the interval 392–537 nm. Furthermore, early experiments on the first-order and second-order Stokes generation in UV OFs has been presented also by Rothschild and Abad [89], and by Pini et al. [90], respectively. In Rothschild and Abad's experiment, a 4-ns nitrogen laser pulse ($\lambda_p = 337.1$ nm) is used for the first-order Stokes generation in a 200 μm-core-diameter 50 m fused silica OF. The output spectrum contains a broad asymmetrical Raman band with a peak at $\lambda_{S1} = 342.3$ nm, corresponding to a Raman shift of 450 cm^{-1}. Pini et al. [90] used both nitrogen ($\lambda_p = 337.1$) and excimer ($\lambda_{p1} = 308$ nm and $\lambda_{p2} = 351$ nm) pump UV lasers. They obtained the first and second Stokes components with Raman shifts between 431 cm^{-1} and 499 cm^{-1}. After these initial experiments, several consequence papers on multiple-order Stokes generation [91] and broadband frequency conversion [92] in multimode UV OFs have been reported. A broadband frequency conversion by SRS in a 200-μm silica core UV OF has been demonstrated by Pini and colleagues [91]. As a pumping laser source, a 4 MW peak-power 7 ns pulse-duration XeCl excimer laser ($\lambda_p = 308$ nm) is used. The broadband UV spectrum contains nine Stokes orders in the 308–350 nm spectral range. Raman spectrum parameters, including the wavelength of the registered Stokes orders and the corresponding Raman shifts, are shown in Table 4. Efficient SRS has been obtained when a 200 μm core-diameter 200 m fused-silica multimode UV OF is pumped by an 8 ns pulse-duration XeCl excimer laser ($\lambda_p = 351$ nm) [92]. Up to 20th Stokes orders are observed, which corresponds to a total frequency shift of about 8800 cm^{-1}. Pini and co-workers have suggested a combined Raman oscillator–amplifier system. In an XeCl ($\lambda_p = 308$ nm) excimer laser-pumped UV multimode OF, a power conversion efficiency of 80% is achieved at a pump power density of 1.1×10^9 W/cm^2.

5. UV Tunable Double-Pass Fiber Raman Laser

In principle, because of the relatively higher total attenuation losses in the UV OFs, one of the main problems in the UV spectral region is that investigations

Table 4 Experimentally Obtained Raman
Spectra Parameters

	Wavelength (nm)	Shift (cm^{-1})
L	307.9–308.1	
S1	312.4	463
S2	316.7	431
S3	320.8	403
S4	325.3	430
S5	329.9	428
S6	334.7	434
S7	339.5	428
S8	344.5	428
S9	349.1	383

Results derived from 105 m long, 100 μm core-
diameter multimode UV OF pumped by the third
harmonic ($\lambda = 355$ nm) of a Q-switched Nd:YAG
laser.

are directed to the development of single-pass FRLs, rather than to fiber Raman
oscillators, has been very successfully realized in the near-IR and visible [76,77].
A new double-pass fiber Raman laser (DFRL) concept was recently suggested
by Ilev and coauthors first in the visible and near-IR [94–96]. This laser provides
a wide tunable range and permits powerful spectral components to be generated,
due to the double passing of the nonlinear converted laser emission through the
OF, the maximum cavity feedback, the use of a multimode OF, and powerful
pumping. Profiting from these important advantages of the DFRL concept, an
alternative all-solid-state UV and blue tunable DFRL was demonstrated in Refs.
97 and 98.

Principle of Operation and General DFRL Properties. The principal optical arrangement of the DFRL is shown in Figure 15a. It is based on the application
of a simple fiber-optic autocollimation technique with Littrow-prism-tuned emission [94–96]. The operating principle of the experimental scheme is more similar
to that for the oscillation regime of the FRL than to the single-pass FRL [76,77].
In fact, this scheme corresponds to a double-pass fiber Raman laser regime, because of both double passing of the nonlinear converted laser emission through
the OF (called Raman OF) and the absence of an output cavity mirror. The double-pass regime of the DFRL is realized experimentally by two sequential stages
(see Fig. 15b). First, an initial broadband single-pass continuum is generated
(single-pass generation) after the forward passing of the pump laser emission
through the OF. Second, the broadband continuum is spectrally dispersed and

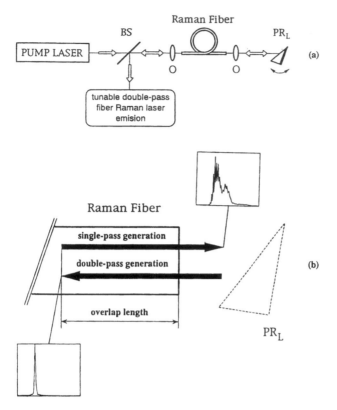

Figure 15 (a) Basic experimental configuration and (b) principal idea of the double-pass fiber Raman laser (DFRL).

reflected from the autocollimation (Littrow) prism. Next, the formatted narrow-band emission is launched again into the OF for the second passing. This emission passes through the forward single-pass emission along a spatial overlap length inside the Raman OF and the useful backward signal is amplified by the powerful forward pumping. Thus, the basic double-pass regimen (double-pass generation, Fig. 15b) of the DFRL is realized and narrow-band tunable spectral components are generated.

The key optical element in the DFRL scheme is the Raman OF used in a multifunctional regime (Fig. 15). It acts simultaneously as a point laser source for the formation of collimated input (to the Littrow prism) emission, as a highly sensitive point receiver for the autocollimation backreflectance signal, and as a medium for nonlinear optical frequency conversion and generation of a broad-band continuum spectrum. In order to be realized as a fiber laser generating high-power and widely tunable emission, whose parameters are comparable with those of a conventional broadband tunable laser source, we used a large-core and low-loss multimode OF. It ensures, compared to small-core single-mode OFs, more efficient launching and supporting of powerful laser emission, higher damage threshold of the fiberoptic material, and higher power of overall Raman Stokes generation [76,77]. A 30 degree Littrow prism (PR_L) with a wideband total reflectance coating is used both to achieve autocollimation back reflectance (maximum cavity feedback) and to disperse the broadband continuum spectrum generated in the OF. In some cases, to increase the spectral resolution for the autocollimation back reflectance after the first passing of the pump laser emission through the OF, additional dispersing prisms in the DFRL cavity may be used. The useful double-pass laser emission is reflected by a beam splitter (BS) that is a dichroic filter with a maximum reflectance for the Stokes part of the initial single-pass continuum.

On the basis of the principal DFRL arrangement presented (Fig. 15), an experimental setup for UV tunable laser generation has been designed (Fig. 16). As a pumping source, the third harmonic ($\lambda_p = 355$ nm) of an 8 ns pulse duration Q-switched Nd:YAG laser is used. The pump pulse duration corresponds to the so-called quasi-cw (or long-time) regime [76] and defines a spatial overlap length (Fig. 15b) between the forward (single-pass) and backward (double-pass) laser pulses exceeding 1.5 m. This overlap length value is close to the experimentally established minimal OF length required to create multicascade stimulated Raman scattering at powerful pumping. It also ensures significant amplification of the output DFRL emission. As an active Raman gain medium, a multimode UV all-silica high-temperature OF is used, which has the following parameters: 100 μm

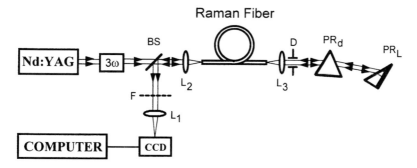

Figure 16 Experimental setup of the UV-tunable DFRL. Nd:YAG, Q-switched Nd:YAG pumping laser; 3ω, third-harmonic generator; BS, beam splitter; L_1, L_2, L_3, lenses; D, diaphragm; PR_d, additional dispersing prism; PR_L, Littrow prism; F, color and neutral filters; CCD, computerized CCD-camera-based spectrometer.

core diameter; 110 μm cladding diameter; 105 m length; NA = 0.22 numerical aperture; 250–1100 nm operating wavelength region; 130 dB/km, 13 dB/km, and 14 dB/km attenuation losses at λ = 355 nm, 488 nm, and 515 nm, respectively. With these features, the OF ensures efficient launching and maintaining of powerful pump laser emission, high damage threshold at power pumping, high power, and multicascade Raman Stokes generation. The 30 degree apex angle Littrow prism (PR$_L$), used both to achieve a maximum laser cavity feedback and to tune the fiber Raman laser emission, as well as the additional dispersing prism (PR$_d$) and focusing lenses L$_1$ and L$_2$ (with focal length 50 mm), are produced from UV-grade fused silica with high UV transmission. The cavity objective (O) is a 13$^\times$ Newport UV microscope objective. The widely tunable laser emission, formed after the second backward pass through the OF, is reflected by the beam splitter (BS; a special dichroic mirror with maximum reflectivity for the Stokes part of the spectrum) and is collected by a fiberoptic spectrometer (Ocean Optics, S1000) with 0.6 nm spectral resolution in the 340–550 nm range.

The double-pass fiber Raman laser generates UV and blue laser emission in two basic regimes: a broad-band multicascade Stokes single-pass generation and a narrow-band tunable double-pass generation. The output laser emission is formed by sequential action of the single-pass and double-pass regimes.

Single-Pass DFRL Generation.   As a result of the single-pass of the pump laser emission through the multimode UV Raman OF, an initial broad-band stimulated spectrum (from 360 nm to 527 nm) is generated, shown in Figure 17.

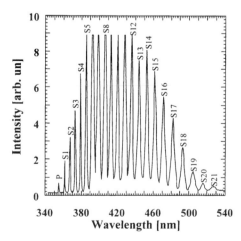

Figure 17 A multi-Stokes Raman spectrum in the 360–527 nm region generated as a result of a single pass of the pump laser emission (λ$_p$ = 355 nm, 150 kW pump power) through a 105 m, 100 μm core-diameter UV Raman OF.

The following features are observed in the spectrum. First, it clearly shows the development of multicascade Raman Stokes nonlinear conversion into the UV Raman OF. As can be seen, up to 21 well-defined discrete maxima corresponding to Raman Stokes orders are obtained. The Raman Stokes shifts are in the interval 412–486 cm^{-1} in agreement with the results obtained by other authors [76,77,91]. Second, the Raman Stokes spectrum has a clear, discrete structure. This result, which was also observed by other authors [91,92,99], represents, in the field of nonlinear fiber optics, an essential difference in the development of nonlinear processes in the near-IR and visible regions (where continuum spectra have been conventionally obtained in many experiments) from that in the UV region. The discrete structure of the registered multicascade Stokes spectrum (Fig. 17) can be explained by the relatively different contributions of basic nonlinear processes that developed in the UV fiber. Under the present experimental conditions (Fig. 16), the dominant nonlinear process in the multimode UV OF is the SRS, which, as a self-phase-matched process, comprises the energy of many fiber modes and leads to formation of the discrete spectrum structure. With regard to self-phase modulation and four-photon mixing, which are the fundamental nonlinear processes in OFs that cause a broadening of the Stokes components and lead to wideband continuum generation, there are no suitable conditions for their development for two basic reasons. One is that it is the nanosecond pump pulse duration (or so-called quasi-cw regimen [76]) that determines that the influence of self-phase modulation on continuum formation is negligible. The other is that the relatively high level of attenuation losses and material dispersion in the UV fibers leads to a considerable reduction in both the typical coherent length required for effective phase-matching parametric processes and the laser power carried by separate OF transverse modes in the framework of which parametric processes occur. Third, although powerful laser pumping (greater than 150 kW peak pump power corresponding to greater than 1.9 GW/cm^2 input pump intensity) near the damage threshold of the fiber-optic material is applied, the experimental conditions for the realization of a coherent regimen of broad-band SRS [100] are not fulfilled. This is due to the high attenuation and dispersion that increase the power threshold for the achievement of this regimen. We conclude that SRS plays a dominant role in the nonlinear processes that developed in the UV Raman OF and is responsible for the formation of the discrete structure of the registered broad-band single-pass multiple-Stokes spectrum.

Double-Pass DFRL Generation. Using the double-pass regimen of the fiber Raman laser, a UV and blue discretely tunable narrow-band emission corresponding to the various Stokes orders over the initial single-pass multicascade Stokes spectrum (see Table 5; Fig. 17) in the range 360–493 nm is generated. Figure 18 shows typical spectral distributions of the double-pass fiber Raman laser emission in the UV regions, where Figure 18a corresponds to $\lambda_{max} = 371$

Table 5 Raman Spectra Parameters

	P	S1	S2	S3	S4	S5	S6	S7	S8	S9	S10
λ(nm)	355	361.2	367.4	373.1	379.2	385.4	392.2	398.8	406.1	413.3	420.5
Shift (cm⁻¹)	0	486	469	412	437	426	445	428	448	430	414

	S11	S12	S13	S14	S15	S16	S17	S18	S19	S20	S21
λ(nm)	428.3	436	444.2	452.9	461.6	471.4	482.2	493	504.2	515.4	527.1
Shift (cm⁻¹)	429	413	425	435	417	449	475	453	452	430	429

Parameters were obtained in a 180 m long UV OF pumped by a 650 kW power XeCl excimer laser (λ = 308 nm).

nm (third Stokes order) and Figure 18b to λ_{max} = 392 nm (sixth Stokes order). The spectral linewidth of the generated components is from 70 cm^{-1} (for the low-order Stokes components) to 380 cm^{-1} (for the high-order Stokes components). A potential method for more effective laser emission generation using the DFRL scheme in a narrower spectral range (in comparison to the spectrum in Fig. 17) involves using a shorter UV Raman fiber. In this case, because of the significant reduction of the total attenuation losses in the OF, the initial single-pass spectrum contains more powerful Stokes components. The Raman amplification of the useful laser emission at the second pass through the OF is also much higher. Thus, we obtain a discrete tunable laser emission with higher (compared with the use of a longer OF, Fig. 18) output power and, in addition, the relative influence of the Raman backscattering and competition between the Stokes orders is strongly reduced. Figure 19 illustrates this scheme's potential by showing typical spectral distributions obtained from a double-pass fiber Raman laser with 25 m 100-μm core-diameter UV fiber at 150 kW peak pump power (the same power level as when the spectra from Figs. 17 and 19 are obtained), where Fig. 19a corresponds to λ_{max} = 361.2 nm (first Stokes order) and Fig. 19b to λ_{max} = 373.1 nm (third Stokes order). However, the use of a shorter OF leads to the generation of a narrower initial single-pass spectrum and, therefore, to a decrease in the laser tunable range.

It can be summarized that the presented UV DFRL possesses several basic advantages. (1) It provides a wide tunable range in the UV-blue spectral region, mainly due to the use of large-core low-loss multimode OFs as well as high pump power level. (2) It permits relatively powerful spectral components to be

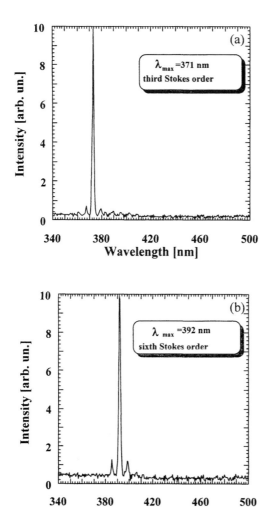

Figure 18 Typical spectral distributions of the UV DFRL generation.

generated, because of the double-pass laser regimen, maximum cavity feedback, use of a multimode OF, and powerful pumping. (3) The principal optical scheme is simplified, compact, and contains very conventional and low-cost optical elements. Moreover, compared to conventional optical parametric oscillator schemes, the principal DFRL scheme excludes critical adjustment and relatively complicated laser cavities and optical elements, as well as strongly temperature- and phase-matching-dependent optical elements such as nonlinear crystals. Also,

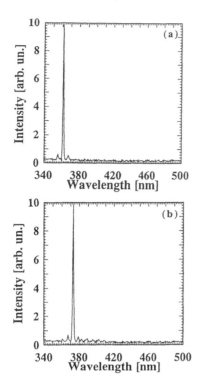

Figure 19 Spectral distributions of UV DFRL at (a) λ_{max} = 361.2 nm and (b) λ_{max} = 373.1 nm when a 25 m, 100 μm core-diameter UV Raman OF is used.

the same Raman OF can be used with a wide range of pump wavelengths at room temperature. The generation of powerful and widely tunable laser emission covering the UV, visible, and near-IR, DFRL can be used for optical signal processing that includes optical pulse shaping and compression, ultrashort pulse generation, and amplification.

To extend the spectral range of UV FRL generation toward the 200 nm range, research efforts are directed to development of low-loss UV-transmitting OFs and high-efficient UV fiber-based delivery systems. UV fibers with attenuation losses below 0.3 dB/m and 1 dB/m for the spectral range around 300 nm and 250 nm, respectively, are already commercially available. Ilev and Waynant [101] recently presented a simple alternative technique for UV laser delivery and high-efficiency direct laser-to-waveguide coupling using an uncoated glass hollow taper. The operating principle of the uncoated hollow taper is based on the light grazing-incidence effect. The grazing-incidence-based taper technique produces a high-quality smooth profile of the output laser beam and ensures high UV laser-to-taper and laser-to-fiber coupling efficiencies [102,103].

III. UV AND VUV GENERATION USING SOLID-STATE RARE-EARTH-ACTIVATED LASER CRYSTALS

In addition to the all-solid-state nonlinear techniques for UV/VUV tunable laser generation considered in the previous section, another promising alternative has been the development of solid-state UV/VUV lasers using rare-earth-activated wide-band-gap fluoride dielectric crystals as the base. The concept of this solid-state UV/VUV laser includes use of the broadband emission features of the inter-configurational d–f transmissions of trivalent rare-earth-activated ions (such as Nd^{3+}, Er^{3+}, Pr^{3+}, Ho^{3+}, Tm^{3+}, Yb^{3+}, Ce^{3+}, Sm^{3+}, Eu^{3+}, Gd^{3+}, Tb^{3+}, and Tb^{3+}). These are doped in dielectric host crystals (such as LaF_3, YF_3, $LiYF_4$, LuF_3, Li-LuF_4, $LiCaAlF_6$, etc.) [104–118]. Figure 20 shows a typical energy-level diagram of Nd^{3+} ions in LaF_3. The excitation spectra centered at $\lambda = 159$ nm originate

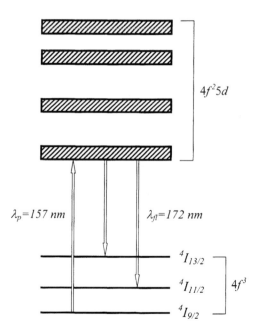

Figure 20   Energy-level diagram of Nd^{3+} ions in LaF_3. The excitation spectra is centered at $\lambda_p = 157$ nm and the VUV $4f^25d \rightarrow 4f^3$ fluorescence is observed at $\lambda_{fl} = 172$ nm.

from transitions from the $4f^3$ ground state to the crystal field split $4f^25d$ levels of the Nd^{3+} ions in LaF_3. The VUV fluorescence observed at $\lambda = 172$ nm originates from the $4f^25d \rightarrow 4f^3$ electric dipole-allowed transition levels and transfers to the $5d \rightarrow {}^4I_{11/2}$ and $5d \rightarrow {}^4I_{13/2}$ interconfigurational transition levels.

In earlier experiments investigating rare-earth-activated fluoride dielectric materials, Yang and DeLuca [104] observed broadband UV/VUV (from 165 to 260 nm) fluorescence in Nd^{3+}-, Er^{3+}-, and Tm^{3+}-doped LaF_3, YF_3, $LiYF_4$, and LuF_3. The first VUV solid-state laser was experimentally constructed by Waynant and Klein in 1986 [105]. The authors obtained VUV laser radiation at $\lambda = 172$ nm produced by pumping of $Nd^{3+}:LaF_3$ crystal with incoherent Kr_2^* radiation at $\lambda = 146$ nm. Figure 21 shows absorption and fluorescence spectra obtained with $Nd^{3+}:LaF_3$ crystals containing Nd^{3+} concentration of 0.1%, 0.5%, and 1.0%. Due mainly to the higher absorption at higher Nd^{3+} concentration, the greatest fluorescence intensity as well as laser action is observed in the lightest (0.1%) doped crystal. As a pumping technique in this experiment, the authors used a source of fast electrons that excite high-pressure Kr gas and generate Kr_2^* dimers.

Several research groups have been working intensively on UV/VUV fluorescence spectroscopy of rare-earth-activated fluoride dielectric materials. Dubinskii, Cefalas, Sarantopoulou, and their coauthors have presented a broad spectrum of experimental realizations of solid-state VUV laser systems using fluoride dielectric crystals doped with various rare-earth-activated ions such as Nd^{3+} [106–

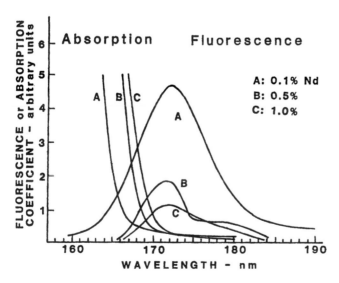

Figure 21   Typical absorption and fluorescence spectra of LaF_3 crystals doped with 0.1, 0.5, and 1.0% Nd^{3+} under pumping by 146 nm radiation from Kr_2^*.

110], Er^{3+} [111], Pr^{3+} [112], Ho^{3+} [113,114], and Tb^{3+} and Tm^{3+} [115,116]. In the usual case, the authors used as a pumping laser source a pulsed-discharged molecular F_2 laser with the following parameters: $\lambda_p = 157$ nm, 12 ns pulse duration, 12-mJ maximum energy, and 120-mJ/cm^2 maximum energy density. Dubinskii et al. obtained an efficient (14% conversion efficiency) laser action at $\lambda = 172$ nm from a Nd^{3+}:LaF$_3$ crystal pumped by a molecular F_2 laser at: $\lambda_p = 157$ nm [106]. VUV laser-induced fluorescent spectra of K_2YF_5:Nd^{3+} (PFYK: Nd) and LiYF$_4$:Nd^{3+} (YLF:Nd) single crystals pumped by the F_2 laser were obtained by Kollia and colleagues [110]. The authors observed eleven and nine dipole transitions, between the $^4I_{9/2}$ ground level of the $4f^3$ configuration and the levels of the $4f^25d$ configuration of the Nd^{3+} ion 9 (see Fig. 1), in YLF and PFYK crystal hosts, respectively. In the most recent papers [113,114], researchers obtained VUV laser-induced fluorescent spectra in the spectral range between 120 and 170 nm using an LiLuF$_4$:Ho^{3+} (LIF) single crystal.

Spectroscopic study of rare-earth-activated fluoride dielectric materials using other doped ions has been reported by another authors. Wegh et al. [117] presented an investigation of the $4f^7$ energy levels of Gd^{3+} in LiF$_4$ in the VUV spectral range. The authors carried out high-resolution excitation measurement using synchrotron radiation. Ehrlich et al. [118] reported on the first observation of stimulated emission from $5d$-$4f$ transition in Ce^{3+} ions in LiYF$_4$. They applied optical pumping using a KrF laser ($\lambda_p = 249$ nm, 25 ns pulse duration) and obtained a Ce^{3+}:LiYF$_4$ stimulated emission at $\lambda_p = 325.5$ nm.

Further research efforts in the field of solid-state rare-earth-activated laser materials for UV and VUV laser generation are focused on two main directions: (1) development of laser materials with improved optical properties for effective UV/VUV stimulated generation; and (2) the search for and development of effective and relatively simple optical techniques for laser pumping. Detailed consideration of the specific features and potential applications of the rare-earth-activated fluoride dielectric materials for UV and VUV laser generation is given in the following chapters.

REFERENCES

1. N Bloembergen. Nonlinear Optics. New York: WA Benjamin, Inc., 1965, pp 1–119.
2. A Yariv. Quantum Electronics 3rd ed New York: John Wiley & Sons, 1989, pp 378–494.
3. DH Auston. Picosecond nonlinear optics. In: SL Shapiro, ed. Ultrashort Light Pulses. New York: Springer-Verlag, 1977, pp 123–201.
4. CL Tang. Nonlinear optics. In: M Bass, ed. Handbook of Optics. 2nd ed. New York: McGraw-Hill, 1995, pp 38.1–38.26.

5. CC Davis. Lasers and Electro-Optics. Cambridge: University Press, 1996, pp 508–560.
6. GL Wood, EJ Sharp. Nonlinear optics. In: RW Waynant, MN Ediger, eds. Electro-Optics Handbook. 2nd ed. New York: McGraw-Hill, 2000, pp 13.1–13.28.
7. A Yariv. Introduction to Optical Electronics. 2nd ed. New York: Holt, Rinehart and Winston, 1976, pp 198–244.
8. AE Siegman. Lasers. Mill Valley, CA: University Science Books, 1986, pp 369–397.
9. RW Waynant, MN Ediger, eds. Selected Papers on UV, VUV, and X-Ray Lasers. Bellingham, WA: SPIE Optical Engineering Press, 1993, pp 253–322.
10. CL Rhodes. Review of ultraviolet laser physics. IEEE J Quantum Electron QE-10: 153–174, 1974.
11. RW Waynant, RC Elton. Review of short wavelength laser research. IEEE J Quantum Electron 64:1059–1092, 1976.
12. F Zernike, JE Midwinter. Applied Nonlinear Optics. New York: John Wiley & Sons, 1973, pp 43–74.
13. WR Hunter. Optics in the vacuum ultraviolet. Electro-Optical Systems Design 5: 16–23, 1973.
14. F Xie, B Wu, G You, C Chen. Characterization of LiB_3O_5 crystal for second-harmonic generation. Opt Lett 16:1237–1239, 1991.
15. TY Fan, CE Huang, BQ Hu, RC Eckardt, YX Fan, RL Byer, RS Feigelson. Second harmonic generation and accurate index of refraction measurements in flux-grown $KTiOPO_4$. Appl Opt 26:2390–2394, 1987.
16. C Chen, Y Fan, R Eckardt, R Byer. Recent developments in barium borate. Proc SPIE 681:12–19, 1986.
17. R Komatsu, T Sugawara, K Sassa, N Sarukura, Z Liu, S Izumida, Y Segawa, S Uda, T Fukuda, K Yamanouchi. Growth and ultraviolet application of $Li_2B_4O_7$ crystals: generation of the forth and fifth harmonics of $Nd:Y_3Al_5O_{12}$ lasers. Appl Phys Lett 70:3492–3494, 1997.
18. N Umemura, K Kato. Phase-matched UV generation at 0.1774 µm in KB_5O_8 $4H_2O$. Appl Opt 35:5332–5335, 1996.
19. C Chen, Y Wu, A Jiang, B Wu, G You, R Li, S Lin. New nonlinear-optical crystal: LiB_3O_5. J Opt Soc Am B 6:616–621, 1989.
20. C Chen, B Wu, G You, A Jiang, Y Huang. High-efficiency and wide-band second-harmonic generation properties of new crystal $\beta\text{-}BaB_2O_4$. Proceedings, International Quantum Electronics Conference, Washington, D.C., 1984, paper MCC5.
21. RC Eckardt, H Masuda, YX Fan, RL Byer. Absolute and relative nonlinear optical coefficients of KDP, KD*P, BaB2O4, LiIO3, MgO:LiNbO3, and KTP measured by phase-matched second-harmonic generation. IEEE J Quantum Electron 26:922–933, 1990.
22. WK Jang, Q Ye, D Hammons, J Eichenholz, J Lim, M Richardson, BHT Chai, EW Vab Stryland. Improved second-harmonic generation by selective Yb ion doping in a new nonlinear optical crystal $Yca_4O(BO_3)_3$. IEEE J Quantum Electron 35:1826–1833.
23. YK Yap, M Inagaki, S Nakajima, Y Mory, T Sasaki. High-power fourth- and fifth-

harmonic generation of a Nd:YAG laser by means of a $CsLiB_6O_{10}$. Opt Lett 21: 1348–1350, 1996.

24. N Umemura, K Kato. Ultraviolet generation tunable to 0.185 μm in $CsLiB_6O_{10}$. Appl Opt 36:6794–6796, 1997.

25. S Imai, T Yamada, Y Fujimori, K Ishikawa. Third-harmonic generation of an alexandrite laser in β-BaB_2O_4. Appl Phys Lett 54:1206–1208, 1989.

26. GC Bhar, S Das, U Chatterjee. Noncollinear phase-matched second-harmonic generation in beta barium borate. Appl Phys Lett 54:1383–1384, 1989.

27. B Wu, N Chen, C Chen, D Deng, Z Xu. Highly efficient ultraviolet generation at 355 nm in LiB_3O_5. Opt Lett 14:1080–1081, 1989.

28. G Feugnet, JP Pocholle. 8-mJ TEM_{00} diode-end-pumped frequency-quadrupled Nd: YAG laser. Opt Lett 23:55–57, 1998.

29. DN Nikogosyan. Beta Barium Borate (BBO): a review of its properties and applications. Appl Phys A52:359–368, 1991.

30. DN Nikogosyan. Lithium Triborate (LBO): a review of its properties and applications. Appl Phys A58:181–190, 1994.

31. S Shapiro. Second-harmonic generation in $LiNBO_3$ by picosecond pulses. Appl Phys Lett 13:19–21, 1968.

32. WH Glenn. Second-harmonic generation by picosecond optical pulses. IEEE J Quantum Electron QE-5:284–290, 1969.

33. JY Huang, YR Shen, C Chen, B Wu. Noncritically phase-matched second-harmonic generation and optical parametric amplification in a lithium triborate crystal. Appl Phys Lett 58:1579–1581, 1991.

34. M Ebrahimzadeh, GJ Hall, AI Ferguson. Singly resonant, all-solid-state, mode-locked LiB_3O_5 optical parametric oscillator tunable from 652 nm to 2.65 μm. Opt Lett 17:652–654, 1992.

35. ST Yang, CC Pohalski, EK Gustafson, RL Byer, RS Feigelson, RJ Raymakers, RK Route. 6.5-W, 532-nm radiation by cw resonant external-cavity second-harmonic generation of an 18-W Nd:YAG laser in LiB_3O_5. Opt Lett 16:1493–1495, 1991.

36. ZY Ou, SE Pereira, ES Polzik, HJ Kimble. 85% efficiency for cw frequency doubling from 1.08 to 0.54 μm. Opt Lett 17:640–642, 1992.

37. H Hemmati. Diode-pumped self-frequency-doubled neodymium yttrium aluminum borate (NYAB) laser. IEEE J Quantum Electron 28:1169–1171, 1992.

38. M Oka, LY Liu, W Wiechmann, N Eguchi, S Kubota. All solid-state continuous-wave frequency-quadrupled Nd:YAG laser. IEEE J Quantum Electron 1:859–866, 1995.

39. DW Anthon, DL Sipes, TJ Pier, MR Ressl. Intracavity doubling of cw diode-pumped Nd:YAG lasers with KTP. IEEE J Quantum Electron 28:1148–1157, 1992.

40. LR Marshall, AD Hays, A Kaz, RL Burnham. IEEE J Quantum Electron 28:1158–1163, 1992.

41. M Oka, S Kubota. Stable intracavity doubling of orthogonal linearly polarized modes in diode-pumped Nd:YAG lasers. Opt Lett 13:805–807, 1988.

42. K Kato. Second-harmonic generation to 2048 Å in β-BaB_2O_4. IEEE J Quantum Electron QE-22:1013–1014, 1986.

43. K Kato. Tunable UV generation to 0.2325 μm in LiB$_3$O$_5$. IEEE J Quantum Electron 26:1173–1175, 1990.

44. WL Glab, JP Hessler. Efficient generation of 200-nm light in β-BaB$_2$O$_4$. Appl Opt 26:3181–3182, 1987.

45. HJ Dewey. Second-harmonic generation in KB$_5$O$_8$4H$_2$O from 217.1 to 315.0 nm. IEEE J Quantum Electron QE-12:303–306, 1976.

46. K Miyazaki, H Sakai, T Sato. Efficient deep-ultraviolet generation by frequency doubling in β-BaB$_2$O$_4$ crystals. Opt Lett 11:797–799, 1986.

47. GA Skripko, SG Bartoshevich, IV Mikhnyuk, IG Tarazevich. LiB$_3$O$_5$: a highly efficient frequency converter for Ti:sapphire lasers. Opt Lett 16:1726–1728, 1991.

48. A Nebel, R Beigang. External frequency conversion of cw mode-locked Ti:Al$_2$O$_3$ laser radiation. Opt Lett 16:1729–1731, 1991.

49. DC Edelstein, ES Wachman, LK Cheng, WR Bosenberg, CL Tang. Femtosecond ultraviolet pulse generation in β-BaB$_2$O$_4$. Appl Phys Lett 52:2211–2213, 1988.

50. F Rotermund, V Petrov. Generation of the fourth harmonic of a femtosecond Ti:sapphire laser. Opt Lett 23:1040–1042, 1998.

51. S Bourzeix, B de Beauvoir, F Nez, F de Tomasi, F Biraben. Ultra-violet light generation at 205 nmby two frequency doubling steps of a cw titanium-sapphire laser. Opt Commun 133:239–244, 1997.

52. S Cussat-Blanc, RM Rassoul, A Ivanov, E Freysz, A Ducasse. Influence of cascading phenomena on a type I second-harmonic wave generated by an intense femtosecond pulse: application to the measurement of the effective second-order coefficient. Opt Lett 23:1585–1587, 1998.

53. Y Wu, T Sasaki, S Nakai, A Yokotani, H Tang, C Chen. CsB$_3$O$_5$: a new nonlinear optical crystal. Appl Phys Lett 62:2614–2615, 1993.

54. Y Mori, I Kuroda, S Nakajima, T Sasaki, S Nakai. New nonlinear optical crystal: cesium lithium borate. Appl Phys Lett 67:1818–1820, 1995.

55. V Petrov, F Rotermund, F Noack, J Ringling, O Kittelmann, R Komatsu. Frequency conversion of Ti:sapphire-based femtosecond laser systems to the 200-nm spectral region using nonlinear optical crystals. IEEE J Sel Topics Quantum Electron 5:1532–1542, 1999.

56. V Petrov, F Rotermund, F Noack, R Komatsu, T Sugawara, S Uda. Vacuum ultraviolet application of Li$_2$B$_4$O$_7$ crystals: generation of 100 fs pulses down to 170 nm. J Appl Phys 84:5887–5892, 1998.

57. C Chen, Z Xu, D Deng, J Zhang, GKL Wong, B Wu, N Ye, D Tang. The vacuum ultraviolet phase-matching characteristics of nonlinear optical KBe$_2$BO$_3$F$_2$ crystal. Appl Phys Lett 68:2930–2931, 1996.

58. K Kato. Temperature-tuned 90° phase-matching properties of LiB$_3$O$_5$. IEEE J Quantum Electron 30:2950–2952, 1994.

59. K Kato. Efficient ultraviolet generation of 2073–2174 A in KB$_5$O$_8$/4H$_2$O. IEEE J Quantum Electron QE-13:544–546, 1977.

60. K Kato. Tunable uv generation in KB$_5$O$_8$4H$_2$O to 1996 A. Appl Phys Lett 30:583–584, 1977.

61. W Muckenheim, P Lokai, B Burghardt, D Basting, L Physik. Attaining the wavelength range 189–197 nm by frequency mixing in β-BaB$_2$O$_4$. Appl Phys B 45:259–261, 1988.

62. A Borsutzky, R Brunger, R Wallenstein. Tunable UV radiation at short wavelengths (188-240 nm) generated by sum-frequency mixing in lithium borate. Appl Phys B 52:380–384, 1991.

63. K Kato. Tunable UV generation to 0.185 µm in CsB_3O_5. IEEE J Quantum Electron 31:169–171, 1995.

64. JA Paisher, ML Spaeth, DC Gerstenberger, IW Ruderman. Generation of tunable radiation below 2000 A, by phase-matched sum-frequency mixing in $KB_5O_84D_2O$. Appl Phys Lett 32:476–478, 1978.

65. RE Stickel Jr, FB Dunning. Generation of tunable coherent vacuum uv radiation in KB5. Appl Opt 17:981–982, 1978.

66. B Wu, F Xie, C Chen, D Deng, Z Xu. Generation of tunable coherent vacuum ultraviolet radiation in LiB_3O_5 crystal. Opt Commun 88:451–454, 1992.

67. GC Bhar, P Kumbhakar, U Chatterjee, AM Rudra, Y Kuwano, H. Kouta. Efficient generation of 200–230-nm radiation in beta barium borate by noncollinear sum-frequency mixing. Appl Opt 37:7827–7831, 1998.

68. A Nebel, R Beigang. Tunable picosecond pulses below 200 nm by external frequency conversion of cw modelocked $Ti:Al_2O_3$ laser radiation. Opt Commun 94:369–372, 1992.

69. J Izawa, K Midorikawa, M Obara, K Toyoda. Picosecond ultraviolet optical parametric generation using a type-II phase-matched lithium triborate crystal for an injection seed of vuv lasers. IEEE J Quantum Electron 33:1997–2000, 1997.

70. J Ringling, O Kittelmann, F Noack, G Korn, J Squier. Tunable femtosecond pulses in the near vacuum ultraviolet generated by frequency conversion of amplified Ti : sapphire laser pulses. Opt Lett 18:2035–2037, 1993.

71. F Seifert, J Ringling, F Noack, V Petrov, O Kittlemann. Generation of tunable fentosecond pulses to as low as 172.7 nm by sum-frequency mixing in lithium triborate. Opt Lett 19:1538–1540, 1994.

72. GPA Malcolm, M Ebrahimzadeh, AI Rerguson. Efficient frequency conversion of mode-locked diode-pumped lasers and tunable all-solid-state laser sources. IEEE J Quantum Electron 28:1172–1178, 1992.

73. MT Reiten, RA Cheville, NJ Halas. Phase matching and focussing effects on non-collinear sum frequency mixing in the near VUV region. Opt Commun 110:645–650, 1994.

74. J Ringling, O Kittelmann, F Noack. Efficient generation of subpicosecond seed pulses at 193 nm for amplification in ArF gain modules by frequency mixing in nonlinear optical crystals. Opt Lett 17:1794–1796, 1992.

75. C Chen, Y Wang, B Wu, K Wu, W Zeng, L Yu. Design and synthesis of an ultraviolet-transparent nonlinear optical crystal $Sr_2Be_2B_2O_7$. Nature 373:322–323, 1995.

76. G Agrawal. Nonlinear Fiber Optics. 2nd ed. London: Academic Press, 1995.

77. C Lin. Nonlinear optics in fibers for fiber measurements and special device functions. J Lightwave Technol LT-4:1103–1115, 1986.

78. S Sudo, H Itoh. Efficient non-linear optical fibres and their applications. Opt Quantum Electron 22:187–212, 1990.

79. F Galleener, J Mikkelsen, R Geils, W Mosby. The relative Raman cross sections of vitreous SiO2, GeO2, B2O3, and P2O5. Appl Phys Lett 32:34–36, 1978.

80. R Smith. Optical power handling capacity of low-loss optical fibers as determined by stimulated Raman and Brillouin scattering. Appl Opt 11:2489–2498, 1972.

81. J AuYeung, A Yariv. Spontaneous and stimulated Raman scattering in long-loss fibers. IEEE J Quantum Electron QE-14:347–352, 1978.

82. J AuYeung, A Yariv. Theory of CW Raman oscillation in optical fibers. J Opt Soc Am 69:803–810, 1079.

83. C Yijiang, A Snyder. Saturation and depletion effect of Raman scattering in optical fibers. J Lightwave Technol 7:1109–1119, 1989.

84. A Chraplyvy, J Stone, C. Burrus. Optical gain exceeding 35 dB at 1.56 μm due to stimulated Raman scattering by molecular D2 in a solid silica optical fiber. Opt Lett 8:415–417, 1983.

85. J Stone, A Chraplyvy, C Burrus. Opt Lett 7:297–299, 1982.

86. C Lin, M Bosch. Large-Stokes-shift stimulated four-photon mixing in optical fibers. Appl Phys Lett 38:479–681, 1981.

87. E Dianov, E Zahidov, A Karasik, P Mamishev, A Prohorov. Stimulated parametric four-photon mixing in glass fibers. Sov Phys JETF Lett 34:38–42, 1981.

88. C Lin, R Stolen. New nanosecond continuum for excited-state spectroscopy. Appl Phys Lett 28:216–218, 1976.

89. M Rothschild, H Abad. Stimulated Raman scattering in fibers in the ultraviolet. Opt Lett 8:653–655, 1983.

90. R Pini, M Mazzoni, R Salimbeni, M Matera, C Lin. Ultraviolet stimulated Raman scattering in multimode silica fibers pumped with excimer lasers. Appl Phys Lett 43:6–8, 1983.

91. R Pini, R Salimbeni, M Matera, C Lin. Wideband frequency conversion in the UV by nine orders of stimulated Raman scattering in a XeCl laser pumped multimode silica fiber. Appl Phys Lett 43:517–518, 1983.

92. K Liu, E Garmire. Role of stimulated four-photon mixing and efficient Stokes generation of stimulated Raman scattering in excimer-laser-pumped UV multimode fibers. J Opt Soc Am 16:174–176, 1991.

93. R Pini, R Salimbeni, M Vannini, AFMY Haider, C Lin. High conversion efficiency ultraviolet fiber Raman oscillator-amplifier system. Appl Opt 25:1048–1050, 1986.

94. I Ilev, H Kumagai, K Toyoda. A widely tunable (0.54–1.01 μm) double-pass fiber Raman laser. Appl Phys Lett 69:1846–1848, 1996.

95. I Ilev, H Kumagai, K Toyoda. A fiber-optic autocollimation refractometric method for dispersion measurement of bulk optical materials using a widely tunable fiber Raman laser. IEEE J. Quantum Electron 32:1897–1902, 1996.

96. I Ilev, H Kumagai, K Toyoda. A powerful and widely tunable double-pass fiber Raman laser. Opt Commun 138:337–340, 1997.

97. I Ilev, H Kumagai, K Toyoda. Ultraviolet and blue discretely tunable double-pass fiber Raman laser. Appl Phys Lett 70:3200–3202, 1997.

98. I Ilev, R Waynant, H Kumagai, K Toyoda. Double-pass fiber Raman laser—a powerful and widely tunable in the ultraviolet, visible and near-infrared fiber Raman laser for biomedical investigations. IEEE J Sel Topics Quantum Electron 5:1013–1018, 1999.

99. K Liu, E Garmire. Understanding the formation of the SRS Stokes spectrum in fused silica fibers. IEEE J Quantum Electron 27:1022–1030, 1991.

100. I Ilev, H Kumagai, K Toyoda, I Koprinkov. Highly-efficient wideband continuum generation in a single-mode optical fiber by powerful broadband laser pumping. Appl Opt 35:2548–2553, 1996.

101. I Ilev, R Waynant, M Ediger, M Bonaguidi. Ultraviolet laser delivery using an uncoated hollow taper. IEEE J Quantum Electron 36:944–948, 2000.

102. I Ilev, RW Waynant. Grazing-incidence-based hollow taper for infrared laser-to-fiber coupling. Appl Phys Lett 74:2921–2923, 1999.

103. I Ilev, R Waynant. Uncoated hollow taper as a simple optical funnel for laser delivery. Rev Sci Instrum 70:3840–3843, 1999.

104. K Yang, J DeLuca. VUV fluorescence of Nd^{3+}-, Er^{3+}-, and Tm^{3+}-doped trifluorides and tunable coherent sources from 1650 to 2600 A. Appl Phys Lett 29:499–501, 1976.

105. RW Waynant, PH Klein. Vacuum ultraviolet laser emission from Nd^{+3}:LaF_3. Appl Phys Lett 46:14–16, 1985.

106. MA Dubinskii, AC Cefalas, E Sarantopoulou, SM Spyrou, CA Nicolaides. Efficient LaF_3:Nd^{3+}-based vacuum-ultraviolet laser at 172 nm. J Opt Soc Am B 9:1148–1150, 1992.

107. MA Dubinskii, AC Cefalas, E Sarantopoulou, R Yu. Abdulsabirov, SL Korableva, AK Naumov, VV Semashko. On the interconfigurational $4f^25d$-$4f^3$ VUV and UV fluorescence fratures of Nd^{3+} in $LiYF_4$ (YLF) single crystals under F^2 laser pumping. Opt Commun 94:115–118, 1992.

108. E Sarantopoulou, AC Cefalas, MA Dubinskii, CA Nicolaides, R Yu, Abdulsabirov, SL Korableva, AK Naumov, VV Semashko. VUV and UV fluorescence and absorption studies of Nd^{3+} and Ho^{3+} ions in $LiYF_4$ single crystals. Opt Commun 107:104–110, 1994.

109. Z Kollia, E Sarantopoulou, AC Cefalas, CA Nicolaides, AK Naumov VV Semashko, R Yu Abdulsabirov, SL Korableva, MA Dubinskii. Vacuum-ultraviolet interconfigurational $4f^3 \rightarrow 4f^25d$ absorption and emission studies of the Nd^{3+} ion in KYF, YF, and YLF crystal hosts. J Opt Soc Am B 12:L782–785, 1995.

110. Z Kollia, E Sarantopoulou, AC Cefalas, AK Naumov, VV Semashko, RY Abdulsabirov, SL Korableva. On the $4f^25d \rightarrow 4f^3$ Interconfigurational transitions of Nd^{3+} ions in K_2YF_5 and $LiYF_4$ crystal hosts. Opt Commun 149:386–392, 1998.

111. E Sarantopoulou, Z Kollia, AC Cefalas, MA Dubinskii, CA Nicolaides, RY Abdulsabirov, SL Korableva, AK Naumov, VV Semashko. Vacuum-ultraviolet and ultraviolet fluorescence and absorption studies of Er^{3+}-doped $LiLuF_4$ single crystals. Appl Phys Lett 65:813–815, 1994.

112. E Sarantopoulou, AC Cefalas, MA Dubinskii, CA Nicolaides, RY Abdulsabirov, SL Korableva, AK Naumov, VV Semashko. VUV and UV fluorescence and absorption studies of Pr^{3+}-doped $LiLuF_4$ single crystals. Opt Lett 19:499–501, 1994.

113. AC Cefalas, E Sarantopoulou, Z Kollia. Vacuum ultraviolet $4f^95d \rightarrow 4f^{10}$ interconfigurational transitions of Ho^{3+} ions in $LiYF_4$ single crystals. J Opt Soc Am B 16:625–630, 1999.

114. E Sarantopoulou, Z Kollia, AC Cefalas. $4f^95d \rightarrow 4f^{10}$ spin-allowed and spin-forbidden transitions of Ho^{3+} ions in $LiYF_4$ single crystals in the vacuum ultraviolet. Opt Commun 169:263–274, 1999.

115. E Sarantopoulou, Z Kollia, AC Cefalas, VV Semashko, RY Abdulsabirov, AK

Naumov, SL Korableva. On the VUV and UV $4f^7(^8S)5d \rightarrow 4f^8$ interconfigurational transitions of Tb^{3+} ions in $LiLuF_4$ single crystal hosts. Opt Commun 156:101–111, 1998.

116. E Sarantopoulou, AC Cefalas, MA Dubinskii, Z Kollia, CA Nicolaides, RY Abdul-sabirov, SL Korableva, AK Naumov, VV Semashko. VUV and UV fluorescence and absorption studies of Tb^{3+} and Tm^{3+} trivalent ions in $LiLuF_4$ single crystal hosts. J Mod Opt 41:767–775, 1994.

117. RT Wegh, H Donker, A Meijerink, RJ Lamminmaki, J Holsa. Vacuum-ultraviolet spectroscopy and quantum cutting for Gd^{3+} in $LiYF^4$. Phys Rev B 56:13841–13848, 1997.

118. DJ Ehrlich, PF Moulton, RM Osgood Jr. Ultraviolet solid-state Ce:YLF laser at 325 nm. Opt Lett 4:184–186, 1979.

8

VUV Laser Spectroscopy of Trivalent Rare-Earth Ions in Wide Band Gap Fluoride Crystals

Evangelia Sarantopoulou and Alciviadis-Constantinos Cefalas
National Hellenic Research Foundation, Theoretical and Physical Chemistry Institute, Athens, Greece

I. INTRODUCTION

The absorption and the excitation spectroscopic characteristics of trivalent rare-earth (RE) ions in the vacuum ultraviolet (VUV) and ultraviolet (UV) spectral regions, activated in wide band gap fluorine dielectric crystals, are due to transitions between the levels of the $4f^n$ single electronic configuration of the trivalent RE ion, and the levels of the $4f^{n-1}5d$ mixed electronic configuration, where a 4f electron is promoted to a 5d localized level [1–3]. The $4f^n \leftrightarrow 4f^{n-1}5d$ electronic transitions are characterized by strong Frank–Condon factors with broadband absorption and emission spectra in the VUV and UV. In contrast, the intraconfigurational $4f^n \rightarrow 4f^n$ transitions are parity forbidden. They are forced by the crystal field configuration mixing and they appear to be weak and sharp.

There are two different experimental methods for exciting a trivalent RE ion in wide band gap dielectric crystals. The first method uses VUV lasers [4] and/or x-ray or synchrotron radiation [5,6]. This pumping arrangement has the advantage of populating the levels of the $4f^{n-1}5d$ electronic configuration directly from the ground state of the trivalent RE ion, via one photon transition only. The subsequent de-excitation mechanisms within the $4f^{n-1}5d$ electronic configuration efficiently populate the levels of the $4f^n$ single electronic configuration of the trivalent RE ion [7–9]. They allow one to study the excitation dynamics and the structure of the levels of the $4f^{n-1}5d$ electronic configuration of the trivalent RE ions. The second method applies upconversion pumping arrangement with all solid state laser elements [10–12], which greatly simplifies the experimental obstacles arising from the VUV pumping. The radiative interconfigurational d–f transitions of the RE activated ions, in the wide band gap of dielectric crystals, offer the possibility that these materials can be used in a variety of applications and for generating coherent VUV or UV light. This is attractive due to its relative simplicity in comparison to existing nonlinear methods using gases and molecules. In the case of optically pumped $LiYF_4:Ce^{3+}$ and $LaF_3:Ce^{3+}$ crystals [13–15], laser emission at 325 and 286 nm was obtained, respectively. Waynant and Klein [16,17] have reported the first laser action in the VUV from solid-state dielectric crystals. They used the $LaF_3:Nd^{3+}$ dielectric crystal to generate laser action at 172 nm when it was optically pumped by incoherent light (emitted from excited Kr_2^* molecules). With a different pumping arrangement using an F_2 pulsed discharge molecular laser operating at 157.6 nm, coherent light from the same crystal at 172 nm was generated [18]. Fluorescent materials based on RE ions can also be used as high quantum efficiency phosphors [19], for plasma display screens and mercury-free light tubes [20–22], fast scintillators [23,24] and for light wave communications [25,26].

In addition to the above-mentioned applications, next-generation microelectronics circuits will have minimum dimensions below 100 nm. It is envisioned that 157 nm laser lithography [27,28,28a] will be the next step in optical lithography. At 157 nm, under VUV illumination of the mask target, lithographic features with dimensions less then 0.10 μm on the photoresist could be achieved. However, there are problems related to the design of the optical projection system, due to the fact that the absorption coefficient of most materials in the VUV is large, yet, their optical properties degrade constantly with time under VUV irradiation. Up to now, only calcium fluoride seems to be promising as optical material for 157 nm photolithography, and the possibility of using wide band gap fluoride dielectric crystals, such as YF_4 and $LiCaAlF_6$, as optical elements for 157 nm photolithography was investigated. These materials could be grown from melts at lower temperature than CaF_2 and they have similar physical properties, despite the fact that they are forming crystals of different symmetry than cubic. Doped or nondoped wide band gap crystals can be used as passive or active optical elements in the VUV: i.e., prisms, lenses, filters of variable and controllable attenuation [28, 28a].

All these applications depend on the development of new dielectric wide band gap materials, and the structure of the levels of the $4f^{n-1}5d$ electronic configuration that specify the d–f transitions of the RE ions. However, despite the early spectroscopic measurements of the f–d transitions [29], only limited information is available in this field. A small amount of data has been taken and analyzed with the use of VUV lasers, x-ray, or synchrotron light sources by exciting a common RE impurity in different dielectric fluoride crystal hosts. Investigation of fundamental physical interactions in the crystal environment is of considerable importance and will determine the response of these materials to VUV light.

II. VACUUM ULTRAVIOLET SPECTROSCOPY: METHODS AND TECHNIQUES

A. Molecular Fluorine Laser

Investigating physical processes in the VUV requires the use of intense sources of photons in this spectral range. For a detailed investigation of the absorption spectrum, the emission characteristics of the sources are such that either the radiation occurs as a continuum or the emission spectrum consists of closely packed lines. In addition, the continuous light sources should be stable with respect to time and the pulsed ones should have good pulse to pulse stability. In most experimental cases, the available continua source suitable for ultraviolet absorption studies is molecular hydrogen. Its molecular spectrum extending from the visible to 100 nm is obtainable from a low-pressure positive column discharge. Continua spectra have likewise been taken from positive column discharges in rare gases, from 60 to 200 nm. At higher pressures, the continuum in helium discharges could be extended up to 400 nm. Tunable VUV radiation can be generated from nonlinear methods using frequency mixing in rare gases and metal vapors [30–33] and nonlinear crystals [34,35]. However, the nonlinear methods are relatively complex and characterized by low conversion efficiency and low-output energy.

The major breakthrough in this direction was made in 1973 with the introduction of the excimer laser sources based on rare gas dimers. Strong dipole transitions between the bound electronic excited state and the dissociative ground state of diatomic clusters of Xe_2, Kr_2, and Ar_2 in liquid, gas, or solid phase [36–39] provided a new powerful radiation source in the VUV. These systems were characterized by high quantum efficiency in the emitted wavelength, high gain, and conversion efficiency. However, limited tunability and e-beam excitation to achieve population inversion imposes serious experimental restrictions on their use for spectroscopic applications.

At the present time, very few laser sources are available in the VUV (Table 1). Among them the ArF excimer laser with limited tunability at 193 nm [40,41], and the molecular fluorine laser at 157 nm [42–50], are the only efficient laser sources available on the laboratory scale. Laser transitions at 156.71, 157.48, and 157.59 nm have been observed for the first time by Woodworth and Rice [42,43] in F_2/He mixtures excited with an electron gun with high efficiency and in electric discharge-pumped lasers. The laser transition has been identified as of the $^3\Pi_{2g} \rightarrow {}^3\Pi_{2u}$ type and is the same as for the laser transition of the 301.5–346 nm band in molecular iodine. The energy difference between the first two transitions

Table 1   VUV Laser Sources

Laser	Emission wavelength (nm)	Excitation
H_2	109.8–164.4	EB
D_2	111.3–161.6	PD
Ar_2	126	EB
Kr_2	145.7	EB
F_2	157.6	PD
ArCl	169, 175	PD
Xe_2	170	EB
Kr^{3+}	175.6	CW
Cu	181	PD
Kr^{4+}	183.2	CW
Ar^{4+}	184.3	CW
CO	187.8, 189.7, 197	PD
ArF	193	PD
Kr^{3+}	195	CW
Kr^{3+}	196	CW

Population inversion is achieved commonly by electron beam (EB) or pulsed (PD) and continued (CW) plasma discharges.

was attributed to de-excitation from two different rotational levels of the same vibrational level of the $^3\Pi_{2g}$ state to the $^3\Pi_{2u}$, and the energy difference between the first and the third level to de-excitation from two different vibrational levels. The energy position of the $^3\Pi_u$ electronic state was found to be 3 eV above the $X^1\Sigma_g$ ground state from electron-scattering experiments [51]. This state is slightly bound by 0.15 eV [52].

The energy position of the lower vibrational level of the $^3\Pi_{2g}$ state was found to be 11.62 eV above the $X^1\Sigma_g$ ground state [53], which is in agreement with theoretical calculations [54].

The kinetic scheme proposed [44] to explain the above transition in the plasma discharge assumes the production of metastable helium atoms He* and e^-/He^+ pairs. The He^+ ions at high pressure are forming He_2^+ and He_3^+ ionized molecular clusters in the ground or excited states, which, finally, through effective collisions, either ionize the fluorine molecules or form atomic species:

$$He_n^+ + F_2 \rightarrow F_2^+ + nHe \ (n = 1,2,3)$$
$$He_n^* + F_2 \rightarrow F_2^+ + nHe \ (n = 1,2) + e$$
$$He_n^* + F_2 \rightarrow F + F^* + nHe \ (n = 1,2,3)$$

Low-energy electrons can efficiently generate atomic fluorine negative ions

$$e + F_2 \rightarrow F + F^- \quad \text{or} \quad e + F_2^+ \rightarrow F + F^*$$

and then

$$F^- + F_2^+ \rightarrow F^* + 2F \text{ or}$$
$$F^- + He_n^+ \rightarrow F^* + nHe$$

The final result of the atomic and molecular collisions is the formation of the F_2 molecules in their excited $^3\Pi_{2g}$ electronic state:

$$F^* + F_2 \rightarrow F_2(^3\Pi_{2g}) + F, \quad F^- + He \rightarrow F_2(^3\Pi_{2g}) + He$$

Once the $F_2(^3\Pi_{2g})$ molecular species have been formed, they can be de-excited by spontaneous and stimulated emission and by collision quenching.

$$F_2(^3\Pi_{2g}) \rightarrow F_2(^3\Pi_{2u}) + h\nu(157.6 \text{ nm})$$
$$F_2(^3\Pi_{2g}) + e^- \rightarrow F_2(^3\Pi_{2u}) + e^-$$
$$F_2(^3\Pi_{2g}) + F_2 \rightarrow 2F_2$$

Electron excitation [42] is a relatively complicated method in comparison to the pulsed fast discharge, using LC-inversion or charge transfer circuits (Fig. 1) together with preionization of the gas mixture. Preionization is achieved by UV light, which is triggered a few nanoseconds before the main discharge [44,55,56]. As soon as the fast switch S (thyratron or spark-gap) closes (Fig. 1), the stored energy in the capacitor C_2 is transferred into the discharge volume through the capacitors C_1. Double preionization (2×80 pins) of the laser volume ensures a uniform and stable main discharge. The electrodes were semicylidrical with round surfaces and no sharp edges, fabricated from stainless steel 316, 1 cm wide and 80 cm long, spaced 2 cm apart.

Good preionization in the circuit is essential to shape together with Rogowski profile electrodes, which are relatively difficult to machine in comparison to the semicylindrical profile. Stability of the discharge main mode was also ensured by the low inductance of the discharge circuit of 60 nH, which gives a fast rise

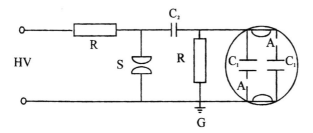

Figure 1   Schematic layout of the electric circuit of the ''charge transfer type'' F_2 laser. HV, high voltage; R, resistive load; S, spark gap; C_1, C_2, capacitors; A, preionization gap; G, ground.

time of the main charging voltage of 50 ns. The discharge volume is $100 \times 1.5 \times 0.3$ cm^3, and this device can deliver 10 mJ at 157 nm per pulse at 2–3 atm total helium pressure. A typical pulse duration is of the order of 10–20 ns, and its spectral linewidth at FWHM is less than 0.05 nm [44]. The small signal gain coefficient of the previous pulsed discharge F$_2$ laser has been measured using the passive cell absorption method at 2 atm of the helium buffer gas. It was found to be 3.2% cm^{-1}. This value was half that predicted by theory [47] considering only dissociative collision of the F$_2$ molecules by either ion–ion recombination or energy transfer reaction and neglecting direct excitation of the F$_2$ molecules by either electron impact or energy transfer from He* and He$_2^*$ molecules.

The small signal gain coefficient using the oscillator–amplifier method was found to be of the same order of magnitude and the saturation intensity was on the order of 4.5 MW/cm^2 [49,50]. Despite the fact that the value of the small signal gain coefficient at 157 nm is comparable to the value of the rare gas halide transitions of the excimer lasers up to 1988, the output energy at 157 nm never exceeded 15 mJ. This was mainly due to the presence of small amounts of organic impurities and air in the commercial gases and the experimental restrictions in raising the total gas pressure above 3 atm. The absorption coefficient of the air at 157 nm is higher than 200 cm^{-1} and therefore the net gain competed with the absorption losses. By increasing the pressure of the helium buffer gas to 8 atm, 112 mJ per pulse were obtained by Yamada et al. [48].

B. VUV Absorption Spectroscopy

Absorption cross-sectional investigations in the VUV require adequate measurement of the ratio of the incident to transmitted radiation. For atomic or molecular systems that exist in the gas state, accurate measurement of pressure and temperature is required. Measurement of the absorption coefficient of metal vapors is made even more difficult by the rapidity with which the reactive vapor attacks the windows of the absorption cell and by the high temperatures often necessary for the production of significant concentrations of the atoms.

For many experimental systems, measurement of the absorption coefficient is prohibited in the wavelength region below the cutoff wavelength of the LiF window (Fig. 2). For other experimental configurations, the source discharge gases are separated from the detector and the sample areas by differential pumping slits. In this case the cutoff wavelength is determined by the efficiency of the gas discharge and the cutoff wavelength of the monochromator and the photon detector (Fig. 3).

The absorption spectra of trivalent RE ions in the VUV in wide band gap dielectric crystals are due to the 4f^n → 4f^{n-1}5d electronic transitions between the ground state of the 4f^n single electronic configuration and the Stark levels of the 4f^{n-1}5d mixed electronic configuration. Experimental apparatus for ob-

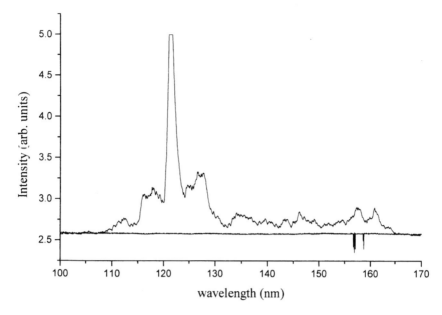

Figure 2 Emission spectrum from a longitudinal stabilized low-pressure hydrogen column recorded with a secondary electron multiplier (SEM), of open configuration. The hydrogen lamp was sealed with an LiF crystal window and it was the only optical element between the VUV lamp and the SEM. With this experimental configuration, the edge of the conduction band of wide band gap dielectric crystals can be determined accurately.

taining the absorption spectra consists mainly of a hydrogen VUV light source, the vacuum chamber where the crystal sample is placed, and the optical and electronic detection equipment (Fig. 4) [7]. The hydrogen light source operates in a longitudinally stabilized discharge mode. The discharge's high stability provides good signal-to-noise ratio (better than 2000). The optical paths of the light source and the fluorescence light beams are inside vacuum lines of stainless steel at 10^{-6} mbar pressure.

The vacuum chamber where the crystal samples are placed is equipped with a cryogenic facility. The optical detection system consists of a VUV monochromator, a solar blind photomultiplier, or a secondary electron multiplier (SEM). When a VUV photon is absorbed within the volume of the crystal sample, new VUV photons are re-emitted within the crystal at the same or different wavelength due to spontaneous emission from the levels of the $4f^n5d$ electronic configuration of the trivalent RE ions. The intensity of the radiation within the crystal volume V depends on the position of the atoms within the crystal and the wavelength [57].

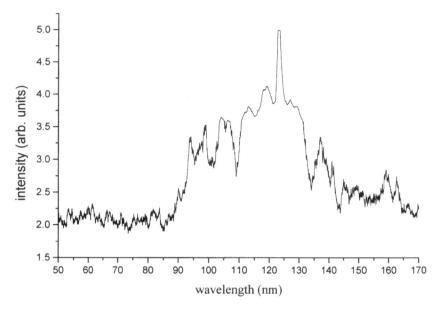

Figure 3 Emission spectrum from a longitudinal stabilized hydrogen lamp. No optical element was placed between the hydrogen discharge column and the secondary electron multiplier (SEM). The hydrogen flow through the discharge cell was controlled with differential pumping at different stages between the discharge and the SEM.

$$\frac{\partial I(\lambda, x)}{\partial x} = [N_2 - N_1]F(\lambda)\left(\frac{Bh\lambda}{Vnc^2}\right) I(\lambda, x) \tag{1}$$

where N_2 and N_1 are the excited and the ground state populations of the atomic transition at λ, $F(\lambda)d\lambda$ is the fraction of the transitions in which the photon frequency lies in a small range $d\lambda$ about the wavelength λ, V is the total crystal volume where the interaction of radiation with matter occurs, B is Einstein's coefficient, and n is the refractive index of the crystal. The solution of this equation is more complicated than it first appears because N_2 and N_1 themselves depend on the intensity of the radiation and hence on the position of the atoms inside the crystal and the wavelength of light. Eq. (1) takes the form

$$\frac{\partial I(\lambda, x)}{\partial x} = [a(\lambda, x) - \beta(\lambda, x)]I(\lambda, x) \tag{2}$$

where $a(\lambda, x)$ is the gain coefficient that describes photon generation through spontaneous emission and $\beta(\lambda, x)$ is the absorption coefficient.

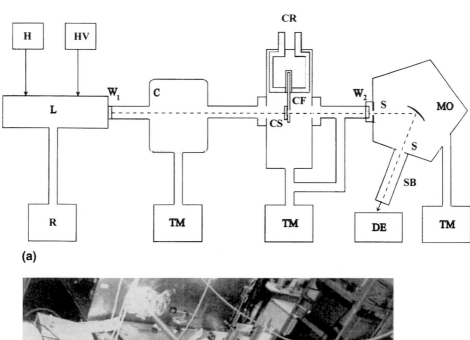

(a)

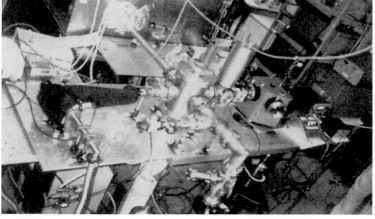

(b)

Figure 4   a. Schematic layout of the VUV absorption spectrometer. H, hydrogen supply; HV, high voltage; L, longitudinal stabilized hydrogen discharge $W_{1,2}$: LiF windows; C, vacuum chamber; CF, cold finger; CS, crystal sample; CR, cryostat; MO, VUV monochromator; S, slits; TM, turbo molecular pump; DE, detection electronics; SB, photomultiplier or secondary electron multiplier. b. X-UV and VUV absorption spectrometer at the National Hellenic Research Foundation (Athens, Greece). Crystal samples could be cooled down to 10K with a closed circle helium cryostat. Further cooling to 4.2 K can be achieved with a specially constructed liquid helium cryostat.

The solution of Eq. (2) is

$$I(\lambda, x) = I(\lambda, x = 0)e^{\int_0^x [a(\lambda,x)-\beta(\lambda,x)]dx} \tag{3}$$

Taking into consideration that the gain coefficient $a(\lambda, x)$ is proportional to the number density of the excited atoms $N_2(x)$ and thus proportional to the intensity of light

$$a(\lambda, x) = \sigma(\lambda)N_2(\lambda, x) \qquad N_2(\lambda, x) = cI(\lambda, x) \tag{4}$$

Equation (3) then becomes

$$I(\lambda, x) = I(\lambda, x = 0) \exp\left\{ \int_0^x [\sigma(\lambda)cI(\lambda, x) - \beta(\lambda, x)] \, dx \right\} \tag{5}$$

This equation can be solved using successive iterations. As zero order approximation we consider the case where there is no secondary photon generation and the absorption coefficient is independent on the position of the interacting atoms inside the crystal. In this case the solution of the equation takes its usual form (Lambert–Beer law)

$$I^{(0)}(\lambda, x) = I(\lambda, x = 0) \exp\left[\int_0^x - \beta(\lambda, x)] \, dx \right] \tag{6}$$

Substituting Eq. (6) into Eq (3) we get $I^{(1)}(\lambda, x)$ in first-order approximation.

$$I^{(1)}(\lambda, x) = I(\lambda, x = 0) \exp\left\{ \int_0^x [\sigma(\lambda)cI(\lambda, x' = 0) \, e^{-\beta(\lambda)x'} - \beta(\lambda)] \, dx \right\} \tag{7}$$

Integrating the above equation and setting

$$F^{(1)}(\lambda, x) = \exp\left[c\sigma(\lambda)\frac{I(\lambda, x = 0)}{\beta(\lambda)} (1 - e^{-\beta(\lambda)x}) \right] \tag{8}$$

Equation (7) becomes

$$I^{(1)}(\lambda, x) = I(\lambda, x = 0) \, F^{(1)}(\lambda, x) \, e^{-\beta(\lambda)x} \tag{9}$$

The $F^{(1)}(\lambda, x)$ term that describes deviation of the absorption processes from the law of Lambert and Beer is a function of wavelength and the crystal thickness. It describes photon emission through spontaneous emission within the crystal volume in first-order approximation. By measuring the photon intensity emitted from the crystal $I^{(1)}(l, x)$ as a function of the crystal thickness, the $F^{(1)}(\lambda, x)$, can be determined.

In the case of "optically thin" samples, $\beta x \ll l$, Eq. (9) takes the form

$$I^{(1)}(\lambda, x) \approx I(\lambda, x = 0) \exp(ax) \approx I(\lambda, x = 0)(1 + ax) \tag{10}$$

Under these experimental conditions, the emitted photon intensity from the crystal is taking place through spontaneous emission processes and varies linearly with crystal thickness. Similarly when $\beta x \gg 1$, "optically thick sample" Eq. (9) takes the form

$$I^{(1)}(\lambda, x) \approx I(\lambda, x = 0) \, e^{[\alpha(\lambda)/\beta(\lambda)]} \approx I(\lambda, x = 0)\left[1 + \frac{a(\lambda)}{\beta(\lambda)} + O\left(\frac{a(\lambda)}{\beta(\lambda)}\right) \right] \quad (11)$$

In this case the emitted intensity from the crystal is independent on its thickness. For cases between Eqs. (10) and (11), the emission is reduced by self-absorption and care has to be taken to ensure that all the terms in the transfer equation are properly allowed for individual cases. When a single RE ion absorbs one VUV photon, the $4f^{n-1}5d$ electronic configuration is populated and competition starts between radiative and nonradiative transitions. The excited ion decays to the different levels of the $4f^{n-1}5d$ electronic configuration and subsequently relaxes to the ground level of the $4f^n$ electronic configuration with VUV or UV photon emission. It is therefore expected that the $F^{(1)}(\lambda, x)$ factor will be equal to unity for the spectral region where there is no photon emission. In this case, absorption processes inside the crystal sample fall within the validity of the Lambert and Beer's law [Eq. (9)]. For example, in the case of the $LiYF_4:Nd^{3+}$ crystal, spontaneous emission modulates the value of the absorption coefficient only in the spectral range from 180 to 182 nm, since the crystal emits strongly in this spectral range when it is excited with VUV laser light at 157 nm [7].

III. OPTICAL AND ELECTRONIC PROPERTIES OF WIDE BAND GAP FLUORIDE DIELECTRIC CRYSTALS DOPED WITH TRIVALENT RE IONS

A. Electronic Properties of Wide Band Gap Materials

The optical properties of the nondoped wide band gap dielectric crystals in the VUV are mainly due to transitions from electronic states between the valance and the conduction band. The high value of the absorption coefficient, for photon energies higher than the crystal's band gap, is due to transitions involving delocalized electronic states in the conduction band, and arise from the crystal symmetry. The crystals' energy bands are formed from the energy levels of atoms when they are brought close together. On the other hand, local lattice imperfections, vacancies and other point defects, and dislocations bring about the formation of allowed states connected with the perturbation area. The electronic wavefunctions of such states are nonzero in approximately the same area where the perturbation exists. In other words, the electron is localized in the perturbation area. The smaller the perturbation energy, the greater the localization area. In this case, transitions are taking place between the valance band and the localized states in the conduction band. The width of the band gap is proportional to the value of the exchange integral, the origin of which is the fact that the electrons may, with some probability, be located near any atom. The exchange integral can be calculated with perturbation theory and accepting electron states of an isolated atom as zero approximation for the solution of the Schrödinger equation for the electron in a periodic crystal field $U(r)$.

$$\hat{H}\psi(r) = E\psi(r)$$

$$\hat{H} = -\frac{h^2}{2m}\Delta + U(r); \ U(r) = U(r + n) \tag{12}$$

denoting the Hamiltonian of an isolated atom by

$$\hat{H}_a = -\frac{h^2}{2m}\Delta + V_a(r)$$

$$\hat{H}_a\psi_a(r) = E_a\psi_a(r) \tag{13}$$

where $V_a(r)$ is the potential energy of the electron in an isolated atom, E_a is the a energy level, and $\psi_a(r)$ is the wavefunction corresponding to E_a. Solution of the equation for the atom is supposed to be known. The zero-order electron wavefunction in the crystal $\psi^{(0)}(r)$ should be the sum of the atomic wavefunctions at the point $r - m$ that satisfy the translational condition at the point $r \pm m$, and n is the coordinate of an ion.

$$\psi^{(0)}(r) = \sum_m \exp(ikm)\psi_a(r - m) \tag{14}$$

Zero approximation of the wavefunction $\psi^{(0)}(r)$ enables the first approximation for the energy of the electron in the crystal field $E^{(1)}$ to be calculated

$$E^{(1)} = \frac{N\left[E_a \sum_p \exp(ikp)S_p + C + \sum_p \exp(ikp)A(p)\right]}{N \Sigma \exp(ikp)S_p} \tag{15}$$

where S_p is the overlapping integral that depends not on the coordinates of two ions, but also on the distance p between them

$$\int \psi_a(r')\psi_a(r' - p) = S_p \tag{16}$$

and $A(p)$ is the exchange energy

$$A(p) = \int \psi_a^*(r') \left[\sum_{n' \neq p} V_a(r' - n) + W(r') \right] \psi_a(r' - p) d\tau' \tag{17}$$

where $W(r')$ is the periodic self-consistent lattice field

$$W(r) = U(r) - \sum_n V_a(r - n) \tag{18}$$

$A(p)$ is made up of the wave functions of two atoms separated by the distance p, with the physical meaning that the two atoms at distance p from each other may exchange electrons. The exchange proceeds through the field of all other atoms and the periodic self-consistent part of the lattice field $W(r)$. Due to the exponential form of the atomic wavefunctions, the exchange takes place mainly between neighboring atoms. Electron exchange between any two atoms of the crystal takes place by way of chain of neighbor exchanges. Therefore, electrons are not localized near individual atoms but move freely through the crystal jumping from one atom to another by the exchange process. From Eq. (15), the energy of the atomic electrons in a crystal field is revealed. When the interaction of the electrons with the ions and with the other electrons is taken into account, the energy levels are lowered by the amount C and split into a band of definite width. Because $S_p \approx 0$ when $p \neq 0$, we may write

$$E^{(1)} = E_a + C + \sum_p \exp(ikp)A_p \tag{19}$$

For a simple cubic lattice like the LiF crystal, every atom has six nearest neighbors with the coordinates $\alpha[(1,0,0), (-1,0,0), (0,1,0), (0,-1,0), (0,0,1), (0,0,-1)]$. Assuming exchange energy to be isotropic and neglecting electron exchange between distant atoms we obtain

$$E^{(1)} = E_a + C + 2A(\cos k_x a + \cos k_y a + \cos k_z a) \tag{20}$$

The energy depends quasicontinuously on the wave vector k and changes from E_{min} to E_{max}

$$\begin{aligned} E_{min} &= E_a + C + 6A \\ E_{max} &= E_a + C - 6A \end{aligned} \tag{21}$$

Thus, as a result of the atomic interaction, the energy level of an isolated atom in a cubic crystal lattice drops by the amount C and splits into a band 12 A wide. The energy band gap depends on the exchange energy A. But the exchange energy itself is dependent on the area of the overlapping of the wave functions: the more the atomic wave functions overlap, the greater will the exchange energy be. It follows from here that energy levels corresponding to inner shell electrons do not split as intensely as those of the outer shells, since the inner shell electrons are localized in smaller areas of space. As the energy increases, bands become wider and the gaps narrower. The higher a level, the lower it drops and the wider it spreads. This result is basic for explaining the dependence of the band gap width on the atomic number of the elements belonging to the same group. For example the width of the forbidden band between the valance and the conduction band for the metals Li, Na, K, Sc, and the halogens F, Cl, Br I should decrease in the same order.

Following the above arguments it is expected that the LiF crystal will have the widest band gap, among all the dielectric crystals, a fact that is verified from experiments. Optical transitions in the VUV might take place between the valance and the core band of the metal cations, which are formed of electron orbitals lying lower in energy than the ones forming the valance band (2p F^-). In this case an electron is excited from the core band to the conduction band. The hole created in the core band relaxes very fast to the band edge and is subsequently annihilated by electron capture from the filled valance band.

The conservation of energy in this case requires emission of a VUV or UV photon. The process is known as cross-relaxation [58] and it has been observed in different wide band gap dielectric crystals [59]. Band-to-band excitonic transitions in the VUV have been observed in various wide band gap materials such as $LiCaAlF_6$ [60] and $LiYF_4$ [61] crystals doped with Nd^{3+} and Ho^{3+} ions, respectively. In these cases, the peaks around 114, 118, and 112, 116, 119 nm might correspond to excitonic transitions between the valence and the conduction band of the host lattice. The exitonic transitions imply a large value of the index of refraction of fluoride crystals in the VUV, in agreement with previous measurements [62–70].

Considering that the cut-off wavelength of the LiF window (which is placed in front of the hydrogen lamp and the solar blind photomultiplier) is at 110 nm (for commercial LiF windows 5 mm thick) (Fig 2.), the absorption spectrum between 113 and 120 nm indicates only the energy of the corresponding transitions. The edge of the conduction band is extended over a relatively wide spectral range (113–120 nm). This situation reflects that the electron wavefunctions of the fluorinc and thc mctal ligands are extended over a large range as is expected from the strong electrostatic nature of the corresponding bonds. The main contribution at the value of the absorption coefficient in the spectral range from 110 to 120 nm is from the absorption of light within the conduction band of the $LiCaAlF_6$ crystal. The absorption within the conduction band of the $LiCaAlF_6$ crystals is at least one order of magnitude higher than the absorption within the $4f^2 5d$ electronic configuration. As a consequence of the large value of the absorption coefficient in this spectral region, it falls out of the dynamic range of the absorption spectrometer and hence deviates significantly from its mean value. Optical transitions in the VUV are likewise taking place between the valance and charge transfer bands situated inside the band gap of the host crystal. They are formed by strong electrostatic interactions between the d electron of the trivalent RE ions and the anions of the host crystal [71–74].

IV. VUV ABSORPTION SPECTROSCOPY OF TRIVALENT RE IONS IN WIDE BAND GAP FLUORIDE CRYSTALS

When a particular atom or ion is surrounded by a symmetrical distribution of atoms or ions (ligands), many new properties emerge that were not present in the free atom case. There are two models for the treatment of such systems. In the first model the central atom is assumed to be surrounded by a distribution of charges that produces an electrostatic field at the position of the central atom. In this model (*crystal field*) there is no sharing of electronic charge between the central atom and the ligands. The second model treats the central atom and its ligands as if they were a single molecule (*molecular–orbital method*). In this case electrons are permitted to overlap. In most cases this model is treated by the methods of molecular–orbital theory. In both cases symmetry considerations are of great importance. The symmetry of the optical site determines the degeneracy of the electronic states and the optical properties of the materials. In the case of wide-band-gap dielectric crystals doped with trivalent RE ions, the crystal field model is usually applied since it is simpler, and the optical properties of the RE ions are determined by the influence of the crystal field on the Hamiltonian of the free ion.

In order to assign the dipole-allowed transitions between the Stark components of the $4f^{n-1}5d$ electronic configuration and the levels of the $4f^n$ electronic configuration, the exact solution of the secular equation of the $4f^{n-1}5d$ electronic configuration in the presence of the crystal field should be known [75,76]. However up to now, there have been no detailed theoretical calculations of the energy position of the levels of the $4f^{n-1}5d$ electronic configuration. Crystal field calculations are carried out only to a low order of perturbation theory where the strongest perturbation is applied first, followed by the next strongest, etc. [77]. The exact theoretical treatment of the energy level problem of the $4f^{n-1}5d$ electronic configuration is difficult to address. Even for the simplest case of the Ce^{3+} ion, the crystal field for a d electron in a cubic symmetry requires 10 parameters for its characterization. In order to identify these from the experimental data, one must resolve and identify at least 11 spectral lines within the configuration. The number of lines within the configuration is large (20 for Pr^{3+}, 3106 for Tb^{3+} ions) and as a consequence attempts to evaluate crystal-field parameters by fitting an experiment are difficult.

In principle, if one knows the radial wave functions of the ion in question, then the crystal field perturbation can be calculated and the energy levels can be compared with the experimental values. However, it is tedious to construct wavefunctions of the electrons in the $4f^{n-1}5d$ mixed configurations, and only qualitative interpretation of the absorption spectra has been available up to now. The early attempt of Dieke [29] to deduce the extent of the $4f^{n-1}5d$ electronic configuration for each of the RE ions was made using experimental data and some observed regularities in the lanthanide spectra. The effect of the crystal field on the $4f^{n-1}5d$ electronic configuration is to lower its energy in comparison to the free ion case (Table 2).

Various empirical methods have been developed [77] to interpret the VUV absorption spectra of the $4f^{n-1}5d$ electronic configuration. These are based on the following principle. The energy difference Δ between two configurations differing by a single electron in the 4f configuration (e.g., $4f^n$ and $4f^{n-1}5d$) is constant for all the lanthanides. This allows the evaluation of the energies of different configurations for free ions with different ionization. Brewer [77] used this to estimate the energy of the lowest energy level of a given configuration, beginning with measured energies for neutral rare earths. For the 5d electronic configuration of the RE ions, the crystal field perturbation is stronger than the electron–electron interaction, due to the fact that the d electron is extended over a large distance and been far apart from the nucleus of the ion. In this case the effect of the crystal field is calculated first or simultaneously with the imposition of the Coulomb interaction followed by the spin–orbit interaction. For example for the D_4 tetragonal symmetry, the energy levels are the following: $\Gamma^{(4)}$, $\Gamma^{(5)}$, $\Gamma^{(3)}$, and $\Gamma^{(1)}$.

When the spin and the spatial parts of the wave functions are coupled within the framework of the symmetry group of the RE ion, the number of the energy levels is described by the decomposition of the product representation of $\Gamma^{(6)}$ (spin part) with $\Gamma^{(4)}$, $\Gamma^{(5)}$, $\Gamma^{(3)}$, and $\Gamma^{(1)}$. The new energy levels now are the following: $\Gamma^{(7)}$ ($^2B_2$), $\Gamma^{(6)}$ (^2E), $\Gamma^{(7)}$ (^2E), $\Gamma^{(7)}$ ($^2B_1$), $\Gamma^{(6)}$ ($^2A_1$). In addition, when only one electron is present in the 5d orbital, as concerns the symmetry of the states produced by the combined action of the crystal field and the spin–orbit interaction, it makes no difference in which order the two perturbations are applied. The symmetry group of the atom is no longer the three-dimensional rotation group because all directions in space are no longer equivalent. Instead, the symmetry is that of a point group containing a finite number of elements and is a subgroup of the three-dimensional rotation group. An irreducible representation of the latter becomes reducible with respect to the lower symmetry point group. The spherical harmonics are not any longer energy eigenstates and are transformed under the irreducible representations of the symmetry operators of the three-dimensional rotation group to its linear combinations. The effect on the states of the free atom is to remove degeneracy. The first-order energies are sensitive to the magnitude of the interactions and hence to the order in which they are applied.

In the case where the Coulomb repulsion of the electrons or the spin–orbit interaction is the strongest perturbation, as in the case of the $4f^n$ or the $4f^{n-1}$ electronic configuration [78], which are well shielded by the 6p and the 6s elec-

Table 2 Energy Position of the Edge of Levels of $4f^{n-1}5d$ Electronic Configuration of Trivalent RE Ions

RE^{3+} ion	Free RE^{3+} ions $\times 10^3$ cm^{-1}	$YF_3 \times$ 10^3 cm^{-1}	$LuF_3 \times$ 10^3 cm^{-1}	$LiYF_4 \times$ 10^3 cm^{-1}	$LaF_3 \times$ 10^3 cm^{-1}	$BaY_2F_8 \times$ 10^3 cm^{-1}	$CaF_2 \times$ 10^3 cm^{-1}	$KY_3F_{10} \times$ 10^3 cm^{-1}
Nd^{3+}	70.1	58.4	57.1	54.6	60.0	53.3		53.8
Er^{3+}	75.4	64.3	63.8	62.5	64.7	61.9		63.3
Tm^{3+}	74.3	63.4	63.0	61.5	64.4		63.0	
Ce^{3+}	49.7			34.4	38.8	34.0	31.9	
Ho^{3+}	58.9			63.3	63.3	63.1	64.1	

Values are given as measured from the ground level of the $4f^n$ electronic configuration in different crystal hosts.

tronic configurations, the transformation properties of the electronic wavefunctions are well described by the three-dimensional rotation group $O^+(3)$ and the spherical harmonics are basic functions for the representations $D^{(1)}(\alpha, \beta, \gamma)$ of the $O^+(3)$. The subsequent application of the crystal field will further split the energy levels.

Vacuum absorption and emission spectra of the trivalent RE ions in wide band gap fluoride crystals are due to the interconfigurational $4f^{n-1}5d \leftrightarrow 4f^n$ transitions of the RE ions. They have been interpreted by fitting the crystal field splitting of the d electronic configuration to the energy gaps in the ground multiplets of the $4f^{n-1}$ core [79]. The high complexity of the energy level systems of the trivalent RE ions in the $4f^{n-1}5d$ electronic configuration makes a detailed interpretation of the observed spectra still impossible. This is mainly due to the fact that the wave functions of the two electronic configurations are transformed differently under the symmetry operations and the classification of spectral terms is rather difficult, since the selection rules for the angular momentum are no longer valid.

A theoretical interpretion of the atomic spectra of the $4f^{n-1}5d$ electronic configuration was based on the finding, that the interaction energy of the crystal field with the d electron exceeds the Coulomb interaction energy between the d and the $4f^{n-1}$ electrons [80]. Energy states of the d electron are characterized by irreducible representations of the corresponding crystal symmetry group $\Gamma^{(2)}$. For the $4f^{n-1}$ electronic configuration, L–S coupling is predominant and further splitting of energy levels is taking place (Table 3). The energy levels are characterized by the $\Gamma^{(1)}$ index of the irreducible representation of the three-dimensional rotation group.

For this method the electronic wave functions can be written in the form $| SLJ\Gamma^{(1)}, \Gamma^{(2)}, \alpha\Gamma\mu>$, where μ is the number of the basis function of the irreducible representation Γ, and the index α identifies the representation Γ that appears several times in the direct product representation $\Gamma^{(1)} \times \Gamma^{(2)}$.

A qualitative interpretation of the $4f^{n-1}5d \leftrightarrow 4f^n$ interconfigurational transitions of trivalent RE ions in wide band gap dielectric crystals was made by Szczurek and Schlesinger [81]. They constructed spectral terms first (the selection rules for angular momentum are valid in this case), and then allowed for crystal field splitting. Because the electronic wave functions of the $4f^{n-1}$ and 5d electronic configurations have different transformation properties, this approximation is expected to be valid for the situation in which a more or less uniform charge distribution of the ligands takes place within the unit crystal cell. This is possible provided that the radius of the d orbital is smaller than lattice constants. In this case, the electric field around the origin for a given distribution of ligands is more or less spherical. This situation partially restores the three-dimensional rotation symmetry of the 5d electronic configuration. As a first approximation, the 5d and $4f^{n-1}$ electronic configurations might well have the same transformation properties under the three-dimensional rotational group.

Table 3 Crystal Field Splitting of the Main Levels of the $4f^2$ Electronic Configuration in the Presence of Crystal Field with Octahedral (O) and Tetragonal (D_4) Symmetry

Configuration	Terms	Levels	O	D_4
f^2	^1S	$^1S_0$	$^1\Gamma_1$	$^1\Gamma_1$
	^3P	$^3P_2$	$^3\Gamma_3 + {}^3\Gamma_5$	$^1\Gamma_1 + {}^1\Gamma_3 + {}^1\Gamma_4 + {}^1\Gamma_5$
		$^3P_1$	$^3\Gamma_4$	$^1\Gamma_2 + {}^1\Gamma_5$
		$^3P_0$	$^3\Gamma_1$	$^3\Gamma_1$
	^1I	$^1I_6$	$^1\Gamma_1 + {}^1\Gamma_2 + {}^1\Gamma_3 + {}^1\Gamma_4 +$ $2{}^1\Gamma_5$	$2{}^1\Gamma_1 + {}^1\Gamma_2 + 2{}^1\Gamma_3 +$ $2{}^1\Gamma_4 + 3{}^1\Gamma_5$
	^1D	$^1D_2$	$^1\Gamma_3 + {}^1\Gamma_5$	$^1\Gamma_1 + {}^1\Gamma_3 + {}^1\Gamma_4 + {}^1\Gamma_5$
	^1G	$^1G_4$	$^1\Gamma_1 + {}^1\Gamma_3 + {}^1\Gamma_4 + {}^1\Gamma_5$	$2{}^1\Gamma_1 + {}^1\Gamma_2 + {}^1\Gamma_3 + {}^1\Gamma_4$ $+ 2{}^1\Gamma_5$
	^3F	$^3F_4$	$^3\Gamma_1 + {}^3\Gamma_3 + {}^3\Gamma_4 + {}^3\Gamma_5$	$2{}^3\Gamma_1 + {}^3\Gamma_2 + {}^3\Gamma_3 + {}^3\Gamma_4$ $+ 2{}^3\Gamma_5$
		$^3F_3$	$^3\Gamma_2 + {}^3\Gamma_4 + {}^3\Gamma_5$	$^3\Gamma_2 + {}^3\Gamma_3 + {}^3\Gamma_4 + 2{}^3\Gamma_5$
		$^3F_2$	$^3\Gamma_2 + {}^3\Gamma_5$	$^3\Gamma_1 + {}^3\Gamma_3 + {}^3\Gamma_4 + {}^3\Gamma_5$
	^3H	$^3H_6$	$^3\Gamma_1 + {}^3\Gamma_2 + {}^3\Gamma_3 + {}^3\Gamma_4 +$ $2{}^3\Gamma_5$	$2{}^3\Gamma_1 + {}^3\Gamma_2 + 2{}^3\Gamma_3 +$ $2{}^3\Gamma_4 + 3{}^3\Gamma_5$
		$^3H_5$	$^3\Gamma_3 + 2{}^3\Gamma_4 + {}^3\Gamma_5$	$^3\Gamma_1 + 2{}^3\Gamma_2 + {}^3\Gamma_3 +$ $^3\Gamma_4 + 3{}^3\Gamma_5$
		$^3H_4$	$^3\Gamma_1 + {}^3\Gamma_3 + {}^3\Gamma_4 + {}^3\Gamma_5$	$2{}^3\Gamma_1 + {}^3\Gamma_2 + {}^3\Gamma_3 +$ $^3\Gamma_4 + 2{}^3\Gamma_5$

A uniform charge distribution could be the result of charge compensation from the F^- ligands, or of the localized character of the d orbitals. For example, for a hydrogen-like atom, the expectation value of the radius of the electronic clound $\langle r \rangle$ is given by the formula

$$\langle r \rangle = \frac{\alpha_0}{2Z}[3n^2 - l(l + 1)] \tag{22}$$

where n is the principal quantum number equal to 5, α_0 is the Bohr radius of the atom, $\alpha_0 = 0.053$ nm, Z is the charge of the nucleus, and $l = 2$. For Nd^{3+}, Tb^{3+}, and Ho^{3+} ions, the calculated values of the ionic radii are $\langle r \rangle \sim 0.028$, 0.027, and 0.030 nm. The experimentally measured values for the three trivalent ions in CaF_2 are 0.11, 0.10, and 0.97 nm. Since the ionic radii of the F^- ions are 0.136 nm, and taking into consideration that the distance between two neighboring atoms is of the order of ~ 0.5 nm (as in $LiYF_4$ crystal), the spherical symmetry is expected to be restored partially by charge compensation from F^- ions and the localized character of the d orbital. Using this approximation, the number of experimentally observed transitions is in excellent agreement with the number of theoretically expected $4f^n \rightarrow 4f^{n-1} 5d$ dipole-allowed transitions for Nd^{3+} and Tb^{3+} ions [82,83]. (See Table 4.)

A. Cerium (Ce³⁺, 4f¹)

The Ce^{3+} ion has only one electron in the d electronic configuration and therefore the spectroscopic assignment of the energy levels is simple. The free Ce^{3+} ion in the 4f electronic configuration has two energy levels—$^2F_{5/2}$, $^2F_{7/2}$,—which are separated by 2253 cm^{-1}. In the d electronic configuration it has also two levels— $^2D_{3/2}$, $^2D_{5/2}$—which are separated by 49,733 and 52,226 cm^{-1} from the ground state, respectively. The crystal field splitting of the energy levels of the 5d electronic configuration depends on the site symmetry of the Ce^{3+} ions. The fivefold degenerate d level split into two levels with degeneracies 2 and 3. The d orbitals $(z^2, d_{x^2} - y^2, d_{yz}, d_{zx}, \text{and } d_{xy})$ are the basis functions for the $D^{(2)}$ representations of $O^+(3)$ group and the reduction of the $D^{(2)}$ into irreducible representations of the cubic group is $E(\Gamma^{(3)}) + T_2(\Gamma^{(5)})$. Since E and T_2 are two- and three-dimensional respectively, the energy levels associated with E and T_2 are doubly and triply degenerate. When the site symmetry of the Ce^{3+} ion is further reduced to a tetragonal D_4 one, degeneracy of the energy levels is removed since the irreducible representations of the cubic group are reducible representations of the D_4 group. The energy levels are split further and the degeneracy is removed: E $(\Gamma^{(3)}) = {}^2A_1$ $(\Gamma^{(1)}) + {}^2B_1 (\Gamma^{(3)})$, $T_2 (\Gamma^{(5)}) = {}^2A_1 (\Gamma^{(1)}) + {}^2B_1 (\Gamma^{(3)})$. The absorption spectra of the Ce^{3+} ions in dielectric crystal hosts are attributed to the dipole-allowed transitions between the $^2F_{5/2}$ ground state of the 4f electronic configuration and the energy levels of the 5d electronic configuration observed early mainly due to experimental reasons, since the Ce^{3+} ions absorb and emit in the UV [84–99]. Six strong interconfigurational dipole-allowed transitions have been observed [92,96]. Four transitions have been assigned to the interconfigurational $4f \rightarrow 5d$ electronic transitions of the Ce^{3+} ions of tetragonal site symmetry, and two to clusters formed by two or more Ce^{3+} ions. One more transition of low intensity corresponds to the e_g level, which is doubly degenerate [97]. The remaining e_g level indicates a strong localized vibrational structure [81]. In this case the electronic wavefunctions of the Ce^{3+} ions do not overlap with the wavefunctions of the ions of the host lattice [99] and the strongest phonon lines are those corresponding to localized phonon frequencies.

B. Praseodymium (Pr³⁺, 4f²)

Energy levels of the free Pr^{3+} ion in various electronic configurations such as $4f^2$, 4f5d, 4f6s, and 4f6p have been analyzed by Sugar [100]. A comparison of all the third and the fourth spectra of the lanthanide ions show a similarity throughout the series [101]. Energy levels of the $4f^2$ electronic configurations have been identified by comparing spectra in the vapor phase and in $LaCl_3:Pr^{3+}$ crystals [102]. The lines arising from the $4f^2 \rightarrow 4f5d$ transitions were identified in a series of exposures of sparks made with successively increasing current. The energy levels were assigned by comparing the level structure with that of the isoelectronic Ce^{2+} ion and from relative line intensities [103]. The free Pr^{3+} ion in the 4f5d electronic configuration has 18 energy levels, the lowest is the $^1G_4$ at 61,171 cm^{-1}. The $^3G_{3,4,5}$, $^3H_{4,5}$ levels are excited through dipole-allowed transitions from the $^3H_4$ ground state of the $4f^2$ electronic configuration. The absorption spectra of the $4f^2$ and the 4f5d electronic configurations of the Pr^{3+} ions were studied previously in a number of crystal lattices [8,104–110]. If the Pr^{3+} ion is located at a C_{4v} symmetry center, the $^3G_{3,4,5}$ and the $^3H_{4,5}$ energy levels are split for a total number of 35 Stark levels [81].

The number of absorption lines in $CaF_2:Pr^{3+}$ crystals is about half that, but some lines may be weak or superimposed on others. If only the strongest lines are considered, an additional selection rule $\Delta J = \Delta L$ is applied in the C_{4v} symmetry, the number of Stark levels is limited to 12, which is close to what was experimentally observed [81]. In the case of a high concentration of the Pr^{3+} ions (higher than 0.01 at.%), the appearance of additional lines in the absorption spectrum is attributed to Pr^{3+} clusters and sites of additional symmetries. A typical VUV absorption spectrum for the $KY_3F_{10}:Pr^{3+}$ crystal for 0.1 at.% concentration of the Pr^{3+} ions at room temperature is indicated in Figure 5.

The 4f5d electronic configuration in the free ion case is extended between 61,170 and 78,776 cm^{-1}. The edge of the levels of the 4f5d electronic configuration in various fluoride dielectric crystals is extended down to 44,440 cm^{-1} and depends on the host material. For example, the edge of the levels of the 4f5d electronic configuration in $LiLuF_4$ crystal hosts is at 45,100 cm^{-1} [8], in CaF_2 at 45,450 cm^{-1} [81], and in LaF_3 at 49,995 cm^{-1} [110]. The $4f^2$ electronic configuration of the Pr^{3+} ions consists of 13 energy levels. $^1S_0$ is the highest one, which lies very close to the 4f5d electronic configuration. The energy position of the $^1S_0$ level is within the 4f5d electronic configuration for most of the cases. For the LaF_3 crystal, the $^1S_0$ level is situated 3000 cm^{-1} below the edge of the band [111], and hence it can be highly populated from the 4f5d electronic configuration. A similar position of the $^1S_0$ energy level is valid for the YF_3 [24] and most of the oxide crystal hosts [112–114].

C. Neodymium (Nd^{3+}, $4f^3$)

The absorption spectrum of the trivalent Nd^{3+} ion in the VUV has been studied previously in different dielectric crystal hosts [4,7,10,61,81,82,109,115,116]. The energy position of the levels of the $4f^25d$ electronic configuration of the free Nd^{3+} ion can be estimated from data available for the isoelectronic Pr^{2+}. According to

Table 4 Number of Electric Dipole-Allowed Interconfigurational Transitions of Trivalent RE Ions Between the Ground State of the $4f^n$ Electronic Configuration and the Excited Terms of the $4f^{n-1}5d$ Electronic Configuration

Ion	n	$4f^n$ configuration		$4f^{n-1}5d$ configuration			Dipole allowed transitions $\Delta S = 0, \Delta L = 0 \pm 1, \Delta L = \Delta J$	
		Number of levels	Ground state	Number of levels	Lowest level	Excited terms	Terms in tetragonal crystal field	Transitions
La^{3+}	0	1	$^1S_0$	0	—	—	—	0
Ce^{3+}	1	2	$^2F_{5/2}$	2	$^2D_{3/2}$	$^2D_{3/2}$	$\Gamma_6 + \Gamma_7$	2
Pr^{3+}	2	13	$^3H_4$	20	$^1G_4$	$^3G_3$ $^3H_4$	$\Gamma_2 + \Gamma_3 + \Gamma_4 + 2\Gamma_5$ $2\,\Gamma_1 + \Gamma_2 + \Gamma_3 + \Gamma_4 + 2\Gamma_5$	5
Nd^{3+}	3	41	$^4I_{9/2}$	107	$^4I_{9/2}$	$^4H_{7/2}(2)$ $^4I_{9/2}$ $^4K_{11/2}$	$2\Gamma_6 + 2\Gamma_7$ $3\,\Gamma_6 + 2\Gamma_7$ $3\Gamma_6 + 3\Gamma_7$	9
Pm^{3+}	4	107	$^5I_4$	386	$^5L_6$	$^5H_3(3)$ $^5I_4(2)$ $^5K_5$	$\Gamma_2 + \Gamma_3 + \Gamma_4 + 2\Gamma_5$ $2\Gamma_1 + \Gamma_2 + \Gamma_3 + \Gamma_4 + 2\Gamma_5$ $\Gamma_1 + 2\Gamma_2 + \Gamma_3 + \Gamma_4 + 3\Gamma_5$	14
Sm^{3+}	5	198	$^6H_{5/2}$	977	$^6L_{11/2}$	$^6G_{3/2}(4)$ $^6H_{5/2}(3)$ $^6I_{7/2}(2)$	$\Gamma_6 + \Gamma_7$ $\Gamma_6 + 2\Gamma_7$ $2\Gamma_6 + 2\Gamma_7$	20
Eu^{3+}	6	295	$^7F_0$	1876	$^7K_4$	$^7D_1(2)$ $^7F_1(3)$ $^7G_1(2)$	$\Gamma_2 + \Gamma_5$ $\Gamma_2 + \Gamma_5$ $\Gamma_2 + \Gamma_5$	7
Gd^{3+}	7	327	$^8S_{7/2}$	2725	$^8H_{3/2}$	$^8P_{9/2}$	$3\Gamma_6 + 2\Gamma_7$	3

Ion						Terms	Decomposition	
Tb^{3+}	8	295	$^7F_6$	3106	$^9D_2$	$^7D_5$ (5) $^7F_6$ (5) $^7G_7$ (5)	$\Gamma_1 + 2\Gamma_2 + \Gamma_3 + \Gamma_4 + 3\Gamma_5$ $2\Gamma_1 + \Gamma_2 + 2\Gamma_3 + 2\Gamma_4 + 3\Gamma_5$ $\Gamma_1 + 2\Gamma_2 + 2\Gamma_3 + 2\Gamma_4 + 4\Gamma_5$	30
Dy^{3+}	9	198	$^6H_{15/2}$	2725	$^8G_{15/2}$	$^6G_{13/2}$ (11) $^6H_{15/2}$ (10) $^6I_{17/2}$ (9)	$3\Gamma_6 + 4\Gamma_7$ $4\Gamma_6 + 4\Gamma_7$ $5\Gamma_6 + 4\Gamma_7$	58
Ho^{3+}	10	107	$^5I_8$	1876	$^7H_8$	$^5H_7$ (18) $^5I_8$ (14) $^5K_9$ (11)	$\Gamma_1 + 2\Gamma_2 + 2\Gamma_3 + 2\Gamma_4 + 4\Gamma_5$ $3\Gamma_1 + 2\Gamma_2 + 2\Gamma_3 + 2\Gamma_4 + 4\Gamma_5$ $2\Gamma_1 + 2\Gamma_2 + 2\Gamma_3 + 2\Gamma_4 + 4\Gamma_5$	79
Er^{3+}	11	41	$^4I_{15/2}$	977	$^6L_{17/2}$	$^4H_{13/2}$ (18) $^4I_{15/2}$ (14) $^4K_{17/2}$ (11)	$3\Gamma_6 + 4\Gamma_7$ $4\Gamma_6 + 4\Gamma_7$ $5\Gamma_6 + 3\Gamma_7$	79
Tm^{3+}	12	13	$^3H_6$	386	$^5L_6$	$^3G_5$ (13) $^3H_6$ (11) $^3I_7$ (9)	$\Gamma_1 + 2\Gamma_2 + \Gamma_3 + \Gamma_4 + 3\Gamma_5$ $2\Gamma_1 + \Gamma_2 + 2\Gamma_3 + 2\Gamma_4 + 3\Gamma_5$ $\Gamma_1 + 2\Gamma_2 + 2\Gamma_3 + 2\Gamma_4 + 4\Gamma_5$	62
Yb^{3+}	13	2	$^2F_{7/2}$	107	$^4H_{9/2}$	$^2D_{5/2}$ (5) $^2F_{7/2}$ (5) $^2G_{9/2}$ (5)	$\Gamma_6 + 2\Gamma_7$ $2\Gamma_6 + 2\Gamma_7$ $3\Gamma_6 + 2\Gamma_7$	25
Lu^{3+}	14	1	$^1S_0$	20	$^3P_2$	$^1P_1$	$\Gamma_2 + \Gamma_5$	1

The number of dipole-allowed transitions is calculated taking into consideration partial restoration of the spherical symmetry for the 5d electronic configuration. The theoretical predictions are in excellent agreement with experimental results.

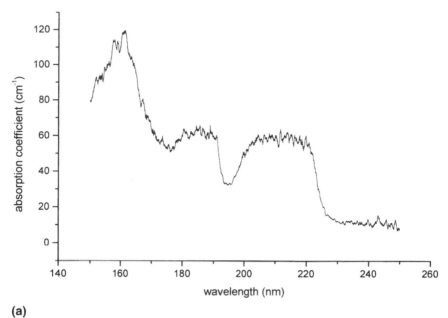

(a)

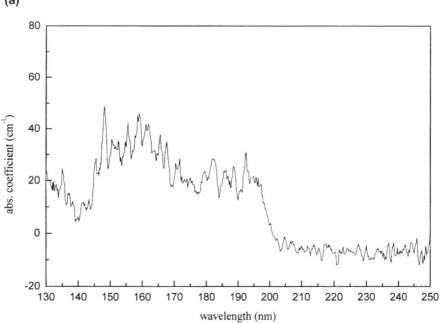

(b)

Figure 5 a. VUV absorption spectrum of the $KY_3F_{10}:Pr^{3+}$ crystal. b. VUV absorption spectrum of $LaF_3:Pr^{3+}$ crystal.

Sugar [117], 107 possible energy levels exist in the free Pr^{2+} ion of the $4f^25d$ electronic configuration. Among them, only nine levels, $4f^2(^3F)5d$ [$^4H_{7/2}$, $^4H_{9/2}$, $^4H_{11/2}$] and $4f^2(^3H)5d$ [$^4H_{7/2}$, $^4H_{9/2}$, $^4H_{11/2}$, $^4I_{9/2}$, $^4I_{11/2}$, $^4K_{11/2}$], are excited from the ground level $^4I_{9/2}$ of the $4f^3$ electronic configuration. In order to assign the dipole-allowed transitions between the Stark components of the $4f^25d$ electronic configuration and the levels of the $4f^3$ electronic configuration, the exact solution of the secular equation of the $4f^25d$ electronic configuration in the presence of the crystal field should be known. However up to now, there have been no theoretical calculations regarding the energy position of the levels of the $4f^25d$ electronic configuration.

To find the total number of interconfigurational dipole-allowed transitions of the Nd^{3+} ions, spectral terms that arise from the 5d and the 4f orbital are formed, followed by calculation of the effect of the crystal field on the system's Hamiltonian. The crystal field divides these nine levels into a total number of 40 Stark components for a C_{4v} or S_4 site symmetry. However if the selection rules for dipole transitions $\Delta S = 0$, $\Delta l = +1$, -1, $\Delta L = 0$, $+1$, -1, $\Delta J = 0$, $+1$, -1 ($J = 0 \rightarrow J = 0$ forbidden) and $\Delta J = \Delta L$ are applied, only 11 transitions remain after excitation from the ground level $^4I_{9/2}$ and the $4f^2(^3F)5d$ [$^4K_{11/2}$] level is the lowest. The $4f^2(^3H)5d$ [$^4H_{7/2}$] level is the highest. In the case of Nd^{3+} ion in $LiYF_4$ crystal hosts (C_{4h}^6 space symmetry group, S_4 point symmetry group of the Nd^{3+} ions), 11 Stark components are populated through dipole-allowed transitions from the ground level $^4I_{9/2}$ of the $4f^3$ electronic configuration, which is in excellent agreement with the experiments [82].

Therefore, the charge compensation and localized character of the d orbitals partially restore the spherical symmetry of angular momentum. In this case, the main effect of a strong crystal field is to lower the energy levels of the $4f^{n-1}5d$ electronic configuration relative to the free ion case. The amount of shifting is directly correlated to the magnitude of the crystal field. The absorption spectrum was slightly broadened at room temperature compared to lower temperature (77 K). The structure of the spectrum verifies the mixed character of the $4f^25d$ electronic configuration and partial restoration of the spherical symmetry. The absorption peak around 66,578 cm^{-1} probably has a weak additional contribution from the highly excited levels of the Nd^{3+} ion of the $4f^3(^2F)$ electronic configuration [29]. In this experiment the band gap of the $LiYF_4$ crystal host was estimated to be 86,900 cm^{-1} wide (for a 0.5 mm thick crystal). However, the value of the band gap is estimated to be wider for a thinner sample. The absorption band from 115 to 118 nm has been observed for various host lattices. It was attributed to transitions between the valence and the conduction band of the host lattice. In condensed matter at room and low temperatures, dipole transitions between the ground level of the Nd^{3+} ion and the levels of the $4f^25d$ electronic configuration usually have a broad band structure, due to the wide spatial distribution of the electronic cloud. A similar response is expected for crystal hosts of different symmetry. For example, in the case of the rhombic PFYK:Nd crystal (space group Pnam) [118], absorption spectrum showed that the $4f^2 5d$ configuration levels split into nine main Stark components with maximum absorption at 56,212; 58,423; 59,180; 62,606; 65,789; 66,906; 70,061; 74,174; and 79,878 cm^{-1}, respectively [82]. The absorption spectrum of the LiCaAlF$_6$:Nd^{3+} crystal in the spectral range from 120 to 200 nm is likewise indicated in Figure 6. The peaks were assigned to strong dipole-allowed transitions between the $^4I_{9/2}$ ground level of the $4f^3$ electronic configuration and the Stark components of the levels of the $4f^2 5d$ electronic configuration.

When the crystal field of trigonal symmetry is applied, on the $^4K_{11/2}$ level of the Nd^{3+} free ion, it splits it into three Stark components: $2(\Gamma^{(4)} + \Gamma^{(5)} + \Gamma^{(6)})$, $\Gamma^{(4)}$, $\Gamma^{(4)}$. The $^4I_{9/2}$ level likewise splits into two Stark components by the trigonal crystal field: $2(\Gamma^{(4)} + \Gamma^{(5)} + \Gamma^{(6)})$, $\Gamma^{(4)}$. The $^4H_{7/2}$ splits into three Stark components: the $\Gamma^{(4)}$, $\Gamma^{(4)}$, $(\Gamma^{(4)} + \Gamma^{(5)} + \Gamma^{(6)})$. Some of these components are degenerate and the degeneracy is removed within the $4f^2 5d$ electronic configuration. In the case of trigonal symmetry, the dipole-allowed transitions between the ground state $^4I_{9/2}$ of the $4f^3$ electronic configuration and the Stark components

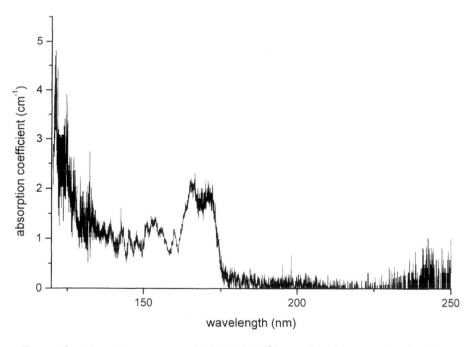

Figure 6 Absorption spectrum of LiCaAlF$_6$:Nd^{3+} crystal in the spectral region from 120 to 250 nm.

of the $4f^25d$ electronic configuration are 20, in agreement with experimental observations. The peak at 115 nm corresponds to excitonic transitions between the valence and conduction band of the host lattice, suggesting a large value of the index of refraction, which is in agreement with previous measurements of fluoride crystals' than films in the VUV.

D. Promethium (Pm^{3+}, $4f^4$)

Promethium is an artificial element that is highly radioactive. The $4f^4$ electronic configuration of Pm^{3+} ions has been investigated in $LaCl_3$ crystals [119,120]. There are 107 and 386 levels of the $4f^4$ and the $4f^35d$ electronic configurations respectively. In the free ion case, $^5L_6$ is the lowest level of the $4f^35d$ electronic configuration. Six terms—the ^5H (3), ^5I, and ^5K—are excited through dipole-allowed transitions from the ground state $^5I_4$ of the $4f^4$ electronic configuration. Fourteen dipole-allowed transitions are expected in the crystal field with tetragonal symmetry. Free-ion calculations put the edge of the $4f^35d$ electronic configuration at 73,300 cm^{-1} and the $4f^34s$ electronic configuration at 111,000 cm^{-1} [2].

E. Samarium (Sm^{3+}, $4f^5$)

The energy position of the levels of $4f^4 5d$ electronic configuration of free Sm^{3+} ion can be estimated based on the data available for isoelectronic Nd^+ [120,121]. The free Sm^{3+} ion has 977 levels of $4f^45d$ electronic configuration. Among them, only nine terms [four ^6G, three ^6H, and two ^6I] are excited from the ground level $^6H_{5/2}$ of the $4f^5$ electronic configuration, giving 20 dipole-allowed transitions excited from the ground state. The lowest level of the $4f^45d$ electronic configuration is the $^6L_{11/2}$ one at 73,700 cm^{-1} above the ground state. The $4f^5$ electronic configuration of the free ion consists of 198 levels [81]. It has been analyzed in crystals [122,123]. Early measurements of crystal field splitting of single and mixed electronic configurations of Sm ions in various dielectric crystal hosts were attributed to the divalent Sm^{2+} ions [124–126]. The edge of the levels of $4f^45d$ electronic configuration in LaF_3 was at 58,000 cm^{-1} and extends well beyond the cut-off of the crystal host. The VUV absorption spectra of Sm ions in various dielectric crystal hosts, CaF_2 [121,127] and $LiYF_4$ [128], were interpreted as $4f^5 \rightarrow 4f^45d$ interconfigurational transitions of the trivalent Sm^{3+} ions. Absorption spectra of $CaF_2:Sm^{3+}$ are unique in that some spectral lines arising from thermally excited Stark levels of the $4f^5$ electronic configuration were observed at low temperatures [81]. The average spacing between zero phonon and vibronic lines is ~486 cm^{-1}. This frequency has been attributed to a local mode between the Sm^{3+} ion and its nearest-neighbor fluorine ions [129].

As already mentioned, the effect of the crystal field on the $4f^{n-1}5d$ electronic configuration energy would be to lower its edge; various empirical methods have been established [77], based on the following principle: The energy difference Δ between two configurations differing by a single electron in the 4f configuration (e.g., $4f^n$ and $4f^{n-1}5d$) is constant for all the lanthanides. Martin [130] estimated the Δ energy difference for neutral monovalent and divalent ions from a few measured values. Sugar and Reader [131] applied this principle to estimate trivalent ion differences using Martin's results. In the free ion case, the energy difference between the $4f^45d$ and the $4f^5$ levels was estimated to be 76,000 ± 2000 cm^{-1} [77,130–131]. In the presence of the CaF_2 crystal field, a depression of the upper configuration by about 18000 cm^{-1} towards the ground state is taking place [92]. Hence the edge of the levels of the $4f^45d$ electronic configuration in the presence of the CaF_2 crystal field is expected to be at 58,000 ± 2000 cm^{-1}, in agreement with the experiments [121].

F. Europium (Eu^{3+}, 4f^6)

The free Eu^{3+} ion has 1878 levels [81] of $4f^55d$ electronic configuration. Among them, seven terms [two ^7G, three ^7F, and two ^7D] are excited from the ground level $^7F_0$ of the $4f^6$ electronic configuration, giving seven dipole allowed transitions excited from the ground state. The lowest level of the $4f^55d$ electronic configuration is the $^7K_4$, at 81,800 cm^{-1} above the ground state $^7F_0$ of the $4f^6$ electronic configuration. The $4f^6$ electronic configuration of the free ion consists of 295 levels [81]. Comparison of the absorption spectrum of CaF_2: Eu^{3+} ion with an energy level system of the free isoelectronic Sm^{2+} has been given by Schlesinger and Szczurek [121]. The so-called system difference $4f^55d$–$4f^6$ in the free Eu^{3+} ions was estimated to be 85,500 ± 1000 cm^{-1}. In the presence of the CaF_2 crystal field, the first level of the $4f^55d$ electronic configuration is expected to be at 67,500 ± 1000 cm^{-1}, which is in agreement with experimental observations [121]. The vibrational structure has been analyzed from the absorption spectrum and the separation from zero-phonon lines. Their vibronics were found to be ~452 cm^{-1} and 486 cm^{-1} [121]. The two values were attributed to the formation of centers of different site symmetry and clustering at higher concentration of the RE ions. Strong absorption of Eu^{3+} ions in LaF$_3$ crystals between 200 and 300 nm has been attributed to interconfigurational transitions from divalent ions [2]. The 5d electronic configuration of the Eu^{2+} ion is calculated to begin at 33,900 cm^{-1} and the 6s electronic configuration at 44,300 cm^{-1} [131].

G. Gadolinium (Gd^{3+}, 4f^7)

Crosswhite and Kielkopf have studied the free ion spectrum and identified several lines from the $4f^65d$ electronic configuration occurring between 104,000 and 122,000 cm^{-1} [132,133]. Despite the fact that Gd^{3+} ions in the $4f^65d$ electronic configuration have 2725 levels [81], only three among them are important if electric dipole-allowed transitions from the ground state are taken into consideration, the $^8P_{5/2}$, $^8P_{7/2}$, and $^8P_{9/2}$. In free Gd^{3+} ions these levels are expected to be separated from the lowest energy level of the $4f^6 5d$ electronic configuration, $^8H_{3/2}$, by about 8000, 9600, and 11,200 cm^{-1}, respectively [132]. These three levels are spread over the energy range of 3200 cm^{-1}. Separation between the extreme nonphonon lines in the absorption spectrum is about the same, namely 3600 cm^{-1} [134]. The lowest energy level of the $4f^6 5d$ electronic configuration $^8H_{3/2}$, in CaF_2 is situated at 77,760 cm^{-1}, in agreement with Loh's estimation at 78,000 cm^{-1} [92]. In this work, the absorption lines at 77,890 and 79,710 cm^{-1} were assigned to spin-forbidden transitions. The spectrum of Gd^{3+} ions in CaF_2 is exceptional in that it does not seem to exhibit the type of vibronic feature at about 490 cm^{-1} from the parent zero phonon line that appears in the $4f \rightarrow 5d$ absorption spectra of all 10 other triply ionized rare earth ions in CaF_2.

H. Terbium (Tb^{3+}, $4f^8$)

The absorption spectrum of the trivalent Tb^{3+} ion has been studied previously in different crystal lattices [2,5,83,135]. The electric crystal field splits all the levels of single and mixed configuration. The number and the spacing of the components depend on the symmetry and intensity of the crystal field. Such splitting has been observed previously for the Tb^{3+} ion in other dielectric crystal host materials as well. For example, regarding the $LiYF_4$:Tb^{3+} crystal (site symmetry C_{4h} or S_4), it was found that the electric crystal field splits the $4f^7 5d$ electronic configuration into five main Stark components with maximum absorption at 46,700, 51,500, 55,000, 60,600, and 80,900 cm^{-1}, respectively [135]. For the LaF_3:Tb^{3+} crystal (C_2 or D_{3h} site symmetry), the electric crystal field likewise splits the levels of the $4f^7 5d$ electronic configuration into seven main Stark components with maximum absorption at 54,000, 55,500, 57,800, 58,800, 61,300, 65,800, and 73,000 cm^{-1} respectively [2]. In the case of the free Tb^{3+} ion, Dieke and Crosswhite indicated that the $4f^7 5d$ electronic configuration splits into two groups of levels, and the energy gap between them extends from 55,700 to 65,000 cm^{-1} [29]. For the LaF_3 crystal, the edge of the levels of the $4f^7 5d$ configuration is at 50,000 cm^{-1} [19]; for the CaF_2 at 46,500 cm^{-1} [19]; and around 44,900 cm^{-1} for the $LiYF_4$ crystal [18]. The edge of the levels of the $4f^7 5d$ electronic configuration in the case of the free Tb^{3+} ion is at 54,900 cm^{-1} [23].

According to Szczurek and Schlesinger [81], the $4f^7$ electronic configuration of the Tb^{3+} ion is half-filled, and the lower energy level of the $4f^7 5d$ electronic configuration is the $^8S_{7/2}$ one, which has zero total angular momentum. Therefore, in a pure L–S coupling the 5d electron does not interact with the $4f^7$ shell. The interaction is taking place only through the crystal field. In this aspect the Tb^{3+} ion is similar to the Ce^{3+} ion, although the last one in the 5d electronic configuration has an empty 4f shell. As a result, the low-energy part of the absorption spectrum of the Tb^{3+} ions in the $4f^7 5d$ electronic configuration should be quite similar to that of Ce^{3+} ions. The absorption spectra of $LiLuF_4$: Tb^{3+} in the 140–220 nm spectral range consists of a band displaying similar structure to the spectrum of Ce^{3+} ions in crystal field in the 180–320 nm spectral range. The separation between the extreme lines in the spectra is about 20,600 cm^{-1} for terbium and about 21,600 cm^{-1} for cerium. This experimental evidence supports the argument that in both cases the lowest energy-level system in the $4f^{n-1}5d$ electronic configuration is formed through the interaction between the 5d electron and the crystal field. Following this argument and taking into consideration that spherical symmetry is partially restored, the $4f^7(^8S)5d$ electronic configuration consists of two terms, the ^9D and the ^7D one, and only the $^7D_5$ is populated through the electric dipole transitions from the $4f^8(^7F_6)$ ground level of the $4f^8$ electronic configuration. The crystal field of tetragonal symmetry should split the $^7D_5$ level into eight Stark components ($^7D_5 = \Gamma_1 + 2\Gamma_2 + \Gamma_3 + \Gamma_4 + 3\,\Gamma_5$).

In order to find the relative energy position of the Stark components, it is necessary to construct the basic functions for the Γ_{1-5} irreducible representations, and then to calculate the matrix elements of the electrostatic potential on the Tb^{3+} ion, for each one of the eight Stark components. In this approximation, the total angular momentum, J, is the good quantum number (J–J coupling). However in order to avoid the lengthy calculations, a different approach can be applied that considers the orbital angular momentum of the d electron ($L = 2$) to be the good quantum number. In this first-order approximation, the Γ_4 irreducible representation has the lowest energy in a tetragonal site symmetry. The next irreducible representation with higher energy to the Γ_4, is the Γ_5 one, followed by the Γ_3 and the Γ_1. The Γ_2 and the Γ_5 Stark components are double and triple degenerate. The degeneracy can be raised when the L–S coupling strength is strong enough within the levels of the $4f^7 5d$ electronic configuration. The splitting of the levels of the $4f^7 5d$ electronic configuration due to the L–S coupling is only few hundred wavenumbers, and this value is considerably smaller than the crystal field splitting. In the case of the $LiYF_4$:Tb^{3+} ions, the eight Stark components of the $^7D_5$ level were assigned to the levels of the $4f^7 5d$ electronic configuration at 47,000, 52,000, 54,400, 55,500, 55,900, 62,200, 64,300, and 67,500 cm^{-1} respectively [135]. We therefore assign the triple degenerate Γ_5 Stark component to the energy levels at 54,400, 55,500, and 55,900 cm^{-1}, and the double degenerate Γ_2 Stark component to the energy levels at 62,200 and 64,300 cm^{-1}, respectively. The remaining three Stark components Γ_1, Γ_3, and Γ_4 were assigned to the energy levels at 67,400, 52,000, and 47,000 cm^{-1}, respectively. The next to the lowest ^8S level of the $4f^7 5d$ electronic configuration is the $4f^7(^6P)5d$ one. The exact value of the separation between the ^6S and the ^6P levels of the $4f^7$ electronic configuration is not known exactly, but it might be assumed to be of the same order of magnitude as in the case of the Gd^{3+} ions, which is 32,000 cm^{-1} [24].

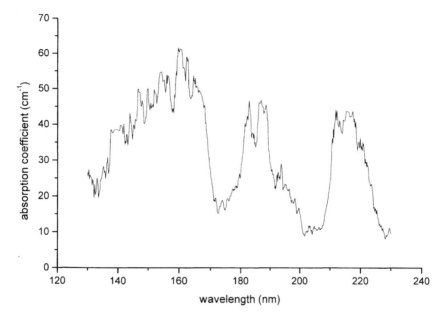

Figure 7   VUV absorption spectrum of K_2YF_5:Tb^{3+}.

The $4f^7(^6P)5d$ electronic configuration has four terms that are excited from the $^7F_6$ ground level of the $4f^8$ electronic configuration: the $^7F_6$, $^7F_5$, $^7G_7$ and the $^7G_6$ ones. The crystal field splits all these levels to a total number of 26 Stark components for a C_{4v} or an S_4 site symmetry. However, if the selection rules for dipole transitions $\Delta S = 0$, $\Delta L = \pm 1$, $\Delta L = 0, \pm 1$, $\Delta J = 0, \pm 1$, ($J = 0 \rightarrow J = 0$ forbidden), and $\Delta J = \Delta L$ applied, only 14 transitions remain after the excitation from the ground level $^7F_6$, which occupies the spectral region from 120 to 240 nm.

A similar structure has been observed previously for the LiYF$_4$: Tb^{3+} crystal in this spectral range [135]. The band gap of the LiLuF$_4$ crystal was estimated to be 77,600 cm^{-1} wide. It is possible the levels of the $4f^7 5d$ electronic configuration overlap with the levels of the $4f^7 6s$ electronic configuration below 120 nm [2]. The VUV spectrum of the Tb^{3+} ions in K$_2$YF$_5$ crystal hosts is indicated in Figure 7. The spectrum is similar to that of the LiYF$_4$:Tb^{3+} crystal [135].

I. Dysprosium (Dy³⁺, 4f⁹)

The $4f^9$ electronic configuration of trivalent dysprosium in LaF$_3$ indicated a complex structure. The edge of the levels of the $4f^8 5d$ electronic configuration of free trivalent dysprosium ions is calculated to be at 66,300 cm^{-1} and at 99,100 cm^{-1} for the 6s electronic configuration. The absorption spectrum of Dy^{3+} ions in LaF_3 indicates a complex structure [136] and accurate assignment of states has not been made above 24,000 cm^{-1}. The VUV absorption spectrum of the Dy^{3+} ions in CaF_2 begins at 175 nm and consists of many sharp zero-phonon lines and elaborately structured phonon side bands. There are two peaks in the VUV absorption spectrum of Dy^{3+} ions in CaF_2 situated at 490 and 100 cm^{-1} from their parent zero-phonon lines [137]. The peak at 490 cm^{-1} is attributed to a localized vibrational mode of the eight nearest fluorine neighbors of the Dy^{3+} ions. The vibration mode at 490 cm^{-1} is present in CaF_2 crystals doped with different RE ions such as Ce^{3+}, Tb^{3+}, and Ho^{3+} ions.

J. Holmium (Ho^{3+}, 4f^{10})

The absorption spectrum of the trivalent Ho^{3+} ion was studied previously in a number of crystal lattices [2,7,81,138,139]. The electric crystal field splits all the levels of the single and the mixed configurations. Three Stark components of the levels of the $4f^95d$ electronic configuration were observed previously in the case of the BaY_2F_8 crystal host, with maximum absorption at 63,900 cm^{-1} (156.4 nm), 67,800 cm^{-1} (147.4 nm), and 71,800 cm^{-1} (139.2 nm). The edge of the levels of the $4f^95d$ electronic configuration was found to be at 63,100 cm^{-1} (158.4 nm) [139]. For the LaF_3 and the CaF_2 crystal hosts, the edges of the levels of the $4f^95d$ electronic configuration for the spin-allowed transitions were found to be at 63,300 cm^{-1} (158.7 nm) and 64,100 cm^{-1} (156.0 nm), respectively [2,138]. For the free ion case, the edge of the levels of the $4f^95d$ electronic configuration was found to be at 58,900 cm^{-1} (169.5 nm) [29]. In the case of the single $LiYF_4$ crystal and for 0.1 at. % concentration of the Ho^{3+} ions, the edge of the levels of the $4f^95d$ electronic configuration for the spin-allowed transitions was found to be at 63,300 cm^{-1} (158 nm) [174]. In the spectral range from 120 to 158 nm, five main Stark components of the $4f^95d$ electronic configuration with maximum absorption at 64,800 cm^{-1} (153.4 nm), 69,300 cm^{-1} (143.5 nm), 73,200 cm^{-1} (136.6 nm), 78,900 cm^{-1} (126.7 nm), and 84,100 cm^{-1} (118.8 nm) were observed.

In the spectral range from 158 to 170 nm, four main Stark components of the $4f^95d$ electronic configuration with maximum absorption at 59,700 cm^{-1} (167.5 nm), 60,200 cm^{-1} (166 nm), 61,200 cm^{-1} (163 nm), and 61,900 cm^{-1} (162 nm) were likewise observed. They were assigned previously to spin-forbidden transitions [128]. This assignment was justified because of the low value of the absorption coefficient, in comparison to the value of the absorption coefficient for wavelengths less than 158 nm, and on the other hand because of the long lifetime (300 ns) of the corresponding transitions. The transitions between 58,800 cm^{-1} (170 nm) and 54,400 cm^{-1} (184 nm) were assigned to the $4f^{10} \rightarrow 4f^{10}$ intraconfigurational dipole-forbidden transitions.

The local maximum around 116 nm [175] is attributed to the transition between the valence and the conduction bands of the host lattice. From the absorption spectrum it is estimated that the band gap of the $LiYF_4$ crystal host is 86,200 cm^{-1} (116 nm) wide. This value is in agreement with the value of the band gap of 10.7 eV (115 nm) of the $LiYF_4$ crystal, estimated previously from reflection spectra [3]. Indeed, the cut-off wavelength of the LiF window, which is placed in front of the hydrogen lamp and the solar blind photomultiplier, is at 110 nm. The transmittance of the LiF window drops successively from its highest value at 120 nm, to zero at 110 nm when it is recorded using a secondary electron multiplier of open geometry without the $LiYF_4$:Ho^{3+} crystal in the optical path. In this spectral region, there is only one local maximum at 112 nm. When the transmittance is recorded with the the $LiYF_4$:Ho^{3+} crystal in the optical path, two new additional local maxima appear in the absorption spectrum from 110 to 120 nm, one at 116 nm and the other at 118.8 nm. A local maximum around 116 nm appears as well in the absorption spectrum of $LiYF_4$ crystals doped with Nd^{3+} ions [3,82]. Hence the position of the local maximum at 116 nm should be independent of the RE dopands in $LiYF_4$ crystal hosts and it is assigned to a transition between the valance and the conduction band of the $LiYF_4$ crystal.

There is not a sharp transition between the valance and the conduction band due to the delocalized nature of the conduction band. From the absorption spectrum, the oscillation strength of the transition between the valance and the conduction band increases as one shift in the wavelength scale from 122 to 116 nm. For wavelengths less than 122 nm, the levels of the $4f^9 5d$ electronic configuration of the Ho^{3+} ions are mixed with the delocalized levels of the conduction band of the crystal host, and the transmittance of the crystal decreases rapidly in this spectral region. Hence, the absorption spectrum between 110 and 122 nm, indicates only the position of the levels of $4f^9 5d$ electronic configuration inside the conduction band. In this case the value of the absorption coefficient cannot be calculated, because of the low value of the signal-to-noise ratio of the light source, which approaches unity. The absorption peak at 122 nm corresponds to the transition between the $^5I_8$ ground level of the $4f^{10}$ electronic configuration and the edge of the conduction band. Therefore the position of the $^5I_8$ ground level of the Ho^{3+} ion is 4200 ± 800 cm^{-1} above the top of the valance band of the $LiYF_4$ crystal host, provided that the band gap of the $LiYF_4$ crystal is 86,200 cm^{-1} wide. The local maximum at 118.5 nm likewise corresponds either to the transition between the ground $^5I_8$ level and a Stark component of the $4f^9 5d$ electronic configuration inside the conduction band or to a transition from the ground $^5I_8$ level to a localized level inside the conduction band.

K. Erbium (Er^{3+}, $4f^{11}$)

The edge of the levels of the $4f^{10} 5d$ electronic configuration of the Er^{3+} ions was calculated to be at 75,400 cm^{-1} [131]. The lowest level of the $4f^{10} 5d$ electronic configuration is the $^6L_{17/2}$ and 79 dipole-allowed transitions are excited from the ground state $^4I_{15/2}$ of the $4f^{11}$ electronic configuration. The absorption spectrum of the trivalent Er^{3+} ion in the VUV has been studied previously in a number of crystal lattices [1–3,9,81,140–141] and it was found to be rather simple, mainly because very few energy levels of the $4f^{10}5d$ electronic configuration are within the band gap of the host material. Intraconfigurational VUV dipole-forbidden $4f^{11} \rightarrow 4f^{11}$ transitions of Er^{3+} ions were observed in various dielectric crystals as well [142]. The edge of the levels of the $4f^{10}5d$ electronic configuration in KY_3F_{10} host was found to be at 60,850 cm^{-1}. Five absorption bands of the $4f^{10}5d$ electronic configuration have been observed in BaY_2F_8 hosts with the maximum of the absorption at 63,700, 67,900, 70,500, 73,400, and 74,600 cm^{-1}. The edge of the levels of the $4f^{10}5d$ electronic configuration in this crystal host was found to be at 61,900 cm^{-1}, 58,000 cm^{-1} in LaF_3 [2], and at 63,000 cm^{-1} in $LiLuF_4$ [9]. Five main Stark components of the $4f^{10}5d$ electronic configuration have been observed as well in $LiLuF_4$ hosts with the maximum of the absorption at 60,980, 69,198, 71,422, 74,343, and 76,916 cm^{-1}. The presence of an additional weak absorption band with maximum of absorption at 163 nm in $LiLuF_4$ host has been observed as well by Devyatkova et al. [141] in KY_3F_{10} matrices at 164.4 nm.

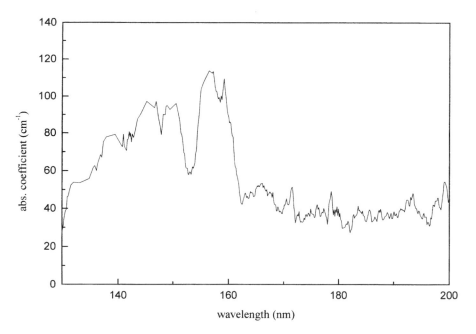

Figure 8 VUV absorption spectrum of $K_2YF_5:Er^{3+}$ crystals.

The VUV absorption spectrum of the Er^{3+} ions in K_2YF_5 crystal hosts is indicated in Figure 8.

L.   Thulium (Tm^{3+}, $4f^{12}$)

The edge of the levels of the $4f^{11}5d$ electronic configuration of the free Tm^{3+} ions was estimated to be at 74,300 cm^{-1}, and that of the $4f^{11}6s$ electronic configuration at 101,300 cm^{-1} [29]. The relative simplicity of the spectrum in various dielectric hosts arises because only some of the levels of the $4f^{11}5d$ electronic configuration lay below the edge of the conduction band. The absorption spectrum of Tm^{3+} ions in $LiLuF_4$ [135] indicates the presence of four bands with maximum of absorption at 63,300, 69,000, 73,000 and 76,500 cm^{-1}, respectively and for the K_2YF_5 host the VUV absorption spectrum of the Tm^{3+} ions is indicated in Figure 9. VUV absorption spectra of Tm^{3+} ions in different fluoride dielectric hosts have been analyzed by Szczurek and Schlesinger [143–145]. The main result of these investigations was that different possible transitions and

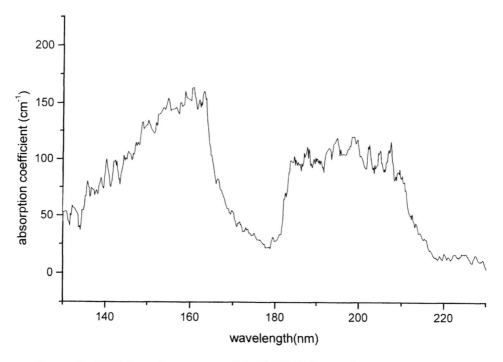

Figure 9   VUV absorption spectrum of the K_2YF_5:Tm^{3+} crystal.

relaxation paths are dependent upon lattice temperature. The temperature-dependent lifetimes can be expected to occur as a result of phonon-assisted radiative transitions.

M. Ytterbium (Yb^{3+}, 4f^{13})

The free Yb^{3+} in the 4f^{12}5d electronic configuration has 107 energy levels. Among these, some 20 levels have been measured from gaseous spectra [146] including the lowest $^4H_{9/2}$ level. This enables one to make some limited comparison of the observed spectra of Yb^{3+} in different crystal hosts with the free Yb^{3+} energy levels [147–149]. The crystal field of the C$_{4v}$ symmetry splits the lowest $^4H_{9/2}$ level into five Stark levels. The separation between the extreme of these five lines is aboute 2100 cm^{-1}. This value is about one order of magnitude smaller than the corresponding 4f^{n-1}5d level splitting observed, for instance, in cerium, praseodymium, or terbium. This suggests that the interaction between the 5d electron and the crystal field is weak.

IV. LASER-INDUCED FLUORESCENCE OF RE IONS IN WIDE BAND GAP FLUORIDE CRYSTALS AT 157 nm

When the RE ion is excited from its ground level of the 4f^n electronic configuration to a given level of the 4f^{n-1}5d electronic configuration, it populates the levels of the 4f^n electronic configuration from the edge and the Stark components of the levels of the 4f^{n-1}5d electronic configuration. This is a rather common response to photoexcitation of the trivalent RE ions in fluoride dielectric crystals, and it has been observed previously for other dielectric crystals doped with RE ions [9,82,150]. In this case competition starts between radiative and nonradiative transitions. The nonradiative transitions are faster and usually the ion will decay to the lowest of the 4f^{n-1}5d levels. The transitions arising from the edge and the levels of the 4f^{n-1}5d electronic configuration and the laser induced fluorescence (LIF) spectrum could be interpreted on the basis of four different processes:

1. The formation of optically active RE centers of different site symmetries, which always are present at high concentration of the RE ions [151].
2. Direct 4f^{n-1}5d → 4f^n emission from the levels of the 4f^{n-1}5d electronic configuration due to large energy separation of the 4f^{n-1}5d levels and/or weak electron [4f^{n-1}5d]–phonon interaction.
3. Emission from the levels of the 4f^{n-1}5d electronic configuration due to the repopulation of these levels via phonon trapping and reabsorp-

tion of the lattice vibrations within the levels of the $4f^{n-1}5d$ electronic configuration.
4. Spin-forbidden transitions.

Direct emission from the Stark components of the levels of the $4f^{n-1}5d$ electronic configuration should be taking place only when: (1) two successive electronic levels are well separated by the combined energy of a few phonons (e.g., the phonon energy of $LiYF_4$ crystal is between 140 and 500 cm^{-1} [152]); (2) the energy separation is larger than kT; and (3) the strength of the electron $[4f^95d]$–phonon interaction is weak.

For example, in the case of the Ho^{3+} ions, there are 1878 electronic levels of the $4f^95d$ electronic configuration in the energy range from 80,000 to 60,000 cm^{-1}. Therefore, two successive electronic levels are separated by 11 cm^{-1} (assuming an even distribution of the electronic energy levels over the spectral range). Because the phonon energies of the $LiYF_4$ crystal are higher than 150 cm^{-1} [153], there is a small probability of direct transitions between the Stark components of the levels of the $4f^95d$ electronic configuration and the levels of the $4f^{10}$ electronic configuration. The interconfigurational $4f^95d \rightarrow 4f^{10}$ transitions in this case are taking place from the edge of the levels of the $4f^95d$ electronic configuration.

For those of the RE ions with weak electron $[4f^{n-1}5d]$–phonon interaction, the coupling strength of the interaction is described within the validity of the Born-Oppenheimer approximation. In this case, radiationless transitions within the levels of the $4f^{n-1}5d$ electronic configuration with a different set of good quantum numbers are forbidden. Therefore, the probability for interconfigurational radiative transitions, which originate directly from the levels of the $4f^{n-1}5d$ electronic configuration, is expected to be high for those of the RE ions where the electron $[4f^{n-1}5d]$–phonon interaction is weak. This situation is confirmed by experimental observations in the case of the Tb^{3+} [83], Gd^{3+} [22], and Er^{3+} [9,128] ions. On the contrary, in the case of strong electron–phonon interaction, as for the Pr^{3+} [8], Nd^{3+} [82], and Ho^{3+} [7] ions, the rate of the internal relaxation within the levels of the $4f^{n-1}5d$ electronic configuration is expected to be high. In this case the dipole-allowed transitions are expected to originate mainly from the edge of the levels of the $4f^{n-1}5d$ electronic configuration.

In addition, for those of the RE ions with strong electron–phonon interaction [5,153–155], dipole-allowed transitions directly from inside the levels (Stark components) of the $4f^{n-1}5d$ electronic configuration could be explained on the basis of the repopulation of these levels through phonon reabsorption and trapping within the levels of the $4f^{n-1}5d$ electronic configuration [82]. The repopulating process will cease when the trapped phonons escape from the active volume by frequency shifting through the dipole-allowed $4f^{n-1}5d \rightarrow 4f^n$ interconfigurational transitions. Finally, the spin allowed and the spin-forbidden [150] interconfigurational transitions can be explained on the basis of selection rules and rates of relaxation within the levels of the $4f^{n-1}5d$ electronic configuration.

A. Cerium

There is much interest in Ce^{3+}-doped ionic crystals for applications in UV scintillators [150–160] and lasers [13–15,26]. Cerium trivalent ions in various dielectric crystal hosts are characterized by excellent photochemical stability, high conversion efficiency, and low excited-state absorption coefficient. They also exhibit broadband tunable positive gain in UV, such as in $LiCaAlF_6$ crystal hosts [161,162]. Polarized luminescence and absorption spectroscopy of Ce^{3+} ions in dielectric fluoride crystals reveal the crystal field splitting of the 5d electronic configuration [93,163]. The luminescence spectra of Ce^{3+} in different dielectric fluoride crystal hosts consists of two bands due to transitions from the lowest Stark component of the 5d electronic configuration to the $^2F_{5/2}$ and $^2F_{7/2}$ levels of the 4f electronic configuration.

B. Praseodymium

The LIF spectrum of Pr^{3+} ions in $LiLuF_4$ single-crystal hosts [8] excited in the 4f5d electronic configuration with the F_2 molecular laser at 157 nm indicated that interconfigurational transitions mainly originate from the edge of the band of the 4f5d electronic configuration at 46400 cm^{-1} and spans the spectral range from 218 to 280 nm. Despite the fact that electron–phonon interaction within the 4f5d electronic configuration is strong, the LIF spectrum indicated the presence of two weak bands around 200 and 170 nm. These bands were assigned to interconfigurational transitions from the Stark components of the 4f5d electronic configuration at 53,200 and 63,400 cm^{-1}, respectively, and taking into consideration the energy gaps of the absorption spectrum (which are similar to the energy gaps in the absorption spectrum of the Tb^{3+} ions [83,135]). The two weak bands have been observed from the same crystal using x-rays as the excitation source [162]. UV fluorescence spectra using two photon excitation techniques have been recorded as well [163]. Luminescence spectra of Pr^{3+} ions in YF_3 and LaF_3 crystals, where the edge of the levels of the 4f5d electronic configuration is at 50,000 cm^{-1}, indicated that radiative interconfigurational transitions in the UV and the visible are taking place between the $^1S_0$ energy level at 48,000 cm^{-1} and the lower energy levels of the $4f^2$ electronic configuration [24,111].

C. Neodymium

In the case of VUV excitation of the YLF:Nd crystal, it was observed experimentally that the emission spectrum depends on the excitation wavelength [115]. Two peaks in this case were situated at 181.5 and 185.3 nm. The two peaks were observed in multiphoton and single photon laser excitation of the YLF:Nd crystals as well [7,10,82], at right angle to the excitation axis. The two peaks were assigned to transitions from the low Stark component of the lower $4f^2 5d$ ($^4K_{11/2}$) level of the Nd^{3+} ion to the levels of the $4f^3$ configuration.

Indeed, when the crystal field is applied on the $^4K_{11/2}$ level of the Nd^{3+} ion it splits it into three Stark components, $\Gamma^{(8)}$, $\Gamma^{(7)}$, and $\Gamma^{(6)}$, for a tetragonal site symmetry of the Nd^{3+} ion. The $\Gamma^{(8)}$ one is quadratic degenerate, $2(\Gamma^{(6)} + \Gamma^{(7)})$ and its degeneracy is removed when the L–S coupling is strong enough within the $4f^2 5d$ electronic configuration. These levels occupy the spectral range from 453 to 1700 cm^{-1} above the low energy limit of the levels of the $4f^2 5d$ electronic configuration, in agreement with findings by Thogersen et al. [10].

The broad band distribution of the electronic cloud near the edge of the band reflects the wide electronic character of the $4f^2 5d$ ($^4K_{11/2}$) $\rightarrow 4f^3$ dipole transition, with the maximum of the emission at 181.5 and 185.3 nm, respectively. The narrow emission peaks of the 180 nm band correspond to transitions from the Stark component of the $4f^2 5d$ ($^4K_{11/2}$) level to the Stark components of the $^4I_{9/2}$, and $^4I_{11/2}$ levels of the $4f^3$ configuration. The $^4I_{9/2}$ level of the $4f^3$ configuration is split into five Stark components at 1, 139, 180, 236, and 529 cm^{-1} [82]. The $^4I_{11/2}$ level of the $4f^3$ configuration is split into five Stark components at 1979, 2015, 2056, 2221, and 2254 cm^{-1}. At room and low temperatures all these levels should be populated from transitions that originate from the $^4K_{11/2}$ electronic level of the $4f^2 5d$ configuration, at 55,648, 55,096, and 54,495 cm^{-1}, respectively. In these experiments the linewidth of the $4f^2 5d \rightarrow 4f^3$ atomic transitions, should be less than 0.1 nm. This fact suggests that the emission cross section of the $4f^2 5d \rightarrow 4f^3$ transitions along the optical axis of the crystal is $\sim 10^{-15}$–10^{-16} cm^2, which is two or three orders of magnitude higher than the emission cross section of the $4f^2 5d \rightarrow 4f^3$ transitions observed along the directions perpendicular to the optical axis of the crystal. In the later case, the linewidth of the $4f^2 5d \rightarrow 4f^3$ transitions was 8.5 nm [10,115], giving a value for the corresponding emission cross section of $\sim 10^{-18}$ cm^2.

A possible mechanism to explain the structure of the LIF spectrum of the 180 nm band could be phonon trapping within the levels of the Stark component of the $^4K_{11/2}$ level within the active volume of the crystal. This assignment is supported by the fact that the energy difference of 552 and 600 cm^{-1} between the Stark component of $^4K_{11/2}$ level reflects the energy difference (within the experimental error) between successive vibronic modes of the YLF:Nd lattice [82].

This process is taking place because of the efficient coupling between sites of the Nd^{3+} ions, through multipole or exchange interactions. The repopulating process of the higher vibronic modes will cease when the trapped phonons escape from the active volume by frequency shifting via the dipole-allowed $4f^2 5d \rightarrow 4f^3$ interconfigurational transitions.

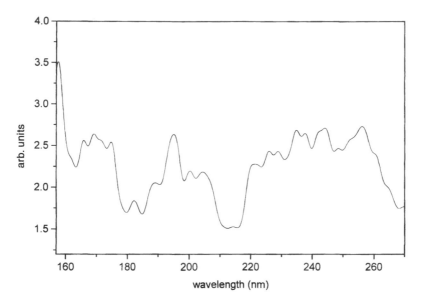

Figure 10   LIF spectrum of $LiLuF_4:Tb^{3+}$ following excitation at 157.6 nm. (From Ref. 83.)

The radiative lifetime of the transitions within the 180 nm band was found to be ~30 ns in agreement with previous measurements [4,6,164], using different crystal samples and measurements on VUV decay and energy transfer dynamics of Nd^{3+} ions [165] in various dielectric crystal hosts under x-ray and synchrotron excitation [166–168] for scintillation and laser [169,170] applications.

D.   Terbium

Terbium trivalent ions are an interesting case since in a pure L–S coupling there is not interaction between the $4f^7$ and the 5d electronic configuration, the interaction is only through the crystal field. On the other hand, partial restoration of the spherical symmetry of the 5d electronic configuration in the crystal field of the surrounding ligands is taking place. The LIF spectrum in the spectral range from 150 to 270 nm under excitation at 157.6 nm, with the F_2 laser, for 0.01 at.% concentration of the Tb^{3+} ion at liquid nitrogen temperature, is in Figure 10. The fluorescence peaks were assigned to the dipole transitions between the levels of the $4f^7$ 5d electronic configuration and the levels of the $4f^8$ electronic configuration of the Tb^{3+} ion [83]. When the Tb^{3+} ion is excited from its $4f^8$ ($^7F_6$) ground level to a given level of the $4f^7 5d$ electronic configuration, the experimental evidences indicate that it populates the levels of the $4f^8$ electronic configuration, from both the levels and the edge of the $4f^7 5d$ electronic configuration. Direct emission from the levels of the $4f^7 5d$ electronic configuration could take place only when two successive electronic levels are well separated by the combined energy of few phonons, as mentioned previously. However, according to Szczurek and Schlesinger [81], there are 3106 electronic levels of the $4f^7 5d$ electronic configuration in the energy range from 80,000 to 45,000 cm^{-1}. This number corresponds to the energy level density of 13 electronic levels per wavenumber (considering an even distribution of the energy levels over the spectral range). Hence there is a small probability of direct transitions between the Stark components of the levels of the $4f^7 5d$ electronic configuration and the levels of the $4f^8$ electronic configuration. Therefore the LIF spectrum can be interpreted provided that repopulation of the levels of the $4f^7 5d$ electronic configuration is taking place. The repopulation occurs via phonon reabsorption and trapping within the levels of the $4f^7 5d$ electronic configuration. These kind of processes have their origin in the strong ionic coupling between two different Tb^{3+} ions through multipole or exchange interactions, and they have been observed previously for different crystal samples as well [172]. The repopulating process will cease when the trapped phonons escape from the active volume by frequency shifting via the interconfigurational transitions. Phonon trapping is usually detected by observing the fluorescent radiation from the excited levels to the ground state. In the case of weak electron $[4f^{n-1}]$–phonon interaction and/or spin-forbidden $4f^{n-1} 5d$ transitions, the phonon trapping and reabsorption processes compete with the nonradiative relaxation process within the levels of the $4f^{n-1} 5d$ electronic configuration. In this case, the interconfigurational transitions from inside the levels of the $4f^{n-1} 5d$ electronic configuration are strong enough, as for example in the case of the Tb^{3+}, Gd^{3+}, and Er^{3+} ions [172]. In the case of strong electron–phonon interaction, for example for the Nd^{3+} ion, the rate of the internal relaxation within the levels of the $4f^{n-1} 5d$ electronic configuration is higher than the rate of the phonon reabsorption within the levels of the $4f^{n-1} 5d$ electronic configuration. In this case, the dipole transitions originate mainly from the edge of the levels of the $4f^{n-1} 5d$ electronic configuration.

As one finds from the LIF spectrum in Figure 10, the dipole transitions, between the $^7D_5$ Stark components of the $4f^7 5d$ electronic configuration and the levels of the $4f^8$ electronic configuration are due to the dipole transitions $^7D_5 \rightarrow {}^7F_{4,5,6}$, which satisfy the selection rules for electric dipole transitions $\Delta S = 0$, $\Delta l = +1, -1$, $\Delta L = 0, +1, -1$, $\Delta J = 0, +1, -1$ and $\Delta J = \Delta L$.

The emission band, which covers the spectral range from 220 to 253 nm, reflects the wide electronic character of the $4f^7 5d(^7D_5) \rightarrow 4f^8 (^7F_{0-6})$ dipole transitions of the Γ_4 Stark component of the levels of the $4f^7 5d$ ($^7D_5$) electronic configuration. The interconfigurational transitions in this spectral range originate from the edge of the levels of the $4f^7 5d$ electronic configuration. From the LIF spectrum of [83], it was found that the energy position with the higher Frank–Condon factor is at $(45.2 \pm 0.2) \times 10^3$ cm^{-1}. In this case the following assignments can be made regarding the observed dipole transitions (Fig. 11).

$4f^7 5d[^7D_5(\Gamma_4), 45.2 \times 10^3$ cm$^{-1}] \rightarrow 4f^8(^7F_6, 0) + h\nu(221.2$ nm)

$4f^7 5d[^7D_5(\Gamma_4), 45.2 \times 10^3$ cm$^{-1}] \rightarrow 4f^8(^7F_5, 1.5 \times 10^3$ cm$^{-1}) + h\nu(228.8$ nm)

$4f^7 5d[^7D_5(\Gamma_4), 45.2 \times 10^3$ cm$^{-1}] \rightarrow 4f^8(^7F_4, 3.1 \times 10^3$ cm$^{-1}) + h\nu(237.7$ nm)

Besides the above-observed dipole-allowed transitions, dipole-forbidden transitions [173] ($\Delta J \neq 0, \pm 1$) were observed from the Γ_4 Stark component of the levels of the $4f^7 5d$ ($^7D_5$) electronic configuration.

$4f^7 5d[^7D_5(\Gamma_4), 45.2 \times 10^3$ cm$^{-1}] \rightarrow 4f^8(^7F_3, 4.0 \times 10^3$ cm$^{-1}) + h\nu(242.5$ nm)

$4f^7 5d[^7D_5(\Gamma_4), 45.2 \times 10^3$ cm$^{-1}] \rightarrow 4f^8(^7F_2, 4.2 \times 10^3$ cm$^{-1}) + h\nu(244.2$ nm)

$4f^7 5d[^7D_5(\Gamma_4), 45.2 \times 10^3$ cm$^{-1}] \rightarrow 4f^8(^7F_1, 5.0 \times 10^3$ cm$^{-1}) + h\nu(248.5$ nm)

$4f^7 5d[^7D_5(\Gamma_4), 45.2 \times 10^3$ cm$^{-1}] \rightarrow 4f^8(^1F_0, 5.6 \times 10^3$ cm$^{-1}) + h\nu(252.7$ nm)

The position of the levels of the $4f^8$ electronic configuration of the Tb^{3+} ion in the LiLuF$_4$, single-crystal host, is not known exactly. The position of these levels in the free ion case, for the $^7F_{0-5}$ terms is at 58,000, 51,000, 48,000, 40,000, 31,000, and 19,000 cm^{-1} respectively. From the LIF spectrum and for the Γ_4 Stark component, the experimental position of the $^7F_{0-5}$ terms in the LiLuF$_4$ crystal was found to be at 56,000, 50,000, 42,000, 40,000, 31,000, and 15,000 cm^{-1}, respectively. Taking into consideration the fact that the electrons of the $4f^8$ electronic configuration are well screened by the electrons of the 5d electronic configuration, the position of the levels of the $4f^8$ electronic configuration inside the LiLuF$_4$ crystal should be shifted relative to the position of the levels in the free ion case, by few wavenumbers, and this situation is reflected well in the experimental data.

The emission spectrum that covers the spectral region from 193 to 220 nm can likewise be assigned to the dipole transitions between the Γ_5 Stark component of the $4f^7 5d(^7D_5)$ electronic configuration, with maximum of absorption at 54,400 cm^{-1} and the levels of the $4f^8$ electronic configuration. In this case the Frank–Condon factor of the corresponding transitions has its highest value at $(52.9 \pm 0.2) \times 10^3$ cm^{-1}. For these transitions, the following assignments can be made within the limits of the experimental error of 200 cm^{-1} for these wavelengths.

$4f^7 5d[^7D_5(\Gamma_5), 52.9 \times 10^3$ cm$^{-1}] \rightarrow 4f^8(^7F_6) + h\nu(188.9$ nm)

$4f^7 5d[^7D_5(\Gamma_5), 52.9 \times 10^3$ cm$^{-1}] \rightarrow 4f^8(^7F_5, 1.5 \times 10^3$ cm$^{-1}) + h\nu(194.7$ nm)

$4f^7 5d[^7D_5(\Gamma_5), 52.9 \times 10^3$ cm$^{-1}] \rightarrow 4f^8(^7F_4, 3.0 \times 10^3$ cm$^{-1}) + h\nu(200.2$ nm)

$4f^7 5d[^7D_5(\Gamma_5), 52.9 \times 10^3$ cm$^{-1}] \rightarrow 4f^8(^7F_3, 4.0 \times 10^3$ cm$^{-1}) + h\nu(204.5$ nm)

From the LIF spectrum [83], it was found that the position of the Stark components of the levels of the $4f^8$ electronic configuration $^7F_{3-5}$ in the LiLuF$_4$ crystal

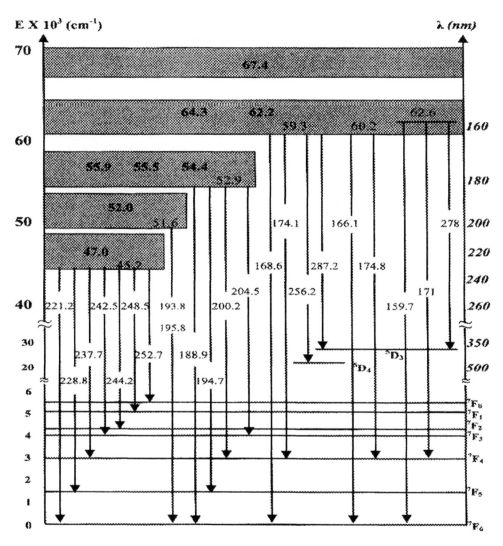

Figure 11 Simplified energy diagram shows various transitions of the LIF spectra of the LiLuF$_4$:Tb^{3+} crystal. The position of levels of the 4f^{7}5d electronic configuration of the Tb^{3+} ion is where the maximum absorption is taking place.

is at 40,000, 30,000, and 15,000 cm^{-1}, respectively, in agreement with the energy position of these levels in the spectral range from 220 to 253 nm.

The transition around 194 nm could originate as well from the Γ_3 Stark component, with maximum of absorption at 52,000 at 51,600 cm^{-1}.

$$4f^75d[^7D_5(\Gamma_3), 51.6 \times 10^3 \text{ cm}^{-1}] \rightarrow 4f^8(^7F_6) + h\nu(193.8-194.7 \text{ nm})$$

The emission spectrum, which covers the spectral range from 159 to 174 nm, can be assigned to the dipole transitions between the Γ_2 Stark component of the $4f^75d(^7D_5)$ electronic configuration, and the levels of the $4f^8$ electronic configuration. However, for this spectral range, it is difficult to make any specific assignment for the $4f^75d \rightarrow 4f^8$ transitions, due to the presence of intense scattered laser light. Taking into consideration the position of the levels of the Tb^{3+} ions of the $4f^8$ electronic configuration in the $LiLuF_4$ crystal host, and the emission spectrum, the following assignments can be made for the $4f^75d \rightarrow 4f^8$ transitions:

$$4f^75d[^7D_5(\Gamma_2), 62.6 \times 10^3 \text{ cm}^{-1}] \rightarrow 4f^8(^7F_6) + h\nu(159.7 \text{ nm})$$
$$4f^75d[^7D_5(\Gamma_2), 62.6 \times 10^3 \text{ cm}^{-1}] \rightarrow 4f^8(^7F_4 \text{ or } ^7F_3) + h\nu(171 \text{ nm})$$
$$4f^75d[^7D_5(\Gamma_2), 59.3 \times 10^3 \text{ cm}^{-1}] \rightarrow 4f^8(^7F_6) + h\nu(168.6 \text{ nm})$$
$$4f^75d[^7D_5(\Gamma_2), 59.3 \times 10^3 \text{ cm}^{-1}] \rightarrow 4f^8(^7F_4) + h\nu(174.1 \text{ nm})$$

The transitions at 278, 256.2, and 287.2 nm were assigned to the spin-forbidden transitions [173], taking into consideration that the energy position of the $^5D_4$ and $^5D_3$ levels of the $4f^8$ electronic configuration in the free Tb^{3+} ion case is at 20,500 and 26,000 cm^{-1}, respectively.

$$4f^75d[^7D_5(\Gamma_1), 60,200 \text{ cm}^{-1}] \rightarrow 4f^8(^5D_3) + h\nu(278 \text{ nm})$$
$$4f^75d[^7D_5(\Gamma_2), 59.3 \times 10^3 \text{ cm}^{-1}] \rightarrow 4f^8(^5D_4) + h\nu(256.2 \text{ nm})$$
$$4f^75d[^7D_5(\Gamma_2), 59.3 \times 10^3 \text{ cm}^{-1}] \rightarrow 4f^8(^5D_3) + h\nu(287.2 \text{ nm})$$

The transitions at 166.1 and 174.8 nm [83] can be assigned to the interconfigurational transitions originating from the energy position at 60,200 cm^{-1}.

$$4f^75d[^7D_5(\Gamma_2), 60.2 \times 10^3 \text{ cm}^{-1}] \rightarrow 4f^8(^7F_6) + h\nu(166.1 \text{ nm})$$
$$4f^75d[^7D_5(\Gamma_2), 60.2 \times 10^3 \text{ cm}^{-1}] \rightarrow 4f^8(^7F_4) + h\nu(174.8 \text{ nm})$$

E. Holmium

The LIF spectrum of the $LiYF_4$:Ho^{3+} crystal, under excitation at 157.6 nm with the F_2 laser, for 0.1 at.% concentration of the Ho^{3+} ion at liquid nitrogen temperature includes both spin-allowed and spin-forbidden interconfigurational transitions and intraconfigurational transitions [7,174,175]. The emission peak at 171.2 nm is assigned to the interconfigurational spin-allowed transition from the lower

Stark component of the $^5H_7$ term at 63,450 cm^{-1} (157.6 nm) to the $^5I_7$ level of the $4f^{10}$ electronic configuration:

$$4f^9 5d(63,450 \text{ cm}^{-1}) \rightarrow 4f^{10}(^5I_7) + h\nu(171.2 \text{ nm})$$

The spin-allowed transition to the ground $^5I_8$ level (which is expected at 157.6 nm) was not observed due to the presence of the intense laser scattered light.

The lowest $^7H_8$ level of the $4f^9 5d$ electronic configuration should likewise split into four doubly degenerate Stark components in the LiYF$_4$ crystal field. The transitions that originate from the Stark components of the $4f^9 5d$ ($^7H_8$) electronic configuration are spin-forbidden and therefore they are expected to be of low intensity. The transitions at 167.5 and 165.6 nm are assigned to the spin-forbidden transitions as follows:

$$4f^9 5d(^7H_8, 60,380 \text{ cm}^{-1}) \rightarrow 4f^{10}(^5I_8) + h\nu(165.6 \text{ nm})$$
$$4f^9 5d(^7H_8, 59,700 \text{ cm}^{-1}) \rightarrow 4f^{10}(^5I_8) + h\nu(167.5 \text{ nm})$$

The LIF transitions from 172.5 to 189 nm were assigned to the $4f^{10} \rightarrow 4f^{10}$ intraconfigurational transitions. From both emission and absorption spectra, the dipole-forbidden transitions at 172.5, 174, 177.5, 178.5, 181, 185, 188, and 189 nm were assigned to the transitions from the high excited levels of the $4f^{10}$ electronic configuration to the ground $^5I_8$ level [174,175].

F. Erbium

The LIF spectrum of the Er^{3+} ions in LiYF$_4$ crystal hosts of Figure 12 includes both spin-allowed and spin-forbidden transitions. The emission bands at 158.6, 157, and 195 nm were assigned to dipole-allowed transitions from the edge of the lower energy level of the $4f^{10} 5d$ electronic configuration to the energy levels of the $4f^{11}$ electronic configuration [9]. Emission at 167–169 nm from an absorption band situated at 163 nm [115] was observed for the first time by Sarantopoulou et al. [9]. It was assigned to spin-forbidden transitions from the Stark component of the $4f^{11}$ electronic configuration to the ground level $^4I_{15/2}$ of the $4f^{11}$ electronic configuration [9,150]. VUV emission of Er^{3+} ions in various dielectric hosts using synchrotron and x-ray excitation has been reported as well [176–179].

G. Thulium

Due to the fact that the laser photons at 157 nm populate the $4f^{11} 5d$ electronic configuration of the Tm^{3+} ions in LiLuF$_4$ crystal hosts just above the edge of the band [135], the LIF spectrum of the LiYF$_4$:Tm^{3+} monocrystals, under the F$_2$ laser excitation, exhibits strong interconfigurational transitions in the VUV (Fig. 13).

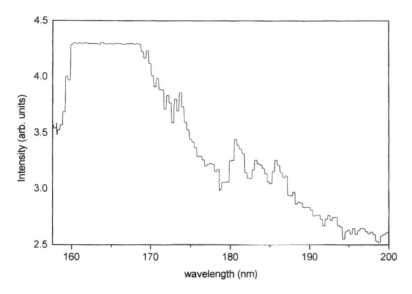

Figure 12   VUV LIF spectrum of LiLuF$_4$:Er^{3+} crystal under F$_2$ laser excitation.

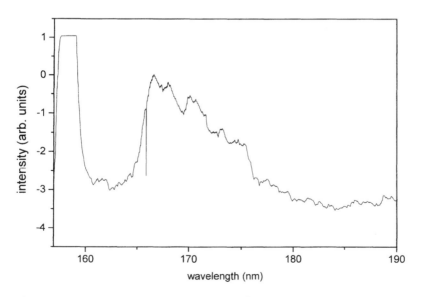

Figure 13   VUV LIF spectrum of LiYF$_4$:Tm^{3+} crystals under F$_2$ laser excitation.

REFERENCES

1. KH Yang, JA DeLuka. VUV fluorescence of Nd^{3+}, Er^{3+} and Tm^{3+} and Tm^{3+} doped trifluorides and tuneable coherent sources from 1650 to 2600 nm. Appl Phys Lett 29: 499–501, 1976.

2. S Heaps, LR Elias, WM Yen. Vacuum-ultraviolet bands of trivalent lanthanides in LaF_3. Phys Rev B 13: 94–104, 1976.

3. KH Yang, JA DeLuca. Vacuum ultraviolet excitation studies of $5d^1 4f^{n-1}$ to $4f^n$ and $4f^n$ to $4f^n$ transitions of Nd^{3+}, Er^{3+} and Tm^{3+}-doped trifluorides. Phys Rev B 17: 4246 4255, 1978.

4. MA Dubinskii, AC Cefalas, E Sarantopoulou, RY Abdulsabirov, SL Korableva, AK Naumov, VV Semashko. On the interconfigurational $4f^2 5d–4f^3$ VUV and UV fluorescence features of Nd^{3+} in YLF single crystals under fluorine laser pumping. Opt Commun 94: 115–118, 1992.

5. JP Krupa, I Gerard, A Mayoler, P Martin. Electronic structure of f-element system in the UV and VUV energy range. Acta Phys Pol A84: 843–848, 1993.

6. R Visser, P Dorenbos, CWE Van Eijk, A Meijerink, HW den Hartog. The scintillation intensity and decay from Nd^{3+} $4f^2 5d$ and $4f^3$ excited states in several fluoride crystals. Phys Cond Matt 5: 8437–8460, 1993.

7. E Sarantopoulou, AC Cefalas, MA Dubinskii, CA Nicolaides, RY Abdulsabirov, SL Korableva, AK Naumov, VV Semashko. VUV and UV fluorescence and absorption studies of Nd^{3+} and Ho^{3+} ions in $LiYF_4$ single crystals. Opt Commun 107: 104–110, 1994.

8. E Sarantopoulou, AC Cefalas, MA Dubinskii, CA Nicolaides, RY Abdulsabirov, SL Korableva, AK Naumov, VV Semashko. VUV and UV fluorescence and absorption studies of Pr^{3+}-doped $LiLuF_4$ single crystals. Opt Lett 19: 499–501, 1994.

9. E Sarantopoulou, Z Kollia, AC Cefalas, MA Dubinskii, CA Nicolaides, RY Abdulsabirov, SL Korableva, AK Naumov, VV Semashko. Vacuum ultraviolet and ultraviolet fluorescence and absorption studies of Er^{3+}-doped $LiLuF_4$ single crystals. Appl Phys Lett 65: 813–815, 1994.

10. J Thogersen, JD Gill, HK Haugen. Stepwise multiphoton excitation of the $4f^2 5d$ configuration in Nd^{3+}:YLF. Opt Commun 132: 83–88, 1996.

11. S Guy, MF Joubert, B Jacquier. Photon avalance and mean-field approximation. Phys Rev B 55: 8240–8248, 1997.

12. M Malinowski, B Jacquier, M Bouazaoui, MF Joubert, C Linares. Laser induced fluorescence and up conversion processes in $LiLuF_4$: Nd^{3+} laser crystals. Phys Rev B 41: 31–40, 1990.

13. DJ Ehrlich, PF Moulton, RM Osgood Jr. Ultraviolet solid state Ce:YLF laser at 325 nm. Opt Lett 4: 184–186, 1979.

14. K-S Lim, DS Hamilton. Optical gain and loss studies in Ce^{3+}: $YLiF_4$. J Opt Soc Am B6: 1401–1406, 1989.

15. DJ Ehrlich, PF Moulton, RM Osgood Jr. Optically pumped Ce:LaF_3 laser at 286 nm. Opt Lett 5: 339–341, 1980.

16. RW Waynant, PH Klein. Vacuum ultraviolet emission from Nd^{3+}: LaF_3. Appl Phys Lett 46: 14–15, 1985.

17. RW Waynant. Vacuum ultraviolet laser emission from Nd^{3+}: LaF_3. Appl Phys B 28: 205, 1982.

18. MA Dubinskii, AC Cefalas, CA Nicolaides. Solid state Nd^{3+} doped LaF_3 VUV laser pumped by a pulsed discharge fluorine-molecular laser at 157 nm. Opt Commun 88: 122–124, 1992.

19. JC van't Spijker, P Dorenbos. CWE van Eijk, JEM Jacobs, HW den Hartog, N Korolev. Luminescence and scintillation properties of $BaY_2F_8:Ce^{3+}$, $BaLu_2F_8$ and $BaLu_2F_8:Ce^{3+}$. J Lumin 85: 11–19, 1999.

20. S Kubodera, M Kitahara, J Kawanaka, W Sasaki, K Kurosava. A vacuum ultraviolet flash lamp with extremely broadened emission spectra. Appl Phys Lett 69: 452–454, 1996.

21. G Blasse, BC Grabmayer. Lumininescent Materials. Berlin: Springer Verlag, 1996.

22. RT Wegh, H Donker, A Meijerink, RJ Lamminmaki, J Holsa. Vacuum-ultraviolet spectroscopy and quantum cutting for Gd^{3+} in $LiLuF_4$. Phys Rev B 56: 13841–13848, 1997.

23. CM Combes, P Dorenbos, CWE Vaneijk, C Pedrini, HW Denhartog, JY Gesland, PA Rodnyi. Optical and scintillation properties of Ce^{3+} doped $LiYF_4$ and $LiLuF_4$ crystals. J Lumin 71: 65–70, 1997.

24. AM Srivastava, SJ Duclos. On the luminescence of YF_3-Pr^{3+} under vacuum ultraviolet and X-ray emission. Chem Phys Lett 275: 453–456, 1997.

25. MJF Digonnet, RW Sadowski, HJ Shaw, RH Pantell. Resonantly enhanced nonlinearity in doped fibers for low power all optical switching. Opt Fib Techn 3: 44–64, 1997.

26. CD Marshall, JA Speth, SA Payne, WF Krupke, GJ Quarles, V Castillo, B Chai. Ultraviolet laser emission properties of Ce^{3+} doped $LiSrAlF_6$ and $LiCaAlF_6$. J Opt Soc Am B 11: 2054–2065, 1994.

27. TM Bloomstein, M Rothschild, RR Kunz, DE Hardy, RB Goodman, ST Palmacci. Critical issues in 157 nm lithography. J Vac Sci Techn B 16: 3154–3157, 1998.

28. AC Cefalas, E Sarantopoulou, P Argitis, E Gogolides. Mass spectroscopic and degassing characteristics of polymeric materials for 157 nm photolithography. Appl Phys A 69: 5229–5234, 2000.

28a. E Sarantopoulou, Z Kollia, AC Cefalas. $YF_3:ND^{3+}$, Pr^{3+}, Gd^{3+} wide band gap crystals as optical materials for 157 nm photolithography. Optical Materials 18: 23–26, 2001.

29. GH Dieke, HM Crosswhite. The spectra of the doubly and triply ionized rare earths. Appl Opt 2: 675–686, 1963.

30. R Hilbic, R Wallenstein. Enhanced Production of Tunable VUV Radiation by phase-matched frequency tripling in krypton and xenon. IEEE J Quantum Electron 17: 1566–1573, 1981.

31. W Jamroz, PE LaRocque. BP Stoicheff. Generation of continuously tunable coherent vacuum-ultraviolet radiation (140–106 nm) in zinc vapor. Opt Lett 7: 617–619, 1982.

32. JC Miller, RN Compton, CD Cooper. Vacuum ultraviolet spectroscopy of molecules using third-harminic generation in rare-gases. J Chem Phys 76: 3967–3973, 1982.

33. Y Hirakawa, K Nakai, T Obara, M Maeda, K Muraoka. Generation of coherent

XUV radiation by 2-photon resonant 4-wave-mixing combined with stimulated Raman-scattering. Opt Commun 92: 215–218, 1992.

34. K Miyazaki, H Sakai, T Sato. Efficient deep-ultraviolet generation by frequency doubling in beta-Bab$_2$O$_4$ crystals. Opt Lett 11: 797–799, 1986.

35. BC Wu, F Xie, C Chen, D Deng, Z Xu. Generation of tunable coherent vacuum ultraviolet-radiation in LiB$_3$O$_5$ crystal. Opt Commun 88: 451–454, 1992.

36. NG Basov, VA Danilychev. Condensed- and compressed-gas lasers. Sov Phys Usp 29:31–56, 1986.

37. RH Lipson, PE Larocque, BP Stoicheff. Vacuum ultraviolet laser-excited spectra of Xe$_2$. Opt Lett 9: 402–404, 1984.

38. T Eftimiopoulos, BP Stoicheff, RI Thomson. Efficient population-inversion in excimer states by supersonic expansion of discharge plasmas. Opt Lett 14: 624–626, 1989.

39. H Nahme, N Schwentner. Luminescence of rare-gas crystals at high-excitation densities for VUV laser applications. Appl Phys B 51: 177–191, 1990.

40. AC Cefalas, TA King. Injection locking of ArF excimer lasers. Appl Phys B37: 159–164, 1985.

41. AC Cefalas, TA King. Phase conjugation by four wave mixing in an ArF excimer amplifier. Opt Commun 51: 105–110, 1984.

42. JR Woodworth, JK Rice. An efficient high-power F$_2$ laser near 157 nm. J Chem Phys 69: 2500–2504, 1978.

43. JR Woodworth, JK Rice. High intensity laser photolysis of OCS at 157 nm. IEEE J Quantum Electron 15: 88D–89D, 1979.

44. H Pummer, OK Hohla, M Digelman, JP Reilly. Discharge pumped F$_2$ pulsed laser at 1580 A. Opt Commun 28: 104–106, 1979.

45. AC Cefalas, C Skordoulis, M Kompitsas, CA Nikolaides. Gain measurements at 157 nm in an F$_2$ pulsed discharge molecular laser. Opt Commun 55: 423–426, 1985.

46. VN Ishchenko, SA Kochubei, AM Razhev. High-power efficient vacuum ultraviolet F$_2$ laser excited by an electric discharge. Sov J Quantum Electron 16:707–709, 1986.

47. M Ohwa, M Obara. Theoretical evaluation of high-efficiency operation of discharge-pumped vacuum ultraviolet F$_2$ lasers. Appl Phys Lett 51: 958–960, 1987.

48. K Yamada, K Miyazaki, T Hasama, T Sato. High-power discharge-pumped F$_2$ molecular laser. Appl Phys Lett 54: 597–599, 1989.

49. C Skordoulis, E Sarantopoulou, S Spyrou, AC Cefalas. Amplification characteristics of a discharge excited F$_2$ laser. J Mod Opt 37: 501–509, 1990.

50. C Skordoulis, S Spyrou, AC Cefalas. Gain and saturation measurements in a discharge excited F$_2$ laser using an oscillator amplifier configuration. Appl Phys B 51: 141–145, 1990.

51. H Nishimura, DC Cartwright, J Trajmar. Electron energy loss spectroscopy of molecular fluorine. J Chem Phys 71: 5039–5041, 1979.

52. DJ Cartwright, PJ Hay. Theoretical studies of the valence electronic states and the $^1\Pi_u \to X\ ^1\Sigma_g^+$ absorption spectrum of the F$_2$ molecule. J Chem Phys 70: 3191–3203, 1979.

53. K Hoshiba, Y Fujita, SS Kano, H Takuma, T Takayanagi, K Wakiya, H Suzuki. Experimantal observation of the F_2 VUV laser levels. J Phys B Atom Mol Phys 18: L875–L879, 1985.

54. RG Wang, ZW Wang, MA Dillon, D Spense. Electron energy loss spectroscopy of molecular fluorine. J Chem Phys 80: 3574–3579, 1984.

55. AC Cefalas, TA King. A doubly preionized ArF laser. J Phys E: Sci Instr 17: 760–764, 1984.

56. RC Caro, MC Gower, CE Webb. A simple tunable KrF laser system with narrow bandwidth and diffraction limited divergence. J Phys D Appl Phys 15: 767–773, 1982.

57. R Loundon. The Quantum Theory of Light. Oxford: Oxford University Press, 1973.

58. NN Ershov, NG Zakharov, PA Rodnyi. Spectral-kinetic study of the intrinsic luminescence characteristics of a fluorite-type crystal. Opt Spectrosk 53: 51–54 1982.

59. CWE van Eijk. Cross-luminescence. J Lumin 60: 936–941, 1994.

60. E Sarantopoulou, Z Kollia, AC Cefalas. LiCaAlF$_6$:Nd^{3+} crystal as optical material for 157 nm photolithography. Opt Commun 177: 377 382, 2000.

61. Z Kollia, E Sarantopoulou, AC Cefalas, CA Nikolaides, AK Naumov, VV Semashko, RYu Abdulsabirov, SL Korableva, MA Dubinskii. Vacuum-ultraviolet interconfigurational $4f^3 \rightarrow 4f^2 5d$ absorption and emission studies of the Nd^{3+} ion in KYF, YF and YLF crystal hosts. J Opt Soc Am B12: 782–785, 1995.

62. V Dauer. Optical constants of lithium fluoride thin films in the far ultraviolet. J Opt Soc Am B 17: 300–303, 2000.

63. ED Palik, ed. Handbook of Optical Constants of Solids. 1st ed. Orlando, FL: Academic Press, 1985.

64. A Milgram, M Parker Givens. Extreme ultraviolet absorption by lithium fluoride. Phys Rev 125: 1506–1509, 1962.

65. P Laporte, JL Subtil, M Courbon, M. Bon. Vacuum ultraviolet refractive index of LiF and MgF$_2$ in the temperature range 80–300K. J Opt Soc Am 73: 1062–1069, 1983.

66. G Stephan, YLe Calvez, JC Lemonier, Mme S Robin. Properties optiques et spectre electronique du MgF$_2$ et du CaF$_2$ de 10 a 48 eV. J Phys Chem Solids 30: 601–608, 1969.

67. OR Wood, HG Craighead, JE Sweeney, PJ Maloney. Vacuum ultraviolet loss in magnesium fluoride films. Appl Opt 23: 3644–3649, 1984.

68. MW Williams, RA MacRae, ET Arakawa. Optical properties of magnesium fluoride in the vacuum ultraviolet. J Appl Phys 38: 1701–1705, 1966.

69. ET Hutcheson, G Hass, JT Cox. Effect of deposition rate and substrate temperature on the vacuum ultraviolet reflectance of MgF$_2$- and LiF-overcoated aluminum mirrors. Appl Opt 11: 2245–2248, 1972.

70. AS Barriere, A Lachter. Optical transitions in disordered thin films of the ionic compounds MgF$_2$ and AlF$_3$ as a function of their conditions of preparation. Appl Opt 16: 2865–2871, 1977.

71. A Mayolet, JC Krupa, I. Gerard, P. Martin. Luminescence of Eu^{3+} doped materials excited by VUV synchrotron radiation. Materials Chem Phys 31: 107–109, 1992.

72. AN Belsky, JC Krupa. Luminescence excitation mechanisms of rare earth doped phosphors in the VUV range. Displays 19:185–196, 1999.

73. G Ionova, JC Krupa, I. Gerard, R. Guillaumont. Systematics in electron-transfer energies for lanthanides and actinides. N J Chem 19: 677–689, 1995.

74. I Gerard, JC Krupa, E Simoni, P Martin. Investigation of charge transfer $O^{2-} \rightarrow Ln^{3+}$ and $F^- \rightarrow Ln^{3+}$ in LaF_3: (Ln^{3+}, O^{2-}) and YF_3: (Ln^{3+}, O^{2-}) systems. J Alloys Comp 207/208: 120–127, 1994.

75. HR Moser, G Wendin. Theoretical-models for intensities of d-f transitions in electron-energy-loss spectra of rare-earth and actinide metals. Phys Rev B 44: 6044–6061, 1991.

76. BF Aull, HP Jenssen. Impact of ion-host interactions on the 5d to 4f spectra of lanthanide rare-earth–metal ions. I. A phenomenological crystal field model. Phys Rev B 34: 6640–6646, 1986.

77. L Brewer. Energies of the electronic configurations of the singly, doubly and triply ionized lanthanides and actinides. J Opt Soc Am 61: 1666–1682, 1971.

78. BR Judd. Optical absorption intensities of rare-earth ions. Phys Rev B127: 750–761, 1962.

79. E Loh. $4f^n \rightarrow 4f^{n-1}5d$ spectra of rare earth ions in crystals. Phys Rev B 175: 533–536, 1968.

80. NV Starostin. On the theory of composite $f^{n-1}d$ configuration of lanthanides in crystalline fields of cubic symmetry. Opt Spec 23: 260–261, 1964.

81. T Szczurek, M Schlesinger. Vacuum ultraviolet absorption spectra of CaF_2:RE^{3+} crystals. In: Proceedings of the International Symposium Rare Earth Spectroscopy. Singapore: World Scientific, pp 309–330, 1985.

82. Z Kollia, E Sarantopoulou, AC Cefalas, AK Naumov, VV Semashko, RYu Abdulsabirov, SL Korableva. On the $4f^2 5d \rightarrow 4f^3$ interconfigurational transitions of Nd^{3+} ions in K_2YF_5 and $LiYF_4$ crystal hosts. Opt Commun 149: 386–392, 1998.

83. E Sarantopoulou, Z Kollia, AC Cefalas, VV Semashko, R Yu Abdulsabirov, AK Naumov, SL Korableva. On the VUV and UV $4f^7(^8S)5d \rightarrow 4f^3$ interconfigurational transitions of Tb^{3+} ions in $LiLuF_4$ single crystal hosts, Opt Commun 156: 101–111, 1998.

84. P Feofilov. On absorption and luminescence spectra of Ce^{3+} ions. Opt Spectr 6: 150–151, 1959.

85. AA Kaplyanski, VN Medvedev, PP Feofilov. The spectra of trivalent cerium ions in alkaline earth fluoride crystals. Opt Spect 14: 351–356, 1963.

86. MH Crozier. Zeeman effect in the $4f \rightarrow 5d$ spectrum of Ce^{3+} in CaF_2. Phys Rev 137: A1781–A1783, 1965.

87. AA Kaplyanski, VN Medvedev. Piezospectroscopic determination of the symmetry of the crystal field acting on trivalent ions of rare-earth in fluorite lattice. Opt Spect 18: 451–456, 1965.

88. AA Kaplyanski, VN Medvedev. Linear Stark effect in the spectra of local centers in cubic crystals. Opt Spect 23: 404–410, 1967.

89. SK Gayen, DS Hamilton. Two photon excitation of the lowest 4f-5d near ultraviolet transition in Ce^{3+}: CaF_2. Phys Rev B28: 3706–3711, 1983.

90. MA Dubinskii, VV Semashko, AK Naumov, R Yu Abdulsabirov, SL Korableva. Spectroscopy of new active medium of a solid-state UV laser with broadband single pass gain. Laser Phys 3: 216–217, 1993.

91. IT Jacobs, GD Jones, K Zdansky, RA Satten. Electron phonon interaction of hydro-

genated, deuterogenated and tritiated crystals of calcium and strontium fluoride containing cerium. Phys Rev B3: 2888–2910, 1971.

92. E Loh. Lowest 4f-5d transition of trivalent rare-earth ions in CaF_2 crystals. Phys Rev 147: 332–335, 1966.

93. M Yamaga, D Lee, B Henderson, T Han, H Gallagher, T Yosida. The magnetic and optical properties of Ce^{3+} in $LiLuCaAlF_6$. J Phys Cond. Matter 10: 3223–3237, 1998.

94. CD Marshall, JA Speth, SA Payne, WF Krupke, GJ Quarles, V Castillo. Ultraviolet laser emission properties of Ce^{3+}-doped $LiSrAlF_6$ and $LiCaAlF_6$. J Opt Soc Am B 11: 2054–2065, 1994.

95. E Loh. Ultraviolet absorption spectra of Ce^{3+} ion in alkaline–earth fluorides. Phys Rev 154: 270–276, 1967.

96. WJ Manthey. Ultraviolet absorption spectra of Ce^{3+} ion in alkaline–earth fluorides. Phys Rev B 8: 4086–4098, 1973.

97. E Loh, G Samoggia, E Reguzzoni, L Nosenzo. Thermaly modulated ultraviolet absorption spectra of CaF_2:Ce^{3+} crystals. Phys Status Solidi B 79: 795–799 (1977).

98. M Schlesinger, PW Whippey. Investigations of 4f-5d transitions of Ce^{3+} in CaF_2. Phys Rev 171: 361–364, 1968.

99. T Szczurek, GWF Drake, M Schlesinger. Vibronic structure in the absorption spectrum of the Ce^{3+} ion in CaF_2. Phys Rev B8: 4910–4912, 1973.

100. J Sugar. Analysis of the spectrum of triply ionized praseodymium (Pr IV). J Opt Soc Am 55: 1058–1061, 1965.

101. GH Dieke, HM Crosswhite, B Dunn. Emission spectra of the doubly and triply ionized rare earths. J Opt Soc Am 51: 820–827, 1961.

102. R Sarup, M Crozier. Analysis of the eigenstates of Pr^{3+} in $LaCl_3$ using the Zeeman effect in high fields. J Chem Phys 42: 371–376, 1965.

103. J Sugar. Description and analysis of the third spectrum of cerium. J Opt Soc Am 55: 33–58, 1965.

104. E Loh. $^1S_0$ level of Pr^{3+} in crystals of fluorides. Phys Rev A 140: 1463–1466, 1965.

105. WT Carnall, PR Fields, R Sarup. $^1S_0$ level of Pr^{3+} in crystal matrices and energy-level parameters for the $4f^2$ configuration of Pr^{3+} in LaF_3. J Chem Phys 51: 2587–2591, 1969.

106. EY Wong, OM Stafsudd, DR Johnston. Absorption and fluorescence spectra of several praseodymium-doped crystals and the change of covalence in the chemical bonds of the praseodymium ion. J Chem Phys 39: 786–793, 1963.

107. HH Caspers, HE Rast, RA Buchanan. Energy levels of the Pr^{3+} in LaF_3. J Chem Phys 43: 2124–2128, 1965.

108. E Loh. Ultraviolet absorption spectra of Pr^{3+} ion in alkaline-earth fluorides. Phys Rev 158: 273–279, 1967.

109. WM S Heaps, LR Elias, WM Yen. Vacuum-ultraviolet absorption bands of trivalent lanthanides in LaF_3. Phys Rev B13: 94–104, 1976.

110. LR Elias, WM S Heaps, WM Yen. Excitation of uv fluorescence in LaF_3 doped with trivalent cerium and praseodymium. Phys Rev B8: 4989–4995, 1973.

111. WW Piper, JA De Luca, FS Ham. Cascade fluorescent decay in Pr^{3+}-doped fluorides: achievement of a quantum yield greater than unity for emission of visible light. J Lumin 8: 344–348, 1974.

112. AM Srivastava, DA Doughty, WW Beers. On the vacuum-ultraviolet excited luminescence of Pr^{3+} in LaB_3O_6. J Electrochem Soc 144: L190–L192, 1997.

113. AM Srivastava, WW Beers. Luminescence of Pr^{3+} in $SrAl_{12}O_{19}$: observation of two photon luminescence in oxide lattice. J Lumin 71: 285–290, 1997.

114. AM Srivastava, DA Doughty, WW Beers. Photon cascade luminescence in Pr^{3+} in $LaMgB_5O_{10}$. J Electrochem Soc 143: 4113–4116, 1997.

115. KM Devyatkova, ON Ivanova, SA Oganesyan, KB Seiranyan, SP Chernov. Luminescent properties of single crystals of $LiYF_4$:Nd^{3+} in the vacuum ultraviolet region. Sov Phys Dokl 35: 56–57, 1990.

116. GJ Quarles, GE Venikouas, RC Powell. Sequential two-photon excitation processes of Nd^{3+} ions in solids. Phys Rev B 31: 6935–6940, 1985.

117. J Sugar. Analysis of the third spectrum of praseodymium. J Opt Soc Am 53:831–839, 1963.

118. MA Dubinskii, NM Khaidukov, IG Garipov, LN Dem'yanets, AK Naumov, VV Semashko, VA Malysov. Spectral–kinetic and laser characteristics of new Nd^{3+}-activated laser hosts of the KF-YF_3 system. J Mod Opt 37: 1355–1360, 1990.

119. W Baer, JG Conway, SP Davis. Crystal spectrum of promethium[3+] in $LaCl_3$. J Chem Phys 59: 2294–2302, 1973.

120. SIJ Weissman. Intermolecular energy transfer. The fluorescence of complexes of Euro J Chem Phys 10: 214–217, 1942.

121. M Schlesinger, T Szczurek. Vacuum ultraviolet spectra of CaF_2:Eu^{3+} and CaF_2:Sm^{3+}. J Opt Soc Am 70: 1025–1029, 1980.

122. HE Rast, JL Fray, HH Caspars. Energy levels of Sm^{3+} in LaF_3. J Chem Phys 46: 1460–1466, 1967.

123. JD Axe, GH Dieke. Calculation of crystal field splittings of Sm^{3+} and Dy^{3+} levels in $LaCl_3$ with inclusion of J mixing. J Chem Phys 37: 2364–2371, 1962.

124. BP Zakharchenya, VP Makarov, A Ya Ruskin. Zeeman effect for d-f transitions in the spectra of Sm^{2+} activated alkali-earth fluoride crystals. Opt Spect 17: 116–120, 1964.

125. ZJ Kiss, HA Weakliem. Stark effect of 4f states and linear crystal field in $BaClF$:Sm^{2+}. Phys Rev Lett 15: 457–460, 1965.

126. BP Zakharchenya, AYa Ruskin. Zeeman effect in the absorption and luminescence spectrum of CaF_2:Sm^{++} and SrF_2:Sm^{++} crystals. Opt Spect 13: 501–502, 1962.

127. E Loh. $4f^n \rightarrow 4f^{n-1}5d$ spectra of rare-earth ions in crystals. Phys Rev 175: 533–536, 1968.

128. RT Wegh, H Donker, A. Meijerink. Vacuum ultraviolet excitation and emission studies of $4f^n \rightarrow 4f^{n-1}5d$ transitions for Ln^{3+} in $LiYF_4$. Proc Elect Soc 97: 284–295, 1988.

129. M Schlesinger, GFW Drake. On the vibronic spectrum of rare-earths in calcium fluoride. Can J Phys 54: 1699–1701, 1976.

130. WC Martin, R Zalubas, L Hagan. Atomic Energy Levels—The Rare Earth Elements. Washington, DC: National Bureau of Standards, 1978.

131. J Sugar, J Reader. Ionization energies of doubly and triply ionized rare earths. J Chem Phys 59: 2083–2089, 1973.

132. JF Kielkopf, HM Crosswhite. Preliminary analysis of the spectrum of triply ionized gadolinium (Gd IV). J Opt Soc Am 60: 347–351, 1970.

133. GH Dieke. In: Crosswhite HM, Crosswhite H, eds. Spectra and Energy Levels of Rare Earth Ions in Crystals, New York: Wiley, 1968, pp 249–253.

134. M Schlesinger, T Szczurek, GWF Drake. The lowest energy 4f → 5d transition of the triply ionized gadolinium in CaF_2. Solid State Commun 28: 165–166, 1978.

135. E Sarantopoulou, AC Cefalas, MA Dubinskii, Z Kollia, CA Nicolaides, RY Abdulsabirov, SL Korableva, AK Naumov, VV Semashko. VUV and UV fluorescence and absorption studies of Tb^{3+} and Tm^{3+} trivalent ions in $LiLuF_4$ single crystal hosts. J Mod Opt 41: 767–775, 1994.

136. JL Fray, HH Caspars, HE Rast, SA Miller. Optical absorption and fluorescence spectra of Dy^{3+} in LaF_3. J Chem Phys 48: 2342–2348, 1968.

137. M Schlesinger, T Szczurek, MCK Wiltshire. 4f → 5d transition studies of Tb^{3+} and Dy^{3+} in calcium fluoride. Can J Phys 54: 753–756, 1975.

138. T Szczurek, M Schlesinger. 4f → 5d transition studies of Ho^{3+} in calcium fluoride. Phys Rev B 9: 3938–3940, 1974.

139. AA Vlasenko, LI Devyatkova, ON Ivanova, VV Mizailin, SP Chernov, T Uvarova, BP Sobolev. Transmission spectra of single crystals of the type $BaLn_2F_8$ in a wide spectral region (from 12 to 0,12 μm). Sov Phys Docl 30: 395–397, 1985.

140. LI Devyatkova, ON Ivanova, VV Mikhailin, SN Rudnev, BP Sobolev, TV Uranova, SP Chernov. Experimental study of 4f-5d transitions in Ho^{3+}, Er^{3+}, Tm^{3+}, and Yb^{3+} in BaY_2F_8. Sov Phys Dolk 30: 687–689, 1985.

141. KM Devyatkova, ON Ivanova, KB Seiranyan, SA Tamazyan, SP Chernov. Vacuum ultraviolet properties of a new fluoride matrix. Sov Phys Dokl 35: 40–41, 1990.

142. LI Devyatkova, ON Ivanova, VV Mikhailin, SN Rudnev, SP Chernov. High energy 4f states of Er^{3+} and Ho^{3+} ions in fluoride crystals. Sov Phys Dolk 62: 275–276, 1985.

143. T Szczurek, M Schlesinger. Spectroscopic studies of excited Tm^{3+} ions in CaF_2 crystals. Phys Rev B34: 6109–6111, 1986.

144. M Schlesinger, T Szczurek. Spectroscopic studies of excited Tm^{3+} ions in alkaline-earth fluorides. Phys Rev B35: 8341–8347, 1986.

145. T Szczurek, M Schlesinger. Temperature dependent rare-earth impurity-site symmetries in CaF_2. Phys Rev B36: 8263–8267, 1987.

146. BW Bryant. Spectra of doubly and triply ionised Ytterbium YbIII and Yb IV. J Opt Soc Am 55: 771–779, 1965.

147. M Schlesinger, T Szczurek, MK Wade, GWF Drake. Anomalies in the vacuum uv absorption spectrum of Yb^{3+} in CaF_2. Phys Rev B18: 6388–6390, 1978.

148. E Loh. Ultraviolet-absorption spectra of europeum and ytterbium in alkaline earth fluorides. Phys Rev B184: 348–352, 1969.

149. E Loh. Strong-field assignment on $4f^{13}5d$ levels of Yb^{2+} in $SrCl_2$. Phys Rev B7: 1846–1850, 1972.

150. RT Wegh, H Donker, A Maijerink. Spin-allowed and spin-forbidden fd emission from Er^{3+} and $LiYF_4$. Phys Rev B57: R2025–R2028, 1998.

151. MB Seelbinder, JC Wright. Site selective spectroscopy of CaF_2:Ho^{3+}. Phys Rev B 20:4308–4320, 1979.

152. E Sarantopoulou, YS Raptis, E Zouboulis and C Raptis. Pressure and temperature-dependent Raman study of $YLiF_4$. Phys Rev B59: 4154–4162, 1999.

153. P Dorenbos, CWE van Eijk, AJJ Bos, CL Melcher. Scintillation and thermolumi-

nescence properties of Lu_2SiO_5:Ce fast scintillation crystals. J Lumin 60&61: 979–982, 1994.

154. O Guillot-Noël, JTM de Haas, P Dorenbos, CWE van Eijk, K Krämer, HU Güdel. Optical and scintillation properties of cerium-doped $LaCl_3$, $LuBr_3$ and $LuCl_3$. J Lumin 85:21–35, 1999.

155. JC van't Spijker, P Dorenbos, CWE van Eijk, K Krämer, HU Güdel. Optical and scintillation properties of Ce^{3+} doped K_2 $LaCl_5$. J Lumin 85: 1–10, 1999.

156. CM Combes, P Dorenbos, CWE van Eijk, K Krämer, HU Güdel. Optical and scintillation properties of pure and Ce^{3+}-doped Cs_2LiYCl_6 and Li_3 Ycl_6:Ce^{3+} crystals. J Lum 82: 299–305, 1999.

157. O Guillot-Noël, JC van't Spijker, JTM de Haas, P Dorenbos, CWE van Eijk, K Krämer, HU Güdel. Scintillation properties of $RbGd_2Br_7$:Ce: advances and limitations. IEEE Trans Nucl Sci 46: 1274–1284, 1999.

158. EG Devitsin, N Yu Kirikova, VA Kozlov, VN Makhov, S Yu Potashov, LN Dmitruk, MA Terekhin, IH Munro, C Mythen, DA Shaw, KW Bell, RM Brown, PS Flower, PW Jeffreys, JM Parker. Time-resolved studies of emission properties of cerium-doped fluoro-hafnate glasses under VUV synchrotron radiation excitation Nucl Instr Methods Phys Res A405: 418–422, 1998.

159. P Dorenbos, JC van't Spijker, OWV Frijns, O Guillot-Noël, CWE van Eijk, K Krämer, HU Güdel, A Ellens. Scintillation properties of $RbGd_2Br_7$:Ce^{3+} crystals; fast efficient and high density scintillators. Nucl Instr Methods Phys Res B132: 728–731 1999.

160. CM Combes, P Dorenbos, CWE van Eijk, C Pedrini, HW Den Hartog, JY Gesland, PA Rodnyi. Optical and scintillation properties of Ce^{3+} doped $LiLuF_4$ and $LiLuF_4$ crystals. J Lumin 71: 65–70, 1997.

161. MA Dubinskii, VV Semashko, AK Naumov, R. Yu Abdulsabirov, S Korableva. Ce^{3+} doped colquirite. A new concept of all-solid state tunable ultraviolet laser. J Mod Opt 40: 1–5, 1993.

162. C Combes. Scintillation properties of ^{6}Li-based materials for thermal-neutron detection. PhD thesis, Technical University of Delft, 1999.

163. S Nicolas, Y Guyot, VV Semashko, R Yu Abdulsabirov, E Descoix, MF Joubert. Spectroscopie de l'ion Pr^{3+} dans $LiLuF_4$. Proceedings, Phenomenes Luminescents des Materiaux Isolants, Lyon, France, 1999.

164. SA Payne, GD Wilke. Transient gratings by 4f to 5d excitation of rare earth impurities in solids. J Lumin 50: 159–168, 1991.

165. RB Barthem, R Buisson, JC Vial, H Harmand. Optical properties of Nd^{3+} pairs in $LiYF_4$-existence of a short range interaction. J Lumin 34: 295–305, 1986.

166. M Gruwe, S Tavernier. Determination of the scintillation light yield of neodymium doped LaF_3 scintillator. Nucl Instr Methods Phys Res A311: 301–305, 1992.

167. P Dorenbos, JTM de Haas, CWE van Eijk. The intensity of the 173 nm emission of LaF_3:Nd^{3+} scintillation crystals. J Lumin 69: 229–233, 1996.

168. AN Belsky, P Chevallier, JY Gesland, N Yu Kirikova, JC Krupa, VN Machov, P Martin, PA Orekhanov, M Queffelec. Emission properties of Nd^{3+} in several fluoride crystals. J Lumin 72/74: 146–148, 1997.

169. MA Dubinskii, R Yu Abdulsabirov, SL Korableva, AK Naumov, VV Semashko.

On the possibility of ultraviolet lasing on f-f transitions in Nd^{3+} ion. Laser Phys 2: 239–241, 1992.

170. MA Dubinskii. Light-driven optical switch, based on excited state absorption in activated dielectric crystals. J Mod Opt 38: 2323–2326, 1991.

171. WM Yen, PM Selzer. Laser Spectroscopy of Solids. Berlin: Springer Verlag, 1981.

172. RT Wegh, H Donker, A Meijerink. Spin allowed and spin-forbidden fd emission from Er^{3+} in $LiYF_4$. Phys Rev B 57: R2025–R2028, 1998.

173. RT Wegh, A Meijerink. Spin-allowed and spin-forbidden $4f^n \rightarrow 4f^{n-1}5d$ transitions for heavy lanthanides in fluoride hosts. Phys Rev B 60: 10820–10830, 1999.

174. AC Cefalas, Z Kollia, E Sarantopoulou. Vacuum Ultraviolet $4f^95d \rightarrow 4f^{10}$ interconfigurational transitions of Ho^{3+} ions in $LiLuF_4$ single crystals. J Opt Soc Am B V16: 625–630, 1999.

175. E Sarantopoulou, Z Kollia, AC Cefalas. $4f^95d \rightarrow 4f^{10}$ spin-allowed and spin-forbidden transitions of Ho^{3+} ions in $LiYF_4$ single crystals in the vacuum ultraviolet. Opt Commun 169: 263–274, 1999.

176. J Becker, JY Gesland, M Yu Kirikova, JC Krupa, VN Makhov, M Runne, M Queffelec, TV Uvarova, G Zimmerer. Fast VUV emission of rare earth ions (Nd^{3+}, Er^{3+}, Tm^{3+}) in wide band gap crystals. J Alloys Comp 275–277: 205–208, 1998.

177. A Meijerink, RT Wegh. VUV spectroscopy of lanthanides: extending the horizon. Materials Sci Forum 315–317: 11–26, 1999.

178. JC Krupa, M Queffelec. UV and VUV optical excitations in wide band gap materials doped with rare earth ions:4f-5d transitions. J Alloys Comp 250: 287–292, 1997.

179. P Dorenbos. The $5d4f^{n-1} \rightarrow 4f^n$ luminescence of trivalent rare earth ions in inorganic crystals. Materials Sci Forum 315–317: 222–227, 1999.

9
Spectroscopy of Broad-Band UV-Emitting Materials Based on Trivalent Rare-Earth Ions

Richard Moncorgé
Université de Caen, Caen, France

I. INTRODUCTION

There are a number of scientific and technological applications for which tunable ultraviolet (UV) solid-state laser sources are already or will soon be well suited. These include in environmental sciences, for the detection of pollutants and biological hazards in the atmosphere as well as measurements and control of various climatic parameters (see Chapter 13). In medicine the applications are in ophthalmology and biotechnologies for which broad-band laser sources can be used for the production of ultra-short excitation light pulses. Other important needs also exist in the fields of photolithography (see Chapter 6); material processing with, for example, the deposition of thin films; and in information technology, with high-density optical storage.

For some of these, such as high-density information storage, compact size and energy efficiency are more important than high power or energy storage. It is thus likely that diode lasers, such as those based on III-N wide bandgap semiconductors, will dominate these applications. For most of the other needs, however, the laser sources based on rare-earth doped materials offer important advantages. We will comment on just two examples.

It is important to monitor some atmospheric pollutants over broad areas, perhaps even globally. Laser techniques such as differential absorption lidar (the optical equivalent of radar) are well suited to this task. In this technique one sends out light beams at two different wavelengths, one of which is absorbed by the chemical species of interest and the other is not. Each is reflected off the ground or some other feature, or is scattered off particles in the atmosphere, such that returning light is detected at the emitting site. The difference in signals at the two wavelengths gives the absorption strength, and hence, the concentration of the species of interest. If distance information is required, the travel time of the laser pulses can provide it. Such applications require substantial pulse energies and high beam quality from a compact system with high efficiency and reliability. Solid-state lasers based on rare-earth-doped materials are exceptionally well suited to such tasks. For example, the National Aeronautics and Space Administration (NASA) uses such systems for ozone detection, which requires UV wavelengths in pairs such as 289 nm for absorption and 300 nm for nonabsorption in the troposphere, or 301 and 311 nm, respectively, in the stratosphere [1]. There has also been interest in UV solid-state lasers for the excitation of tryptophan fluorescence to detect biological hazards in the atmosphere at a distance [2].

As the efficiency and lifetime of features in semiconductor devices continue to improve and their size to shrink, photolithography at shorter wavelengths is also becoming very important [3]. Considerable effort has already been put into 193 nm technology [4], with the expectation of using XeF excimer lasers. However, the greater reliability and potentially lower life-cycle costs of solid-state lasers have attracted interest in the development of rare-earth-doped crystalline systems to reach the same wavelength [5]. This approach combines an Nd laser, frequency multiplication, and optical parametric oscillator (OPO) technology to achieve the desired wavelength and beam quality [6]. It is believed that the requisite average power can be achieved. This is an application potentially of very great importance.

II. STATE OF THE ART FOR BROAD-BAND UV LASER SYSTEMS

Most of the tunable laser sources operating in the near UV that are now commercially available remain either very complicated and/or very expensive. As mentioned above, these are based on frequency upconversion: frequency-doubling, tripling, and/or mixing in nonlinear crystals of tunable visible or near-infrared laser sources. These sources can be dye lasers pumped by frequency-doubled or tripled Nd:YAG lasers, which have good efficiencies but pose those problems of maintenance, lifetime, dimensions, and toxicity. They can be solid-state Ti-sapphire or Cr:LiSAF lasers themselves, pumped by flash-lamps, frequency-doubled Nd:YAG, or laser diodes. Finally, they can be OPOs based on nonlinear crystals such as beta-baryum-borate (BBO). However, in this case, one comes up against a problem of cost, since visible OPOs necessitate high-beam-quality frequency-tripled Nd:YAG laser sources, which often means injection by diode-pumped minilasers. The global optical conversion efficiency does not exceed a few percent and the quality of the resulting laser beam is often very disappointing. In addition, there are problems of maintenance and dimensions and, more scientifically, of wavelength adjustment and versatility, and of spectral bandwidth.

For these reasons, the most attractive solution for obtaining a tunable UV laser source that is efficient, reliable, compact, and of reasonable cost is to take advantage of the 5d $\rightarrow$ 4f interconfigurational broad-band laser transitions of rare-earth ions such as Ce^{3+} in crystals.

This also appears to be the only way to obtain a really good-quality laser beam, controlled adjustment of the spectral width of the laser emission (which is useful for some application), and a multiwavelength operation from only one laser oscillator.

However, because of serious problems of solarization, due to the formation of transitory or permanent color centers consecutive to the absorption of UV excitation and laser radiation in the rare-earth-ion-emitting state, this type of solid-state laser has been discarded for many years. Only recently, with the discovery [7] of the Ce^{3+} doped $LiCaAlF_6$ (LiCAF) laser crystal (see Chapter 11), has this type of laser really been understood. It was demonstrated, for example, that starting from a careful spectroscopic analysis (that of the excited state absorption as a function of polarization) one could deduce optimal operating conditions and obtain a spectacular improvement in the laser performance of the materials [8,9]. It was even shown that solarization could be reduced and laser efficiencies increased by pumping crystals with the frequency-quadrupled radiation of a pulsed Nd:YAG laser, and exposing them simultaneously to residual green light resulting from frequency-doubling of the same pump laser. The laser conversion efficiency of this type of material now reaches nearly 30–40 % with respect to the UV pump radiation, and about 10% with respect to the infrared (IR) radiation at 1.064 μm of the Nd:YAG pump laser, which is quite good. The tunability domain is moderately wide (285–310 nm) but it already allows several applications and a complete commercial system is now available from Lambda-Physics [10]. It includes a 1 kHz, frequency-quadrupled diode-pumped Nd:YAG (Star Line series) pumping a Ce:LiCAF laser that delivers 500 mW average power at gain maximum.

Research is now very active in this field, including investigations of several other materials doped with Ce^{3+} and other rare-earth ions such as Pr^{3+} and Nd^{3+}, which present similar broad-band UV emissions in complementary wavelength domains. For example, $Ce:LiLuF_4$ is now competing seriously with Ce:LiCAF, since the former recently proved to have exceptional laser efficiencies (around 45–50%) and a very interesting wavelength tunability extending from 305 up to 335 nm [11–13]. In the case of the other ions, however, the situation seems less favorable since, up to now, only one really led to laser action: $Nd:LaF_3$ [14,15]. This crystal was pumped directly at 157 nm with the aid of an F_2 molecular laser into the first absorption band of the $4f^2 5d$-excited electronic configuration of the Nd^{3+} ions. Laser action occurred at 172 nm ($4f^2 5d \rightarrow 4f^3$ transition), which makes $Nd:LaF_3$ the solid-state laser with the shortest laser wavelength demonstrated so far. Attempts have been made with several Pr^{3+}-doped materials, by using pump-probe techniques, to detect gain at particular emission wavelengths [16]. None of these attempts have been successful [16,17]. However, considering the complexity of the problem and referring to the case of Ce^{3+}, this does not mean that it cannot be resolved. Materials doped with ions such as Pr^{3+}, Nd^{3+}, or Tm^{3+} might be more interesting, in the end, than Ce^{3+}-doped ones, because of their energy level structures (see in Fig. 1). Indeed, they are characterized not only by $4f \leftrightarrow 5d$ interconfigurational optical transitions in the UV domain but also by $4f \leftrightarrow 4f$ intraconfigurational transitions in the visible. This means that their UV emissions can be excited via multiple absorptions of visible photons within their 4f energy levels. This mechanism has only been studied so far as an optical loss for laser transitions in the visible and the near-infrared between the 4f energy levels. It has not yet been really studied as a population mechanism of the UV emission bands in view of their potentials as tunable laser sources. Such a mechanism, however, could be very interesting, since multistep excitation

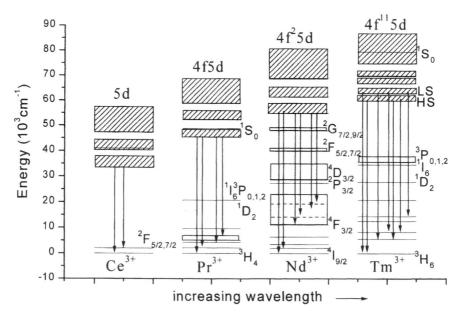

Figure 1 Approximate positions of $4f^n$ and $4f^{n-1}5d$ energy levels Ce^{3+}, Pr^{3+}, Nd^{3+}, and Tm^{3+} in $YLiF_4$.

with the aid of visible and, eventually, infrared photons would likely reduce the solarization effects associated with a UV pumping light.

The following sections examine various mechanisms that can be involved in the excitation and emission processes of these rare-earth doped materials. Based on our present understanding of the question, we then derive some guidelines for future work, with particular emphasis on the spectroscopy and dynamics of materials doped with Pr^{3+}.

III. INTEREST AND LIMITATIONS OF Pr^{3+}, Nd^{3+}, Er^{3+}, AND Tm^{3+} COMPARED TO Ce^{3+}

The positions of the $4f^n \leftrightarrow 4f^{n-1}5d$ absorption and emission transitions do not only depend on the type of rare-earth ion to be considered, but also on the host matrix. This is due to the strong crystal field effect on the 5d outer-shell orbital, which splits into various components depending on the symmetry of the crystalline environment around the rare-earth active ion. This can be used to advantage since emissions can occur in very different, and thus complementary, spectral domain. This is shown in Figure 2 for Ce^{3+} and Pr^{3+} in various materials.

A strong crystal field may also lead to strong electron–phonon interaction and to increased Stokes shifts between the potential curves associated with energy

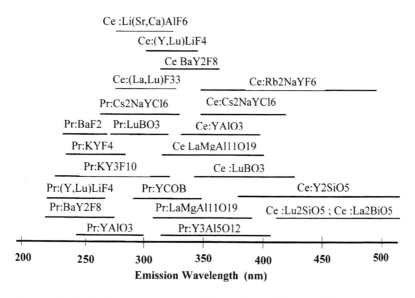

Figure 2 Emission domains spanned by various Ce^{3+} and Pr^{3+}-doped materials.

levels of various electronic configuration. It can thus lead to increased nonradiative relaxations, which has been studied in great detail by researchers investigating these materials for phosphor and scintillator applications.

Depending on the host materials, sharp line emissions may occur rather than broad bands. Strong ligand-to-metal charge transfer absorption transitions (LMCT) may overlap with the $4f^{n-1}5d \leftrightarrow 4f^n$ interconfigurational absorption and emission transitions.

Even in the case of strong $4f^{n-1}5d \rightarrow 4f^n$ emission transitions, laser gain may not be observed because of excited-state absorption losses occuring in the wavelength domain of excitation or of UV emission, and/or the subsequent transitory or permanent formation of absorbing color centers due to so-called solarization effects.

This means that several parameters have to be considered before one may discover a new laser material.

A. Position and Nature of the $4f^{n-1}5d$-Emitting Level

As seen in Figures 1 and 2, in the same crystal host, the Pr^{3+} emission bands lie at higher energies (shorter wavelengths) than the Ce^{3+} ones. They are shifted from each other by some 12,000 cm^{-1}. This agrees with numerous previous estimations [18]. For example, Pr:YLF (YLiF$_4$) emission starts around 220 nm, while Ce:YLF starts around 300 nm, which leads to a difference in energy of 12,100 cm^{-1}. In the case of YAP (YAlO$_3$) and YAG (Y$_3$Al$_5$O$_{12}$), which give very distinct emission bands, they are found as 11,500 and 12,250 cm^{-1}, respectively. This also agrees with the observation reported recently by Dorenbos [19] that knowledge of the position of the lowest energy level of the $4f^{n-1}5d$ configuration of a particular rare-earth ion in a specific host allows one to predict the position of the lowest energy level of the other 13 Ln^{3+} ions in that same host. This can be of considerable help in predicting the position of the emission bands.

According to the positions of the free ion energy levels of the trivalent rare-earth ions and their positions in a fluoride host such as LaF$_3$, as reported by Dieke and Crosswhite and Heaps et al. [20,21] as well as by deVries and Blasse [18], and based on results collected in the literature of the last 20 years, in addition to Ce^{3+} and Pr^{3+}, only Nd^{3+}, Er^{3+}, and Tm^{3+}-doped compounds would lead to efficient $4f^{n-1}5d \rightarrow 4f^2$ emissions in the near UV [see 22–25]. Among these, only Tm^{3+} would lead to emission bands in oxide hosts; Nd^{3+} and Er^{3+} would only emit in wide-bandgap materials such as fluorides. In particular, Er^{3+} and Tm^{3+} are the only ions having more than half-filled 4f shells ($n = 11$ and 12, respectively) for which there might be no close $4f^n$ energy levels lying below the $4f^{n-1}5d$-emitting one, and thus for which nonradiative quenching might be negligible.

According again to the same literature, the emission bands of Nd^{3+} would start some 10,000 cm^{-1} above those for Pr^{3+}, thus some 22,000 cm^{-1} above those for Ce^{3+}. In the case of Nd:YLF that would mean an emission spectrum starting around 180 nm, which is in perfect agreement, for example, with the spectrum reported by Thogersen et al. [23]. Using the same rule of thumb and data reported by Yang and Deluca [26], the $4f^{n-1}5d \rightarrow 4f^n$ emission spectra for Er^{3+} and Tm^{3+} should start 8300 cm^{-1} above those for Nd^{3+} and 17,850 cm^{-1} above those for Pr^{3+}. This would be at around 158 nm in the case of YLF and 200 nm in the case of YAG, for example. These results are in good agreement with the spectra of Er:YLF and Tm:YLF reported by Wegh and Meijerink [25] and the spectrum of Tm:YAG reported by Lips et al [22].

When there are no spectroscopic data available in the literature, it is worth noting that a rough estimate of the energy of the lowest 5d absorption band can be obtained by calculating the change of the 4f–5d centroid energy difference from its free-ion value due to the polarizability of the ligands. This was found for Pr^{3+}, Nd^{3+}, Er^{3+}, and Tm^{3+} to be roughly equal to the amount of energy by which the crystal field depresses the $4f^{n-1}5d$ configuration. The procedure was introduced by Morrison [27], based on a Judd suggestion and used, for example, by Aull and Jenssen [17] and Merlde et al [28]. According to these authors, the 4f–5d centroid energy difference is given by

$$(E_{5d} - E_{4f})_{\text{solid}} \cong (E_{5d} - E_{4f})_{\text{free-ion}} - \sigma_2 \sum_i (\alpha_i e^2 / R_i^6) \tag{1}$$

where α_i is the polarizability of the ith ligand, R_i is the RE–ligand distance, and σ_2 is given by

$$\sigma_2 = \langle r^2 \rangle_{5d} - \langle r^2 \rangle_{4f} \tag{2}$$

that is, the difference in expectation values of r^2 between the 5d and 4f wave functions.

By treating σ_2 as an adjustable parameter, it was found to be a constant if slightly decreasing value (lanthanide contraction effect) ranging from 2.3 to 1.9 (Å)^2 for Nd^{3+}, Er^{3+}, and Tm^{3+}-doped fluorides such as $YLiF_4$ and LaF_3, and about 3.1 (Å)^2 for Pr^{3+}-doped oxides such as YAG, YAP, and SAM ($SrAl_{12}O_{19}$). These values assume polarizabilities of about 1 and 3.8 (Å)^3 for fluorine and oxygen ions, respectively. Using, for example, the calculated and measured $4f^2$–$4f5d$ energy differences for the free Pr^{3+} ion and for Pr:YLiF$_4$ (i.e., 61,170 cm^{-1} [29] and 45,450 cm^{-1}, respectively), a σ_2 value of 2.3 is found by using the same other parameters used by Morrison [27] and, thus in good agreement with the above values. This would mean, for example, that a σ_2 value of about 3 should be used for Tm^{3+}-doped YAG. According to Morrison [27] and Merkle et al. [28], it would lead to a $4f^{12} - 4f^{11}5d$ minimum energy gap of about 49,500 cm^{-1}, which is in perfect agreement with the onset of the $4f^{11}5d \rightarrow 4f^{12}$ UV fluorescence observed in this material around 200 nm and reported by Lips et al. [22].

For the more compact (tighter) octahedral complexes such as elpasolite Cs_2NaYF_6 and borate $ScBO_3$, however, the situation is more complex. This is due to a stronger crystal field as well as more significant metal–ligand covalency effects. It partially results in greater effective ligand polarizabilities, which further shrink the 4f–5d centroïd energy difference. In the case of the $Ce:Rb_2NaYF_6$ fluoro-elpasolite, for example, an effective fluoride polarizability $\alpha = 1.83$ (Å)^3 was reported [17] and the value $\sigma_2 = 1.25$. These values give, according to Eqs. (1) and (2), a 4f–5d centroïd shift of 14,000 cm^{-1}. With a free-ion centroïd difference of 50,000 cm^{-1}, a 5d crystal-field splitting of the order of 18,000 cm^{-1}, and a 4f spin-orbit splitting of 2000 cm^{-1}, this locates the onset of UV emission around 350 nm, which is in good agreement with the so-called blue-site emission reported by Aull and Jenssen [17]. In the case of the chloro-elpasolites [30–32], because of the greater polarizability of the chlorine ions ($\alpha \approx 3.6$ according to [33]), the 4f–5d centroïd shift should be even larger than in fluorides and lead to much lower emission bands. This is not as dramatic, however, because of the smaller crystal field experienced by the rare-earth dopant. The onset of the emission band in a Ce^{3+}-doped chloro-elpasolite such as Cs_2NaYCl_6 (onset at about 360 nm [30–32]) is only shifted down by about 800 cm^{-1}.

It is worth noting that for materials with the same type of ligands, oxygen O^{2-} for oxides and fluor F^- for fluorides, for example, the higher the coordination number (CN) the longer the metal–ligand distances and the weaker the local crystal field [34]. With CN = 8 for YAG and CN = 12 for YAP, the crystal field is stronger in the case of YAG. It has an emission spectrum starting at much lower energy than that of YAP, because of a larger crystal field splitting of the 5d bands. The same type of argument is applicable in the case of materials with different types of ligands but the same coordination number, such as YAP and YLF (or BYF) for both of which CN = 8. In this case, the higher the ligand valency the stronger the crystal field and the lower the emitting state.

The question then arises as to the nature of the lowest energy-emitting level of the $4f^{n-1}5d$ electronic configuration. This is of great importance, since it governs the radiative emission probability of the emission transition and thus the value of the stimulated emission cross section. To discuss that question we refer to two very recent works, by Wegh and Meijerink [25] and by Laroche et al., [35] which are discussed in more detail below. According to these authors, the case of the more than half-filled 4f shell ions such as Er^{3+} and Tm^{3+} might be very different from that of the less than half-filled 4f shell ions, such as Ce^{3+}, Pr^{3+}, and Nd^{3+}. Indeed, because of Hund's rule, if spin still can be considered as a good quantum number (which is not proven), the high-spin states of the ground and first-excited configurations $4f^n$ and $4f^{n-1}5d$ of more than half-filled 4f shell ions have the lowest energies but their spins differ by unity. Moreover, the high-spin state in the first excited configuration could lie just below a low-spin one. For Er^{3+}, for example, the high-spin ground state is a quartet ($^4I_{15/2}$) whereas in the $4f^{10}5d$-excited configuration the lowest energy level is likely to come from a high-spin sextet. The next higher-lying one could be again a quartet. This means that all the emitting transitions from the sextet energy level to the lower energy levels of the $4f^{11}$ ground state configuration (levels $^4I_{15/2}$, $^4I_{13/2}$, etc. . . .) will have a spin-forbidden character. All the emitting transitions coming from the above-lying quartet (provided that radiative probability competes favorably with nonradiative decay) are spin-allowed. This is what was observed and reported recently for Er^{3+} and Tm^{3+}-doped fluorides [24,25], especially at low temperature. The first consequence is a fluorescence lifetime of a few μs. It was 3 and 8 μs in the case of Er^{3+} and Tm^{3+}-doped YLF, respectively, which is two orders of magnitude longer than what is usually found with Ce^{3+} or Pr^{3+}. This is not good for scintillator or laser applications, since it means much smaller stimulated emission cross sections, and thus smaller laser gains. Wegh and Meijerink [25] compared this to the situation encountered in organic molecules; it occurs with the singlet and triplet states of organic dyes. The situation would be more favorable if the energy mismatch between high and low spin-excited states (3000 cm^{-1} in the case of $Er:YLiF_4$ [25]) could be reduced to a few hundred cm^{-1}. Thermalization could then occur at room temperature, resulting in a situation comparable to that found with the $^2E \rightarrow {}^4A_2$ and $^4T_2 \rightarrow {}^4A_2$ spin-forbidden and spin-allowed transitions of Cr^{3+} in alexandrite ($BeAl_2O_4$) and other medium-field materials [36].

In the case of the less than half-filled 4f shell ions such as Ce^{3+}, Pr^{3+}, and Nd^{3+}, the situation is different. Assuming, as previously, that LS coupling holds and that the 4f–5d electrostatic interaction dominates over spin-orbit coupling, the ground and first-excited electronic configurations would have the same high-spin states: doublets, triplets, and quartets in the case of Ce^{3+}, Pr^{3+}, and Nd^{3+}, respectively. As a consequence, for all three ions, the $4f^{n-1}5d \rightarrow 4f^n$ emission transitions should be parity as well as spin-allowed transitions. Although the real situation is likely to be more complex, at least in the case of Pr^{3+} and Nd^{3+}, as was proved recently with $Pr:YLiF_4$ [35], this is consistent with measured fluorescence lifetimes, which usually range between 10 and 40 ns, depending on the ion and the crystal host. For fluorescence quantum efficiencies nearly equal to one, such fluorescence lifetimes also mean stimulated emission cross sections on the order of 10^{-18} cm^2.

This is illustrated in Figure 3, in which we have reported the stimulated emission spectra of Pr^{3+}-doped $LiYF_4$, BaY_2F_8, KY_3F_{10}, and Cs_2NaYCl_6. It also accounts for the various wavelength domains and emission band shapes that can be observed by changing host materials. In the case of Pr^{3+} (and the same observation could be made with Nd^{3+}), the most important emission peaks correspond to the lowest Stark levels $^3H_{4,5,6}$ and $^3F_{2,3,4}$ of the $4f^2$ ground-state configuration

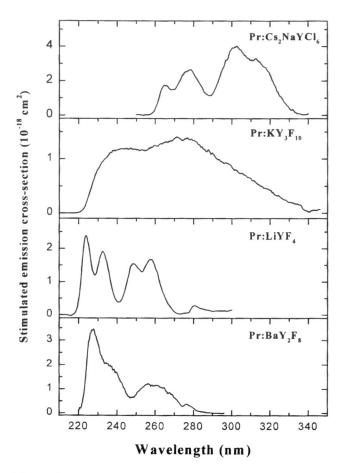

Figure 3   Stimulated emission spectra of some Pr^{3+}-doped crystals.

(see Fig. 1). These levels are more or less split by the crystal field and the spin-orbit interaction, and the transitions are more or less sensitive to electron–phonon coupling. In the case of Ce^{3+}, observed emission peaks come from transitions to the two spin-orbit levels $^2F_{5/2}$ and $^2F_{7/2}$ that are split and eventually mixed by the local crystal field, depending on its characteristics (strength and degree of distortion).

B.   Fluorescence Quantum Efficiency

Fluorescence quantum efficiency is a very important parameter that might depend on various nonradiative channels. An extensive literature has been produced on the subject. However, the situation is far from satisfactory, probably because the measurement of absolute fluorescence quantum efficiencies is not simple, and interpretation of the observed phenomena varies according to the type of host.

For example, in the case of Pr^{3+}-doped materials, the authors often refer to work by Piper et al. [37], according to which the fluorescence quantum efficiencies of the $4f^2 5d \rightarrow 4f^2$ UV emissions of Pr^{3+}-doped KYF_4, KY_3F_{10}, BaY_2F_8, and YPO_4 would be considerably lower than that of Pr^{3+} : $YLiF_4$. However, nothing has been said about the quality of the samples, so it would be educative to do these measurements again with high-quality and well-characterized samples. It was demonstrated in our laboratory, and also reported by Dubinskii et al. [38], that Ce^{3+} : $LiLuF_4$ was much more fluorescent than Ce^{3+} : $YLiF_4$. We do not know yet for sure if this is due to the quality of the crystals or to more physical reasons, such as the smaller size and the larger weight of Lu^{3+} compared to that of Y^{3+}. It may also be related to excitation or relaxation processes involving the top of the valence band, with its $4f(Lu^{3+})$ character on the one hand and its $2p(F^-)$ character on the other.

Three types of mechanisms are usually involved in the nonradiative relaxation of the $4f^{n-1}5d$ excited configuration and subsequent reduction of fluorescence quantum efficiency (see in [39,40], for example). There are photoionization (or metal $\rightarrow$ ligand charge/electron transfer), when the lowest $4f^{n-1}5d$ levels lie just inside or just below the conduction band (CB) of the host material; absorption by overlapping charge transfer (ligand $\rightarrow$ metal charge/electron transfer) bands; and $4f^{n-1}5d \rightarrow 4f^n$ intersystem crossing and/or multiphonon relaxation.

Of course, the former cause of nonradiative quenching can be avoided easily by working with host materials such as fluorides, which only absorb at high energy in the VUV spectral domain. However, for different reasons, including material fabrication and robustness, oxides may be more suitable. This is because, despite their lower optical absorption edge, many produce very efficient UV luminescence. Ce^{3+} and Pr^{3+} : YAG [41,42], with an intrinsic absorption edge around 52,000 cm^{-1} and 5d and 4f5d lowest levels around 20,500 and 32,000 cm^{-1}, respectively, are good examples.

In many materials, however, it is clear that direct photoionization will be more of a problem with Pr^{3+} and Tm^{3+} than with Ce^{3+}, because of higher-energy $4f^{n-1}5d$ absorption bands. The other ions of interest here (i.e., Nd^{3+} and Er^{3+}), never led to $4f^{n-1}5d$ UV luminescence other than in a fluoride crystal. Even though photoionization prevents UV luminescence in Ce^{3+}-doped materials such as nitrates, vanadates, and carboxylates with absorption edges below 40,000 cm^{-1}, many other Ce^{3+}-doped oxides, borates, phosphates, silicates, and aluminates with absorption edges above 50,000 cm^{-1} give intense luminescences. According to Yen et al. [43], the crossover from nonluminescent to luminescent Ce^{3+}-doped systems would occur for host materials with intrinsic absorption energy ranging between 45,000 and 50,000 cm^{-1}. In the case of Pr^{3+}, the number of UV-emitting oxide materials is not only more restricted but the interpretation of the quenching mechanism may be more subtle.

The case of Pr^{3+}-doped $(Sc, Lu, Y)_2O_3$, discussed by Blasse et al. [39] and Aumuller et al. [44], is educative. The fast nonradiative $^3P_0 \rightarrow {}^1D_2$ relaxation as well as the 4f 5d UV luminescence quenching observed in these materials are accounted for by 4f 5d–4f^2 intersystem crossing. The explanation is based on the fact that the Pr^{3+} ion substitutes for smaller-size cations, such that the Pr-O distance is usually short and the covalent character of the metal–ligand bonding is increased. This causes the 4f 5d configuration to shift to lower energies, close to the $^3P_0$ level of the underlying 4f^2 configuration, together with a strong crystal field splitting. This is in accordance with the observations made in the Pr^{3+}-doped $(Sc, Lu, Y, La)BO_3$ series [45]. The 4f5d lowest emitting state was observed to shift to lower energies with decreasing substituted ion radius (Pr^{3+}:0.99 Å, Sc^{3+}: 0.74 Å, Lu^{3+}:0.86 Å, Y^{3+}:0.90 Å, for coordination 6 [46]), an effect sometimes termed nephelauxetic [47]. There is, however, some disagreement as to the effect of the increased metal–ligand covalent character resulting from tighter metal–ligand distances, and thus from a stiffer crystalline environment, on the Stokes shift. Usually in ionic crystals, the tighter the crystal environment, the smaller the Stokes shift, which is in good agreement with observations reported [45] concerning Pr^{3+}-doped borates. DeMello Donega and Colleagues [40], however, argue that the fast $^3P_0 \rightarrow {}^1D_2$ relaxation observed even at low temperature in the Pr^{3+}:Ln_2O_3 series could not be accounted for. In case of increased overlap between the metal and the ligand wavefunctions, the 4f5d state would acquire some charge-transfer character. Thus as the 5d wavefunction becomes more delocalized, the 4f5d state would be more contracted (greater force constant leading to a steeper potential curve), thus resulting in a larger Stokes shift. This apparent discrepancy was already noted in the case of the Ce^{3+}-doped fluoro-elpasolites [17] mentioned above, since only a relatively strong covalent character (compared to the other strongly ionic fluorides) could account for both a 4f–5d band at low energy (strong crystal field) and a large Stokes shift.

Some attention can be paid to LMCTs such as $F^- \rightarrow Ln^{3+}$ and $O^{2-} \rightarrow Ln^{3+}$ in fluoride and oxide host lattices containing Ln^{3+} trivalent lanthanides. According to Jorgensen [47], the position of an LMCT transition can be estimated by:

$$E_{CT} \approx 3 \times 10^4[\chi(L) - \chi(M)] \text{ cm}^{-1} \tag{3}$$

where χ stands for the electronegativities of the considered ligand and metal ions. According to work by Jorgensen, van Vught et al., Ropp and Carrol, and Gérard et al. [47–50],

$$\chi(Cl^-) \approx 3, \qquad \chi(O^{2-}) \approx 3.2, \qquad \chi(F^-) \approx 3.9$$

and

$$\chi(Ce^{3+}) \approx 2, \quad \text{and} \quad \chi(Pr^{3+}) \approx 1.1$$

so that it can be found that

$$E_{CT}(F^- \rightarrow Ce^{3+}) \approx 57,000 \text{ cm}^{-1} \quad \text{and} \quad E_{CT}(F^- \rightarrow Pr^{3+}) \approx 84,000 \text{ cm}^{-1}$$

and

$$E_{CT}(O^{2-} \rightarrow Ce^{3+}) \approx 36,000 \text{ cm}^{-1} \quad \text{and} \quad E_{CT}(O^{2-} \rightarrow Pr^{3+}) \approx 63,000 \text{ cm}^{-1}$$

which also means

$$E_{CT}(\text{fluorides}) \approx E_{CT}(\text{oxides}) + 21,000 \text{ cm}^{-1}$$

Experimentally, it has been found, for example, that

$$E_{CT}(YLF) < E_{CT}(YPO_4, YAlO_3, YBO_3) < E_{CT}(YAG)$$

Based on Figures 2 and 3, it is worth noting here that LMCT transitions should not be a problem for any Ce^{3+} or Pr^{3+}-doped materials of good quality. There might be some problem, however, if oxygen traces are present, for example, in Ce^{3+}-doped fluoride materials (which is a topic often discussed in the literature). Indeed, in this hypothesis, the Ce^{3+} emission bands of fluorides such as $(La,Lu)F_3$ might overlap with a $O^{2-} \rightarrow Ce^{3+}$ LMCT absorption band located around 275 nm.

To summarize, it is likely that photoionization and intersystem crossing are more serious problems than LMCT absorption, especially in the case of the Pr^{3+}-doped oxide materials. Among all the emitting ones, it is clear that those with the higher optical absorption edge, such as Pr^{3+} : $YAlO_3$ and Pr^{3+} : (Sc, Y, Lu)BO_3, might be the best candidates for UV laser operation. The other Pr^{3+}-doped materials are subject either to nonradiative relaxations or excited-state absorption (ESA).

IV. $4f^2 \rightarrow 4f5d$ ESA TRANSITIONS AND $4f5d$ ENERGY LEVELS OF Pr^{3+}

Because of the spin-forbidden character of emission transitions, there seems to be little hope of finding a UV or VUV laser system based on Er^{3+} or Tm^{3+}-doped materials. Apart from Ce^{3+}, which will not be discussed here, and for Nd^{3+}, which already lased in LaF_3, only Pr^{3+} may lead to laser action and probably also in a fluoride host.

For that reason we will concentrate on Pr^{3+} by presenting the results of a recent experimental and theoretical analysis [51,52] evaluating the possibility of

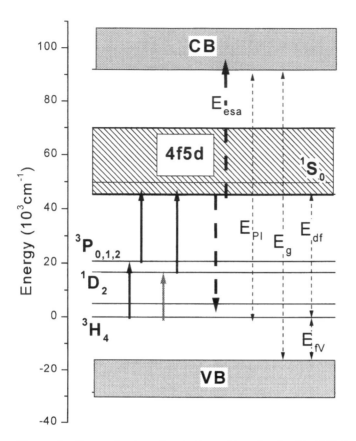

Figure 4 Energy levels and optical transitions involved in Pr³⁺-doped materials.

pumping the 4f5d UV emission band with the aid of two photons (see Fig. 4). One would bring the Pr^{3+} ion into its $^3P_0$ or $^1D_2$ excited state. Another photon would bring the system from this excited level into the 4f5d absorption band, following work done in the past with Pr^{3+}:YAG [53,54].

The materials used for these investigations included Perovskite Pr:YAlO₃, because good-quality single crystals with various dopant concentrations were available. Among the oxide systems, Pr:YAlO₃ is presently the most serious candidate material for UV laser emission. Also used were the fluorides Pr:YLiF₄, LiLuF₄, BaY₂F₈, KY₃F₁₀, and KYF₄, all grown in our laboratory. Although few data can be found on Pr:KY₃F₁₀ and KYF₄, most of the basic optical properties of the other selected materials have been reported in the literature.

Positions of energy levels of the 4f² ground configuration for Pr^{3+}-doped yttrium orthoaluminate (YAlO₃) are available [55]. The emission spectrum of the

4f5d first excited configuration exhibits two broad bands at room temperature centered at 247 and 282 nm [56,57]. Their excitation spectrum exhibits, at energies below the band gap energy (8 eV [58]), two groups of bands located around 46,080 and 55,550 cm^{-1}, respectively, assigned to the $4f^2 \rightarrow 4f5d$ transitions [56].

The basic luminescence properties of Pr:YLiF$_4$ ($4f^2$ energy level positions and intraconfigurational transition intensities) can be found elsewhere [59–61]. Unpolarized data on $4f^2$-$4f5d$ optical transitions were reported by Lawson and Payne [16] and Piper et al. [37]. In the case of Pr:BaY$_2$F$_8$, $4f^2$ luminescence and stimulated emission data have been reported [62,63], as have $4f^2$-$4f5d$ unpolarized spectra [16,64]. Information on Pr:KY$_3$F$_{10}$, is also available [51,65].

In the case of Pr:KYF$_4$, no information is available on the positions of the $4f^2$ energy levels, and on its optical transitions between $4f^2$ and $4f5d$ configurations.

We report in Table 1 the positions of $^1D_2$ and $^3P_0$ energy levels and their fluorescence lifetimes measured at room temperature. Based on these data, absorption in the spin-triplet state is more efficient (spin-allowed transition) but more energy can be stored into the singlet. Both metastable levels can be interesting as intermediate states in the two-step pumping process mentioned above.

A. Excited-State Absorption Data

Excited-state absorption (ESA) measurements were performed using a pump-probe experimental set-up previously described [51]. The pump source was provided by a broad-band optical parametric oscillator (OPO GWU model C355) widely tunable in the visible and infrared domains and delivering pump pulses of about 10 mJ with a pulse duration of 10 ns at a repetition rate of 10 Hz. The probe beam was a continuous wave (CW) high pressure Xe arc-lamp (Osram 100 W). The probe light propagated colinearly to the pump beam through the sample, but in the opposite direction. It was collimated and focused successively onto the sample and the entrance slit of a monochromator with the use of concave mirrors to avoid chromatic aberrations.

Table 1 Approximate Positions and Fluorescence Lifetimes of $^1D_2$ and $^3P_0$ Metastable Levels in Pr-Doped YAlO$_3$, KY$_3$F$_{10}$, YLiF$_4$, BaY$_2$F$_8$, and KYF$_4$

Materials	Pr:YAlO$_3$	Pr:KY$_3$F$_{10}$	Pr:LiYF$_4$	Pr:BaY$_2$F$_8$	Pr:KYF$_4$
$v(^3P_0)$	20,410 cm^{-1}	20,730 cm^{-1}	20,860 cm^{-1}	20,840 cm^{-1}	21,000 cm^{-1}
$\tau(^3P_0)$	11 μs	33.5 μs	43.5 μs	42.5 μs	63 μs
$v(^1D_2)$	16,380 cm^{-1}	16,670 cm^{-1}	16,740 cm^{-1}	16,650 cm^{-1}	17,540 cm^{-1}
$\tau(^1D_2)$	165 μs	92.5 μs	205 μs	175 μs	300 μs

Polarized and calibrated ESA spectra could be recorded from 450 to 220 nm (22,000–45,000 cm^{-1}). Those obtained in the case of Pr:YLiF$_4$ are shown in Figures 5a and 6a. Those for the other materials can be found elsewhere [51,52]. These spectra are reported in units of wavenumbers (cm^{-1}). They can be shifted by the energies of the respective $^3P_0$ and $^1D_2$-absorbing excited levels (see Table 1) to yield the positions of the involved 4f5d levels:

$$\bar{v}(4f5d) = \bar{v}(^3P_0) + \bar{v}(^3P_0 \rightarrow 4f5d)$$
$$\bar{v}(4f5d) = \bar{v}(^1D_2) + \bar{v}(^1D_2 \rightarrow 4f5d) \tag{4}$$

As can be seen in Figures 5 and 6, ESA cross sections are on the order of 10^{-18} cm^2, which is typical for 4f-5d parity-allowed electric dipole transitions (see, for example, the absorption cross section of Ce^{3+} ions in LiCa(Sr)AlF$_6$ [8]).

Inspection of these $^3P_J \rightarrow 4f5d$ and $^1D_2 \rightarrow 4f5d$ ESA bands and comparison, especially in the case of Pr:YLiF$_4$, with the corresponding $^3H_4 \rightarrow 4f5d$ ground-state absorption (GSA) or excitation spectra reported elsewhere [51,52] leads to the following conclusions:

ESA spectra strongly depend on polarization.
ESA spectra from the ^3P_J and $^1D_2$ levels strongly differ from each other.
$4f^2(^3P_J, {}^1D_2) \rightarrow 4f5d$ ESA spectra are much more structured than $4f^2(^3H_4)$
$\rightarrow 4f5d$ GSA spectra.

The first observation is to be expected but this is the first time it has been demonstrated. The usually reported $4f^2 \rightarrow 4f5d$ UV and vacuum ultraviolet (VUV) spectra are not polarized, for technical reasons, and their spectral resolution is poor. For that reason, $4f^2 \rightarrow 4f5d$ ESA spectra, which are recorded in the near-UV domain, should supply more information on the structure of the 4f5d levels. They show first that the 4f5d level structure is much more complicated than the usually reported 5d splitting into e$_g$ and t$_{2g}$ levels, for example, in the case of cubic crystal-field symmetry, and that these observed 4f5d levels should not be interpreted solely on the basis of the local crystal-field environment. They should also most likely be interpreted in terms of electrostatic interaction and spin-orbit coupling. On the other hand, these levels should not be interpreted considering electrostatic interaction as the major source of splitting, as for the free-ion, and then selecting the only levels for which selection rule $\Delta S = 0$ is verified [68,69]. A better method, as suggested previously [70] for the $4f^65d$ configuration of Eu^{2+} and the $4f^75d$ configuration of Tb^{3+} [71], would be to consider the $4f^{n-1}$ electron core and 5d electron as separate systems with internal interactions (electrostatic + spin-orbit interactions for the former, and crystal-field for the latter) that exceed their mutual electrostatic and spin-orbit interactions.

This is relatively easy to do in the case of Tb^{3+} since the $4f^7$ core leads to the isolated orbital singlet ^8S (ground state of Gd^{3+}). In the case of Pr^{3+}, which immediately follows the relatively trivial case of Ce^{3+}, the levels of the $4f5d$ excited configuration could be constructed from the separate states of $4f$ and $5d$ electrons (i.e., by coupling $^2F_{5/2}(4f)$ and $^2F_{7/2}(4f)$ states). This presupposes negligible crystal-field splitting of these $4f$ states (which might not be accurate in certain strong-field systems such as the fluoro-elpasolites [17]), and the $5d$ states resulting from crystal field splitting (E_g and T_{2g} in case of T_d or O_h symmetry, for example). In the case of Nd^{3+}, the levels of the first excited electronic configuration $4f^25d$ could be constructed from the separate states of $4f^2$ and $5d$ electrons, i.e., by coupling (since we are more interested in the lower energy levels) the states $4f^2(^3H_J, ^3F_J, ^1G_4)$ of lower energies with the $5d$ states resulting from crystal-field splitting. This is not done exactly the same way as in previous works [69]. An important question, in particular, is that concerning spin–orbit coupling: Should we treat spin–orbit coupling in the $4f^{n-1}$ separate state before or on the same level as the $5d$ crystal-field splitting and $4f-5d$ electrostatic interaction? The answer was given recently in the case of the $4f5d$ configuration of Pr^{3+} [35].

B. 4f5d Energy Level Analysis

Regarding the energy levels of the $4f5d$ electronic configuration for the free ion [72], the energy spacings (a few thousands cm^{-1}) between the different spectral terms resulting from the $4f-5d$ electrostatic interaction have comparable magnitude to that usually produced by the local crystal field on the $5d$ orbitals. It is likely as a consequence that the electrostatic and local distortion hamiltonians should be considered together. Without complete calculations or a knowledge of their respective energies, we can only speculate on the nature and the symmetry of the resulting states.

Full calculation of the $4f5d$ sublevels and of the electric dipole transitions from different states of the $4f^2$ configuration towards these sublevels were recently performed for $Pr:YLiF_4$. In addition to very detailed polarized ESA spectra (from $4f^2$ states $^1D_2$ and $^3P_0 + ^1I_6$) recorded at low as well as room temperature [35,51], complete ground state absorption data were available [59,66]. Calculations were made by choosing pure LS coupling wavefunctions $/f^2SLJM>$ and $/fdLSJM>$ and the hamiltonian $H = H_{ee} + V_{SO} + V_{cryst}$, sum of the electrostatic, spin–orbit, and crystal-field hamiltonians, respectively. For the free-ion parameters, Sugar's data were used [72] and some small correction factors introduced. The $4f$ crystal field parameters were those previously determined experimentally [59]. Among the three $5d$ crystal-field parameters, two of them were taken as adjustable parameters, the third one $B_{44}(5d)$ being linked to the others. It was also assumed that the ratio $B_{kq}(5d)/B_{kq}(4f)$ should be independent on q, but not on k, as assumed by Apaev et al. [73]. The $\langle r \rangle_{fd}$ radial integral involved in the electric dipole transition matrix elements was adjusted to yield the best overall agreement between observed and calculated cross sections at absorption maxima.

The results of a simulation of the observed ESA spectra of Pr:YLiF$_4$ based on these calculations are reported in the Figures 5b and 6b. A very good agreement is thus obtained with the experimental data, at least in the wavelength domains in which they have been recorded.

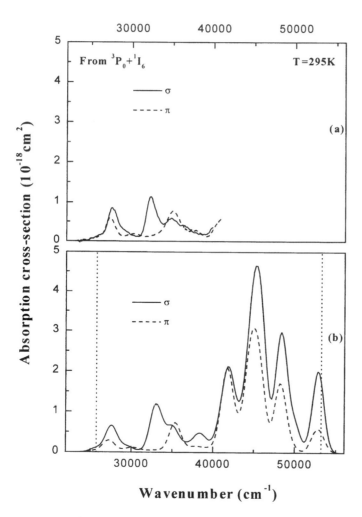

Figure 5 $(^3P_{0,1,2}, {}^1I_6) \rightarrow$ 4f 5d polarized ESA spectra of Pr:YLiF$_4$ (a) recorded and (b) calculated at room temperature (after M. Laroche, J. Margerie, et al.).

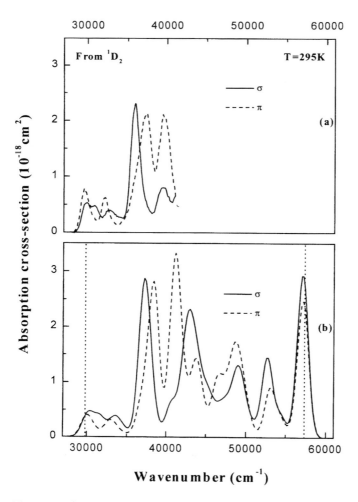

Figure 6 $^1D_2 \rightarrow$ 4f 5d polarized ESA spectra of Pr : YLiF$_4$ (a) recorded and (b) calculated at room temperature.

These calculations show clearly that the main interaction within the 4f 5d configuration is the crystal-field splitting of the 5d orbital into four levels with crystallographic quantum numbers [74] $\mu_d = 0$, $\mu_d = \pm 1$ (doubly degenerate), and twice $\mu_d = 2$ labeled 2s and 2a, for symmetrical and antisymmetrical, respectively (or Γ_2, $\Gamma_{3,4}$, Γ_2, and Γ_1 respectively, in Bethe notation [75]). More exactly (Fig. 7), the calculations show that the local crystal-field of symmetry S$_4$ in YLiF$_4$ creates three groups of sublevels (and not four) because $/\mu_d = \pm 1>$ and $/\mu_d = 2s>$ are too close to be resolved. This is in very good agreement, for example,

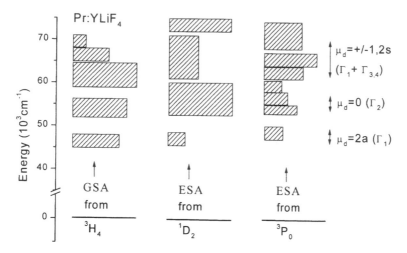

Figure 7   Position and overall nature of the 4f 5d energy levels in Pr:YLiF₄.

with the three peaks observed in the ground-state absorption spectrum [67] around 215 nm (46,500 cm^{-1}), 185 nm (54,000 cm^{-1}), and 160 nm (62500 cm^{-1}) for transitions to states $/\mu_d = 2a>$, $/\mu_d = 0>$, and $/\mu_d = \pm 1,2s>$, respectively. This splitting is very similar to that found in the case of Ce:LiLuF₄ [73].

   The calculations also show that the 4f crystal-field parameters are more than one order of magnitude smaller than the 5ds, that spin–orbit coupling of the 5d electron is quenched, and that at least in the domain below 60,000 cm^{-1}, which was more closely investigated via ESA, only three potentially significant interactions remain. These three are the crystal field on the 5d electron, the electrostatic interaction of the 4f and 5d electrons, and the spin–orbit coupling of the 4f electron. All of these are necessary to yield a reasonable description of the GSA and of the polarized ESA spectra. It was also demonstrated, however, that although the electrostatic interaction is important to account for the description of the high-energy part of the spectra ($>$58,000 cm^{-1}), the two other interactions were sufficient to explain roughly the two groups of bands of lower energies ranging between 45,000 cm^{-1} and 50,500 cm^{-1} (222–198 nm) and between 51,000 cm^{-1} and 58,000 cm^{-1} (196–172 nm). Moreover, especially for the first group, the observed splitting of about 3200 cm^{-1} (seen more clearly on the $^1D_2 \rightarrow$ 4f 5d ESA spectra) would be predominantly due to the spin–orbit coupling $7\xi_{4f}/2$ of the 4f electron with j = 5/2 and 7/2 values. The difference in the shapes of the absorption spectra, with much more complicated ones derived from ESA than GSA, is thus explained by the percentages of pure j – j states in the starting multiplets. For example, the only multiplet for which the j monoelectronic quantum number has a strong preference for a particular value (either 5/2 or 7/2) is the $^3H_4$ ground state, with 75% (5/2, 5/2) pure character and 22% mixed (5/2, 7/2) character.

These unexpected results should be related to the observation made by Lawson and Payne [76], referring to other authors [77,78], following which interactions between 5d and 4f electrons can be small due to a fortuitous cancellation of the multipole–multipole Coulomb interactions and the anisotropic exchange interactions.

At least for Pr^{3+}, the interpretation is far from that consisting of separating the energy levels of the 4f 5d configuration in purely low- and high-spin states.

V. $4f^3 \rightarrow 4f^25d$ ESA TRANSITIONS OF Nd^{3+}

As shown in Figure 1, the lowest energy levels for the first excited configuration $4f^25d$ of Nd^{3+} are relatively close to the higher-lying ones $^2G_{7/2,9/2}$ of the $4f^3$ fundamental configuration. They become closer as the crystal field becomes stronger. In the case of oxide materials, this means very reduced energy gaps, and, due to phonon energies of the order of 600–900 cm^{-1}, this also means strong nonradiative relaxations. On the other hand, in the case of fluorides, with energy gaps of the order of 7500 cm^{-1} for $YLiF_4$ and BaY_2F_8 and of 14,500 cm^{-1} for LaF_3 and KY_3F_{10} [79], and phonon energies of the order of 450 cm^{-1}, nonradiative relaxations are strongly reduced and efficient $4f^25d \rightarrow 4f^3$ UV emissions occur corresponding to transitions ending on levels $^4I_{9/2,11/2,13/2}$ (between 0 and 5000 cm^{-1}), $^4F_{3/2}-^4F_{9/2}$ (between 11,000 and 14,500 cm^{-1}), and $^4G_{5/2,7/2}-^4G_{7/2,9/2}$ (between 17,500 and 20,500 cm^{-1}). In the case of $Nd:YLiF_4$ for example, this means three groups of bands located around 180–200 nm, 220–240 nm, and 255–280 nm [23].

Two-step and three-step excitation pumping were considered (see in Fig. 8). Three-step excitation pumping was successfully realized, for example, to excite the VUV emissions of $Nd:YLiF_4$ [23,79] by using the same pump photons at 532 nm of a frequency-doubled $Nd:YAG$ laser. The pumping sequence was

$$^4I_{9/2} \rightarrow \ ^4G_{7/2,9/2}, \ ^2K_{13/2}(\sim 20000 \ cm^{-1}) \rightarrow \ ^2F_{5/2}(\sim 38000 \ cm^{-1}) \rightarrow 4f\,5d$$

As in the case of Pr^{3+}, two pathways were considered for two-step excitation [79]. Pumping was produced either into the $^4D_{3/2}$ multiplet with the frequency-tripled radiation of a $Nd:YAG$ laser at 355 nm or into the $^4F_{3/2}$ metastable level around 800 nm with an OPO, a Ti:Sa, or a diode laser. Then the excited ions were brought into the $4f^25d$ excited configuration bands with the aid of a second photon. In this case, the pumping sequences were as follows:

$$^4I_{9/2} \rightarrow \ ^4D_{3/2}(\sim 28500 \ cm^{-1}) \rightarrow 4f^25d$$

$$\text{or} \quad ^4I_{9/2} \rightarrow \ ^4F_{3/2}(\sim 11000 \ cm^{-1}) \rightarrow 4f^25d$$

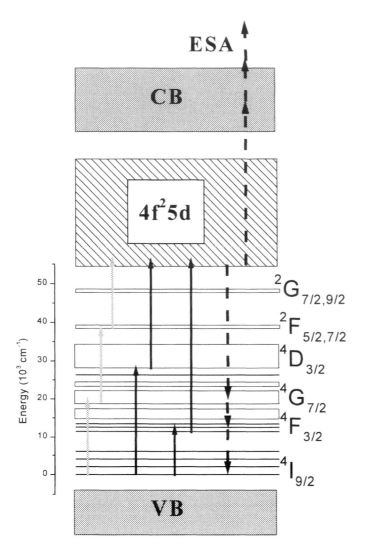

Figure 8 Energy levels and optical transitions involved in Nd^{3+}-doped materials.

Whatever the excitation process is, only a detailed knowledge of the relaxation times of the involved intermediate excited levels and of the ESA cross section spectra can determine the more efficient one. Table 2 shows characteristic lifetimes of the levels involved in some fluorides. From these values, it is clear that three-step excitation pumping involving short-lived excited states such as the $^4G_{7/2}$ multiplet [80,81] is likely to be more efficient by using short, nanosec-

Table 2 Lifetimes of Some Energy Levels of Nd-Doped Fluoride Crystals [80,81]

Materials	$Nd:LaF_3$	$Nd:YLiF_4$	$Nd:BaY_2F_8$	$Nd:NaYF_4$
$\tau(^2F_{5/2})$	23 μs	14.5 μs	13 μs	6μs
$\tau(^4G_{7/2})$	56 ns	8.5 ns		
$\tau(^4D_{3/2})$	13.5 μs	1.3 μs	2.7μs	7μs
$\tau(^2P_{3/2})$	225 μs	35 μs	50 μs	62 μs
$\tau(^4F_{3/2})$	700 μs	550 μs	660 μs	305 μs

Source: Refs. 80, 81.

ond-type, laser excitation pulses. For two-step pumping, involving intermediate long-lived levels with lifetimes of several microseconds, longer excitation pulses may allow larger energy storage and larger excitation pump powers.

The preliminary experiments mentioned above [79] have shown that ESA into the lowest-energy $4f^2 5d$ absorption bands is much more efficient after pumping around 800 nm (i.e., after populating the $^4F_{3/2}$ metastable level) than after pumping around 355 nm in the $^4D_{3/2}$. Indeed, ESA spectra obtained from these two excited states exhibit cross sections on the order of 10^{-18} cm^2 and 10^{-20} cm^2, respectively. Such a difference indicates, as in the case of Pr^{3+}, that selection rules apply in some way and that they are substantially different, at least in the ESA wavelength domain investigated, for levels such as $^4D_{3/2}$ and $^4F_{3/2}$, although they are both spin-quartets.

As in the case of Pr^{3+}, it would be worth performing the calculations to clarify this situation. At the moment, without adopting Szczurek and Schleisinger's procedure [68] in its entirety, the Russel-Saunders approximation scheme can be used to give a crude interpretation of the above-mentioned observation. Let us assume that the $4f^2 5d$ levels resulting from spin-orbit, electrostatic, and crystal field interactions keep some memory of their free-ion structure and positions [82]. In this case, if we use selection rules $\Delta S = 0$, $\Delta L = 0$, ± 1, $\Delta J = 0$, ± 1 (except $0 \leftrightarrow 0$) for transitions between $^{2S+1}L_J$ levels, the $4f^3 \rightarrow 4f^2 5d$ allowed transitions of lower energy from the above considered energy levels would be as follows:

$^4I_{9/2} \rightarrow ^4K_{11/2}$, $^4I_{9/2}$, and $^4H_{7/2}$, up to levels located at frequencies $\sigma_0 + 500$ cm^{-1}, $\sigma_0 + 1700$ cm^{-1}, and $\sigma_0 + 2600$ cm^{-1}, respectively
$^4F_{3/2} \rightarrow ^4G_{5/2}$ at $\sigma_0 + 2200$ cm^{-1}
$^4D_{3/2} \rightarrow ^4F_{3/2}$ at $\sigma_0 + 7300$ cm^{-1}

where σ_0 stands for the energy of the lowest level $^2H_{9/2}$ of the $4f^2 5d$ configuration.

This is very crude, but begin to interpret the ESA results and the observation of strong ESA from state $^4F_{3/2}$ and weak ESA from state $^4D_{3/2}$ in the investigated wavelength domain, corresponding to the lower energy levels of the $4f^25d$ electronic configuration.

Consequently, as expected, efficiency of multistep excitation of the VUV emission bands of Nd^{3+} should be strongly dependent on the wavelength (and polarization) of the pumping photons.

VI. SIMULATION OF TWO-STEP EXCITATION PUMPING AND TENTATIVE GAIN MEASUREMENTS

The different ESA spectra reported above suggest that two-step excitation, because of a second step with a very high cross section, should be an efficient upconversion process to populate $4f5d$ energy levels in the Pr^{3+}-doped fluoride and oxide crystals.

This was demonstrated by Laroche et al. [51] in the case of $Pr:KY_3F_{10}$. Two-step excitation efficiency was investigated by estimating theoretically the single-pass gain that should be obtained after excitation with 5 ns pump pulses at 445 and 355 nm for the first and second steps $^3H_4 \rightarrow {}^3P_2$ and $^3P_0 \rightarrow 4f5d$, respectively (see in Fig. 4). Using the fluorescence lifetimes of 30 µs and 26 ns of the $^3P_0$ and $4f5d$-emitting states and the adequate rate equation model, 20% gain per pass was found with pump fluences of 0.5 J/cm^2. Such a result is certainly too optimistic, considering that the simulation was made by assuming negligible losses due to ESA of the pump or of the UV-emitted photons in the $4f5d$ emitting state and/or due to solarization effects (see Sec. VII.).

Measurements of UV laser gain were also attempted in all the materials at a wavelength of 266 nm and by pumping them, as shown in Figure 4, with two consecutive photons at around 470 nm (to excite the $^3P_{0,1,2}$, $^1I_6$ multiplets) and 355 nm (to reach the $4f5d$ band from the $^3P_{0,1,2}$, $^1I_6$-excited multiplet). Gain was attempted at 266 nm because it was the only probe wavelength that was not too difficult to produce with our Nd:YAG-pumped OPO laser (by frequency doubling the residual of the 532 nm radiation). This probe was at a longer wavelength than that already tried unsuccessfully [16]: around 225 nm in $Pr:YLiF_4$ and $Pr:BaY_2F_8$ (but also in $Pr:CaF_2$ and several other similar fluorides). Unfortunately, either the emission cross section was too weak at this probe wavelength (in the case of $Pr:LiYF_4$ or $Pr:BaY_2F_{10}$) or this wavelength was well adapted (in the case of $Pr:KY_3F_{10}$ or $Pr:YAlO_3$), but the crystals solarized so much that it was impossible to detect any laser gain.

VII. SOLARIZATION EFFECTS

As mentioned previously, solarization (the creation of color centers under UV light excitation) is a critical problem in most of these UV-emitting materials.

However, it has been demonstrated, at least in the case of the now famous Ce: LiCAF and Ce:LiSAF laser systems, that these solarization effects depend not only on the characteristics of the optical transitions such as polarization but also on the purity and the dopant substitution process of the crystals. Concerning the first aspect, solarization was attributed to ESA (of the pump and/or the emitted photons) from the 5d UV-emitting state up into the conduction band of the crystals) followed by trapping of the resulting free electrons by impurity traps (traps resulting from incomplete charge compensation in the doping process, in the case of Ce:LiCAF and Ce:LiSAF). We have not yet tried to pump our Pr doped crystals directly into their 4f5d band to check because the 213 nm Nd:YAG laser pump source necessary was not available. However, we could observe some solarization effects in the course of the two-photon excitation experiments as well as during the tentative gain measurements reported in the previous section [83]. These solarization effects manifest themselves by the appearance of broad absorption bands in the blue/UV domain extending, for example, from 230 to 320 nm and from 350 to 550 nm in the case of $Pr:KY_3F_{10}$ (see Fig. 9) and from 250 to 650 nm in the case of $Pr:YAlO_3$ (Fig. 10). Concerning these systems, it is worth noting the two following points. Solarization effects in $Pr:KY_3F_{10}$ depend strongly on crystal quality and, in the case of $Pr:YAlO_3$, on dopant concentration. In the case of $Pr:KY_3F_{10}$, poor optical quality also means a larger number of

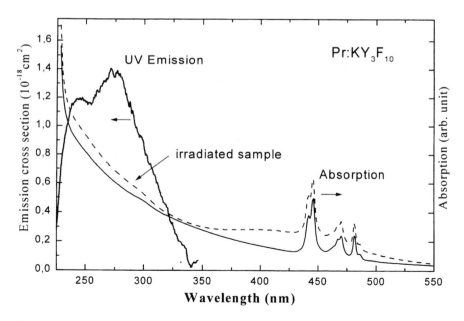

Figure 9 Color center absorption bands (dashed curve) produced in a Pr^{3+}-doped KY_3F_{10} single crystal.

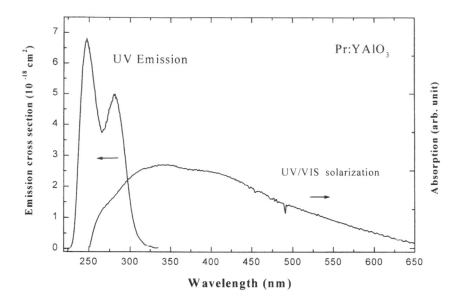

Figure 10 Deconvoluted color center absorption band (UV/visible solarization) produced in a Pr^{3+}-doped $YAlO_3$ single crystal.

lattice defects, and therefore a larger number of potential trap centers for the free electrons produced in the conduction band. In the case of $Pr:YAlO_3$, solarization effects could be reduced significantly by increasing the Pr dopant concentration. This is in agreement with previous work [84] showing that the color center formation in $YAlO_3$ comes mainly from the formation of electron-hole pairs followed by trapping of holes by Pr^{3+} ions, thus by the formation of Pr^{4+} ($Pr^{3+} + e^+ \rightarrow Pr^{4+}$). However, the reverse electron capture process at Pr^{4+} ions also occurs ($Pr^{4+} + e^- \rightarrow Pr^{3+}$). So, Pr^{3+} ions, serving as additional traps, change the direction of processes of charge carriers' release and capture, which results in an increase of the material's stability to short wavelength radiation.

These results both mean that better-quality crystals need to be grown and that dopant concentration can be a very important parameter.

VIII. DISCUSSION AND CONCLUSION

As mentioned above and already discussed in the cases of Ce^{3+} [42,85–88], Pr^{3+} [16,54] and Eu^{2+} and Sm^{2+} [76] doped materials, solarization effects would likely result from ESA from the 5d emitting level to the host conduction band (CB). According to Griffith [75], the large oscillator strength of this localized-to-delocalized 5d → CB transition would be due to an "intensity borrowing"from the valence band (VB) to the conduction band (CB) transition. This would be due to the large spatial extent of the 5d orbital and its mixing, in case of fluorides for example, with the 2p orbitals of the F⁻ ligands. Such an intensity borrowing mechanism was also involved recently to account for the large oscillator strength of the Nd^{3+} absorption transitions found in YVO_4 [89]. It turns out that the position and strength of this 5d → CB ESA transition depend on several parameters, and that only the position of this band can be estimated correctly at the moment.

This can be made in two ways, depending on the available data: first via photoionization threshold measurements and using the electrostatic model developed in the past by Pedrini et al. [90]; and, second, via XPS measurements.

Photoionization measurements give E_{PI}, (the energy difference between the $4f^n$ ground state of the rare-earth dopant and the conduction band [see Fig. 4 for Pr^{3+}]) and, according to the electrostatic model, E_{PI} can be also obtained from the ionization potential I_p of the free ion by using the relation:

$$E_{pi} = I_p - C \tag{5}$$

where C is a correction factor due to the crystal environment. Knowing the energy of the 5d-emitting level with respect to the 4f ground-state, defined as E_{df}, the 5d → CB ESA threshold would occur at the energy:

$$E_{esa} \leq E_{pi} - E_{df} = I_p - C - E_{df} \tag{6}$$

On the other hand, XPS data give the approximative position of the 4f ground state within the bandgap above the top of the valence band, which we denote as E_{fv}. In this case, the 5d → BC ESA threshold would be obtained from:

$$E_{esa} \geq E_g - E_{df} - E_{fV} \tag{7}$$

Using the data of Pedrini et al. [90] for divalent rare-earth ions in fluorides and assuming, according to Sugar and Reader [91], that the ionization energies for the trivalent ions have the same relative values, it means that for Ce^{3+}, Pr^{3+}, and Nd^{3+}, photoionization energies can be deduced from each other by using the respective ratios $1:1.2:1.31$.

Take $YLiF_4$, for example, for which the top of the valence band VB is predominantly made up of $2p_F$ orbitals and the bottom of the BC, of $4d_Y$, with an energy gap of about 84,500 cm⁻¹ (10.1 eV). According to the XPS results of Guillot-Noel et al. [89], $E_{fV}(Nd^{3+}) \approx 9700$ cm⁻¹ ≈ 1.1 eV, thus $E_{PI}(Nd^{3+}) \approx$ 74,800 cm⁻¹. This means that $E_{PI}(Pr^{3+}) \approx 74,800 \times (1.2/1.31) \approx 68,500$ cm⁻¹ and $E_{PI}(Ce^{3+}) \approx 74,800 \times (1/1.31) \approx 57,000$ cm⁻¹, thus $E_{fV}(Pr^{3+}) \approx 16,000$ cm⁻¹ ≈ 1.9 eV, and $E_{fV}(Ce^{3+}) \approx 27,500$ cm⁻¹ ≈ 3.3 eV. These results are in perfect agreement with the XPS measurements made recently on CeF_3 and PrF_3 crystals using synchrotron radiation [92].

The same procedure can be used in the case of $YAlO_3$, for which XPS spectra indicate E_{fv} values of $E_{fv}(Pr^{3+}) \approx 0$ and $E_{fv}(Ce^{3+}) \approx 1$ eV [93], the top of the valence band being made predominantly of $2p_0$ orbitals. Indeed, with an energy gap $E_g \approx 61,700$ cm^{-1}, it is found $E_{Pf}(Pr^{3+}) \geq 61,700$ cm^{-1} and $E_{Pf}(Ce^{3+}) \geq 61,700 \times (1/1.2) \approx 51,400$ cm^{-1} thus $E_{fv}(Ce^{3+}) \leq 10,300$ cm^1 (1.2 eV).

Consequently, it seems reasonable to use the above E_{fv} values along with the known E_g and E_{df} values to give estimates of the positions of the 5d $\rightarrow$ CB ESA band thresholds, at least in fluorides and oxides. A complication may occur, however, in the case of the Lu-based fluorides such as $LiLuF_4$, since XPS measurements [94] proved that the 4f orbital of Lu^{3+} lies just above the 2p for F$^-$. In this case, the above E_{fv} values for fluorides would be reduced by about 1 eV, giving values of about 0, 1, and 2 eV for Nd^{3+}, Pr^{3+}, and Ce^{3+}, respectively. Once we are able to estimate the onset for the 5d $\rightarrow$ CB ESA transition, we are faced with the question of the width of the CB (Δ_{CB}). To answer that question, knowledge of the band structure of the material is necessary, which is not always easy to determine.

Let us see the consequences of the above considerations in the case of the known laser materials: first $Ce:LiCaAlF_6$ and $Ce:YLiF_4$ with emissions in the ranges 280–320 nm and 300–340 nm, respectively. With 5d metastable emitting levels located around 280 and 300 nm (i.e., $E_{fd} \approx 35,700$ and 33,300 cm^{-1}) and 4f ground states around 3.3 eV above the top of the valence band (i.e., $E_{fv} \approx 27,500$ cm^{-1}) ESA of the laser emissions in the 5d levels could occur between 91,700 and 98,500 cm^{-1}, in the case of Ce:LiCAF, and between 90,200 and 94,100 cm^{-1}, for Ce:YLF (i.e., between 11 and 12 eV and between 10 and 11 eV, respectively, above the top of the valence band). These energies coincide with the edge of the VB–CB energy gaps E_g in these materials. It explains why ESA-induced solarization effects were observed in these systems, although they are reduced compared to other lower-energy-gap materials. This may explain why solarization effects seem slightly more important in the case of Ce:LiSAF than in the case of Ce:LiCAF, since optical absorption edges occur around 10.4 and 10.7 eV [95], respectively. This also agrees well with our own observation in the case of $Ce:LiLuF_4$. No color center formation was ever observed in our samples by irradiating them at 248 nm with a KrF laser producing up to 4J/cm^2 at 100 Hz repetition rate, whereas weak and strong parasitic absorption bands could be easily observed, in the same pumping conditions, in the case of Ce doped $YLiF_4$ and LuF_3, respectively (see Figure 11).

From these observations, it can be concluded that $Ce:YLiF_4$ solarizes more efficiently than $Ce:LiLuF_4$ for two reasons. Band calculations [96] have shown that the band gap energy E_g of $LiLuF_4$ is larger than that of $LiYF_4$ by a factor of 1.2, which means some 12.5 eV values instead of 10.5 eV. Moreover, according to XPS measurements [94], the top of the valence band of $LiLuF_4$ would

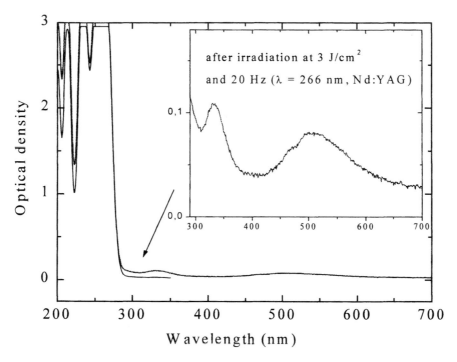

Figure 11   Color center absorption bands produced in a Ce^{3+}-doped LuF_3 single crystal.

have a $2p_F4f_{Lu}$ mixed orbital character, which means that the 4f ground state of the Ce^{3+} dopant ion should be closer to the top of the valence band. Both effects thus act favorably to relegate the edge of the 5d → CB ESA to higher energies, at least beyond the emission domain of this Ce^{3+}-doped material.

Based on these results, some prospective arguments can be derived for the search for new Ce^{3+} laser systems and a good Pr^{3+} candidate.

Let us consider first the case of Pr^{3+} in YAP ($YAlO_3$) and LuAP ($LuAlO_3$) with $E_g \approx 7.4$ eV, in YPO_4 and $LuPO_4$ with $E_g \approx 8.1$ and 8.3 eV, and in $YLiF_4$ and $LiLuF_4$ with $E_g \approx 10.5$ and 12.5 eV. In the case of Pr:YAP and LuAP and of Pr:YPO_4 and $LuPO_4$, emissions occur between about 240 and 300 nm. With a 4f ground-state located near the top of the VB, ESA in the emission domain would thus occur between about 9 and 10 eV (i.e., directly from the 5d-emitting level into the conduction band). In the case of Pr:$LiYF_4$ and $LiLuF_4$, with E_{fV} values of about 1.9 and 0 eV, respectively, and emissions ranging between 220 and 270 nm, ESA would occur between about 10.7 and 11.7 eV, which is somewhat more favorable. Consequently, it seems difficult to find a Pr-doped material for which ESA in the region of the emission does not fall within the conduction band. This does not mean that laser operation in these Pr-doped crystals is hope-less. Indeed, some good conditions (CB structure, polarization of the transitions) may be encountered for which positive gain may be obtained. Such a hope comes from the fact, as mentioned by Lawson and Payne [16], that the ESA cross sections seem barely to exceed the emission cross sections (e.g., $YLiF_4$), which means that some portion of the emission band could exhibit laser action.

In the case of Ce^{3+} and Pr^{3+}, another alternative is to search for materials with large bandgaps that induce large crystal-field splittings (from a strong local crystal-field). With a Ce^{3+} emission extending from about 320 up to 400 nm in Rb_2NaYF_6, for example [17], and thus a Pr^{3+} emission starting around 230 nm and extending to about 300 nm, ESA in the emission domain would reach levels located somewhere between 76,800 and 87,000 cm^{-1} (9.2 and 11.5 eV). Elpasolite systems such as $(Rb,Cs,K)_2Na(Y,Sc,Lu)F_6$ would be interesting to study.

The last question concerns the potentiality of new Nd^{3+}-doped materials. In this case, ESA in the emission domain of these materials would occur, for example, in the case of $Nd:YLiF_4$ (see Fig. 8), with a value $E_{fv} \approx 1.1$ eV and an emission domain ranging between 180 and 280 nm, between about 12 and 14.5 eV. This would thus eventually be above the CB of the material [97], in a region where levels should be found from the $4f^26s$ second excited electronic configuration, the next higher-lying one $4f^26p$ being located around 17.5 eV, and thus at much higher energies [98]. However, because of the Laporte selection rule, ESA transitions from the emitting level of the $4f^25d$ electronic configuration to the next $4f^26s$ should not be very intense, and most probably less intense than the parity-allowed emission itself.

In conclusion, several directions of research still need to be explored. These include efforts to find new crystalline materials with good enough optical quality, characterized by large bandgaps of (10–12 eV), a strong local crystal field (tight lattices), and reduced Stokes shifts (especially in the case of Pr^{3+}) to avoid nonradiative quenching (stiff lattices). Much attention must be payed to the spectroscopic aspects, such as the structure of the conduction band of the materials and the intensity and state of polarization of the $4f^{n-1}5d \to CB$ ESA transitions, which will require further calculations and experiments.

ACKNOWLEDGMENTS

Most of the work reported in this Chapter and quoted in Refs. 13, 28, 32, 50, 51, 78, and 81 was done under the auspices of the US Army European Research Office (contract N68171-97-M-5764) and of CNRS (PICS + GDR Lasmat), and with the collaboration of colleagues from the University of Caen (Prof. J. Margerie, Dr. S. Girard, and PhD student M. Laroche), the University of Houston-Downtown (Dr. L. D Merkle), and the Universities of Lyon and St Etienne (Drs

Y. Guyot, M.F. Joubert, S. Guy, C. Pédrini, Prof. E. Descroix, and PhD student S. Nicolas). Many thanks are expressed to all of them.

REFERENCES

1. EV Browell. In: BHT Chai, SA Payne, eds. OSA Proceedings on Advanced Solid-State Lasers. Washington, DC: Optical Society of America, 1995, vol 24, pp 2–4.
2. J Fox, C Swim. In: BHT Chai, SA Payne, eds. OSA Proceedings on Advanced Solid-State Lasers. Washington, DC: Optical Society of America, 1995, vol 24, pp 13–15.
3. J Wallace. Laser Focus World 35(8): 139–142, 1999.
4. Special Report. Laser Focus World. July: 82–96, 1997.
5. RD Mead, CE Hamilton, DD Lowenthal. Proc SPIE 3051: 882, 1997.
6. AA Buj, AV Kachinsky, AV Ermakov, AS Grabchikov, IA Khodasevich, VA Orlovich. Conference on Lasers and Electro-Optics CLEO/Europ'2000. Nice, France, 2000.
7. MA Dubinskii, VV Semashko, AK Naumov, R Yu. Abdulsabirov, SL Korableva. Laser Physics 3: 216–217, 1993.
8. CD Marshall, JE Speth, SA Payne, WF Krupke, GJ Quarles, V Castillo, BHT Chai. J Opt Soc Am B11 2054–2065, 1994. AJ Bayramian, CD Marshall, JH Wu, JA Speth, SA Payne, GJ Quarles, V Castillo. In: SA Payne, C Pollock, eds. OSA Trends in Optics and Photonics. Washington, DC: Optical Society of America, 1: 60–65, 1996.
9. JF Pinto, L Esterowitz, GJ Quarles. Electron Lett 31: 2009–2011, 1995.
10. SV Govorkov, AO Wiessner, Th Schroder, U Stamm, W Zschocke, D Bastings. OSA Trends Optics Photonics 19: 2–5, 1998.
11. N Sarukura, MA Dubinskii, Z Liu, V Semashko, AK Naumov, SL Korableva, RY Abdulsabirov, K Edamatsu, Y Suzuki, T Itoh, Y Segawa. IEEE J Select Topics Quant Electron 1: 792–804, 1995.
12. P Rambaldi, R Moncorgé, JP Wolf, C Pédrini, JY Gesland. Opt Commun 146: 163–166, 1998. Laser Focus World 33: 26, 1997.
13. AJS McGonigle, S Girard, DW Coutts, R Moncorgé. Electron Lett 35: 1640–1641, 1999.
14. RW Waynant, Ph H Klein. Appl Phys Lett 46: 14–16, 1985.
15. MA Dubinskii, AC Cefalas, E Sarantpoulou, SM Spyrou, CA Nicolaides, R Yu. Abdulsabirov, SL Korableva, VV Semashko. J Opt Soc Am B9: 1148–1150, 1992.
16. JK Lawson, SA Payne. Opt Mater 2: 225–232, 1993.
17. BF Aull, HP Jenssen. Phys Rev B34: 6640–6646, 1986. Phys Rev B34: 6648–6655, 1986.
18. AJ deVries, G Blasse. Mater Res Bull 21: 683–694, 1986.
19. P Dorenbos. International Conference on Luminescence, (Osaka, Japan 1999), paper BO4–5.
20. GH Dieke, HM Crosswhite. Appl Opt 2: 675–686, 1963.
21. WS Heaps, LR Elias, WM Yen. Phys Rev B 13: 94–104, 1975.

22. HV Lips, N Schwentner, G Sliwinski, K Petermann. J Appl Spectr 62: 803–814, 1995.

23. J Thogersen, JD Gill, HK Haugen. Opt Comm 132: 83–88, 1996.

24. J Becker, JY Gesland, N Yu. Kirikova, JC Krupa, VN Makhov, M Runne, M Queffelec, TV Uvarova, G Zimmerer. J Lumin 78: 91–96, 1998.

25. RT Wegh, A Meijerink. Phys Rev B60: 10820–10830, 1999.

26. KH Yang, JA Deluca. Phys Rev B 17: 4246–4255, 1978.

27. CA Morrison. J Chem Phys 72: 1001–1002, 1980.

28. LD Merkle, B Zandi, Y Guyot, HR Verdun, B McIntosh, BHT Chai, JB Gruber, M Seltzer, CA Morrison, R Moncorgé. In: TY Fan, BHT Chai, eds. OSA Proceedings on Advanced Solid State Lasers. Washington, DC: OSA, 20: 361–366, 1994.

29. J Sugar. J Opt Soc Am 55: 1058–1061; 1965. Phys Rev Lett 14: 731–732, 1965.

30. S Mroczkowski, P Dorain. J Less Common Met 110: 259–265, 1985.

31. JC van't Spijker, P Dorenbos, CWE van Eijk, MS Wickleder, HU Gudel, PA Rodnyi. J Lumin 72–74: 786–788, 1997.

32. M Laroche, M Bettinelli, S Girard, R Moncorgé. Chem Phys Lett 311: 167–172, 1999.

33. L Pauling. Proc Soc (London) A114: 181, 1927.

34. GA Slack, SL Dole, V Tsoukala, GS Nolas. J Opt Soc Am B11: 961–974, 1994.

35. M Laroche, JL Doualan, S Girard, J Margerie, R Moncorgé. J Opt Soc Am B17: 1291–1303, 2000.

36. JC Walling, OG Peterson, HP Jenssen, RC Morris, EW O'Dell. IEEE J Quant Electron 16: 1302–1315, 1980.

37. WW Piper, JA DeLuca, FS Ham. J Lumin 8: 344–348, 1974.

38. MA Dubinskii, VV Semashko, AK Naumov, R Yu. Abdulsabirov, SL Korableva. Laser Physics 4: 480–484, 1994.

39. G Blasse, W Schipper, JJ Hamelink. Inorg Chim Acta 189: 77–85, 1991.

40. C De Mello Donega, A Meijerink, G Blasse. J Phys Cond Matter 4: 8889–8902; 1992. J Phys Chem Sol 56: 673–685, 1995.

41. MJ Weber. Solid Stat Commun 12: 741–744, 1973.

42. DS Hamilton, SK Gayen, GJ Pogatshnik, RD Ghen, WJ Miniscalco. Phys Rev B39: 8807–8815, 1989.

43. WM Yen, M Raukas, SA Basun, W van Schaik, U Happek. J Lumin 69: 287–294, 1996.

44. GC Aumüller, W Köstler, BC Grabmaier, R Frey. J Phys Chem Sol 55: 767–772, 1994.

45. G Blasse, JP van Vliet, JWM Verwey, R Hoogendam, M Wiegel. J Phys Chem Sol 50: 583–585, 1989.

46. RD Shannon. Acta Crystal A 32: 75, 1976.

47. CK Jorgensen. Modern Aspects of Ligand-Field Theory. Amsterdam: North-Holland, 1971; R Reisfeld, CK Jorgensen. Lasers and Excited State of Rare-Earths. Berlin: Springer, 1977.

48. N van Vugt, T Wigmans, G Blasse. J Inorg Nucl Chem 35: 2602, 1973.

49. RC Ropp, B Carrol. J Phys Chem 81: 1699–1700, 1977.

50. I Gérard, JC Krupa, E Simoni, P Martin. J Alloys Comp 207/208: 120–127, 1994.

51. M Laroche, A Braud, S Girard, JL Doualan, R Moncorgé, M Thuau, LD Merkle. J Opt Soc Am B 16: 2269–2277, 1999.
52. S Nicolas, M Laroche, S Girard, R Moncorgé, Y Guyot, MF Joubert, E Descroix, AG Petrosyan. J Phys Cond Matter 11: 7937–7946, 1999.
53. J Ganem. WM Dennis, WM Yen. J Lumin 54: 79–87, 1992.
54. YM Cheung, SK Gayen. Phys Rev B 49: 14827–14835, 1994.
55. M Malinowski, C Garapon, MF Joubert, B Jacquier. J Phys Condens Matter 7: 199, 1995.
56. EG Gumanskaya, MV Korzhik, SA Smirnova, VB Pavienko, AA Fedorov. Opt Spectrosc 72: 86–88, 1991.
57. C Pédrini, D Bouttet, C Dujardin, B Moine, I Dafinei, P Lecoq, M Koselja, K Blazek. Opt Mat 3: 81–88, 1994.
58. T Tomiki, F Kaminao, M Fujisawa, Y Tanahara. J Phys Soc Japan 55: 2090; 1986. T Tomiki, F Kaminao, M Fujisawa, Y Tanahara and T Futemna. J Lumin 40&41: 379, 1988.
59. L Esterowitz, FJ Bartoli, RE Allen, DE Wortman, CA Morrison, RP Leavitt. Phys Rev B 19: 6442–6455, 1979.
60. GM Renfro, JC Windscheif, WA Sibley, RF Belt. J Lumin 22: 51–68, 1980.
61. JL Adam, WA Sibley, DR Gabbe. J Lumin 33: 391–407, 1985.
62. BM Antipenko. Opt Spektrosc 56: 72–77, 1984.
63. AA Kaminskii, SE Sarkisov. Phys Stat Sol 97: K163–168, 1986.
64. SP Chernov, LI Devyatkova, ON Ivanova, AA Kaminskii, VV Mikhailin, SN Rudnev, TV Uvarova. Phys Stat Sol 88: K169–173, 1985.
65. CA Morrison, DF Wortman, RP Leavitt, HP Jenssen. Harry Diamond Labs report, HDL-TR 1897, 1980.
66. E Sarantopoulou, AC Cefalas, MA Dubinskii, CA Nicolaides, R Yu. Abdulsabirov, SL Korableva, AK Naumov, VV Semashko. Opt Lett 19: 499–501, 1994.
67. JC Krupa, M Queffelec. J Alloys Comp 250: 287–292, 1997.
68. T Szczurek, M Schleisinger. In: B Jerowska-Trzebiatowska, J Legendziewicz, W Strek, eds. Rare Earth Spectroscopy, Proceedings, International Symposium on RE Spectroscopy, Wroclaw, Poland, Sept. 1984, pp 309–330.
69. Z Kollia, E Sarantopoulou, AC Cefalas, AK Naumov, VV Semashko, RY Abdulsabirov, SL Korableva. Opt Comm 149: 368–392, 1998.
70. AA Kaplyanski, AK Przhevuskii. Opt Spectrosc 19: 331–338, 1965.
71. T Hoshina, SZ Kuboniwa. J Phys Soc Jpn (171): 828–840.
72. HM Crosswhite, GH Dieke, WJ Carter. J Chem Phys 43: 2047–2054, 1965.
73. RA Apaev, MV Eremin, AK Naumov, VV Semashko, R Yu. Abdulsabirov, SL Korableva. Opt Spectrosc 84: 735–737, 1998.
74. KH Hellwege. Ann Phys 4: 95, 1948.
75. JS Griffith. The Theory of Transition-Metal Ions. Cambridge, UK: Cambridge University Press, 1964.
76. JK Lawson, SA Payne. Phys Rev B47: 14003–14010; 1993. J Opt Soc Am 8: 1404–1411, 1991.
77. MJ Freiser, S Methfessel, F Holtzberg. J Appl Phys 39: 900–902, 1968.
78. A Yanase, T Kasuga. Prog Theor Phys Suppl 46: 388–410, 1970.

79. Y Guyot, S Guy, MF Joubert. J Alloys Comp 323/324: 722, 2001.
80. TT Basiev, A Yu. Dergachev, YV Orlovski, AM Prokhorov. In: A Dubé, L Chase, eds. OSA Proceedings, Advanced Solid State Lasers. Washington, DC: Optical Society of America, vol 10, 358–362, 1991.
81. MF Joubert, B Jacquier, C Linares, R Macfarlane. J Lumin 53: 477–482, 1992.
82. J Sugar. J Opt Soc Am 53: 831–839, 1963.
83. M Laroche. University of Caen, private communication, 2000.
84. TI Butaeva, KL Ovanesyan, AG Petrosyan. Cryst Res Tech 23: 849, 1988.
85. KS Lim, DS Hamilton. J Opt Soc Am 6: 1401, 1989.
86. RR Jacobs, WF Krupke, MJ Weber. Appl Phys Lett 33: 410, 1978.
87. WJ Miniscalco, JM Pellegrino, WM Yen. J Appl Phys 49: 6109, 1978.
88. JF Owen, PB Dorain, T Kobayasi. J Appl Phys 52: 1216, 1983.
89. O Guillot-Nöel, B Bellamy, B Viana, D Gourier. Phys Rev B60: 1668, 1999.
90. C Pédrini, F Rogemond, DS McClure. J Appl Phys 59: 1196; 1986. C Pédrini, DS McClure, CH Anderson. J Chem Phys 70: 4959, 1979.
91. J Sugar, J Reader. J Chem Phys 59: 2083–2089, 1973.
92. D Bouttet, C Dujardin, C Pédrini, W Brunat, D Tran Minh Duc, JY Gesland. Proceedings, International Conference on Inorganic Scintillators and their Applications. Delft, pp 111–113, 1996.
93. C Pédrini, D Bouttet, C Dujardin. Proceedings, International Conference on Inorganic Scintillators and their Applications, Delft 106–108, 1996.
94. B Moine, C Dujardin, H Lautesse, C Pédrini, CM Combes, A Belski, P Martin, JY Gesland. Mater Sci Forum 239–241: 245, 1997.
95. K Shimamura, SL Baldochi, N Mujilatu, K Nakano, Z Liu, N Sarukura, T Fukyda. J Cryst Growth 2000.
96. A Hairie, M Laroche, S Girard, R Moncorgé (unpublished).
97. RA Heaton, CC Lin. Phys Rev B22: 3629–3638, 1980.
98. L Brewer. J Opt Soc Am 61: 1666–1682, 1971.

10

Generation of Coherent Ultraviolet and Vacuum Ultraviolet Radiation by Nonlinear Processes in Intense Optical Fields

A. L. Oldenburg and J. G. Eden
University of Illinois, Urbana, Illinois

> *For the rest of my life, I will reflect on what light is.*
> A. Einstein (ca. 1917)

I. INTRODUCTION

The year 2001 marks the 200th anniversary of the discovery of the ultraviolet (UV). In 1801, Johann Wilhelm Ritter reported [1] in *Annalen der Physik* that, on the 22nd of February of that year, he had detected solar radiation ''on the side of the violet of the color spectrum, *outside of the same* . . .'' (emphasis added). A more detailed paper, published by Ritter in 1803, states: ''Dafs auch aufserhalb des Violett des Newtonschen Spectrums unsichtbare strahlen anzutreffen find . . .'' (roughly translated: ''That outside of the violet of Newton's spectrum, invisible radiation is found . . .'') and concludes that '' . . . these chemically active rays must be completely different from the colored ones'' [2]. Tousey [3] has recounted the pioneering work of Stokes and Schumann, who discovered the vacuum ultraviolet (VUV) later in the 19th century and extended the frontiers of spectroscopy to wavelengths below 130 nm.

One hundred and sixty-two years following Ritter's announcement, H. G. Heard ushered in a new era for optical physics in the UV and beyond when he stated, in a 30 line letter published in *Nature*, that "Ultraviolet coherent light has been generated directly at room temperature in a pulsed nitrogen-gas laser . . . Despite the short pulse duration, the output spot of the laser is exceedingly brilliant" [4]. The peak power generated at 337.1 nm in these first experiments was a modest 10 W but D. A. Leonard [5] succeeded less than 2 years later in producing 200 kW pulses from an oscillator–amplifier system employing a segmented transverse discharge. In the early 1970s, the discharge-pumped N_2 laser was the only widely available coherent UV source for the laboratory. Capable of producing single-pulse energies of several mJ at repetition frequencies of tens of Hz, commercial N_2 laser systems were indispensable for pumping dye lasers. Nevertheless, because their peak (instantaneous) pulse powers are on the order of 1 MW, the N_2 laser proved to be of limited value in exploring nonlinear phenomena.

The discovery of the rare gas–halide lasers in 1975 introduced a family of oscillators offering peak powers of tens of MW at several discrete wavelengths in the UV and VUV (193–353 nm) [6]. Despite the enormous impact that these lasers have had in several areas such as photochemistry and medical procedures, the difficulty in generating laser pulse widths below ~0.5 ps, in addition to the beam quality, limits the peak intensities readily attainable from a system of acceptable complexity. Four decades after the advent of the laser, therefore, few primary sources offering peak powers above 1 MW exist in the UV and VUV. Those systems that are available are characterized by limited tunability (typically <150 cm^{-1}) and are incapable of generating intensities above ~10^{12} W-cm^{-2} in a compact (tabletop), high pulse repetition frequency (PRF) system.

The second avenue to generating coherent UV/VUV radiation, nonlinear frequency upconversion from a fixed or tunable primary source, has a history that also extends back prior to 1970. Over the past 30 years, a wide variety of nonlinear processes have been demonstrated to yield coherent UV or VUV radiation (tunable single frequency *or* broadband) and it is this subject to which the present chapter is devoted. For the sake of brevity, we choose to limit our remarks to three processes that have proven particularly valuable in the advance of spectroscopy and optical physics in atoms and small molecules: continuum generation, four-wave mixing, and a source of more recent origin—high-order harmonic generation.

II. CONTINUUM GENERATION

Continuum generation is a convenient and versatile process by which UV radiation can be produced and, accordingly, has been applied to time-resolved absorption spectroscopy and the production of ultrashort laser pulses. Because of the large body of literature devoted to this subject, the present review will discuss only the fundamental mechanisms of continuum generation and the practical as pects of the production of broad continua while maximizing efficiency.

When a sufficiently intense laser pulse excites a nonabsorbing medium, it will invariably, in the absence of competing processes, exhibit continuum generation. Continuum generation is characterized by spectral broadening of the incident laser light and, near threshold, has the important feature of preserving the temporal coherence of the laser. Broadening occurs as a result of the nonlinear refractive index (n_2) of the medium, which gives rise to self-phase modulation (SPM) and higher-order processes. As a result, continuum generation is typically achievable only with laser pulses $\lesssim 10$ ps in duration and $\gtrsim 10^{10}$ W/cm^2 in intensity.

A. Theory and Experimental Considerations

To understand the mechanisms that drive continuum generation in a particular medium, it is useful to write the refractive index in terms of the average linear refractive index n_0 and the nonlinear refractive index n_2:

$$n(t) = n_0 + n_2|A(t)|^2 \tag{1}$$

where A is the complex electric field envelope. We have assumed here that the response time of n_2 is sufficiently fast to follow the pulse envelope, but not on the same time scale as an optical cycle. Contributions to the magnitude and response time of n_2 (which is $\propto \chi^{(3)}$ in lowest order) can come from various physical processes such as electronic polarization, molecular orientation, and electrostriction. For practical purposes, it is a general rule that vapors have a lower n_2 than do condensed media such as liquids and glasses.

To illustrate the SPM process, one can write the solution for a linearly polarized laser pulse propagating in a nonlinear medium (neglecting processes such as Raman scattering and absorption) as follows [7]:

$$A(z, t) = a_0 F(t)\exp\left[i\frac{\omega_0 n_2}{2c} a_0^2 F^2(t)z\right] \tag{2}$$

where $F(t)$ is the laser pulse envelope, a_0 is its amplitude, and ω_0 is the fundamental frequency. As the laser pulse arrives, it induces a change in the refractive index of the medium (as quantified by n_2), which, in turn, induces a phase change in the pulse envelope that is proportional to the field strength. This phase change causes the instantaneous frequency ($\omega(t) = \omega_0 - d\phi(t)/dt$) to be first red-shifted and subsequently blue-shifted, giving rise to an upchirp in the central portion of the pulse. Computing the spectrum from this solution illustrates that not only spectral broadening but also spectral modulation occurs, both of which are spectrally symmetrical about ω_0 in the pure SPM case. It is also important to note that in this ideal case, the temporal coherence of the pulse is preserved. Although the temporal pulse width may increase following SPM, it is theoretically possible to recompress the pulse to durations shorter than that for the pre-SPM pulse by compensating for the spectral phase.

By assuming a gaussian temporal pulse envelope with a FWHM of τ_p, we can estimate the maximum spectral extent of the SPM-broadened pulse to be:

$$\Delta\omega_{max} \approx \sqrt{\frac{8 \ln 2}{e}} \frac{\omega_0 n_2 a_0^2 z}{c \tau p} \tag{3}$$

Although the broadening is proportional to the interaction length z and $1/\tau_p$, for femtosecond driving pulses one must also consider the detrimental effects of group velocity dispersion (GVD) which increase with z and the laser bandwidth (typically, $\propto 1/\tau_p$). GVD gives rise to temporal broadening (in the form of laser chirp), which reduces the peak intensity, thereby quenching the SPM effect. For this reason, vapors may be more desirable, due to their reduced GVD, than condensed media.

Another result of the above expression [Eq. (3)] is that broadening is proportional to the laser intensity, given by a_0^2. A second limitation, aside from the availability of intense laser sources, is breakdown of the medium, which occurs at much lower intensities for condensed media than for vapors. Also, vapors and liquids are self-healing, whereas solid samples generally are not.

It should be noted that, in the above treatment, other nonlinear effects such as self-focusing and self-steepening, which are often inextricably linked with continuum generation, were neglected. Self-focusing arises as a result of the change in index over the spatial profile of the electric field $E(r)$ associated with the laser pulse laser; this mechanism facilitates the SPM process by confining the laser pulses to a smaller cross-sectional area of higher intensity. However, self-focusing also gives rise to potentially undesirable effects such as conical emission (see color insert, Fig. 1A), which appears in the far-field pattern, and the breakup of the laser profile into microfilaments within the interaction medium. Both are capable of destroying the spatial and temporal coherence of the radiation. The process of self-steepening is characterized by a time delay in the peak of the pulse with respect to the edges, caused by the larger index of refraction induced by the peak intensity. GVD and self-steepening can be responsible for asymmetrics in the spectrum of an otherwise symmetrical, purely SPM-broadened continuum.

B. Review of Experimental Results

Although self-phase modulation was observed in liquid CS_2 in 1967 by Shimizu [8], the first demonstration of continuum generation over thousands of cm^{-1}(later dubbed supercontinuum generation) was reported by Alfano and Shapiro in 1970 [9–11]. These supercontinua, which in some cases extended from 400 to 700

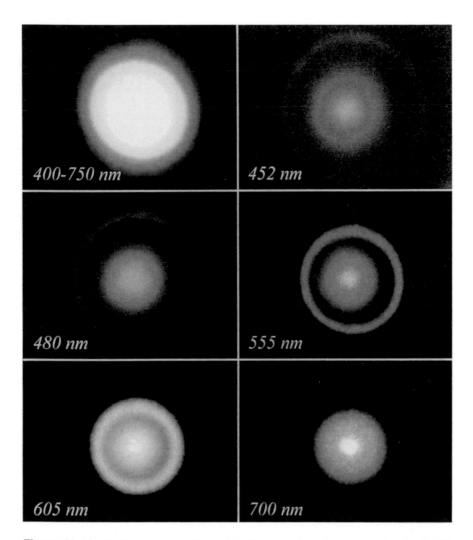

Figure 10.1A Conical emission observed in the generation of continua when 1 mJ, 120 fs pulses at 780 nm are focused into a ~1.2 cm thick disc of fused silica. For illustrative purposes, only spectral slices in the visible portion of the continuum are presented and each photograph shows the profile recorded after the beam passes through a bandpass filter having the indicated center wavelength.

Figure 10.5 Photograph of the violet emission at 420 nm produced when 2 Torr of Rb vapor (in the heat pipe at right) is irradiated with 1 mJ, 120 fs pulses at 770 nm produced by a Ti:sapphire laser. The pump radiation is incident on the heat pipe from the right and the coherent violet/UV radiation emerges from the left end of the heat pipe. The violet emission is the result of FWPO in the vicinity of the 6p → 5s transition of Rb and the divergence of the coherent radiation matches that of the incident (pump) laser radiation.

nm, were produced in such media as glass, crystals, and condensed rare gases with 4 ps laser pulses at 530 nm. Microfilaments 5–20 μm in diameter were observed and credited with increasing the laser power density to $\sim 10^{13}$ W/cm^2.

Since, ideally, the continuum spectrum is symmetrical about the pump frequency, high-intensity ($>10^{13}$ W/cm^2) femtosecond laser pulses are necessary when one is working with visible or near-infrared pump sources to produce experimentally useful continua extending into the ultraviolet. Since stimulated Raman scattering (SRS) becomes of increasing importance with excitation of increasing duration, [7] femtosecond lasers are the logical choice for realizing the required pumping intensities and minimizing the detrimental effects of SRS.

To the authors' knowledge, Fork et al. were the first to report, in 1982, the generation of continua extending significantly into the ultraviolet, [12,13] using a colliding pulse mode-locked (CPM) laser generating 65 fs pulses centered at 627 nm. By focusing ($f/12$) the laser pulses into an ethylene glycol jet, a supercontinuum having a spectral range from 190 nm to 1.6 μm was produced. In this experiment, the effect of chirp due to GVD was essentially negligible, owing to the small thickness of the jet (500 μm), giving rise to delays as low as 10 fs/100 nm of spectral width.

Because of the lower nonlinear index n_2 generally exhibited by gases and vapors as compared to condensed media (typically differing by four orders of magnitude), the former were not considered seriously as a source of efficient continuum generation until 1985, when Glownia et al. [14] found that intense pulses gently focused into air exhibited a surprising degree of SPM. Using XeCl amplifier modules to produce 10 mJ, 350 fs pulses at 308 nm, it was found that pulses exiting the first amplifier with an energy of 1.5 mJ were spectrally broadened in air from 60 cm^{-1} into a continuum of ~ 1000 cm^{-1} breadth. It was found independently [15] that supercontinua extending over 10,000 cm^{-1} could easily be produced in many high-pressure (1–40 atm) gases, such as most of the rare gases, H_2, N_2, and CO_2. In retrospect, gases are the medium of choice for most applications in the UV, due to their low GVD and, in many cases, transparency in the UV.

Upon further study of continuum generation in gases, [15,16] it was suggested by Corkum and co-workers [15] that the threshold for continuum generation is closely associated with that of self-focusing. In support of this concept, Glownia et al. [17,18] reported that the onset of continuum generation coincides with the formation of filaments within the medium. As a result, the continuum-generation process cannot be properly described by the simplified theory presented above. The reader is referred to several theoretical papers [19–22] which take self-focusing and higher-order effects into account. In fact, these papers illustrate that GVD or another process, which quenches self-focusing, must be present to prevent total collapse of the light filaments to a diameter on the order of a wavelength.

Subsequent to the discovery of continuum generation in gases and vapors, a number of experiments were reported. The studies carried out by Corkum et al. [15,16,23] utilized a 600 nm laser producing pulses with energies of 0.5 mJ and temporal widths of 70 fs and 2 ps (Table 1). For the gases studied (Ar, Kr, Xe, H_2, N_2, and CO_2), the "cutoff" wavelength on the blue side ranged from 290 nm in Xe to 330 nm in CO_2, and was independent of pump intensity, but did increase slightly with pressure. In general, the position of the blue edge of the spectral profile appeared to be fixed, independent of the gas used and the intensity applied above threshold. Glownia and co-workers [17,18] utilized a 308 nm source generating 350 fs pulses (as mentioned above) to produce a continuum with a spectral range from 235 nm to 450 nm, for the purposes of time-resolved absorption spectroscopy experiments. Their best results were achieved in 33 atm of argon by focusing the pump light with a 1.5 m focal length lens to a point 20 cm before the exit window of a cell, 1 m in length. Approximately 10 filaments were observed in the near field. Several single-shot spectra were obtained that exhibited a widely varying frequency of modulation, illustrating the problem of

Table 1   Summary of the Characteristics of Continua Generated in Several Selected Experiments

Pump laser characteristics	Medium	Continuum spectral range	Continuum spectral density	References
627 nm/70 fs/250 μJ	Ethylene glycol	190 nm–1.6 μm	~2 nJ/nm @ 400 nm	12, 13
600 nm/2 ps/500 μJ and 600 nm/70 fs/350 μJ	Xe CO_2 Ar, Kr, N_2, H_2	290 nm-IR 330 nm-IR ~300 nm-IR	NA	15, 16, 23
308 nm/300 fs/2.5 mJ	Ar H_2 CO_2	235–450 nm mostly SRS 295–360 nm	~2 μJ/nm @ 280 nm	17, 18
248 nm/700 fs/10 mJ	Ne Ar Kr	190–400 nm 200–340 nm 200–320 nm	NA	24
405 nm/165 fs/5 mJ	Liquid H_2O	200–600 nm	~5 μJ/nm @ 300 nm	25
790 nm/125 fs/167 mJ	Ne Ar Kr Xe	700 nm-IR 150 nm-IR 220 nm-IR 280 nm-IR	— ~20 μJ/nm @ 300 nm ~100 μJ/nm @ 300 nm ~2 μJ/nm @ 300 nm (see Fig. 2)	26

The spectral densities given are order-of-magnitude values for the sake of comparison.

shot-to-shot noise for this highly nonlinear experiment. Also, it is interesting to note that, in H_2, both SRS and continuum generation were present, with the continuum generation contributing less to the spectrum at lower pressures.

At shorter wavelengths, continuum generation has been studied by Gosnell et al. [24] in Ne, Ar, and Kr at high pressures (≥ 40 atm). With 700 fs pump pulses at 248 nm, they found the threshold energy for continuum generation to be as low as 60 µJ. They also determined that the shot-to-shot noise can be reduced to a reasonable level (0.2% absorption sensitivity when averaging 100 shots) by using Kr at 10 atm. Figure 1 illustrates the results obtained for neon, argon, and krypton, which produce continua spanning from roughly 200 nm to 320 nm. Neon is recommended for shorter-wavelength applications because it exhibits the broadest spectrum, extending from the violet to 190 nm.

Although the feasibility of using gases for ultraviolet continuum generation was accepted and generally preferred by the early 1990s, an experiment by Rodriguez and co-workers [25] illustrates the potential advantages of a liquid medium. The frequency-doubled output of a Ti:sapphire laser, yielding 5 mJ, 165 fs pulses, was used to excite H_2O. The continuum generated was extremely broad ($\sim 33{,}000$ cm^{-1}), extending from 200 nm (limited by the cell transmission) to 600 nm. A

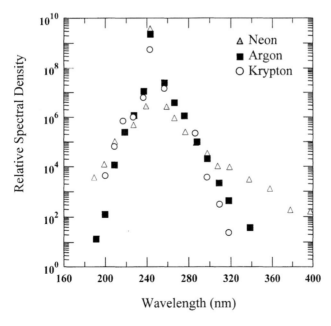

Figure 1 Supercontinuum spectra obtained with neon, argon, and krypton at 40 atm. The input pulse energy is approximately 10 mJ. (From Ref. 24, reprinted by permission.)

surprising finding was that several SPM signatures such as microfilamentation, conical emission, and spectral modulation were not observed in this experiment, resulting in good shot-to-shot stability and spatial coherence.

As the availability of strong optical fields (intensities $> 10^{13}-10^{15}$ W/cm²) produced by near-IR femtosecond lasers became more commonplace in the 1990s, XUV production through high-order harmonic generation in the rare gases (see Sec. IV) drew considerable attention. However, using one of these highly intense lasers strictly for continuum generation can also produce interesting results. In 1995, for example, Nishioka et al. [26] demonstrated that a pulse at 790 nm, having a peak power of 1.6 TW, will break down 1 atm of rare gas, even if focusing of the beam is weak ($f = 3$ m). For a focal length of 5 m, however, fine trapping channels formed, which were measured in air to be 3 m in length. The resulting spectra spanned the region from the IR down to 150 nm for argon (see Fig. 2). These are, to our knowledge, the broadest continua reported to date. The threshold for continuum generation (i.e., at which a single filament would form) occurred at 10 mJ. As the laser power was increased by an order of magnitude, the spatial mode pattern actually improved, possibly due to the presence of a higher number of microfilaments.

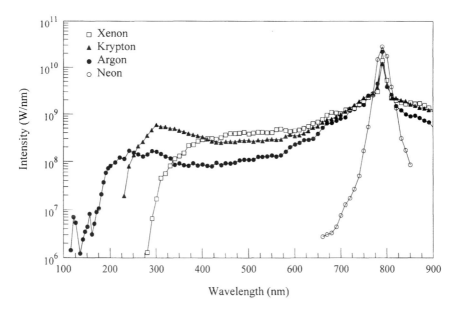

Figure 2   Spectral intensity as a function of wavelength for continua generated in atmospheric-pressure rare gases. The laser power is 1.6 TW. (From Ref. 26, reprinted by permission.)

For those not having access to a terawatt laser, however, the next best option for UV continuum generation might be the use of a gas-filled hollow fiber. It combines the advantages of the gaseous medium (low GVD, UV transmittance) with a waveguide that maintains the optical field intensity over an interaction region of extended length. For applications in femtosecond pulse generation, the hollow fiber is also desirable because it improves the spatial coherence of the output radiation, which, in turn, minimizes loss of the temporal coherence. Within the femtosecond laser community, it is well known that preserving temporal coherence (for the purpose of pulse compression) requires that continua be generated near threshold so as to minimize the number of microfilaments. The hollow fiber, therefore, affords a means of increasing the field–gaseous medium interaction length without the loss of coherence.

Recent experiments that took advantage of the hollow fiber were conducted by Nibbering et al. [27,28] to produce continua spanning the 360–430 nm spectral region. The emitted light was subsequently phase compensated to produce pulses as short as 8 fs with energies of 15 µJ. The hollow fiber used had a bore diameter of 220 µm, and was filled with Ar. Argon was chosen specifically because it balances the ionization threshold with a moderate nonlinear susceptibility. In contrast, neon and helium have larger ionization thresholds, but the nonlinear susceptibilities are considerably smaller than that for Ar. Sokolowski-Tinten et al. [29] have also shown that the hollow fiber provides a degree of spectral broadening comparable to that obtainable for glass. Their experiments, conducted with 1.2 mJ pulses at 800 nm launched into a 70 cm long fiber of 170 µm bore diameter filled with up to 5 atm of Ar, also demonstrated that the spatial quality of the radiation emerging from a hollow fiber is preferable to that generated in BK7 glass.

III. FOUR-WAVE MIXING

Four-wave mixing (FWM) is a parametric process that owes its origin to the third-order nonlinear susceptibility ($\chi^{(3)}$) of the medium. In essence, a coherent fourth wave is generated with a frequency (ω_4) that is one of the following linear combinations of the input waves: $\omega_4 = 3\,\omega_1$ (third harmonic generation; THG), $\omega_4 = 2\omega_1 + \omega_2$, and $\omega_4 = \omega_1 + \omega_2 + \omega_3$ (sum-frequency generation; SFG), and $\omega_4 = 2\omega_1 - \omega_2$ and $\omega_4 = \omega_1 - \omega_2 \pm \omega_3$ (difference-frequency generation; DFG). As outlined below, four-wave mixing can be used for efficient frequency conversion (>50%) into the near UV as well as the VUV and XUV. This can be achieved by optimizing the effects of phase matching in combination with resonant enhancement of the third-order susceptibility ($\chi^{(3)}$).

A comprehensive treatment of FWM theory and an overview of the considerable corpus of experimental work in the literature would itself require several chapters. Rather, the fundamental theoretical principles and a sampling of experimental techniques and results are presented here with the intention of providing an introduction to the subject. For a more detailed discussion, the reader is encouraged to consult Refs. 30–33.

A. Background

Before delving into the experimental details of FWM, it may be useful to review the basic concepts and theoretical considerations associated with this process. If ground-state depletion is ignored, three sufficiently intense electromagnetic waves will induce a third-order polarization (in a nonlinear medium) that oscillates at one of the mixing frequencies of the input waves. This polarization facilitates the growth of a fourth wave over the length of the medium. For incident plane waves, the intensity of the generated fourth wave can be written as [30]

$$I_4 \propto \frac{|\chi^{(3)}|^2}{n_1 n_2 n_3 n_4 \lambda_4^2 c^2} I_1 I_2 I_3 \frac{L^2 \sin^2(\Delta k L/2)}{(\Delta k L/2)^2} \tag{4}$$

where the I_j are the intensities of the incident waves, n_j are the indices of refraction at the wave frequencies ω_j, L is the length of the medium, and Δk is the "phase mismatch" of the four wave vectors [$\Delta k = k_4 - (k_1 \pm k_2 \pm k_3)$]. The last term in the above expression is essentially a *sinc* function, which is responsible for a significant (if not total) loss of power delivered to the fourth wave, unless the phase-matching condition $\Delta k = 0$ is satisfied. For sum-frequency generation, for example, the phase-matching condition can be written as:

$$2\pi c \Delta k = n(\omega_4)\omega_4 - [n(\omega_1)\omega_1 + n(\omega_2)\omega_2 + n(\omega_3)\omega_3] = 0 \tag{5}$$

Since, in this instance, $\omega_4 = \omega_1 + \omega_2 + \omega_3$, the presence of positive or negative dispersion in the medium ($dn(\omega)/d\omega > 0$ or < 0, respectively) gives rise to a positive or negative Δk, respectively.

The phase-matching condition, presented above in the plane-wave approximation, is sensitive to the tightness of focusing, as well as the positioning of the focal point within the medium. Table 2 summarizes the optimum value of the vector Δk (which has been computed for gaussian beams by Bjorklund [34]) in terms of the confocal parameter $b = 2\pi\omega_0^2 n/\lambda$ (where ω_0 is the beam-waist radius) of the incident pump waves. As illustrated by Table 2, for certain cases a "perfect" phase-match ($\Delta k = 0$) may actually result in no output when the beams are tightly focused ($b \ll L$). Also, the desired sign of Δk depends on the type of FWM process. Specifically, a normally dispersive medium induces a positive Δk and is, therefore, well suited for DFG of the type $\omega_4 = \omega_1 - \omega_2 - \omega_3$.

Table 2 Optimal Values for the Wave-Vector Mismatch Δk in Terms
of the Confocal Parameter b for Various Focusing Schemes and FWM Processes

	Optimum Δk			
	Focus in center of medium		Plane wave	Focus far outside medium
FWM process	Tight focusing $(b \ll L)$	Loose focusing $(b \gg L)$		
THG and SFG	$\Delta k = -2/b$ (No FWM for $\Delta k \geq 0$)	$\Delta k = -4/b$		
$\omega_4 = \omega_1 + \omega_2 - \omega_3$ and $\omega_4 = 2\omega_1 - \omega_2$	$\Delta k = 0$		$\Delta k \to 0$ as $b \to \infty$	$\Delta k \to 0$
$\omega_4 = \omega_1 - \omega_2 - \omega_3$ and $\omega_4 = \omega_1 - 2\omega_2$	$\Delta k = 2/b$ (No FWM for $\Delta k \leq 0$)	$\Delta k = 4/b$		

Source: Adapted from Ref. 34.

This is in contrast to THG or SFG, for which a negatively dispersive medium is desirable.

Various techniques can be used to achieve optimal phase matching. In solid media, the length of the medium (and, if the medium is a crystal, the angle of orientation) is often optimized for the wavelengths of interest. Also, birefringence (the dependence of n on the angle of polarization) can be exploited when the generated wave is driven with a polarization perpendicular to one or more of the incident waves. In isotropic media where birefringence does not occur, however, the dispersion is typically positive in the photon energy region below the first excited state, with anomalous dispersion occurring at each single-photon resonance. It is often possible to achieve phase-matching by simply tuning or choosing the input frequencies carefully, in order to take advantage of the anomalous dispersion effect. One technique for phase-matching in vapors involves using a mixture of positively and negatively dispersive gases, in which the partial pressure of each gas can be tuned so as to achieve phase-matching. Typically, one constituent of the vapor mixture will actually contribute to the interaction (through $\chi^{(3)}$), while both contribute to the dispersion $n(\omega)$. Of course, since $n(\omega)$ is also dependent upon the number density, pressure and temperature tuning, in general, are important to the FWM process in gases and vapors. Finally, the angle of the incident beams can be varied, which takes advantage of the vector nature of Δk. This is particularly useful in liquids and solids, where pressure tuning is not possible.

Again considering Eq. (4), it is evident that the FWM process is critically dependent upon the third-order susceptibility, $\chi^{(3)}$. The lowest-order nonlinear term for isotropic media, $\chi^{(3)}(\omega)$, is greatly enhanced at frequencies near resonance. However, tuning an input wave near a single photon resonance is accompanied by resonant absorption, which quenches FWM. Also, anomalous dispersion near a resonance becomes steep (i.e., $|dn/d\omega|$ is large), so that the detuning requirements for phase-matching can be quite strict. This is also true for three-photon resonances, where the output wave is resonant with the medium (and so is absorbed by the medium). Tuning near a two-photon resonance, however, is often ideal in that it effectively enhances $\chi^{(3)}$ without exhibiting anomalous dispersion (again assuming negligible depletion of the ground state). In this way, $\chi^{(3)}$ can be enhanced without affecting the phase-matching condition.

B. Experiments

1. FWM in Alkali Metal Vapors

Because of the large dipole moments associated with optical transitions in the alkali atoms, one expects $\chi^{(3)}$ to be large and conducive to efficient FWM. Indeed, Harris and Miles [35] in 1971 predicted $\chi^{(3)}$ for the alkali atoms to be larger than that for the rare gases by six orders of magnitude. In experiments reported that year, Young and co-workers [36] demonstrated the potential of FWM in the alkali vapors by frequency-tripling the output of the Nd:YAG laser to 354.7 nm. Their results showed for the first time that FWM could be a tool for generating coherent radiation at new wavelengths and for upconverting fixed wavelength primary sources, in particular. A key aspect of these experiments was the introduction of mixing two (or more) vapors to realize phase-matched third harmonic generation. A factor of 33 improvement in the efficiency for generating UV photons was obtained when a 412:1 Xe/Rb vapor mixture (rather than pure Rb vapor) served as the conversion medium. The Xe/Rb ratio was chosen such that the index of refraction at the fundamental wavelength (1.064 µm) was the same as that at the third harmonic. As one might expect, phase matching is equally sensitive to temperature. Optimum conversion of power at the fundamental into the third harmonic was measured for a temperature of 265°C. Deviations of the temperature from this value as small as ± 10°C yielded virtually no output.

The experiments of Young and colleagues [36] were designed such that, as illustrated in Figure 3a, the THG process in Rb vapor is enhanced by a three-photon near-resonance. That is, the combined energies of three photons, at frequency ω_1, differs from the energy of a three photon-accessible state by the detuning $\delta \sim 300$ cm^{-1}. However, as discussed earlier, a more favorable situation for FWM is that in which a *two*-photon excitation step in the overall process is resonantly enhanced. As an example, Leung et al. [37] tripled the frequency of the 694 nm output of a ruby laser in Cs vapor by a three-photon process having a

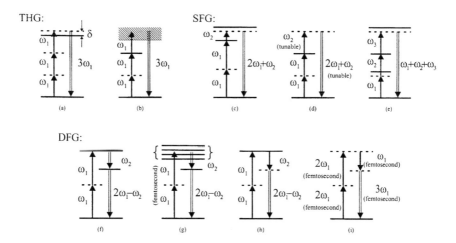

THG: SFG:

DFG:

Figure 3 Generalized atomic energy level diagrams illustrating a variety of four wave-mixing processes that have been employed in metal vapors and gases. Solid and dashed horizontal lines denote stationary and virtual states, respectively, and the shaded region in (b) indicates the ionization continuum. An arrow with a heavy line indicates a photon incident on the nonlinear medium; the double-line arrows represent photons generated within the nonlinear medium. Examples of experiments that have demonstrated each of these processes include (a) Refs. 36, 52; (b) Refs. 37, 38, 53; (c) Ref. 39; (d) Refs. 42, 50; (e) Refs. 48, 49; (f) Refs. 38, 39; (g) Ref. 40; (h) Refs. 55, 57–60; (i) Refs. 61, 62.

two-photon intermediate resonance (cf. Fig. 3b). Although $\chi^{(3)}$ for this process is roughly a factor of 70 larger than that for the frequency-tripling experiments of Young et al., [36] the combined energies of the three-pump photons exceed the ionization potential for Cs, with the result that the conversion efficiency is insensitive to phase matching with the rare gases [38].

An alternative, and considerably more efficient, process was reported by Taylor [38] whose experiments in sodium vapor demonstrated an $\omega_3 = 2\omega_1 - \omega_2$ DFG process (schematically illustrated in Fig. 3f). Occurring simultaneously with THG, the DFG process produced 330.2 nm photons in the UV—in the vicinity of the second principal series of Na ($4^2P_J \rightarrow 3^2S$, $J = 1/2, 3/2$)—with an efficiency of $5 \times 10^{-2}\%$, or a factor of ~ 500 greater than that for the THG process at the same fundamental wavelength. In this process, also known as four-wave parametric oscillation (FWPO), only ω_1 is supplied externally whereas ω_2 and ω_3 are produced within the medium. Two possible mechanisms for generation of the idler wave are stimulated electronic Raman scattering (SERS), and optically pumped stimulated emission (OPSE), which can occur as a result of the initial two-photon excitation process. The idler then acts in concert with the pump radiation (ω_1) to generate the signal wave near the np $\rightarrow$ ground transition by

FWM. An interesting set of experiments carried out by Smith and Ward [39] examined FWPO in Cs vapor. With pulses from a ruby laser, they were able to excite near-resonantly the $9\,^2D_{3/2} \leftarrow\leftarrow 6\,^2S_{1/2}$ two-photon transition and observed coherent emission in the vicinity of the $np\,^2P_{1/2}\,(7 \leq n \leq 10) \rightarrow 6\,^2S_{1/2}$ transitions. As much as 3 W of peak power generated at the signal wavelengths was observed.

The comparatively recent development of femtosecond laser systems offers a new tool with which to examine FWM. Because of the large spectral bandwidths associated with $\lesssim 100$ fs pulses, for example, it is now possible to drive simultaneously more than one two-photon transition in the alkalis (see Fig. 3(g)). Figure 4 displays the near-UV spectrum observed when Rb vapor is excited by ~620 nm, 120 fs pulses from a colliding pulse, mode-locked (CPM)-dye amplifier system [40]. Phase matching dictates that coherent emission is produced just to the "blue" side of the Rb $np\,^2P \rightarrow 5s\,^2S_{1/2}$ transitions. Pump-probe experiments [40] demonstrate that this emission serves as a means of detecting an atomic wavepacket. A photograph of the violet signal radiation produced by two-photon excitation of Rb vapor with ~100 fs pulses at 770 nm $(Ti:Al_2O_3$ laser system) is shown in Figure 5.

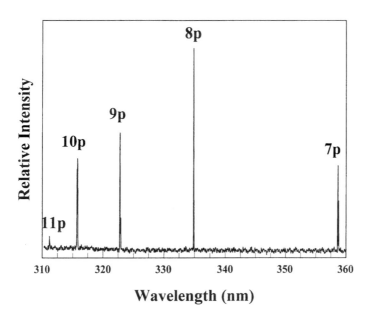

Figure 4   UV spectrum of the coherent emission produced by Rb vapor when excited by 620 nm, ~120 fs pulses. Each of the sharp features observed arises from four-wave parametric oscillation near the terms of the $np\,^2P_J \rightarrow 5s\,^2S_{1/2}\,(7 \leq n \leq 11)$ principal series of Rb. (Data from Ref. 40.)

Figure 5 Photograph of the violet emission at 420 nm produced when 2 Torr of Rb vapor (in the heat pipe at right) is irradiated with 1 mJ, 120 fs pulses at 770 nm produced by a Ti:sapphire laser. The pump radiation is incident on the heat pipe from the right and the coherent violet/UV radiation emerges from the left end of the heat pipe. The violet emission is the result of FWPO in the vicinity of the 6p → 5s transition of Rb and the divergence of the coherent radiation matches that of the incident (pump) laser radiation. (See also color insert.)

2. Other Metal Vapors

Although the highest conversion efficiencies for FWM have generally been achieved with the alkali metal vapors, the available wavelengths are typically constrained to narrow spectral regions in the near-UV. The desire to generate tunable VUV radiation prompts us to turn to other nonlinear media and the group II metal vapors, in particular. In the early 1970s, Kung et al. [41] demonstrated the ability of parametric processes in metal vapor/rare gas mixtures to produce coherent VUV. With phase-matched mixtures of Cd vapor and Ar, they generated radiation at 118, 152, and 177 nm and peak powers as high as 7 kW. Various SFG processes were invoked, the most efficient of which was tripling frequency-doubled Nd:YAG radiation (532 nm) to yield 177 nm photons.

Hodgson, Sorokin, and Wynne [42] subsequently reported that resonant enhancements of the FWM process of $\sim 10^4$ could be realized by tuning a dye laser near an atomic two photon-allowed transition. Utilizing two dye lasers, they were able to generate, in Sr vapor, coherent radiation over broad regions (1200–3500 cm^{-1}) in the VUV by the $2\omega_1 + \omega_2$ SFG process illustrated in Figure 3d. McKee and colleagues [43] extended this approach to the important spectral region encompassing the Lyman-α transition of hydrogen (121.6 nm). A photon flux $\geq 10^8$ photons/pulse was produced at the Lyman-α wavelength with two KrF laser (248 nm)-pumped dye lasers to excite Mg vapor, yielding a narrow-linewidth source that is well suited for the detection of H atoms.

Four-wave mixing in Hg vapor was reported in 1981 by Mahon and Tompkins [44,45], who showed that conversion efficiencies and peak powers up to 1% and 10 kW, respectively, were obtainable but the tuning range was initially limited to 2–8 Å wide windows in the 121.9–125.5 nm region. Hilbig and Wallenstein [46] subsequently extended the lower limit of the tuning range for four-wave sum mixing in Hg vapor to 109.0 nm. Tunability down to the LiF cutoff (~ 104.5 nm) was achieved by Herman and co-workers [47] by using the 6s 8s ^{1}S state of Hg for the two-photon intermediate resonance. Unprecedented conversion efficiencies as high as 5% were reported by Muller et al. [48] in the late 1980s for the generation of 130 nm photons in Hg vapor using a collimated beam geometry and optical field-nonlinear medium interaction lengths greater than 1 m. Recently, FWM in Hg vapor has successfully produced CW radiation at the hydrogen Lyman-α wavelength [49]. Because of the narrower linewidths offered by CW FWM, this process is attractive for high-resolution spectroscopy. First observed by Freeman et al. [50] in 1978, the CW generation of coherent VUV radiation by FWM places stringent requirements on the frequencies of the pump lasers. As illustrated qualitatively by Figure 3e, efficient energy conversion dictates that both ω_1 and $\omega_3 + \omega_2 + \omega_1$ be nearly resonant with the atomic medium *and* $\omega_1 + \omega_2$ must be on resonance.

3. The Rare Gases

Although the initial experiments in THG [51] were performed in the rare gases, it was the pioneering work of Harris, Young, and colleagues in the alkalis [35,36] that considerably heightened interest in FWM as a practical technique for wavelength conversion. However, the corrosive properties of the alkalis prompted a re-examination of the rare gases. In 1973, Kung and co-workers [52] reported an efficiency approaching 3% for the conversion of 355 nm energy into its third harmonic (118 nm) in phase-matched Xe/Ar gas mixtures. Subsequent work in the 1970s extended the limit of nonlinear frequency upconversion to 57 nm, well into the extreme ultraviolet (XUV) [33,53]. Most of these experiments were fixed-frequency THG or SFG performed in the rare gases, using either an excimer laser pump or the frequency-quadrupled output of an Nd:YAG or Nd:glass laser [33].

It was found that the rare gases are also suitable for generating intense Lyman-α radiation. Langer and colleagues [54] demonstrated conversion efficiencies up to $1.4 \cdot 10^{-4}$ in a mixture of Kr and Ar, by tripling the frequency-doubled output of a dye laser. Recently, Meyer and Faris [55] were able to produce 7 μJ of Lyman-α radiation in pulses having a peak power of 1.3 kW. They chose a phase-matched gas mixture of Kr and Ar as the nonlinear medium and generated the desired wavelength by DFG with an ArF (193 nm) excimer laser and a dye laser.

Broadly tunable coherent radiation in the VUV has also been produced by FWM in the rare gases and Kr and Xe, in particular [56]. Marangos et al. [57] demonstrated one practical scheme for generating tunable VUV radiation spanning the entire 120–200 nm spectral region. They locked the frequency of one dye laser onto a two-photon resonance of Kr to take advantage of the resonance enhancement, while tuning a second dye laser in order to achieve DFG of the form $2\omega_1 - \omega_2$ (illustrated schematically in Fig. 3(h)).

Finally, with the development of femtosecond lasers in the last decade, the generation of ultrafast pulses in the UV and VUV has become a reality. One scheme advanced by Glownia et al. [58] produced pulses in the VUV at 156 nm having energies as high as 60 μJ. An amplified KrF laser served as the source for ω_1, a femtosecond Ti:Al$_2$O$_3$ laser generated ω_2, and the VUV radiation was produced in Xe gas by the $2\omega_1 - \omega_2$ DFG process of Figure 3h. It is interesting to note that they also observed simultaneous FWPO at 148 nm, both in the presence and absence of pumping at ω_2. Subsequent work [59,60] using techniques similar to those introduced in Ref. 58 but with other rare gases and different excimer laser wavelengths, have produced femtosecond pulses elsewhere in the VUV. Another route was taken by Durfee and co-workers [61,62], who introduced a hollow core fiber to facilitate phase-matching and increase the interaction length. In these experiments, femtosecond pulses in the deep UV were produced from pulses at 800 nm and 400 nm, the fundamental and frequency-doubled outputs of a Ti:sapphire amplifier system, by the process $3\omega_1 = 2\omega_1 + 2\omega_1 - \omega_1$ (Fig. 3i). Experiments performed in rare gases both with and without the fiber produced $3\omega_1$ light with conversion efficiencies of 13% and 0.1%, respectively, illustrating the substantial improvement achieved with the hollow fiber. The UV light generated within the hollow fiber was subsequently phase-compensated to produce 8 fs pulses, centered at 270 nm, with energies of 1 μJ. Although this DFG experiment was nonresonant, the use of the hollow fiber might be expected to achieve similarly spectacular results for resonant FWM.

Before concluding this section, we would be remiss in not mentioning the recent experiments conducted at Stanford [63] in which VUV radiation was produced efficiently by adapting electromagnetically induced transparency (EIT) to enhance the FWM process resonantly, *without* the need to resort to phase-matching. Merriam and co-workers reported that 233 nm photons could be converted in lead vapor to a wavelength of 186 nm with a photon conversion efficiency approaching unity. The application of EIT to short-wavelength generation is in its infancy but, when coupled with existing nonlinear frequency upconversion processes, is likely in the future to yield sources of significantly improved efficiencies and beam quality.

IV. HIGH-ORDER HARMONIC GENERATION

Over the past 10–15 years, experimentalists have observed a remarkable variety of nonlinear effects that are a result of the optical field intensities and, hence, field strengths that can now be generated. The impact of the mode-locked Ti-doped sapphire ($Ti:Al_2O_3$) laser, in particular, on atomic and optical physics over the past several years rivals that of the tunable dye laser more than two decades ago. From its discovery in 1982, the $Ti:Al_2O_3$ laser has been developed rapidly and commercially available, mode-locked oscillator/regenerative amplifier systems are now capable of generating sub-20 fs pulses having energies beyond 1 mJ at kHz pulse repetition frequencies (PRF). Consequently, this laser offers access to a temporal and intensity ($>10^{16}$ W-cm^{-2}) realm likely to yield a number of surprising nonlinear optical effects. One such phenomenon is high-order harmonic generation (HG), the production of odd harmonics when an intense optical field interacts with an atom or molecule in the gas phase.

A. Experimental Observations

It has been known for more than three decades that odd harmonics of the optical field frequency are generated when an intense laser pulse interacts with an atom in the gas phase. In 1967, New and Ward [51] reported producing the third harmonic of the ruby laser in the rare gases. The ensuing years have witnessed steady improvements in the number of harmonics produced. Early experiments by Reintjes and colleagues [64] in He and Ne produced harmonics having wavelengths as short as 38 nm with ~30 ps pulses at 266 nm (fourth harmonic of Nd:YAG) and, following the introduction of the excimer laser, several studies [65,66] were carried out at 308 nm (XeCl) and 248 nm (KrF) with pulse widths as low as 15 ps [66] and, in one experiment, coherent radiation at the seventh harmonic of 248 nm (35.5 nm) was observed.

A turning point in the field occurred when McPherson et al. [67] detected radiation corresponding to the *17th harmonic* of 248 nm when Ne was exposed

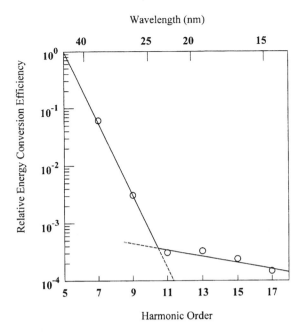

Figure 6 Relative efficiency for the conversion of 248 nm fundamental radiation to the odd harmonics $5 \leq N \leq 17$. The data shown are for Ne and fundamental intensities of $10^{15} - 10^{16}$ W-cm^{-2}. (Adapted from Ref. 67.)

to driver field intensities in the $10^{15}–10^{16}$ W-cm^{-2} range. More importantly, the rate of decline of the relative intensity of the harmonics slowed dramatically beyond the 11th harmonic (in Ne). Figure 6 shows the data reported in Ref. 67 in 1987. Similar results were observed for the other rare gases. Ferray and co-workers [68] and Li et al. [69] of Saclay subsequently confirmed the presence of higher harmonics (harmonic order $N \leq 33$ for Ne) for optical fields of longer wavelength as well ($\lambda = 1.06$ μm). An intriguing characteristic of harmonic spectra is a region, known as the plateau, in which the harmonic intensity is relatively insensitive to harmonic order N, followed by cutoff, a rapid decline in the harmonic response. For the *lowest* harmonics ($N \lesssim 7$), intensity also falls rapidly with increasing harmonic order (as had been observed previously), prompting Li et al. [69] to comment that "The first harmonics behave . . . according to theoretical predictions derived from lowest-order perturbation theory. . . ." However, "neither the extended plateau . . . nor the cutoff at high order can be explained by the traditional theory of nonlinear optics in [a] weak field. . . ." [69] Therefore, the experimental results demonstrate that HG cannot be described by conventional perturbation theory. A vivid illustration of the nonper-

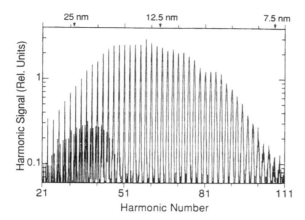

Figure 7 Harmonic spectrum reported by Macklin et al. for 125 fs, 15 mJ pulses from a Ti:Al$_2$O$_3$ laser ($\lambda \sim$ 800 nm) focused to an intensity of $\sim$1.3 × 10^{15} W-cm^{-2} in Ne gas. (From Ref. 70, reprinted by permission.)

turbative behavior of HG is provided by the data of Macklin et al. [70] illustrated in Figure 7. With 125 fs, 15 mJ pulses produced by a Ti:sapphire laser ($\lambda \cong$ 800 nm), they produced a spectrum in Ne extending out to the 109th harmonic. The authors noted that the roll-off in intensity for the lower harmonics is an artifact of the spectral response of the monochromator. In reality, the harmonic response declines slowly (with increasing harmonic order) up to approximately $n = 85$ and then falls rapidly. In addition to the harmonic spectra reported by several groups over the past decade [71–84], extensive measurements of the polarization characteristics [85–93], temporal coherence [94], and spatial properties [95] of the harmonic radiation have appeared in the literature.

Perhaps the primary motivation for the experimental and theoretical interest in HG is its potential for providing a table-top source of coherent XUV or soft X-ray radiation. Recent experiments [70–84] have produced coherent radiation at wavelengths at the edge of the UV to below 5 nm (soft x-ray region) from both atoms and molecules for instantaneous optical intensities from 10^{15} to $\sim$10^{18} W-cm^{-2}. With pulses as short as 5 fs produced by mode-locked Ti:sapphire lasers, harmonics having photon energies beyond 500 eV have been generated at kHz repetition frequencies. In the UV and VUV, average powers in the µW range can now be readily produced for individual harmonics, given the current capabilities of commercial Ti:Al$_2$O$_3$ oscillator/amplifier systems. This reliable source of coherent short wavelength radiation is a tremendous asset for experimentalists and promises future applications in fields such as biochemistry and medical diagnostics as the pulse energy per harmonic continues to rise with improvements in optical driver technology.

B. Theoretical Treatment of Harmonic Generation

Not only is HG of interest for the generation of short-wavelength coherent radiation but it also offers a unique tool for exploring the interaction between an atom and a strong optical field. Two competing views of the physical processes responsible for HG have emerged and will be briefly described here. For further detail, the reader is urged to consult the references cited.

Describing the harmonic response of the atom at high optical fields, and the presence and characteristics of the higher harmonics comprising the plateau and cutoff, in particular, requires (as noted earlier) nonperturbative theoretical methods. Several approaches have been explored [84,96–115]. By direct numerical integration of the time-dependent Schrödinger equation, Krause, Schafer and Kulander [98–100] calculated the response of a single atom to a strong field, assuming the single electron approximation to be valid. However, the computational capacity required by this approach and the desirability for insight into the physical processes underlying HG have spawned the pursuit of other nonperturbative theoretical models. Several groups [92,97,102–107] have solved Schrödinger's equation by approximating the atomic potential, the number of bound atomic states, or ground state depletion in different ways. In each instance, the description of the optical field and its interaction with an electron were treated classically.

The physical picture that best describes these calculations and the underlying assumptions is, perhaps, the two-step model proposed by Kulander et al. [100] and Corkum [101]. In this view, the atom is first ionized and the photoelectron is accelerated by the optical field. When the electric field reverses direction in the next half-cycle, the electron is accelerated back toward the parent atom where recombination occurs and radiation is produced. The role of the optical field in this semiclassical model, therefore, is the ionization of the atom and acceleration (or deceleration) of the free electron. For this reason, the common theme of semiclassical theoretical treatments of HG is the goal of determining the temporal behavior (and, hence, dipole moment) of the electron oscillating in the optical field. The semiclassical formulation proposed by Lewenstein et al. [102] has been widely applied to describing HG experimental results, including emission spectra, and polarization and harmonic linewidth data. It has been successful in describing the general characteristics of HG data acquired with both long (>100 fs) and short (<10 fs) optical pulse widths. A detailed description of semiclassical theory, the single atom response to an intense optical field and the propagation of the resulting spectrum through a plasma channel, can be found elsewhere [84,102–111].

Recently, a second theoretical view of HG was introduced [112–115]. This theory quantizes the optical field and makes use of the stationary solutions to Schrödinger's equation for an electron interacting with an optical field [116]. Known as quantized field (quantum) Volkov states, these stationary states are separated by integral multiples of the fundamental photon energy. An analytical expression for the matrix element for transitions between Volkov states was determined [112,114]. Consequently, the picture proposed is analogous to that for the generation of radiation by an atom: namely, transitions between stationary states. In this case, however, the eigenstates arise from the interaction of an electron with an optical field as opposed to an electron interacting with an ion core.

The second theoretical picture of HG, therefore, suggests that once an electron is released from the atom via ionization, the electron interacts with the fundamental photon field, resulting in the formation of quantum Volkov states. Optical transitions between Volkov states generate harmonic radiation, and the relative intensity of each harmonic is governed by the transition matrix element. The quantum theory also predicts that the velocity of the photoelectron is unchanged by Volkov-to-Volkov state transitions. Comparisons of theoretical predictions with a variety of experimental data show that the quantum theory also describes experiments well [112–115]. To differentiate between the semiclassical and quantum models of HG will require additional experiments over a range of parameters and, possibly, including multicolor driving fields or an x-ray prepulse to provide a background of energetic photoelectrons.

V. CONCLUSIONS

The two centuries following the discovery of the UV have witnessed enormous progress in both the development and application of new sources in the UV and VUV. Today, primary and nonlinear (secondary) sources of coherent radiation allow one to access any portion of the UV and VUV and to do so with pulses of at least a few kW of peak power. At several discrete wavelengths in this spectral region, peak powers beyond tens of MW are available from the F_2 (157 nm) and rare gas–halide lasers as well as high-order harmonic generation and wave-mixing processes in the rare gases and alkali vapors.

Challenges for future work continue to be the conception and demonstration of CW sources in the VUV as well as pulsed sources of higher efficiency and yet reduced cost. If the events of the past 40 years are any guide to the future, we can expect major advances to be made as the intricacies of the interaction of radiation with atoms, molecules, and artificial (synthetic) structures are probed to ever deeper levels.

ACKNOWLEDGMENTS

This work was supported by the U.S. Air Force Office of Scientific Research.

REFERENCES

1. JW Ritter. Ann Phys 7: 527, 1801.
2. JW Ritter. Ann Phys 12: 409, 1803.
3. R Tousey. Appl Opt 1: 679, 1962.
4. HG Heard. Nature 200: 667, 1963.
5. DA Leonard. Appl Phys Lett 7: 4, 1965.
6. CK Rhodes, ed. Excimer Lasers, 2nd ed. Berlin: Springer-Verlag, 1984.
7. RR Alfano, ed. The Supercontinuum Laser Source. New York: Springer-Verlag, 1989.
8. F Shimizu. Phys Rev Lett 19: 1097, 1967.
9. RR Alfano, SL Shapiro. Phys Rev Lett 24: 592, 1970.
10. RR Alfano, SL Shapiro. Phys Rev Lett 24: 584, 1970.
11. RR Alfano, SL Shapiro. Phys Rev Lett 24: 1217, 1970.
12. RL Fork, CV Shank, C Hirlimann, R Yen, WJ Tomlinson. Opt Lett 8: 1, 1983.
13. RL Fork, CV Shank, RT Yen, C Hirlimann, WJ Tomlinson. In: KB Eisenthal, RM Hochstrasser, W Kaiser, A Laubereau, eds. Picosecond Phenoma III. New York: Springer-Verlag, 1982, pp 10–13.
14. JH Glownia, G Arjavalingam, PP Sorokin, JE Rothenberg. Opt Lett 11: 79, 1985.
15. PB Corkum, C Rolland, T Srinivasan-Rao. Phys Rev Lett 57: 2268, 1986.
16. PB Corkum, C Rolland, T Srinivasan-Rao. In: GR Fleming, AE Siegman, eds. Ultrafast Phenomena V. New York: Springer-Verlag, 1986, pp 149–152.
17. JH Glownia, J Misewich, PP Sorokin. J Opt Soc Am B 3: 1573, 1986.
18. JH Glownia, J Misewich, PP Sorokin. In: GR Fleming, AE Siegman, eds. Ultrafast Phenomena V, New York: Springer-Verlag, 1986, pp 153–156.
19. JE Rothenberg. Opt Lett 17: 583, 1992.
20. JE Rothenberg. Opt Lett 17: 1340, 1992.
21. D Strickland, PB Corkum. J Opt Soc Am B 11: 492, 1994.
22. AA Zozulya, SA Diddams, AG Van Engen, TS Clement. Phys Rev Lett 82: 1430, 1999.
23. PB Corkum, C Rolland, IEEE J Quantum Electron 25: 2634, 1989.
24. TR Gosnell, AJ Taylor, DP Greene. Opt Lett 15: 130, 1990.
25. G Rodriguez, JP Roberts, AJ Taylor. Opt Lett 19: 1146, 1994.
26. H Nishioka, W Odajima, K Ueda, H Takuma, Opt Lett 20: 2505, 1995.
27. ETJ Nibbering, O Duhr, G Korn. Opt Lett 22: 1335, 1997.
28. O Duhr, ETJ Nibbering, G Korn, G Tempea, F Krausz. Opt Lett 24: 34, 1999.
29. K Sokolowski-Tinten, M Werner, P Zhou, D von der Linde. In: Conference on Laser and Electro-Optics. OSA Technical Digest. Washington, DC: Optical Society of America, pp 324–325, 1999.

30. JF Reintjes. Nonlinear Optical Parametric Processes in Liquids and Gases. New York: Academic Press, 1984.
31. CR Vidal. In: LF Mollenauer, JC White, eds. Tunable Lasers. Berlin: Springer-Verlag, pp 57–113, 1987.
32. R Hilbig, G Hilber, A Lago, B Wolff, R Wallenstein. In: P Yeh, ed. Nonlinear Optics and Applications. Bellingham, WA: SPIE, pp 48–55, 1986.
33. J Reintjes. Appl Opt 19: 3889, 1980.
34. GC Bjorklund. IEEE J Quantum Electron QE-11: 287, 1975.
35. SE Harris, RB Miles. Appl Phys Lett 19: 385, 1971.
36. JF Young, GC Bjorklund, AH Kung, RB Miles, SE Harris. Phys Rev Lett 27: 1551, 1971.
37. KM Leung, JF Ward, BJ Orr. Phys Rev A 9: 2440, 1974.
38. JR Taylor. Opt Comm 18: 504, 1976.
39. AV Smith, JF Ward. IEEE J Quantum Electron QE-17: 525, 1981.
40. HC Tran, PC John, J Gao, JG Eden. Opt Lett 23: 70, 1998.
41. AH Kung, JF Young, GC Bjorklund, SE Harris. Phys Rev Lett 29: 985, (1972).
42. RT Hodgson, PP Sorokin, JJ Wynne. Phys Rev Lett 32: 343, 1974.
43. TJ McKee, BP Stoicheff, SC Wallace. Opt Lett 3: 207, 1978.
44. FS Tomkins, R Mahon. Opt Lett 6: 179, 1981.
45. R Mahon, FS Tomkins. IEEE J Quantum Electron QE-18: 913, 1982.
46. R Hilbig, R Wallenstein. IEEE J Quantum Electron QE-19: 1759, 1983.
47. PR Herman, PE LaRocque, RH Lipson, W Jamroz, BP Stoicheff. Can J Phys 63: 1581, 1985.
48. CH Muller III, DD Lowenthal, MA DeFaccio, AV Smith. Opt Lett 13: 651, 1988.
49. KSE Eikema, J Walz, TW Hänsch. Phys Rev Lett 83: 3828, 1999.
50. RR Freeman, GC Bjorklund, NP Economou, PF Liao, JE Bjorkholm. Appl Phys Lett 33: 739, 1978.
51. GHC New, JF Ward. Phys Rev Lett 19: 556, 1967; see also JF Ward, GHC New, Phys Rev 185: 57, 1969.
52. AH Kung, JF Young, SE Harris. Appl Phys Lett 22: 301, 1973.
53. MHR Hutchinson, CC Ling, DJ Bradley. Opt Comm 18: 203, 1976.
54. H Langer, H Puell, H Rohr. Opt Comm 34: 137, 1980.
55. SA Meyer, GW Faris. Opt Lett 23: 204, 1998.
56. R Hilbig, R Wallenstein. Appl Opt 21: 913, 1982.
57. JP Marangos, N Shen, H Ma, MHR Hutchinson, JP Connerade. J Opt Soc Am B 7: 1254, 1990.
58. JH Glownia, DR Gnass, PP Sorokin. J Opt Soc Am B 11. 2427, 1994.
59. O Kittelmann, J Ringling, G Korn, A Nazarkin, IV Hertel. Opt Lett 21: 1159, 1996.
60. M Wittmann, MT Wick, O Steinkellner, P Farmanara, V Stert, W Radloff, G Korn, IV Hertel. Opt Comm 173: 323, 2000.
61. CG Durfee III, S Backus, MM Murnane, HC Kapteyn. Opt Lett 22: 1565, 1997.
62. CG Durfee III, S Backus, HC Kapteyn, MM Murnane. Opt Lett 24: 697, 1999.
63. AJ Merriam, SJ Sharpe, H Xia, DA Manuszak, GY Yin, SE Harris. IEEE J Select Topics Quantum Electron 5: 1502, 1999.
64. J Reintjes, C-Y She, RC Eckardt. IEEE J Quantum Electron QE-14: 581, 1978.
65. J Reintjes, LL Tankersley, R Christensen. Opt Comm 39: 334, 1981.

66. J Bokor, PH Bucksbaum, RR Freeman. Opt Lett 8: 217, 1983.
67. A McPherson, G Gibson, H Jara, U Johann, TS Luk, I McIntyre, K Boyer, CK Rhodes. J Opt Soc Am B 4: 595, 1987.
68. M Ferray, A L'Huillier, XF Li, LA Lompré, G Mainfray, C Manus. J Phys B 21: L31, 1988.
69. XF Li, A L'Huillier, M Ferray, LA Lompré, G Mainfray. Phys Rev A 39: 5751, 1989.
70. JJ Macklin, JD Kmetec, CL Gordon III. Phys Rev Lett 70: 766, 1993.
71. N Sarukura, K Hata, T Adachi, R Nodomi, M Watanabe, S Watanabe. Phys Rev A 43: 1669, 1991.
72. K Miyazaki, H Sakai. J Phys B 25: L83, 1992.
73. JK Crane, MD Perry, S Herman, RW Falcone. Opt Lett 17: 1256, 1992.
74. A L'Huillier, Ph Balcou. Phys Rev Lett 70: 774, 1993.
75. A L'Huillier, M Lewenstein, P Salières, Ph Balcou, M Yu Ivanov, J Larsson, CG Wahlström. Phys Rev A 48: R3433, 1993.
76. Y Nagata, K Midorikawa, M Obara, K Toyoda. Opt Lett 21: 15, 1996.
77. I Mercer, E Mevel, R Zerne, A L'Huillier, Ph Antoine, CG Wahlström. Phys Rev Lett 77: 1731, 1996.
78. Z Chang, A Rundquist, H Wang, MM Murnane, HC Kapteyn. Phys Rev Lett 79: 2967, 1997.
79. Ch Spielmann, NH Burnett, S Sartania, R Koppitsch, M Schnürer, C Kan, M Lenzner, P Wobrauschek, F Krausz. Science 278: 661, 1997.
80. Ch Spielmann, C Kan, NH Burnett, T Brabec, M Geissler, A Scrinzi, M Schnürer, F Krausz. IEEE J Select Topics Quantum Electron 4: 249, 1998.
81. M Schnürer, Ch Spielmann, P Wobrauschek, C Streli, NH Burnett, C Kan, K Ferencz, R Koppitsch, Z Cheng, T Brabec, F Krausz. Phys Rev Lett 80: 3236, 1998.
82. A Rundquist, CG Durfee III, Z Chang, C Herne, S Backus, MM Murnane, HC Kapteyn. Science 280: 1412, 1998.
83. C Altucci, R Bruzzese, C deLisio, M Nisoli, S Stagira, S DeSilvestri, O Svelto, A Boscolo, P Ceccherini, L Poletto, G Tondello, P Villoresi. Phys Rev A 61: 021801, 1999.
84. P Villoresi, P. Ceccherini, L Poletto, G Tondello, C Altucci, R Bruzzese, C deLisio, M Nisoli, S Stagira, G Cerullo, S DeSilvestri, O Svelto. Phys Rev Lett 85: 2494, 2000.
85. KS Budil, P Salières, A L'Huillier, T Ditmire, MD Perry. Phys Rev A 48: R3437, 1993.
86. NH Burnett, C Kan, PB Corkum. Phys Rev A 51: R3418, 1995.
87. FA Weihe, SK Dutta, G Korn, D Du, PH Bucksbaum, PL Shkolnikov. Phys Rev A 51: R3433, 1995.
88. K Miyazaki, H Takada. Phys Rev A 52: 3007, 1995.
89. FA Weihe, PH Bucksbaum. J Opt Soc Am B 13: 157, 1996.
90. M Yu Ivanov, T Brabec, N Burnett. Phys Rev A 54: 742, 1996.
91. M Kakehata, H Takada, H Yumoto, K Miyazaki. Phys Rev A 55: R861, 1997.
92. Ph Antoine, A L'Huillier, M Lewenstein, P Salières, B Carré. Phys Rev A 53: 1725, 1996.

93.   Ph Antoine, B Carré, A L'Huillier, M Lewenstein. Phys Rev A 55: 1314, 1997.
94.   M Bellini, C Lyngå, A Tozzi, MB Gaarde, TW Hänsch, A L'Huillier, C-G Wahl-ström. Phys Rev Lett 81: 297, 1998.
95.   P Salières, T Ditmire, KS Budil, MD Perry, A L'Huillier. J Phys B 27: L217, 1994, and refs. cited therein.
96.   JH Eberly, Q Su, J Javanainen. J Opt Soc Am B 6, 1289 1989; Phys Rev Lett 62: 881, 1989.
97.   W Becker, S Long, JK McIver. Phys Rev A 41: 4112, 1990.
98.   JL Krause, KJ Schafer, KC Kulander. Phys Rev Lett 68: 3535, 1992; Phys Rev A 45: 4998, 1992.
99.   KC Kulander, KJ Schafer, JL Krause. In: M Gavrila, ed. Atoms in Intense Radiation Fields. New York: Academic Press, 1992.
100. KC Kulander, KJ Schafer, JL Krause. In: B Piraux, A L'Huillier, K Rzążewski, eds. Super-Intense Laser-Atom Physics. NATO ASI Ser B, Vol 316. New York: Plenum, 1993.
101. PB Corkum. Phys Rev Lett 71: 1994, 1993.
102. M Lewenstein, Ph Balcou, M Yu Ivanov, A L'Huillier, PB Corkum. Phys Rev A 49: 2117, 1994.
103. W Becker, S Long, JK McIver. Phys Rev A 50: 1540, 1994.
104. SC Rae, K Burnett, J Cooper. Phys Rev A 50: 3438, 1994.
105. M Lewenstein, KC Kulander, KJ Schafer, P Bucksbaum. Phys Rev A 51: 1495, 1995.
106. M Lewenstein, P Salières, A L'Huillier. Phys Rev A 52: 4747, 1995.
107. W Becker, A Lohr, M Kleber, M Lewenstein. Phys Rev A 56: 645, 1997.
108. C Kan, NH Burnett, CE Capjack, R Rankin. Phys Rev Lett 79: 2971, 1997.
109. MB Gaarde, Ph Antoine, A L'Huillier, KJ Schafer, KC Kulander. Phys Rev A 57: 4553, 1998.
110. W Becker, S Long, JK McIver. Phys Rev A 50: 1540, 1994.
111. E Priori, G Cerullo, M Nisoli, S Stagira, S DeSilvestri, P Villoresi, L Poletto, P Ceccherini, C Altucci, R Bruzzese, C deLisio. Phys Rev A 61: 063801, 2000.
112. J Gao, F Shen, JG Eden. Phys Rev Lett 81: 1833, 1998.
113. J Gao, F Shen, JG Eden. J Phys B 32: 4153, 1999.
114. J Gao, F Shen, JG Eden. Phys Rev A 61: 043812, 2000.
115. J Gao, F Shen, JG Eden. Int J Mod Phys B 14: 889, 2000.
116. D-S Guo, GWF Drake. J Phys A 25: 3383, 1992.

11

All-Solid-State, Short-Pulse, Tunable, Ultraviolet Laser Sources Based on Ce^{3+}-Activated Fluoride Crystals

Zhenlin Liu
Japan Science and Technology Corporation, Kawasaki, Kanagawa, Japan

Nobuhiko Sarukura
Okazaki National Research Institutes, Myodaiji, Okazaki, Japan

Mark A. Dubinskii
Magnon, Inc., Reisterstown, Maryland

I. INTRODUCTION

Ultraviolet (UV) tunable lasers have become the most important tool in many fields of science and technology. Their most impressive applications include environmental sensing, engine combustion diagnostics, semiconductor processing, micromachining, optical communications, and medicinal and biological applications.

For example, the behavior of trace constituents in the earth's upper atmosphere, governed by chemical, dynamic, and radiative processes, is of particular importance for the overall balance of the stratosphere and mesosphere. In particular, ozone plays a dominant role by absorbing short-wavelength UV radiation that might damage living organisms and by maintaining the radiative budget equilibrium. Measurement of the total ozone column content and vertical profile by a ground-based UV spectrometer network or by satellite-borne systems remains the fundamental basis for global observations and trend analysis. Remote measurements of trace constituents using an active technique such as lidar have been made possible by the rapid development of powerful tunable laser sources. These have opened a new field in atmospheric spectroscopy by providing sources that can be tuned to characteristic spectral features of atmospheric constituents [1].

Recently, National Aeronautics and Space Administration (NASA) missions used tunable UV laser sources for atmospheric differential absorption lidar (DIAL) measurements from the airplanes to analyze the global distribution of O_3 radicals, which is directly relevant to the "ozone hole" and global climate formation problems [2]. In the airborne UV DIAL system, two frequency-doubled Nd:YAG lasers are used to pump two high-conversion-efficiency, frequency-doubled, tunable dye lasers. NASA used the UV lasers in the wavelength region from 289 to 311 nm.

Considerable effort has been devoted since the early 1970s to the development of new vibronic crystals for tunable solid-state lasers. However, almost all the efficient and successful vibronic materials, including Ti:sapphire [3], alexandrite [4] and, more recently, Cr^{3+}:LiCaAlF$_6$ (Cr:LiCAF) [5] and Cr^{3+}:LiSrAlF$_6$ (Cr:LiSAF) [6] are emitting in the same near-infrared (IR) spectral region (700–1000 nm).

A. UV Laser Systems Using Frequency Conversion

The existing commercially available tunable UV laser sources (comprising subsequent steps of nonlinear frequency conversion: doubling, tripling, and/or mixing of tunable radiation obtained from the primary traditional tunable visible or near infrared lasers) have been extremely complicated and expensive (also bulky, inefficient, inconvenient, and unreliable for airborne measurements in flight or aboard spacecraft applications [2] up to now. For example, UV-tunable lasers based on dye-laser tunability, in addition to a pumping source (usually an Ar^+ ion laser or an Nd:YAG laser with an attached nonlinear frequency-doubler), need a tunable dye laser in the orange–red region of the spectrum. The laser should be provided with a dye circulation system as well as a system for frequency-doubling of the dye laser radiation with a servotuning system (often called "Autotracker") to follow the wavelength changes, and also a system for separation of the visible and UV beams emanating from it. Solid-state tunable UV lasers using Ti:sapphire lasers with subsequent mixing of its tunable output with 532-nm pumping radiation should also be supplied by "Autotracker" [7].

Recent development of frequency-tripled flash lamp-pumped Q-switched Cr:LiSAF tunable laser holds promise for devising more reliable tunable solid-state UV lasers. Cr:LiSAF is of particular interest due to its strong absorption in the 670 nm wavelength region, allowing for potential pumping with AlGaInP/GaAs diode lasers. Harmonic generation in the UV region from 260 to 320 nm was demonstrated using an LiB_3O_5 (LBO)/β-BaB_2O_4 (BBO) sum-frequency mixing scheme. The maximum output was 6 mJ at the wavelength of 300 nm [8].

On the other hand, remarkable progress in high-power ultrashort pulse lasers has been made since chirped-pulse amplification [9] was applied to solid-state lasers such as Ti:sapphire lasers along with the generation of ultrashort pulses. These success, however, did not extend to UV-wavelength region, because the frequency upconversion limits the bandwidth, resulting in a long pulse width and low conversion efficiency. For higher peak power or high average power in the UV region, UV gain media with broad gain bandwidth are necessary.

B. Direct Generation of UV Coherent Light from Lasers

To provide tunable or ultrashort UV laser radiation in a reliable and efficient way, the version with the best prospects at the moment would be to use directly pumped solid-state UV active media, based on the electrically dipole-allowed interconfigurational 5d–4f transitions of rare-earth ions in wide band-gap fluoride crystals: $YLiF_4$ (YLF) [10], LaF_3 [11], and more recently $LuLiF_4$ (LLF) [12,13], $LiCaAlF_6$ (LiCAF) [14–16], and $LiSrAlF_6$ (LiSAF) [17,18]. In fact, this is the only version that also allows independent control of tunable radiation bandwidth, or when necessary, even provides multiwavelength UV output in one laser beam from the oscillator [12,13]. This version was proposed in 1977 from purely spectroscopic considerations [19] and then confirmed experimentally on Ce^{3+} ion in UV region by Ehrlich et al. [10,11] and on Nd^{3+} ion in the VUV region by Waynant [20].

As early as 1977, Yang and Deluca proposed a simple way of implementing a tunable laser capable of producing radiation directly in the UV and even VUV spectral ranges [19]. For this purpose, they proposed to use interconfiguration 5d–4f transitions of rare-earth ions in wide band-gap dielectric crystals. Because of the strong lattice interaction with 5d electrons, the fluorescence that results from 5d to 4f transitions of trivalent rare-earth ions in solid hosts is characterized by broad bandwidths and large Stokes shifts. Such fluorescence is particularly attractive for the development of tunable lasers. Powder samples of $Ce^{3+}:LaF_3$ and $Ce^{3+}:LuF_3$ were excited by the 253.7 nm radiation transmitted through a narrow-band interference filter inserted in front of an Hg lamp source. Broadband UV fluorescence was reported for $Ce^{3+}:LaF_3$ (276–312 nm) and $Ce^{3+}:LuF_3$ (288–322 nm). The fluorescence quantum yields account for the fact that not all of the atoms raised to the pump bands subsequently decay to the upper laser level. Some of these atoms can in fact decay from the pump bands straight back to the ground state or perhaps to other levels, which are not useful. The pump quantum efficiency or fluorescence quantum yields $\eta_q(\lambda)$ are defined as the ratio of the number of atoms that decay to the upper laser level to the number of atoms raised to the pump band by a monochromatic pump at wavelength λ. The fluorescence quantum yields of LaF_3:1%Ce^{3+} and LuF_3:0.1%Ce^{3+} are 0.9 and 0.82, respectively. Estimates of the threshold power for lasing action suggested that a laser system tunable from 276 to 322 nm is feasible with noble-gas–halide lasers as pumping sources [19].

After that, Ce^{3+}:$Y_3Al_5O_{12}$ (YAG) was investigated as a model system for a 5d–4f solid-state tunable laser [21]. This system was chosen since YAG had been extensively studied as a laser host and good-quality crystals were readily available. Despite providing apparently adequate conditions to achieve stimulated emission, it was unable to detect laser action in Ce^{3+}:YAG. It was found that there was strong excited-state absorption (ESA) in this material at the wavelengths of its fluorescence. The ESA was sufficiently strong to quench completely any possible laser action. The crystal showed a net optical loss instead of optical gain at the wavelength of the fluorescence transition. This self-absorption may explain the failure of all attempts to obtain stimulated emission in this material.

A laser of the type, which has a 5d–4f transition, was originally implemented with Ce^{3+}:$YLiF_4$ (Ce:YLF) as a laser medium [10]. It should be noted that Ce^{3+} ions are the most promising activators for the UV spectral range. However, in spite of a large number of studied Ce-activated materials [22,23], the investigations performed before 1992 revealed only two laser-active media [10,11].

In 1979, Ehrlich et al. reported the first observation of stimulated emission from a 5d–4f transition in triply ionized rare earth-doped crystal Ce:YLF, optically pumped at 249 nm, and emitted at 325.5 nm [10]. Since the Ce^{3+} ion has only one electron in the 4f state, the impurity energy levels of Ce-doped crystals are particularly simple. The ground state is split into a $^2F_{5/2}$ and a $^2F_{7/2}$ levels by the spin-orbit interaction. The first excited state is a 5d state, which interacts strongly with the host lattice because of the large spatial extent of the 5d wavefunction. Thus the crystal-field interaction dominates over the spin-orbit interaction and the 5d state splits into four levels as a result of the S_4 site symmetry for the rare-earth ion in YLF.

The broad absorption bands that peak at 195, 205, 240, and 290 nm result from transitions from the 4f ground state to the crystal-field split 5d levels of the Ce^{3+} ion. The fluorescence spectrum has two peaks, the result of transitions from the lowest 5d level to the two spin-orbit ground states $4f(^2F_{5/2})$ and $4f(^2F_{7/2})$. The 40 ns radiative lifetime of the 5d level results from the electric-dipole-allowed character of the 5d $\rightarrow$ 4f transition. The potential tuning range of the Ce:YLF laser estimated from the half-power points of the fluorescence spectrum is from 305 to 335 nm. The maximum output energy observed was ~1 µJ in a pulse width of 35 ns for an absorbed pump energy of 300 µJ.

However, the operation of the Ce:YLF laser was hampered by several poor performance characteristics. These include an early onset of saturation and rolloff in the above-threshold gain and power output as well as a drop in the output for pulse repetition rates above 0.5 Hz. It has been shown that an excited-state absorption of the UV pump light is responsible for a photoionization of the Ce^{3+} ions, which in turn leads to the formation of transient and stable color centers. The color centers have a deleterious effect on the lasing characteristics of Ce: YLF since they absorb at the cerium emission wavelengths. The growth and relaxation of these centers influence the gain saturation and pump rate limitation of the Ce: YLF laser [24,25]. This experiment is of historic significance, but it is short of practical use due to the existence of solarization.

In 1980, the same group mentioned above reported the operation of an optically pumped Ce^{3+}:LaF_3 laser [11]. Because of the rapid internal relaxation to the lowest 5d state, the fluorescence spectrum was not noticeably different for ArF (193 nm), KrF (248 nm), or frequency-doubled Ar^+ ion laser (257 nm) pumping. The fluorescence lifetime was likewise identical for 248 nm or 193 nm excitation. For the 0.05%-doped crystal, the lifetime was 18 ± 2 ns. The approximate potential tuning range is from 275 to 315 nm. The output of a small commercial excimer laser, producing 40 mJ at 248 nm or 10 mJ at 193 nm in a 25 ns full width at half maximum (FWHM) pulse, was used for optical pumping. The primary difficulties encountered (i.e., low output power and high threshold) can probably be ascribed to initial difficulties in crystal growth.

Subsequent studies showed that the attempts to find laser-active media among Ce-activated materials failed because of absorption from the excited state of the 5d configuration of Ce^{3+} ions in Ce^{3+}:YAG [26,27] and Ce^{3+}:CaF_2 [28], formation of stable or transient color centers in Ce-activated samples [23,24,28], and other complicated processes occurring in such media under the high-power UV pumping.

The spectrally broad vibronic emission bands in impurity-doped solids serve as the basis for wavelength-tunable laser operation in these materials. Because of the broad emission and absorption bands, however, these materials are susceptible to ESA, which can significantly reduce the performance characteristics of the laser materials. The ESA is a two-step process, by which the first photon absorbed promotes an electron from the 4f ground state to the lowest 5d state of the trivalent cerium ion. Within the lifetime of this excited 5d state, a second photon is absorbed, which then photoionizes the ion by promoting that electron to the conduction band. The free electron subsequently traps out at an electron acceptor site, forming a stable color center. These color centers are absorptive at the wavelengths for stimulated emission of the trivalent cerium ions, and hence they serve as a quenching mechanism for laser gain in this crystal. The color centers produced are photochromic in that they can be optically bleached [23]. Over a long period of time, difficulties in overcoming these problems, which are inherent in well-known materials used for producing UV light

[10,11], made investigators believe that this scheme of UV lasers was of little promise.

However, recent investigations showed that by an appropriate choice of activator–matrix complexes and active medium–pump source combinations, one can create efficient tunable lasers using d–f transitions in Ce^{3+} ion in the UV spectral range [12–16] and in Nd^{3+} ion in the VUV range [29]. Furthermore, such lasers proved to be stable under the pumping.

In 1992, Dubinskii et al. reported UV laser medium, $Ce^{3+}:LuLiF_4$ (Ce: LLF) crystal, which can be pumped by KrF excimer laser [12,13]. This crystal has almost the same optical properties as Ce:YLF. But Ce:LLF has a smaller solarization effect, and it shows more promise in practical use. In 1993, the same group reported $Ce^{3+}:LiCaAlF_6$ (Ce:LiCAF) crystal, which can be pumped by the fourth harmonic of Nd:YAG lasers. No solarization effect of this crystal was observed [14–16].

$Ce^{3+}:LiSrAlF_6$ (Ce:LiSAF) crystal was reported in 1994, and it also can be pumped by the fourth harmonic of Nd:YAG lasers [17,18]. It has similar laser properties to Ce:LiCAF crystal.

Since these new Ce-doped crystals were reported, studies on the solid-state ultraviolet tunable lasers have become popular again. All these new Ce-doped crystals have a broad gain-bandwidth in the ultraviolet region, which is especially attractive for ultrashort-pulse generation and amplification. Figure 1 shows the tunable wavelength regions of the five known Ce-doped laser crystals.

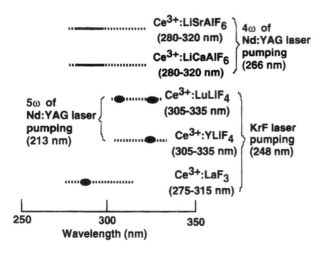

Figure 1 Various tunable lasers in ultraviolet region. Solid lines and dots indicate the confirmed tunable wavelength region; dotted lines show potential tunable wavelength region.

II. NEW UV LASER MATERIALS: Ce:LLF AND Ce:LiCAF CRYSTALS

A. Basic Properties of Ce:LLF Laser Medium

Ce:LLF is a tunable solid-state laser material in the UV region that was first reported in 1992 [12,13]. The choice of this material for the experiments in UV is more reasonable because of its structural and chemical similarity to comprehensively studied and commercially grown YLF single crystal. The LLF crystals belong to the scheelite structural type. Similar to all rare-earth ions, Ce^{3+} ions take part in activation, substituting for Lu^{3+} ions in the position with the point group S_4.

Ce:LLF has a potential tuning region of 305–340 nm, so it is especially attractive for use in spectroscopy of wide band-gap semiconductors for blue laser diodes, such as GaN [30]. The fluorescence spectrum of Ce:LLF crystal has two peaks at 311 nm and 328 nm (Fig. 2). KrF excimer laser (248-nm) can be used as the pumping source for Ce:LLF crystal. The fluorescence lifetime was found to be about 40 ns.

B. Gain Spectrum of Ce:LLF crystal

A Ce:LLF single crystal was grown from a carbon crucible using the Bridgman-Stockbarger method in a fluorinated atmosphere by Dubinskii et al. [31,32]. A

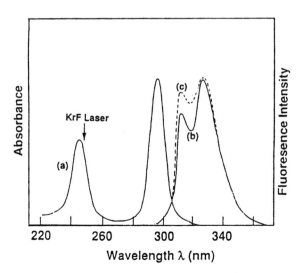

Figure 2   (a) π-Polarized absorption and normalized fluorescence (b: π-polarization, c: σ-polarization) spectra of Ce:LLF single crystal.

seeded crucible was used to obtain oriented boule. The seed orientation was such that the c-axis was perpendicular to the cylindrical axis of the crucible. The corresponding orientation of c-axis of grown crystal was checked by observing the conoscopic picture of the boule. The sample obtained contains about 0.5 at.% Ce^{3+} ions.

The sample for the amplification experiment (1 cm long and 5 mm diameter) was cut cylindrically. A flat window was polished on the side of the cylinder parallel to its axis. The end surfaces were polished parallel within 5 arcmin. The sample was oriented in such a manner that the c-axis of the crystal was parallel to the side window and perpendicular to the axis of the cylinder. The pump and probe experiment was done in order to measure the gain spectrum for the 1 cm long Ce:LLF crystal under the randomly polarized KrF excimer laser excitation with side-pumping geometry (Fig. 3).

The probe beam was the second harmonic of a tunable mode-locked DCM (4-Dicyanmethylene-2-methyl-6-(p-dimethylaminostyryl)-4H-pyran) dye laser, which was synchronously pumped by the second harmonic of a cw mode-locked Nd:YAG laser. The tuning range of the DCM dye laser is from 645 to 680 nm. The second harmonic of the DCM laser was generated using an LBO nonlinear crystal in type I phase-matching condition. When the wavelength of the DCM dye laser was changed, it was necessary to rotate the LBO crystal to satisfy the phase-matching condition. The pumping source to excite Ce:LLF crystal was a conventional randomly polarized KrF excimer laser. The pumping pulses were focused onto the side window of the Ce:LLF crystal by a cylindrical lens.

A biplanar phototube was used to detect the signal. A Tektronix 7934 storage oscilloscope was used to display the gain traces. The data were taken for two times (with and without probe beam) and saved on the oscilloscope screen. Then the amplified pulse train and the fluorescence pulse train could be seen at

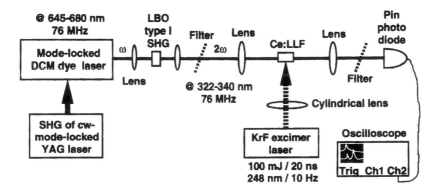

Figure 3 Gain spectrum measurement using the second harmonic of a mode-locked DCM dye laser as a probe.

the same time. The difference between the upper signal and the lower signal will be the net gain. With the use of the mode-locked pulse train as a probe beam, the fluorescence background is easily identified.

The probe beam was σ-polarized (the polarization is perpendicular to the optical axis or c axis of the Ce:LLF crystal). In order to check the gain-polarization dependence, the polarization direction of the DCM dye laser was changed by rotating a half-wave plate. No noticeable anisotropy of the gain was observed within the accuracy of the measurements (at least for 325 nm probe radiation). The gain was demonstrated from 323 nm to 335 nm (Fig. 4), due to the available probe laser tunability. This gain bandwidth is large enough for amplification of tunable femtosecond pulses.

C. Small-Signal Gain and Saturation Fluence of Ce:LLF Crystal

Small-signal gain and saturation fluence are very important parameters for designing lasers [31]. The small-signal gain and saturation fluence of Ce:LLF were evaluated using the second harmonic (325 nm) of a nanosecond DCM-dye laser as a probe as shown in Figure 5. A KrF excimer laser was used to pump the Ce:LLF from its side window. The frequency conversion scheme used an LBO nonlinear crystal to double the dye laser output. The DCM dye laser was pumped by the second harmonic of a Q-switched Nd:YAG laser. The relative timing of the probe dye laser and the pumping excimer laser were controlled by a synchronizer to obtain the largest possible gain. The single-pass gain in the small-signal region ($\sim$1 mJ/cm^2) was over 6-dB/cm (4.3 times) with 0.5-J/cm^2 pumping flux (Fig. 6).

The 10 ns pulsewidth of the probe light was short enough in comparison with a 40 ns fluorescence lifetime of Ce:LLF. Therefore, the evaluation of the

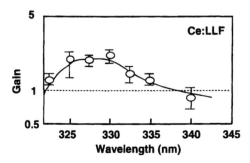

Figure 4 Gain spectrum of Ce:LLF pumped by randomly polarized KrF laser using the second harmonic of a picosecond CW mode-locked DCM dye laser as a σ-polarized probe (pumping fluence $\sim$ 0.1 J/cm^2).

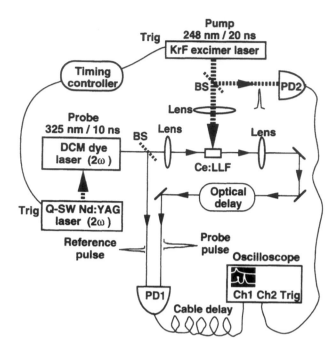

Figure 5 Small-signal gain and saturation fluence were evaluated using the second harmonic (325 nm) of a nanosecond DCM dye laser as a probe.

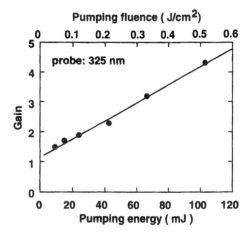

Figure 6 Gain dependence of Ce:LLF on pumping power at 325 nm. The single-pass gain in the small-signal region was over 6 dB/cm (4.3 times) with 0.5 J/cm^2 excitation fluence.

saturation fluence and emission cross section is possible under the assumption of the Frantz-Nodvik relation modeled for slow-decay gain medium [33].

$$G = E_s \log_e\{1 + G_0[\exp(E_i/E_s) - 1]\}/E_i$$

where E_i is the input fluence, E_s is the saturation fluence, G_0 is the single-pass small-signal gain, and G is the gain corresponding to E_i. The saturation fluence of Ce:LLF, assuming the Frantz-Nodvik relation, was estimated to be 50 mJ/cm^2 at 325 nm (Fig. 7), which is about two orders of magnitude higher than that of organic dyes ($\sim$1 mJ/cm^2). The emission cross section estimated from this saturation fluence was 10^{-17} cm^2 by the relation of $E_s = h\nu/\sigma$. These results indicate a high potential for Ce:LLF as a power amplification medium.

D. Basic Properties of Ce:LiCAF Laser Medium

In contrast to excimer laser-pumped UV solid-state laser media, such as Ce:LLF and Ce:YLF, the Ce:LiCAF crystal is the first known tunable UV laser directly pumped by the fourth harmonic of a standard Nd:YAG laser [14–16].

The samples of Ce:LiCAF were also grown in a fluorinated atmosphere using the Bridgeman-Stockbarger technique from carbon crucibles by Dubinskii et al. The crystals grown had the colquiriite structure and the space group P31c. The nonpolarized absorption spectrum of a 2.3 mm thick Ce:LiCAF sample, containing about 0.1% of Ce^{3+} ions, is shown in Figure 8a. Due to the first 4f–5d absorption peak with a half width of 3000 cm^{-1} centered at about 37,000

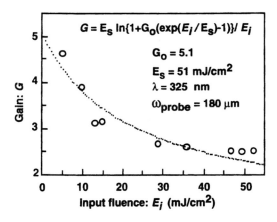

Figure 7 Ce:LLF gain dependence on input fluence of 10-ns, 325-nm probe pulses. The dotted line indicates the gain saturation curve assuming the Frantz-Nodvik relation, which was fitted by the least square method. The saturation fluence and the small signal gain were fitted to be 50 mJ/cm^2 and 5.1 (pumping fluence $\sim$ 0.5 J/cm^2).

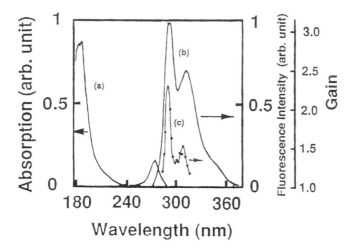

Figure 8 (a) Nonpolarized absorption spectrum of a Ce:LiCAF sample (0.1 at.%; 2.3 mm in length). (b) Nonpolarized fluorescence spectrum of Ce:LiCAF (0.9 at.%). (c) Single-pass small signal gain dependence on the probe beam wavelength for a Ce:LiCAF sample (0.9 at.%; 2.3 mm in length). (From Ref. 34.)

cm^{-1}, Ce:LiCAF can be pumped by the fourth harmonics of various commercially available Nd-lasers (e.g., YAG, YAP, YLF, and GSGG). The Ce:LiCAF fluorescence spectrum (Fig. 8b) displays the nearly two-humped shape characteristic of Ce^{3+} ions in most known hosts, due to the allowed 5d–4f transitions terminating at the $^2F_{7/2}$ and $^2F_{5/2}$ components of the spin–orbit split ground term. Ce:LiCAF has a potential tuning range from 280 to 320 nm. Ce:LiCAF has sufficiently higher effective gain cross section (6.0×10^{-18} cm^2) compared with Ti:sapphire [3], which is favorable for designing laser oscillators. Ce:LiCAF has also larger saturation fluence (115 mJ/cm^2) [18] than organic dyes, which is attractive for designing power amplifiers. The fluorescence lifetime was reported to be 30 ns, which is too short for constructing regenerative amplifiers. However, it is long enough for designing multipass amplifiers. The nonpolarized small-signal single-pass gain dependence on the probe beam wavelength for a 2.3-mm thick Ce:LiCAF sample with Ce^{3+} ion concentration of 0.9 at.% is shown in Figure 8c. The sample optical axis orientation with respect to the direction of observation in this experiment was the same as for obtaining the fluorescence spectra (Fig. 8b). The pump fluence of 0.3 J/cm^2 was used to obtain the above dependence. The probe fluence was less than 1 mJ/cm^2. From a comparison of Figure 8b and 8c, it is evident that the small-signal gain curve shape is similar to the fluorescence spectrum observed. This means that induced absorption is small in the gain spectral region. The small-signal gain reaches a value of 2.5 in the vicinity of the main fluorescence peak [34].

III. GENERATION OF SUBNANOSECOND PULSES FROM Ce³⁺-DOPED FLUORIDE LASERS

Direct ultrashort pulse generation has not been obtained from ultraviolet solid-state lasers as it has for near-infrared tunable laser materials like Ti:sapphire and Cr:LiSAF crystals. This is due to the difficulty in obtaining continuous-wave (cw) UV laser operation, which is required for Kerr lens mode-locking (KLM) schemes utilizing spatial or temporal Kerr type nonlinearity [35,36].

A. Short Pulse Generation with Simple Laser Scheme

A general technique for subnanosecond pulse generation from laser-pumped dye laser has been described [37]. The technique makes use of the resonator transients, are in the form of damped relaxation oscillation or "spiking." These resonator transients are the consequences of the interaction between the excess population inversion and the photons in the cavity. Their durations can be controlled by proper choices of photon cavity decay time and pumping level. The decay lifetime (photon lifetime) t_c of a cavity mode is defined as

$$dE/dt = -E/t_c \tag{1}$$

where E is the energy stored in the mode. If the fractional (intensity) loss per pass is L and the length of the resonator is l_c, then the fractional loss per unit time is cL/nl_c; therefore,

$$t_c = -nl_c/cL \tag{2}$$

For the case of a resonator with mirrors' reflectivities R_1 and R_2,

$$t_c \approx \frac{nl_c}{c(1 - \sqrt{R_1 R_2})} \tag{3}$$

The quality factor of the resonator is defined universally as:

$$Q = \omega E/P = -\omega E/(dE/dt) \tag{4}$$

where E is the stored energy, ω is the resonant frequency, and $P = -dE/dt$ is the power dissipated. By comparing Eqs. (4) and (1), we obtain

$$Q = \omega t_c \tag{5}$$

Let us consider a laser resonator with two mirrors pumped by a Q-switched laser. Normally the pulse duration from the Q-switched laser is approximately 10 ns. During the pump pulse, inversion will build up in the laser active medium; after threshold inversion is reached, a delayed laser pulse will develop. The proposed method of short-pulse generation is based on the fact that this pulse may be shorter than the pump pulse, due to the transient characteristics of the laser oscillator.

The transient behavior of such a laser can be understood by treating numerical examples with a computer. The laser is adequately described by the well-known rate equations for four-level systems as follows:

$$dn/dt = W(t) - Bnq \qquad (6)$$

$$dq/dt = Bnq - q/t_c \qquad (7)$$

where n is the population inversion ($n_2 - n_1$) and nearly equal to the upper state population n_2, $W(t)$ is the pumping rate, B is the Einstein B coefficient for stimulated emission, q is the total number of photons in the cavity, and t_c is the resonator (photon) lifetime.

The pumping rate was assumed to have the form of a gaussian pulse with full half-width T_1 and integrated photon number N. A large reduction of the overall pulse duration demands a high value of T_1/t_c. The ratio T_1/t_c can easily be made large, since the laser resonator needs no switching elements and therefore need not be longer than the material itself. Furthermore, it is possible to use a laser resonator of low mirror reflectivity, whose resonator lifetime is not markedly longer than the single-pass transit time l_c/c. To obtain short single pulses, it is necessary to use resonator lifetimes that are small compared to the pump duration in combination with controlling the level of pumping [38].

B. Subnanosecond Ce:LLF Laser

1. Subnanosecond Ce:LLF Laser Pumped by KrF Excimer Laser

Based on the technique mentioned above, a subnanosecond Ce:LLF laser was made with a low-Q, short-cavity configuration [39]. The experimental setup is simple. A Ce:LLF sample (2.5 cm long, 5 mm diameter) was cut cylindrically. A flat window was polished on the side of the cylinder parallel to its axis. The sample was oriented so that the c-axis of the crystal was parallel to the side window and perpendicular to the axis of the cylinder. No dielectric coatings were deposited on the polished surfaces.

The resonator was formed by a 30%-transmission flat output coupler and the surface reflection of the crystal ($\sim 5\%$ reflectivity) as shown in Figure 9 ($t_c \sim 0.2$ ns $\ll \tau_p = 10$ ns). The cavity spacing is 3 cm. KrF laser pumping pulses of 40 mJ, 10 ns, 1 Hz were focused on the side window by a 20 cm focal-length cylindrical lens. A single 3 μJ pulse from the side of 30% transmission output coupler was obtained at the sacrifice of the extracted energy. The pulse duration observed by a streak camera was 950 ps. The pulse did not have any satellite pulses because the pulse duration was longer than the cavity round trip time (200 ps).

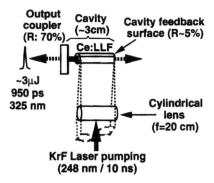

Figure 9 Low-Q, short-cavity Ce:LLF laser. The cavity is formed by a 30% transmission flat output coupler and the surface reflection of the crystal. The cavity spacing is 3 cm.

At higher pumping levels, the pulse develops double or triple peaks in the form of damped relaxation oscillation. The pumping energy dependence of the output performance has been investigated. As shown in Figure 10, single-pulse operation of 1 ns duration was observed for a sufficiently wide pumping energy region from the lasing threshold up to 60 mJ pumping energy. Above this pump-

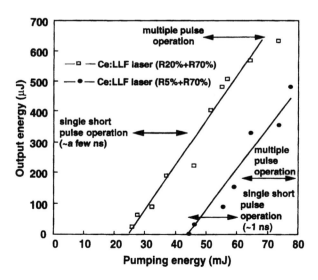

Figure 10 Pumping-energy dependence of output performance for two Ce:LLF lasers with a 70% reflection end mirror (squares: 20%-reflection output coupler; circles: 5% crystal-surface reflection coupler). Single-pulse operation was achieved in wide pumping-energy region.

ing level, multiple-pulse operation was observed. For the higher-Q cavity configuration (the cavity is formed by a 70% reflection end mirror and a 20% reflection output coupler), a wide single-pulse operation region was obtained with a pulse duration of few nanoseconds. In these experiments, the output energy increases linearly with the pumping energy.

Figure 11 shows a typical single-pulse operation under the appropriate pumping energy (Fig. 11a) and a typical multiple pulse operation under higher pumping energy (Fig. 11b). The laser pulse was observed using a Tektronix 7934 storage oscilloscope with a biplanar phototube as the detector.

A 3 cm cavity-length oscillator composed of a flat high-reflection mirror and a 70% transmission flat output coupler was also tested. KrF laser pumping pulses 50 of mJ, 10 ns, 1 Hz were focused on the side window by a 20 cm focal-length cylindrical lens up to 200 mJ/cm^2 pumping fluence. A 2 mJ output at 325 nm was obtained. The obtained pulse had multiple temporal spikes.

From these results, it can be concluded that the conditions for single-pulse operation are not critical, either to the pumping energy or to the coupling constant. In this way, a single short pulse can be easily generated from a low-Q, short-cavity Ce:LLF laser pumped by a standard 10 ns KrF laser.

2. Subnanosecond Ce:LLF Laser Pumped by the Fifth Harmonic of Nd:YAG Laser

Figure 12 indicates the absorption spectrum (nonpolarized) of a Ce:LLF sample containing 1 at.% of Ce^{3+} ions (in the melt). Considering the Ce:LLF absorption, one can see that while this material does not noticeably absorb at 266 nm, it will absorb pumping radiation at the fifth harmonic of Nd:YAG laser wavelength

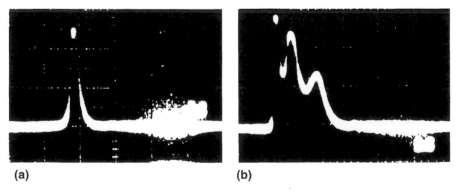

(a)   **(b)**

Figure 11   (a) Typical single pulse operation under appropriate pumping energy. (b) Typical multiple pulse operation under higher pumping energy.

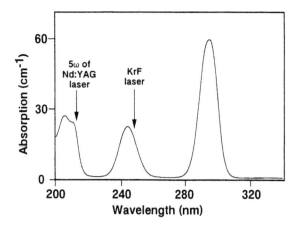

Figure 12 Nonpolarized absorption spectrum of Ce^{3+} ions in $LiLuF_4$ single crystal (Cerium concentration in melt: 1 atomic %).

(213 nm) almost as efficiently as at the KrF-excimer pumping wavelength of 248 nm [40]. With the recent development of new nonlinear borate crystal materials such as $CsLiB_6O_{10}$ [41] and $Li_2B_4O_7$ [42], the fifth harmonic of Nd:YAG lasers has become usable and practical due to the significantly improved efficiency and stability, close to those typical of the fourth harmonic of Nd:YAG laser.

From Fig. 12, it is evident that even the off-peak absorption at 213 nm is strong enough to consider the crystal suitable for an all-solid-state approach using the fifth harmonic of an Nd:YAG laser for side pumping. To prove this, the Ce:LLF was tested under side pumping conditions by the fifth harmonic of an Nd:YAG laser. The 213 nm, 25 mJ, 5 ns, horizontally polarized pulses (σ-pumping) for pumping Ce:LLF laser were stably obtained in $Li_2B_4O_7$ crystal using the type-I sum frequency generation process between 1064 nm and 266 nm pulses from a conventional Q-switched (Q-sw) Nd:YAG laser [42]. The optical layout for a short-cavity Ce:LLF laser is shown schematically in Figure 13. A Ce:LLF single crystal with 0.2 at.% doping level was cut to form a cylinder (5 mm diameter and 25 mm length) with a flat polished window on the side. The sample was oriented so that its optical axis was parallel to the side window and perpendicular to the cylinder longitudinal axis. No antireflection coatings were applied to the rod ends for this experiment. To increase the efficiency of side pumping, the effective pumping penetration depth was geometrically reduced using the novel tilted-incidence-angle side-pumping scheme instead of conventional normal-incidence side pumping. Using a 20 cm focal length cylindrical lens that was also tilted to be parallel to the side window of the laser crystal, the pumping pulse was focused down to a 1.2×0.15 cm^2 line-shaped area to provide the 140 mJ/

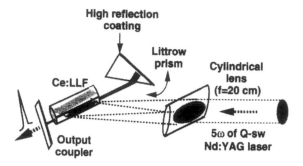

Figure 13 Short-cavity, tilted-incidence-angle (approximately 70 degrees) side-pumping, tunable Ce:LLF laser layout. The pumping pulse was focused by a 20 cm focal length cylindrical lens tilted so that the cylinder axis is parallel to the side window of the laser crystal.

cm^2 pumping fluence at nearly 70 degree incidence angle. Observed efficient penetration depth was estimated to be ~1 mm. In a conventional normal-incidence side pumping scheme, the pumping penetration depth was over 3 mm for the Ce:LLF crystal used here, which is too deep to obtain a good output beam pattern. Obvious advantages of the above-mentioned tilted-incidence-angle, side-pumping scheme are very simple: focused geometry, reduced pumping fluence at the rod surface, reduced risk of damaging optics, and better matching of the excited rod volume and the laser cavity mode volumes (similar to the coaxial pumping scheme). This better matching resulted in a better output beam quality. The low-Q, short-cavity Ce:LLF laser consisted of a Littrow prism used as an end mirror, and a low-reflection (20%) flat output coupler. The total length of the laser cavity was 6 cm. Typical spectrally and temporally resolved streak-camera images of the Ce:LLF laser output pulse are shown in Figure 14. Using the 213 nm, 5 ns, 16 mJ pumping pulses, we were able to obtain 880 ps, 77 µJ, σ-polarized, and satellite-free reproducible pulses at 309 nm. It is worth mentioning here that pumping at 213 nm does not cause noticeable laser rod solarization or laser performance degradation during several hours of continuous operation at a 10 Hz repetition rate.

C.   Short Pulse Generation from Ce:LiCAF Laser

1.   Nanosecond Pumping

Ce:LiCAF is a tunable UV laser medium that can be directly pumped by the fourth harmonic of a standard Nd:YAG laser. A 1% doped, 5 mm cubic Ce:LiCAF sample was used without any dielectric coatings on the polished surfaces

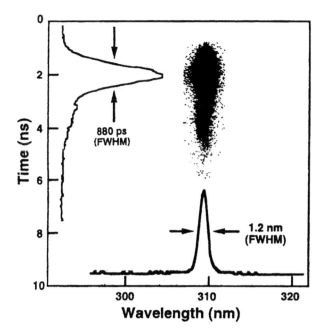

Figure 14 Temporally and spectrally resolved streak camera image of a UV short pulse from the low-Q, short-cavity Ce:LLF laser.

[43]. The experimental setup of the subnanosecond Ce:LiCAF laser is shown in Figure 15. The 1.5 cm long laser cavity was formed by a flat high-reflection mirror and an 80% transmission flat output coupler. The 20 mJ, 10 ns, 1 Hz, 266 nm, horizontally polarized pumping pulses (the fourth harmonic of a conventional 10 ns Q-sw Nd:YAG laser) were focused longitudinally from the high-reflection

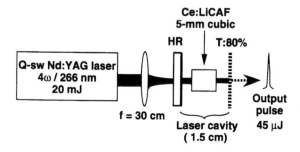

Figure 15 Experimental setup of the low-Q, short-cavity Ce:LiCAF oscillator with nanosecond pumping.

mirror side by a 30 cm focal-length lens with ~300 mJ/cm^2 pumping fluence inside the active medium. The c-axis of the Ce:LiCAF laser crystal sample was parallel to the direction of the pumping polarization. The absorbed energy was 5 mJ. The single-output pulse has energy of 45 μJ. The pulse duration was measured to be 600 ps using a streak camera. In this simple way, subnanosecond pulse can be generated from an all-solid-state Ce:LiCAF laser pumped by a 10 ns Q-sw Nd:YAG laser.

2. Picosecond Pumping

Six hundred picosecond pulses have been generated from a low-Q, short-cavity Ce:LiCAF laser pumped by the fourth-harmonic of a Q-sw 10 ns Nd:YAG laser above. For shorter pulse generation, a shorter pumping source was tried: the fourth harmonic of a mode-locked Nd:YAG oscillator and regenerative amplifier system operated with the repetition rate of 10 Hz [44]. A low-Q, short-cavity Ce:LiCAF oscillator was formed by a flat high-reflection mirror and a 30% reflection flat output coupler. The cavity length was 1.5 cm. A 10 mm Brewster-cut, 1% doped (in the melt) Ce:LiCAF crystal was used. A typical spectrally and temporally resolved streak camera image of the output pulse of the Ce:LiCAF laser is shown in Figure 16. The Ce:LiCAF laser pulse width was measured to be 150 ps; the pumping pulse width was 75 ps.

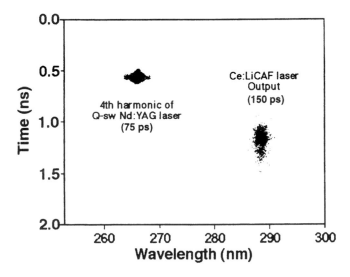

Figure 16 Streak camera trace of the pulse from the low-Q, short-cavity Ce:LiCAF laser with picosecond pumping.

IV. PASSIVE SELF-INJECTION-SEEDED PULSE TRAIN LASERS

Liu [45,46] and, independently, Ewart [47] presented a novel and useful technique for the production of nanosecond and subnanosecond high-power laser pulses in a stable, reproducible, and efficient way: the so-called self-injection technique. Brito Cruz et al. obtained the single or double, highly stabilized nanosecond pulses with large efficiency by the application of self-injection and cavity dumping techniques [48]. They also demonstrated the high-power (0.9 GW) picosecond pulse generation by intracavity nonlinear compression in a self-injected Nd:YAG laser [49]. The technique employs the principle of regenerative amplification of a seed pulse. It differs from previous work [50] using a regenerative amplifier in that it does not require an external oscillator to produce the seed pulse. As a result, the operation is greatly simplified.

A. Passive Self-Injection-Seeded Pulse Train Laser Scheme

Passive self-injection-seeded pulse train laser (SSPT laser) employs the principle of subnanosecond pulse generation, making use of the resonator transient described in Sec. III, and the self-injection, regenerative amplification process. There is no Pockels cell in our laser cavity and no timing controller, so it is completely passive.

Figure 17 shows conceptual design of the passive SSPT laser. The laser consists of a pulse-seeding laser cavity and a feedback laser cavity. In the pulse-

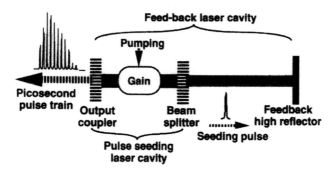

Figure 17 Conceptual design for the self-injection-seeded pulse train laser. The laser consists of a pulse-seeding laser cavity and a feedback laser cavity. A single short pulse is generated in the pulse-seeding laser cavity. The short pulse is injected into the feedback laser cavity and reflected from the end mirror, and then amplified regeneratively in the same gain medium.

seeding laser cavity, a single short pulse is generated without any deep gain quenching. The seeding laser can generate short pulse even without any coupling with the feedback laser cavity part. This is an advantage over the previously reported active self-injection seeding scheme, which cannot operate without the coupled cavity. The short pulse is injected into the feedback laser cavity, reflected from the end mirror, and then amplified in the gain medium. Some portion of the amplified pulse will be coupled out; the other part will be injected into the feedback laser cavity again. This regenerative amplification continues until the gain is quenched completely. Such a cavity is, therefore, expected to generate pulse trains similar to those from Q-switched mode-locked lasers, even though the generation mechanism differs.

There is no requirement for CW-operation capability or an external short-pulse seeding laser [51] to generate such short pulse train. The pulse duration is determined by the seeding pulse, and the pulse separation can be controlled by the length of the feedback laser cavity as in a mode-locked laser (Fig. 18).

The operation conditions for the passive SSPT scheme are as follows. The effective lifetime (T_{Geff}) of the gain medium, considering the excitation pulsewidth (T_{ex}) and relaxation time (T_1), should exceed the round-trip time (T_{rFB}) of the feedback laser cavity. A single pulse should be generated from the pulse-seeding laser cavity without any deep gain quenching. The intracavity fluence of the seeding pulse (F_{SP}) should be much smaller than the saturation fluence (F_{sat}) of the gain medium. The round-trip time (T_{rPS}) of the pulse-seeding laser cavity should be shorter than the duration of seeding pulse (T_p), to avoid multiple satellite pulses. Under this condition ($T_{rPS} \ll T_p$), the reflected portion of the main pulse will overlap the trailing edge of the main pulse itself. The limiting

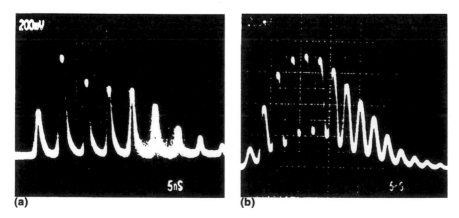

Figure 18   Output of passive SSPT Ce:LLF laser for different feedback cavity length (a) L = 115 cm and (b) L = 65 cm. The separation of pulses can be changed as expected.

factor of pulse duration in this passive SSPT laser is round-trip time or cavity length of the pulse-seeding laser. For an active self-injection seeding or cavity-flipping technique [45–49], it is limited by the transit time of the Pockels cell. All of these requirements ($T_{rPS} \ll T_p \ll T_{rFB} \ll T_{Geff}$, $F_{SP} \ll F_{sat}$) can be easily satisfied in any Ce-doped tunable laser material, and will be adequate for other solid-state laser media, organic dyes, or semiconductors if appropriate cavity parameters are selected.

B. Self-Injection-Seeded Pulse Train Ce:LLF Laser

The experimental setup of the passive SSPT Ce:LLF laser is shown in Figure 19. The seeding laser cavity was the same as the low-Q, short-cavity Ce:LLF laser described in Sec. III.B.1 (the resonator was formed by a 30% transmission flat output coupler and the end surface reflection of the Ce:LLF crystal) [39]. A confocal lens pair (f = 10 cm) with antireflection coating was inserted in the feedback laser cavity for better mode matching with the pulse-seeding laser cavity. The lens pair was adjusted to obtain the maximum output energy. The pumping source was a KrF excimer laser. The 3 µJ, 325 nm, 950 ps seeding pulse was amplified up to 60 µJ energy in the total pulse train with ~200 mJ/cm² pumping fluence.

A spectrometer was inserted before the streak camera to measure the spectrum. The spectrum scale was calculated using a CW He-Cd laser with wavelength at 325 nm. Observation with a streak camera during this self-injection seeding amplification process revealed that neither the pulsewidth nor the spectrum width changed, as shown in Figure 20. This result shows that the output properties of this laser can be totally controlled by the pulse-seeding laser.

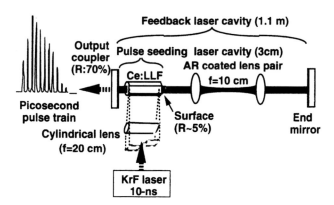

Figure 19   Experimental setup of the passive self-injection-seeded pulse train UV Ce:LLF laser.

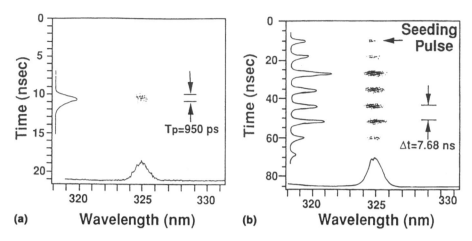

Figure 20 (a) Spectrally and temporally resolved streak camera image of the injection seeding pulse. The pulse duration in FWHM was 950 ps, and the center wavelength was 325 nm. (b) Spectrally and temporally resolved streak camera image of pulse train. This image shows that neither the pulsewidth nor the spectrum width changed during this self-injection seeding amplification process.

The parameters for the passive SSPT Ce:LLF laser in this case are T_{rPS} (200 ps) $\ll T_p$ (950 ps) $\ll T_{rFB}$ (7.7 ns) $\ll T_{Geff}$ (~50 ns), F_{SP} (~1 mJ/cm^2) $\ll F_{sat}$ (~50 mJ/cm^2) (the spot size in the gain medium was ~1 mm). The performance of this laser can be improved after these parameters are optimized. In this way, UV short-pulse trains were directly generated from a Ce:LLF laser pumped by a standard 10 ns KrF excimer laser using the simple passive self-injection-seeded pulse train laser scheme.

C. Self-Injection-Seeded Pulse Train Ce:LiCAF Laser

To obtain a short pulse train, the passive SSPT laser scheme was employed in the Ce:LiCAF laser as it was for the Ce:LLF laser described above [43].

The experimental setup of the SSPT Ce:LiCAF laser is shown in Figure 21. The 1.5 cm long seeding laser cavity was formed by a flat high-reflection mirror and an 80% transmission flat output coupler. The 20 mJ, 10 ns, 1 Hz, 266 nm, horizontally polarized pumping pulses (the fourth harmonic of a conventional 10 ns Q-sw Nd:YAG laser) were focused longitudinally from the high-reflection mirror side by a 30 cm focal-length lens with ~300 mJ/cm^2 pumping fluence inside the active medium. The c-axis of the Ce:LiCAF laser crystal sample was parallel to the direction of the pumping polarization. The absorbed energy was 5 mJ. In the feedback laser cavity, a confocal lens pair (f = 10 cm) with antire-

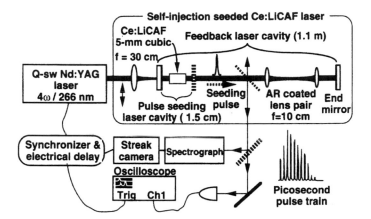

Figure 21 Experimental setup of the passive self-injection-seeded pulse train UV Ce: LiCAF laser.

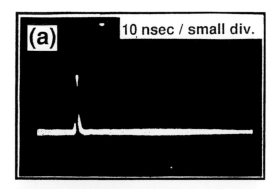

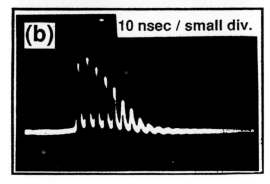

Figure 22 (a) Oscilloscope trace of seeding pulse. (b) Oscilloscope trace of pulse train.

flection coatings was inserted for better mode matching with the pulse-seeding laser cavity. A 1 mm thick fused silica plate as an intracavity output coupler was placed at an angle of 45 degrees to the resonator axis. The estimated coupling constant was 3.6%.

As can be seen from the oscilloscope traces (Fig. 22) of the single seeding pulse and the pulse train, this SSPT scheme is quite efficient with a Ce:LiCAF laser. The ~45 µJ, 290 nm, horizontally polarized, 600 ps seeding pulse was amplified up to ~230 µJ in the total pulse train inside the cavity, and 8 µJ was coupled out.

This Ce:LiCAF laser meets all the requirements for the efficient passive SSPT scheme operation as follows: T_{rPS} (100 ps) $\ll$ T_p (600 ps) $\ll$ T_{rFB} (7.4 ns) $\ll$ T_{Geff} (~40 ns), F_{SP} (~5 mJ/cm^2) $\ll$ F_{sat} (~100 mJ/cm^2). Because of the shorter length of the gain medium (0.5 cm) and pulse-seeding laser cavity (1.5 cm), a shorter pulse duration (600 ps) from the Ce:LiCAF laser than that from the SSPT Ce:LLF laser (crystal length: 2.5 cm; cavity length: 3 cm; pulse duration: 950 ps; see Sec. IV.B) was obtained.

V. ULTRAVIOLET-TUNABLE SUBNANOSECOND PULSE GENERATION

In most lasers, all of the energy released via stimulated emission by the excited medium is in the form of photons. Tunability of the emission in solid-state lasers is achieved when the stimulated emission of photons is intimately coupled to the emission of vibrational quanta (phonons) in a crystal lattice. In these "vibronic" lasers, the total energy of the lasing transition is fixed, but can be partitioned between photons and phonons in a continuous fashion. The result is broad wavelength tunability of the laser output. In other words, the existence of tunable solid-state lasers is due to the subtle interplay between the Coulomb field of the lasing ion, the crystal field of the host lattice, and electron–phonon coupling permitting broadband absorption and emission. Therefore, the gain in vibronic lasers depends on transitions between coupled vibrational and electronic states; that is, a phonon is either emitted or absorbed with each electronic transition.

Rare earth ions doped in appropriate host crystals exhibit vibronic lasing. The main difference between transition metal and rare earth ions is that the former is crystal-field-sensitive and the latter is not. As distinct from transition metal ions, the broad-band transitions for rare earth ions are quantum mechanically allowed and therefore have short lifetimes and high cross sections. The Ce^{3+} ion laser, using a 5d–4f transition, has operated in the host crystals. Such a system would be an alternative to the excimer laser as a UV source, with the added advantage of broad tunability.

Due to the vibronic nature of the Ce^{3+} ion laser, the emission of a photon is accompanied by the emission of phonons. These phonons contribute to thermalization of the ground-state vibrational levels. The laser wavelength depends on which vibrationally excited terminal level acts as the transition terminus; any energy not released by the laser photon will then be carried off by a vibrational phonon, leaving the Ce^{3+} ion at its ground state. The terminal laser level is a set of vibrational states well above the ground state. So the Ce^{3+} ion lasers belong to four-level lasers.

A. Tunable Ce:LLF Laser

The fluorescence spectrum of Ce:LLF crystal excited by the fifth harmonic of an Nd:YAG laser is shown in Figure 23 [40]. The Ce:LLF crystal was the one used in the experiment described in Sec. III.B.2. Two peaks around 308 nm and 325 nm are seen from the fluorescence spectrum. The fluorescence lifetime of the Ce:LLF crystal was estimated to be 40 ns. To test the Ce:LLF laser tunability under 213 nm pumping conditions, we employed a high-Q cavity by replacing the output coupler shown in Figure 12 with an 80% reflection flat mirror. Tunable operation was realized by rotating the prism about the vertical axis. The Ce:LLF laser tunability obtained at the pumping energy level of 22 mJ at 213 nm is shown

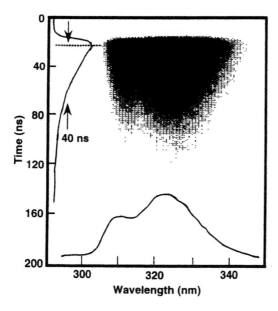

Figure 23 Spectrum of Ce:LLF pumped by the fifth harmonic of an Nd:YAG laser. There are two peaks around 308 nm and 325 nm. The fluorescence lifetime is 40 ns.

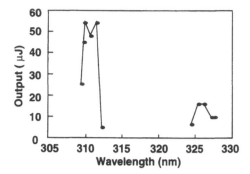

Figure 24 Short-cavity, high-Q Ce:LLF laser tunability obtained at a pumping energy of 22 mJ at 213 nm.

in Figure 24. In a subnanosecond pulse regimen, the tuning was achieved from 309.5 nm to 312.3 nm and from 324.5 nm to 327.7 nm. The gap in the tuning curve is attributed to the close-to-the-threshold operation regimen necessary to maintain a single subnanosecond pulse operation. This tuning behavior resembles that for Ce:YLF laser [52].

B. Tunable Ce:LiCAF Laser

The experimental setup of the tunable, short-cavity Ce:LiCAF laser is shown schematically in Figure 25. The cavity length was 25 mm. To study the tuning performance of this laser without consideration of its temporal characteristics, a high-Q laser with a low-transmission, flat-output coupler (T = 20%) was designed [53,54]. The laser consisted of a Littrow prism with a high reflection coating at one face used as the end mirror and a 10 mm long Brewster cut at the

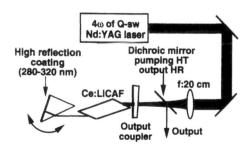

Figure 25 Experimental setup of the tunable, short-cavity Ce:LiCAF oscillator pumped by the fourth harmonics of a Q-switched Nd:YAG laser.

end faces, 1% doped (in the melt) Ce:LiCAF crystal used as the gain medium without any special cooling. The 266 nm, horizontally polarized pumping pulses from a Q-sw Nd:YAG laser in short-pulse operation mode were focused longitudinally from the output mirror side using a 20 cm focal-length lens. In a single pumping shot, there were four short pulses with pulse durations of about 1 ns separated by about 5 ns as shown in Figures 28 and 29. In most cases, the laser operated at 2 Hz to reduce any possible thermal problems in the Ce:LiCAF crystal without mandatory cooling, and to obtain higher extraction efficiency. The c-axis of the Ce:LiCAF crystal was parallel to the direction of the pumping polarization. The tuning operation was realized by rotating the prism horizontally. Tuning using a Ce:LiCAF crystal with Brewster-cut end faces is very efficient because it can increase the dispersion of the laser beam. The output pulse was separated from the pumping beam by a dichroic mirror, which has high reflection in the region of 280–320 nm at a 45 degree incidence angle and high transmission for the pumping beam (266 nm). The output pulse energy tuning curve obtained is shown in Figure 26, with the pumping energy of 8 mJ. As illustrated in Figure 26, the pulses were multipeak at some points in the form of damped relaxation oscillation, and the main peaks had pulse durations of 2–3 ns. The corresponding tuning was accomplished from 282 nm to 314 nm, sacrificing the capability of single short-pulse generation.

To obtain shorter pulses and a single-pulse output, a low-Q, short-cavity Ce:LiCAF laser was constructed by changing the output coupler of the laser mentioned above to a 75% transmission flat coupler. The output pulse energy dependence on wavelength is shown in Figure 27. The pumping energies for different wavelengths were varied between 2 and 4 mJ to obtain single-pulse generation. The demonstrated tuning range was 281 nm to 297 nm. The pulse

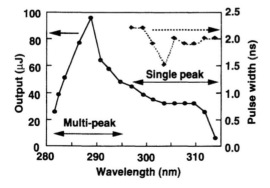

Figure 26   For the high-Q ($T_c = 20\%$), short-cavity Ce:LiCAF laser, tunability between 282 and 314 nm was obtained.

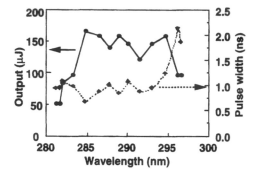

Figure 27   From the low-Q (T_c = 75%), short-cavity Ce:LiCAF laser with a Littrow-prism end mirror, tunability from 281 to 297 nm was achieved while maintaining single short-pulse properties.

durations at different wavelengths observed by a streak camera are also shown in Figure 27. Under the appropriate pumping fluence control, no satellite pulse was observed because the pulse duration exceeded the cavity round trip time (170 ps). The typical spectrally and temporally resolved streak camera image of the output pulse at a wavelength of 289 nm, including the image of the pumping pulse, is given in Figure 28. They were taken in a single shot. Figure 29 was

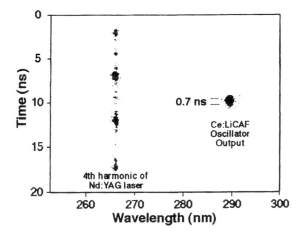

Figure 28   Spectrally and temporally resolved streak camera image of the output pulse at 289 nm from the low-Q, short-cavity Ce:LiCAF laser. The image of the pumping pulse can also be observed.

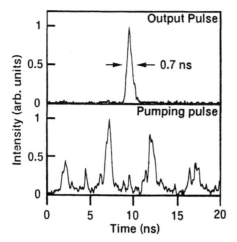

Figure 29 Temporally resolved streak camera traces of the output pulse and the pumping pulse processed from the image in Figure 28. The pulse width (at half maximum) of the output pulse was 0.7 ns, and the center wavelength was 289 nm with a spectrum width (at half maximum) of about 2 nm.

derived from Fig. 28 and describes the temporally resolved streak camera traces of the output pulse and the pumping pulse [53].

To increase the output energy while maintaining the single short output pulse, a 55% reflection flat-output coupler was used. A single-pulse output was generated with the pumping energy of 15 mJ. The demonstrated tuning range was 282–314 nm (Fig. 30). The maximum single subnanosecond pulse output

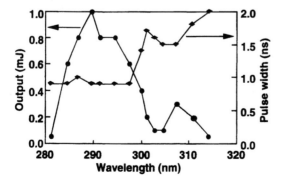

Figure 30 Tuning curve for the short-cavity Ce:LiCAF laser with 55% reflection output coupler.

was 1 mJ with the pulse width of 0.9 ns observed by a streak camera. In this way, single short pulses with broad tuning regions can be easily generated from a low-Q, short-cavity, tunable Ce:LiCAF laser pumped by the fourth harmonic of a Q-sw Nd:YAG laser.

C. Tunable Subnanosecond Pulse Generation Around 230 nm by Sum-Frequency Mixing

To obtain tunable subnanosecond UV pulses around 230 nm, subnanosecond pulses from a tunable Ce:LiCAF master oscillator and power amplifier (MOPA) system were mixed with the fundamental radiation of a Q-sw Nd:YAG laser [54]. The spectral region covered by this sum-frequency generation lies between the fourth and fifth harmonics of Nd lasers and also between the third and fourth harmonics of tunable near-infrared lasers such as Cr:LiSAF. The spectral region near 230 nm is important practically because it spans the signature absorption features of chemical and biological species such as nitric oxide and tryptophan.

The experimental setup is shown schematically in Figure 31. The oscillator stage of the MOPA system is the same as mentioned above (see Sec. V. B). The double-side-pumping, double-pass-amplification configuration was applied in the amplifier stage. A Q-sw Nd:YAG laser was used to pump the oscillator and

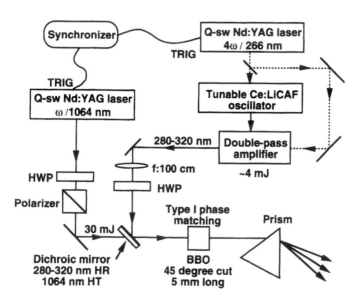

Figure 31 Experimental setup of sum-frequency mixing of Ce:LiCAF and Nd:YAG laser beams through a BBO crystal in type I phase matching condition.

amplifier stages with optimized optical delay. The 266 nm pulses with 12 mJ and 15 mJ energies from each side were slightly focused on the Ce:LiCAF crystal of the amplifier. The pump energy for the Ce:LiCAF oscillator was adjusted to obtain a better beam pattern so as to increase the efficiency of the amplifier. An energy-tuning curve for the Ce:LiCAF MOPA system is shown in Figure 32. Continuous tunability of the MOPA system was obtained from 284 to 299 nm with the maximum pulse energy up to 4 mJ.

A BBO nonlinear crystal was utilized for frequency mixing of the Ce: LiCAF MOPA system output with the fundamental output of another Q-sw Nd: YAG laser synchronized to the Nd:YAG laser for pumping the Ce:LiCAF MOPA system. The BBO crystal (5 mm × 5 mm × 5 mm) was cut at $\theta = 45$ degree for type I phase matching. The Ce:LiCAF laser beam was focused with a 100 cm focal-length lens. Using a dichroic beamsplitter, the Ce:LiCAF and Nd:YAG laser beams were spatially overlapped and subsequently input into the BBO crystal. The delay between the Ce:LiCAF and Nd:YAG laser beams was electrically controlled to ensure temporal overlapping of the two input beams. Pulse energy of the Nd:YAG laser fundamental beam was 30 mJ. The tunability after the sum frequency mixing is shown in Figure 33. The obtained tuning region was from 223 to 232 nm. Sum-frequency generation was optimized at each wavelength by tuning the BBO crystal to maintain the proper phase matching. The peak of the tuning occurred at 227 nm with 0.5 mJ of output energy. The spectrally and temporally resolved streak camera image of the output pulse at this wavelength is shown in Figure 34, and the pulse width was measured to be 1.0 ns.

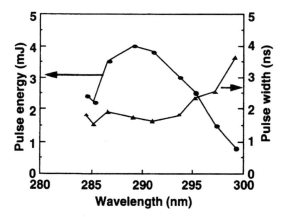

Figure 32 Tuning curve for the Ce:LiCAF MOPA system for the sum frequency mixing with Nd:YAG laser.

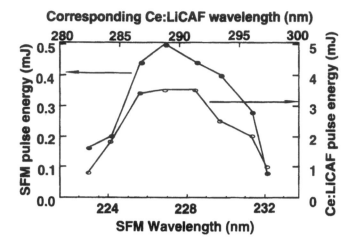

Figure 33 Tuning curve for sum-frequency mixing of Ce:LiCAF and Nd:YAG laser beams.

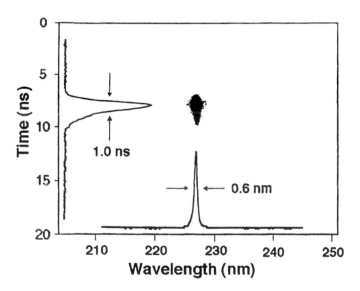

Figure 34 Spectrally and temporally resolved streak camera image of the sum-frequency generation at the peak point of the tuning curve of Figure 33.

VI. EFFICIENT UV LASER AMPLIFICATION USING Ce:LLF AND Ce:LiCAF CRYSTALS

The generation of high-energy pulses is based on the combination of a master oscillator and multistage power amplifiers. In an oscillator–amplifier system, pulse width, beam divergence, and spectral width are primarily determined by the oscillator, whereas pulse energy and power are determined by the amplifier. Operating an oscillator at relatively low energy levels reduces beam divergence and spectral width. Therefore, from an oscillator-amplifier combination one can obtain either a higher energy than is achievable from an oscillator alone or the same energy in a beam with smaller beam divergence and narrower linewidth. Generally speaking, the purpose of adding an amplifier to a laser oscillator is to increase the brightness B [W cm^{-2} sr^{-1}] of the output beam

$$B = P/A\Omega$$

where P is the power of the output beam emitted from the area A and Ω is the solid-angle divergence of the beam.

A. Confocal Four-Pass CW Ce:LLF Amplifier

To demonstrate a practical device with high small-signal gain, the Ce:LLF amplifier was designed with a side-pumped confocal four-pass configuration as shown in Figure 35. The Ce:LLF sample used for the amplifier contains about 0.5 at.%

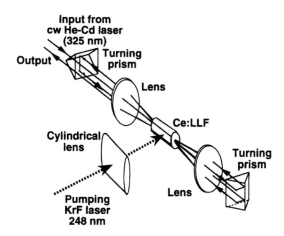

Figure 35 Side-pumped confocal four-pass Ce:LLF amplifier. The pumping area was 3 mm wide and 10 mm long. A confocal lens pair and prisms with antireflection coatings were used to reduce the loss.

Ce^{3+} ions [31]. The sample (1 cm long and 5 mm diameter) for the amplification experiment was prepared for the transverse pumping. The amplifier consisted of a gain medium located at the beam waist of a confocal lens pair (15 cm focal length) and two turning prisms for a small displacement of each pass. This configuration allowed the beams from different passes to overlap well in a small pumped region. Side pumping is possible due to the high absorption (3 cm^{-1}) at the pumping wavelength. KrF-laser pulses (10 mJ, 10 ns) were weakly focused by a 10 cm focal-length cylindrical lens. The pumping laser operated at a 3 Hz repetition rate to reduce the possible influence of color-center formation [24].

A Cw He-Cd laser was used as the probe light with the wavelength at 325 nm. A biplanar phototube was used as the detector. The oscilloscope trace of the gain profile is shown in Figure 36. The upper trace is with the Cw He-Cd laser probe; the lower trace is only the fluorescence without the probe. The difference between the two traces is the real amplification. Considering this difference and the probe level measured separately, the four-pass differential gain was estimated to be 20 dB for the transmitted probe light. This small signal gain is large enough for any practical use.

B.　Picosecond Pulse Ce:LLF Amplifier

Using the above laser material and the same four-pass amplifier configuration, a solid-state UV picosecond laser system was designed (Fig. 37). The seeding

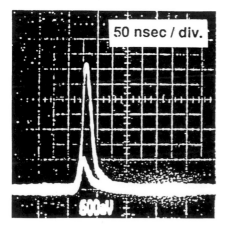

Figure 36　Temporal gain profile of the 4-pass amplifier. The upper trace is with the cw He-Cd laser probe beam, and the lower trace is only fluorescence without the probe. Considering the difference of the two traces and the probe level without pumping measured separately, the gain is estimated to be 20 dB for the transmitted probe light.

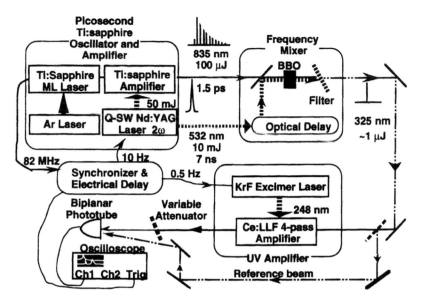

Figure 37 Experimental set-up of solid-state UV picosecond laser system. This system consisted of a Ti: sapphire oscillator and amplifier system, a frequency mixer, and a Ce: LLF amplifier.

picosecond UV pulses at 325 nm were generated by the sum-frequency mixing of the 835 nm picosecond pulses from a Ti: sapphire oscillator and amplifier system and the second harmonic of an Nd: YAG laser [31]. The 10 Hz Q-sw Nd: YAG laser was synchronized to an 82-MHz mode-locked Ti: sapphire laser oscillator, and delivered 50 mJ pulse energy (532 nm) for pumping the Ti: sapphire amplifier and 10 mJ pulse energy for frequency mixing. The 10 nJ, 1.5 ps pulses in the mode-locked pulse train from the Ti: sapphire oscillator without any single-pulse selection were amplified typically up to a 100 μJ level by a double-side-pumped confocal four-pass Ti: sapphire amplifier. For efficient sum-frequency generation, the peak position of the 532 nm, 7 ns pulse was adjusted to the largest pulse among the amplified picosecond pulse train by an optical delay (40 ns). In this way, a single 1 μJ, 325 nm seeding pulse was generated at a 10 Hz repetition rate through a 5 mm thick BBO crystal cut at 35 degrees. The duration of the UV pulses was evaluated to be less than 3 ps using a streak camera.

The confocal four-pass Ce: LLF amplifier as described in Sec. VI. A was pumped by 100 mJ KrF-laser pulses at a 0.5-Hz repetition rate synchronized to the 10 Hz UV seed pulses with 0.5 J/cm^2 pumping fluence. To calibrate the gain factor, a reference pulse separated from the input pulse was detected together

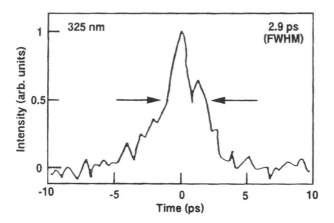

Figure 38 Streak camera trace of the amplified ultraviolet pulse. The pulse duration was measured to be 2.9-ps (resolution limit).

with the attenuated amplified pulse. A 17 dB gain was demonstrated for the transmitted 325 nm ps pulses.

Using a streak camera, the duration of amplified UV pulses was evaluated to be less than 2.9 ps, but the measurement was limited by the resolution of the apparatus (Fig. 38). This result shows that there is no significant pulse width broadening in the amplifier under the conditions of the experiments.

C. Efficient UV Short-Pulse Amplification in a Ce:LiCAF MOPA System

The experiment setup of a Ce:LiCAF MOPA system is shown in Figure 39. A low-Q, short-cavity Ce:LiCAF master oscillator with 15 mm cavity length was formed by a flat high-reflection mirror and a 30% reflection flat output coupler [55]. A 10 mm Brewster-cut, 1% doped (in the melt) Ce:LiCAF crystal was used as the oscillator gain medium. The 15 mJ, 10 ns, 266 nm, horizontally polarized pumping pulses from a Q-sw Nd:YAG laser were focused longitudinally from the high-reflection mirror side by a 20 cm focal-length lens to obtain a 1 J/cm² pumping fluence inside the active medium. In most cases, the laser operated at 2 Hz to avoid possible thermal problems in the Ce:LiCAF crystal and to obtain higher extraction efficiency. The c-axis of the Ce:LiCAF crystal was parallel to the direction of the pumping polarization. A single 1 mJ, horizontally polarized pulse at 289 nm was obtained at the sacrifice of the extracted energy from this master oscillator. The pulse duration observed by a streak camera was 1.0 ns as shown in Figures 40 and 41a. Under the appropriate pumping fluence control,

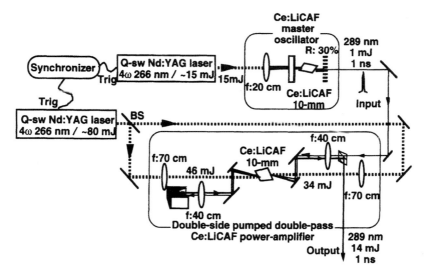

Figure 39 Experiment setup of the Ce:LiCAF MOPA system pumped by fourth harmonics of two conventional 10-ns Q-switched Nd:YAG lasers. This system consists of a low-Q, short-cavity master oscillator, and a confocal double-pass power amplifier.

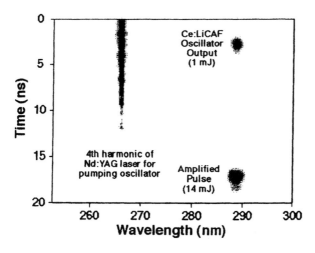

Figure 40 Spectrally and temporally resolved streak-camera image of the UV short pulses from the oscillator stage and after the amplifier stage. A part of the oscillator-pumping pulse image can also be observed. All these images are taken in a single shot.

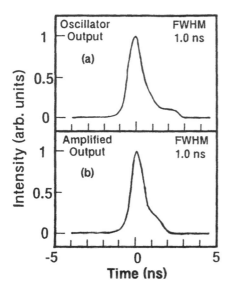

Figure 41 Temporally resolved streak-camera traces of the pulses processed from the image in Figure 40. (a) The temporal pulse shape from the low-Q, short-cavity master oscillator. The pulse duration (FWHM) was 1.0 ns, and the center wavelength was 289 nm. (b) The temporal pulse shape after the amplifier. The pulse duration (FWHM) was also 1.0 ns.

no satellite pulse was observed because the pulse duration exceeded the cavity round trip time (100 ps).

Another 10 mm, 1% doped Brewster-cut Ce:LiCAF crystal was employed in the power amplifier stage. The amplifier was designed with a confocal double-pass configuration similar to the double-side, coaxially pumped, confocal multipass configuration of a Ti:sapphire amplifier [56]. Another 10 ns, Q-sw Nd:YAG laser for pumping the amplifier operated with variable Q-switch delay. The 266 nm pulses with 46 mJ and 34 mJ energies from each side were slightly focused by 70 cm focal-length lens down to ~1 mm beam diameter (~5 J/cm²). The amplifier consisted of a gain medium located at the beam waist of a confocal lens pair (40 cm focal length) and a roof reflector with dielectric coating for a small vertical displacement of each pass (Fig. 39). The signal passes coincide with the pumped region with a small angular separation from the pumping beams (less than a few degrees). This configuration allowed the signal beams with different passes to overlap almost completely in a small pumped region. The output energy of the amplifier was measured for different pumping delays as shown in Figure 42. This result indicated that the relative timing between two Nd:YAG lasers should be controlled within an accuracy of a few nanoseconds. Here the

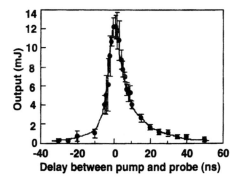

Figure 42   Output energy of the amplifier for different delay timings of the probe pulse from the oscillator and the amplifier-pumping pulse from the Nd:YAG laser.

optimum relative delay timing between the Q-switch trigger signals for the two Nd:YAG lasers was 15 ns. A single-pass gain of over 10 times was observed. The double-pass differential small-signal gain reached 100 times, as shown in Figure 43. The 1 mJ input pulse was amplified up to 14 mJ with 14 MW peak power at a 2 Hz repetition rate. The amplified pulse duration observed by a streak camera was 1.0 ns as shown in Figure 41b. There was no noticeable pulse broadening accompanying the amplification process. The output gain dependence on different input energy of the amplifier is shown in Figure 43. This result clearly shows that the amplification saturation is reached with sufficient input flux. The energy extraction efficiency in the amplifier stage exceeded 18%, which was

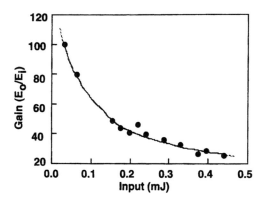

Figure 43   Output gain dependence on input energy of the amplifier. This result clearly shows that the amplification saturation is reached with sufficient input flux.

sufficient for practical use. Even at a 10 Hz operation, 100 mW average power after amplifier was obtained.

VII. HIGH-PULSE-ENERGY UV LASERS USING LARGE Ce:FLUORIDE CRYSTALS

A short-pulse (1 ns) Ce:LiCAF MOPA system and an all-solid-state Ce:LLF laser have been demonstrated. Due to the limited size of the available crystals, it was difficult to obtain high-energy outputs directly from Ce:LiCAF and Ce: LLF lasers. Recently, the growth of large Ce:LiCAF and Ce:LLF crystals has become successful [57], making possible high-energy pulse generation from Ce: LiCAF and Ce:LLF lasers.

A. High-Energy Pulse Generation from a Ce:LiCAF Oscillator

The schematic diagram of a Ce:LiCAF laser resonator using the large Ce:LiCAF crystal is shown in Figure 44. The laser resonator is established by a flat high reflector and a flat output coupler with 30% reflection for 290 nm and 75% transmission for 266 nm separated by 4 cm [58]. The large Ce:LiCAF crystal (18 mm in diameter, clear aperture 15 mm, length 10 mm) is doped with 1.2 mol% Ce^{3+} ions. There is no coating on the parallel end faces of the crystal that are perpendicular to the optical axis of the resonator. The fourth harmonic of a Q-sw Nd:YAG laser is used as the pumping source. The quasilongitudinal pumping method was used to obtain a high-quality laser beam. Because it is difficult to fabricate an end mirror with high reflection for 290 nm and high transmission

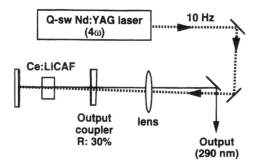

Figure 44 Experimental setup of the Ce:LiCAF laser oscillator pumped by the fourth harmonic of a Q-switched Nd:YAG laser using a quasilongitudinal pumping scheme.

for 266 nm pump beams while maintaining a high damage threshold, the Ce: LiCAF crystal was pumped from the output coupler side that has almost the same transmission for the pump and output wavelengths. To obtain high-energy output without damage to the crystal and optics in the cavity, a large pump beam cross section is necessary. The horizontally polarized pump beam is focused with a 40 cm focal-length lens to produce a 4 mm diameter spot at the surface of the Ce: LiCAF crystal without any damage to the crystal. To reduce the diffraction effects and disturbance to the beam uniformity, it is better to choose a ratio of crystal radius to beam radius of 2 or more. Therefore, a much larger crystal diameter than the pump beam diameter is preferred. More than 85% of the incident pump pulse energy is absorbed by the crystal, so the crystal is long enough for practical use. Figure 45 presents the obtained output energies at 290 nm as a function of the absorbed 266 nm pump energy. The laser oscillation threshold is 12 mJ, which corresponds to a threshold fluence of approximately 100 mJ/cm^2. The measured output energy remained linear with pump fluence with a slope efficiency of 39%. The efficiency can be improved by using a Brewster-cut crystal with antireflective coating. The highest pulse energy was 21 mJ at 10 Hz and 290 nm. The satellite-free single pulse had a pulse duration of 3 ns.

Because the loss of the pumping pulse energy through the output coupler was still large, a noncollinear pumping scheme was tried, as shown in Figure 46. The large pumping beam cross section made it possible to achieve efficient pumping. The angle between the pumping beam and the output beam was about 5 degrees. The obtained output energies at 290 nm as a function of the absorbed 266-nm pump energy are shown in Figure 47. In this case, the output pulse energy was improved up to 30.5 mJ, and the slope efficiency was also 39%.

To demonstrate the tunability of the Ce:LiCAF laser consisting of the large

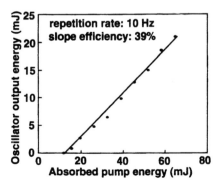

Figure 45 Laser output energy as a function of absorbed pump energy. The measured output energy remained linear with pump fluence with a slope efficiency of 39%.

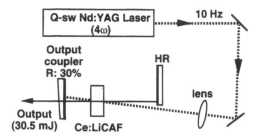

Figure 46 Experimental setup of the Ce:LiCAF laser oscillator using a noncollinear pumping scheme.

Ce:LiCAF crystal, a grating was used as the tuning element which also acted as the end mirror of the oscillator, as shown in Figure 48. The grating worked in the Littrow condition. The incidence of the grating and its first-order diffraction overlapped. The grating used here was blazed for 500 nm wavelength. The obtained tuning range was from 284 to 294 nm, as shown in Figure 49, and maximum output was 6 mJ. The limited tunable range was due to the low diffraction efficiency of the grating for the 290 nm wavelength. A broader tunable range can be expected using a grating blazed for 290 nm wavelength. The pulse widths were approximately 4 ns, and the spectrum widths were about 0.7 nm.

To generate pulses with much higher energy, the fourth harmonics of two simultaneous Q-sw Nd:YAG lasers were used as the pumping sources (Fig. 50). The three pump beams were focused with a 40 cm focal-length lens to produce a spot size of 6 mm in diameter at the surface of the Ce:LiCAF crystal. With

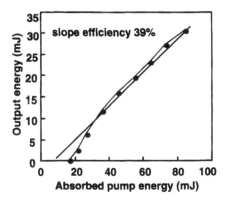

Figure 47 Laser output energy as a function of absorbed pump energy. The output pulse energy was improved up to 30.5 mJ, and the slope efficiency was also 39%.

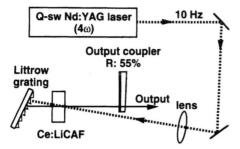

Figure 48 Experimental setup of tunable Ce:LiCAF laser. A grating was used as the tuning element, which also acted as the end mirror of the oscillator. The grating worked in the Littrow condition.

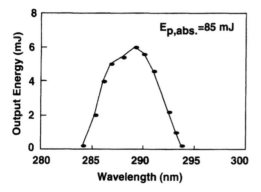

Figure 49 Tuning curve for the tunable Ce:LiCAF laser. The limited tunable range was due to the low diffraction efficiency of the grating for the 290 nm wavelength.

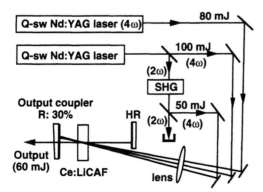

Figure 50 Experimental setup of the Ce:LiCAF laser oscillator pumped by the fourth harmonics of two Q-switched Nd:YAG lasers. The pulse energy as high as 60 mJ was achieved at 290 nm at 10 Hz repetition rate.

the total pumping energy of 230 mJ, the output pulse energy as high as 60 mJ was achieved at 290 nm at 10 Hz repetition rate, which is the highest performance reported for a Ce:LiCAF oscillator until now. In this way, high-energy pulses at 290 nm were generated very easily and efficiently.

B. Ce:LiCAF Oscillator with Noncollinear Brewster Angle Pumping

To scale the output energy of a Ce:LiCAF laser, a new pumping scheme permitting efficient, high-energy pumping is necessary [59]. A pumping scheme with pumping at Brewster angles on a disk Ce:LiCAF crystal is proposed, shown schematically in Figure 51. The disk Ce:LiCAF crystal (20×20 mm^2 in cross section, 5 mm in length) was placed at the 289 nm wavelength Brewster angle relative to the oscillator axis. The pumping pulses (the fourth harmonics of two Q-sw Nd:YAG lasers) were directed to the crystal in the direction near the reflection of the Ce:LiCAF laser pulse on the crystal surface. In this case, the pumping pulses at 266 nm wavelength could also be at the Brewster angle. Thus, the pumping energy of p-polarization pulses could be efficiently coupled into the crystal from the two sides of the crystal without the risk of possible cavity mirror damage, as in the collinear pumping scheme [58]. The laser resonator is constructed simply, using a flat, high reflector and a flat output coupler with a transmission of 45% (Fig. 51). The pumping pulses with horizontal polarization from the two Q-sw Nd:YAG lasers operated at 1 Hz were focused softly on the Ce:LiCAF crystal by two 70 cm focal-length spherical lenses. The delay between the pumping pulses from the two Nd:YAG lasers was controlled by an electric delay controller. The pumping pulse energies of the two Nd:YAG lasers were 80 mJ and 120 mJ, and approximately 50% of these were absorbed by the Ce:

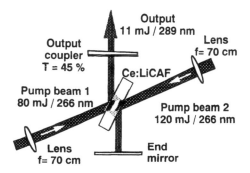

Figure 51 Experimental setup for disk Ce:LiCAF laser with noncollinear, near-Brewster-angle pumping scheme.

LiCAF crystal. A maximum output energy of 11 mJ was obtained from the Ce: LiCAF laser with Brewster-angle pumping when the pumping pulses from the two Nd : YAG lasers arrived on the crystal simultaneously. Even though the output is presently not very high, this noncollinear Brewster-angle-pumping disk oscillator scheme will make it possible to scale the output energy significantly by using larger pumping sources and crystals with higher absorption due to the higher pumping efficiency. The potential advantages of this scheme are the reduced risk of possible mirror damage, ease of pumping beam multiplexing, and better gain profile compared with the side-pumping scheme.

C. Ce:LLF Laser Using Large Ce:LLF Crystal

A Ce:LLF laser resonator is established by using a flat high reflector and a flat output coupler [60]. The length of the laser cavity was 6 cm. The layout is shown schematically in Figure 52. High-quality, large Ce:LLF crystals (ϕ 18 mm $\times$ 10 mm in length) were grown successfully at Tohoku University, Japan by the Czochralski (CZ) method. The Ce:LLF sample used here was obtained by cutting the grown Ce:LLF crystal in the middle along its axis (a half-cut cylinder). The side window and two end surfaces were polished. No dielectric coatings were deposited on the polished end surfaces and the side window. The pumping pulses from a randomly polarized KrF excimer laser operated at 1 Hz were focused softly on the side window of the Ce:LLF crystal using a 50 cm focal-length spherical lens under a normal-incidence side-pumping condition. Because the output pulse of the KrF excimer laser has a rectangular shape, it is not difficult to make a near-line-shaped pumping area on the Ce:LLF crystal through a spherical lens. Almost all of the pumping pulse energy (maximum: 230 mJ) was absorbed by the Ce:LLF crystal. To obtain high-output energy from the Ce:LLF

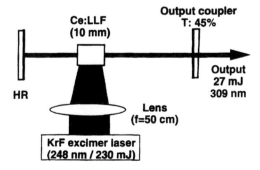

Figure 52 Experimental setup of the high-power Ce:LLF laser pumped by a randomly polarized KrF excimer laser operated at a repetition of 1 Hz.

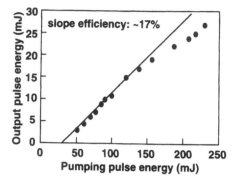

Figure 53 Input–output dependence curve for the Ce:LLF laser. The maximum output energy reaches 27 mJ at the wavelength of 309 nm.

laser, we tested some output couplers with different transmissions. The best result was obtained with the coupler with 45% transmission. Figure 53 shows the input–output dependence curves for the Ce:LLF laser. The maximum output pulse energy reached 27 mJ with the pumping pulse energy of 230 mJ, and the corresponding pumping fluence was approximately 0.6 J/cm^2. This is the highest-output pulse energy ever achieved from this laser medium. The free-running Ce:LLF laser operated at the wavelength of 309 nm. The slope efficiency was approximately 17%.

VIII. CONCLUSIONS AND PROSPECTS

As mentioned above, tunable lasers in the ultraviolet region with tunability centered around 290 nm are of special interest for applications relating to remote sensing. The simple, compact, all-solid-state Ce:LiCAF (282–314 nm) laser can generate coherent radiation in this wavelength. Ce:LLF has a potential longer-wavelength tuning region of around 305–340 nm, so it is especially attractive for use in spectroscopy of wide band-gap semiconductors for blue laser diodes, such as GaN. Their broad gain bandwidth corresponding to a few femtoseconds is extremely attractive for short-pulse applications.

The new solid-state, tunable, ultraviolet crystals Ce:LLF and Ce:LiCAF are proving to be very efficient and reliable for realizing UV lasers. Gain for the Ce:LLF crystal was demonstrated from 323 to 335 nm. The gain bandwidth makes generation of 10 fs pulses possible in the transform limit condition. The saturation fluence of the Ce:LLF crystal was estimated to be 50 mJ/cm^2 through a pump-probe experiment [31]. Subnanosecond ultraviolet coherent pulses were generated directly from solid-state lasers simply for the first time using low-Q, short-cavity Ce:fluoride lasers pumped by KrF laser [39], the fifth [40] and fourth [53] harmonics of Nd:YAG lasers.

For such new laser materials, a passive self-injection seeding scheme was proposed for the direct generation of short-pulse trains, which did not require CW operation capability or an external short-pulse seeding laser. Using this simple scheme, UV subnanosecond pulse trains are directly and passively generated from Ce:LLF pumped by a standard 10 ns KrF excimer laser [39] and Ce:LiCAF pumped by the fourth harmonic of a conventional 10 ns Q-sw Nd:YAG laser [43].

To prove the tunability of these new laser materials, a tunable all-solid-state Ce:LLF laser was made with the fifth harmonic of an Nd:YAG laser as the pumping source [40]. A tunable Ce:LiCAF laser was also demonstrated with broad tuning region from 282 to 314 nm [53]. Furthermore, tunable UV pulses around 230 nm were obtained by the sum frequency mixing of Ce:LiCAF and Nd:YAG lasers [54].

For these new laser materials with large bandwidth, short pulse applications have been investigated. UV picosecond-pulse amplification with large small signal gain as high as 17 dB was demonstrated using Ce:LLF in a confocal four-pass amplifier [31]. The direct generation and efficient amplification of UV short pulses have been demonstrated from the simplest, all-solid-state, UV short-pulse, MOPA system composed of Ce:LiCAF crystals and conventional Q-sw Nd:YAG lasers [55]. In this way, it has been proven that the Ce:LiCAF-based MOPA system is as effective and practical as other UV short-pulse systems.

Large Ce:LiCAF crystals with 15 mm diameter were successfully grown by the Czochralski method recently [57]. Due to the available large Ce:LiCAF crystal, 60 mJ output energy was obtained from the Ce:LiCAF laser. This is the highest output directly from a Ce:LiCAF laser reported until now. A much higher output can be expected by fully utilizing the crystal cross size while using a larger pumping source. This suggests that Ce:LiCAF is a promising material for high-energy ultraviolet pulse generation combined with a high-power, Q-sw Nd:YAG laser [58].

Ce:fluoride laser has been used to environmental sensing [61]. Narrow lasing linewidth (<0.1 nm) operation was demonstrated in distributed-feedback, tunable Ce^{3+}-doped colquiriite lasers [62]. Ce:LiCAF laser worked efficiently at 20 kHz repetition rate [63]. A Ce:LiCAF laser pumped by the sum-frequency-mixing (271 nm) of the green and yellow fundamental lines (511 and 578 nm) from a copper-vapor laser was recently reported [64]. The basic optical properties of Ce^{3+} ions in both oxide and fluoride hosts were investigated [65]. The room temperature fluorescence spectrum of Ce^{3+}-doped $LiBaF_3$ (Ce:LBF) spans the spectral region of 300–450 nm [66].

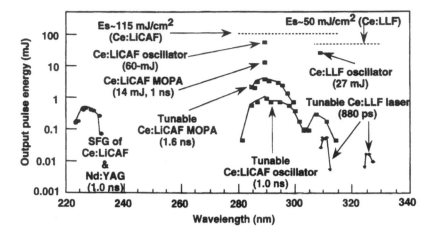

Figure 54 Tuning curves for Ce^{3+} ion-activated laser systems.

With the development of the new laser media and nonlinear crystals, solid-state tunable ultraviolet short-pulse lasers have been realized, as shown in Figure 54 [34,44]. It is reasonable to expect CW solid-state ultraviolet lasers in future with the improvement of Ce:fluoride crystal qualities and high-power ultraviolet CW pumping sources.

It is believable that, with improvement in the quality and size of UV laser crystals and nonlinear crystals (which are important for the CW UV pumping sources), all-solid-state, compact, ultrashort pulse, ultraviolet, tunable Ce:fluoride lasers will be possible in the near future. Further development of laser systems using these new laser media will open up new possibilities for simple and compact tunable UV ultrashort-pulse laser light sources.

REFERENCES

1. GJ Megie, G Ancellet, J Pelon. Lidar measurements of ozone vertical profiles. Appl Opt 24: 3454–3463, 1985.
2. EV Browell. Applications of lasers in remote sensing. Advanced Solid-State Lasers, OSA Technical Digest, 1995, pp 2–4.
3. PF Moulton. Spectroscopic and laser characteristics of Ti:Al_2O_3. J Opt Soc Am B 3: 125–133, 1986.
4. JC Walling, OG Peterson, HP Jenssen, RC Morris, EW O'Dell. Tunable alexandrite lasers. IEEE J Quantum Electron. QE-16: 1302–1315, 1980.
5. SA Payne, LL Chase, HW Newkirk, LK Smith, WF Krupke. LiCaAlF$_6$:Cr^{3+}: a

promising new solid-state laser material. IEEE J Quantum Electron. 24: 2243–2252, 1988.

6. SA Payne, LL Chase, LK Smith, WL Kway, HW Newkirk. Laser performance of $LiSrAlF_6:Cr^{3+}$. J Appl Phys 66: 1051–1056, 1989.

7. WL Pryor. Laser Focus World: Buyers' Guide. Tulsa, OK: PennWell, 1994.

8. JF Pinto, L Esterowitz, GH Rosenblatt. Tunable UV source based on tripled Cr:LiSAF. Advanced Solid-State Lasers, OSA Technical Digest, 1995, pp 279–281.

9. D Strickland, G Mourou. Compression of amplified chirped optical pulses. Opt Commun 56: 219–221, 1985.

10. DJ Ehrlich, PF Moulton, RM Osgood, Jr. Ultraviolet solid-state Ce:YLF laser at 325 nm. Opt Lett 4: 184–186, 1979.

11. DJ Ehrlich, PF Moulton, RM Osgood, Jr. Optically pumped $Ce:LaF_3$ laser at 286 nm. Opt Lett 5: 339–341, 1980.

12. MA Dubinskii, RY Abdulsabirov, SL Korableva, AK Naumov, VV Semashko. New solid-state active medium for tunable ultraviolet lasers. 18th International Quantum Electronics Conference, OSA Technical Digest. Washington, DC Optical Society of America, 1992, pp 548–550.

13. MA Dubinskii, RY Abdulsabirov, SL Korableva, AK Naumov, VV Semashko. A new active medium for a tunable solid-state UV laser with an excimer pump. Laser Phys 4: 480–484, 1994.

14. MA Dubinskii, VV Semashko, AK Naumov, RY Abdulsabirov, SL Korableva. Active medium for all-solid-state tunable UV laser. In: AA Pinto, TY Fan, eds. OSA Proceedings on Advanced Solid-State Lasers. Washington, DC: Optical Society of America, 1993, vol 15, pp 195–198.

15. MA Dubinskii, VV Semashko, AK Naumov, RY Abdulsabirov, SL Korableva. Spectroscopy of a new active medium of a solid-state UV laser with broadband single-pass gain. Laser Phys 3: 216–217, 1993.

16. MA Dubinskii, VV Semashko, AK Naumov, RY Abdulsabirov, SL Korableva. Ce^{3+}-doped colquiriite, a new concept of all-solid-state tunable ultraviolet laser. J Mod Opt 40: 1–5, 1993.

17. JF Pinto, GH Rosenblatt, L Esterowitz, GJ Quarles. Tunable solid-state laser action in $Ce^{3+}:LiSrAlF_6$. Electron. Lett 30: 240–241, 1994.

18. CD Marshall, SA Payne, JA Speth, WF Krupke, GJ Quarles, V Castillo, BHT Chai. Ultraviolet laser emission properties of Ce^{3+}-doped $LiSrAlF_6$ and $LiCaAlF_6$. J Opt Soc Am B 11: 2054–2065, 1994.

19. KH Yang, JA Deluca. UV fluorescence of cerium-doped lutetium and lanthanum trifluorides, potential tunable coherent sources from 2760 to 3220 A. Appl Phys Lett 31: 594–596, 1977.

20. RW Waynant. Vacuum ultraviolet laser emission from $Nd^{+3}:LaF_3$. Appl Phys B 28: 205, 1982.

21. WJ Miniscalco, JM Pellegrino, WM Yen. Measurements of excited-state absorption in $Ce^{3+}:YAG$. J Appl Phys 49: 6109–6111, 1978.

22. RR Jacobs, WF Krupke, MJ Weber. Measurement of excited-state absorption loss for Ce^{3+} in $Y_3Al_5O_{12}$ and implications for 5d-4f rare-earth lasers. Appl Phys Lett 33: 410–412, 1978.

23. DS Hamilton. Trivalent cerium doped crystals as tunable system. Two bad apples. In: P Hammerling, AB Budgor, A Pinto, eds. Tunable Solid State Lasers. Berlin: Springer, 1985, pp 80–90.

24. KS Lim, DS Hamilton. Optical gain and loss studies in Ce^{3+}:$YLiF_4$. J Opt Soc Am B 6: 1401–1406, 1989.

25. KS Lim, DS Hamilton. UV-induced loss mechanisms in a Ce^{3+}:$YLiF_4$ laser. J Luminescence 40 & 41: 319–320, 1988.

26. JF Owen, PB Dorain, T Kobayasi. Excited-state absorption in Eu^{2+}:CaF_2 and Ce^{3+}:YAG single crystals at 298 and 77 KJ Appl Phys 52: 1216–1223, 1981.

27. DS Hamilton, SK Gayen, GJ Pogatshnik, RD Ghen. Optical-absorption and photo-ionization measurements from the excited states of Ce^{3+}:$Y_3Al_5O_{12}$. Phys Rev B 39: 8807–8815, 1989.

28. GJ Pogatshnik, DS Hamilton. Excited-state photoionization of Ce^{3+}ions in Ce^{3+}:CaF_2. Phys Rev B 36: 8251–8257, 1987.

29. MA Dubinskii, AC Cefalas, E Sarantopoulou, SM Spyrou, CA Nicolaides, RY Abdulsabirov, SL Korableva, VV Semashko. Efficient LaF_3:Nd^{3+}-based vacuum-ultraviolet laser at 172 nm. J Opt Soc Am B 9: 1148–1150, 1992.

30. S Nakamura, M Senoh, S Nagahama, N Iwasa, T Yamada, T Matsushita, Y Sugimoto, H Kiyoku. Continuous-wave operation of InGaN multi-quantum-well-structure laser diodes at 233 K Appl Phys Lett 69: 3034–3036, 1996.

31. N Sarukura, Z Liu, Y Segawa, K Edamatsu, Y Suzuki, T Itoh, VV Semashko, AK Naumov, SL Korableva, RY Abdulsabirov, MA Dubinskii. Ce^{3+}:$LiLuF_4$ as a broad band ultraviolet amplification medium. Opt Lett 20: 294–296, 1995.

32. RY Abdulsabirov, MA Dubinskii, BN Kazakov. Sov Phys Crystallogr 32: 559, 1987.

33. LM Frantz, JS Nodvick. Theory of pulse propagation in a laser amplifier. J Appl Phys 34: 2346–2349, 1963.

34. N Sarukura, MA Dubinskii, Z Liu, VV Semashko, AK Naumov, SL Korableva, R Yu. Abdulsabirov, K Edamatsu, Y Suzuki, T Itoh, Y Segawa. Ce^{3+} activated fluoride crystals as prospective active media for widely tunable ultraviolet ultrafast lasers with direct 10-nsec pumping. IEEE J Select Topics Quant Electron 1: 792–804, 1995.

35. DE Spence, PN Kean, W Sibbett. 60-fsec pulse generation from a self-mode-locked Ti:sapphire laser. Opt Lett 16: 42–44, 1991.

36. N Sarukura, Y Ishida, and H Nakano. Generation of 50-fsec pulses from a pulse-compressed, cw, passively mode-locked Ti:sapphire laser. Opt Lett 16: 153–155, 1991.

37. C Lin, CV Shank. Subnanosecond tunable dye laser pulse generation by controlled resonator transients. Appl Phys Lett 26: 389–391, 1975.

38. D Roess. Giant pulse shortening by resonator transients. J Appl Phys 37: 2004–2006, 1966.

39. N Sarukura, Z Liu, Y Segawa, VV Semashko, AK Naumov, SL Korableva, RY Abdulsabirov, MA Dubinskii. Direct and passive sub-nanosecond pulse-train generation from a self-injection-seeded ultraviolet solid-state laser. Opt Lett 20: 599–601, 1995.

40. N Sarukura, Z Liu, S Izumida, MA Dubinskii, RY Abdulsabirov, SL Korableva. All-solid-state tunable ultraviolet sub-nanosecond laser with direct pumping by the fifth harmonic of an Nd:YAG laser. Appl Opt 37: 6446–6448, 1998.

41. Y Mori, S Nakajima, A Taguchi, A Miyamoto, M Inakaki, W Zhou, T Sasaki, S Nakai. Nonlinear optical properties of cesium lithium borate. Jpn J Appl Phys 34: L296–L298, 1995.

42. R Komatsu, T Sugawara, K Sassa, N Sarukura, Z Liu, S Izumida, S Uda, T Fukuda, K Yamanouchi. Growth and ultraviolet application of $Li_2B_4O_7$ crystals: generation of the fourth harmonic and fifth harmonics of Nd: $Y_3Al_5O_{12}$ lasers. Appl Phys Lett 70: 3492–3494, 1997.

43. N Sarukura, Z Liu, Y Segawa, VV Semashko, AK Naumov, SL Korableva, RY Abdulsabirov, MA Dubinskii. Ultraviolet subnanosecond pulse train generation from an all-solid-state Ce:LiCAF laser. Appl Phys Lett 67: 602–604, 1995.

44. Z Liu, N Sarukura, MA Dubinskii, RY Abdulsabirov, SL Korableva. All-solid-state subnanosecond tunable ultraviolet laser sources based on Ce^{3+}-activated fluoride crystals. J Nonlinear Opt Physics Mater 8: 41–54, 1999.

45. YS Liu. Nanosecond pulse generation from a self-injected laser-pumped dye laser using a novel cavity-flipping technique. Opt Lett 3: 167–169, 1978.

46. YS Liu. Generation of high-power nanosecond pulses from a Q-switched Nd : YAG oscillator using intracavity-injecting technique. Opt Lett 4: 372–374, 1979.

47. P Ewart. Frequency tunable, nanosecond duration pulses from flashlamp pumped dye lasers by pulsed Q-modulation. Opt Commun 28: 379–382, 1979.

48. CH Brito Cruz, E Palange, F De Martini. High power subnanosecond pulse generation in Nd-YAG lasers. Opt Commun 39: 331–333, 1981.

49. CH Brito Cruz, F De Martini, HL Fragnito, E Palange. Picosecond pulse generation by intracavity nonlinear compression in self-injected Nd-YAG laser. Opt Commun 40: 298–301, 1982.

50. EG Erickson, LL Harper, in Digest of Technical Papers, OSA/IEEE Conference on Laser Engineering and Applications. New York: (Institute of Electrical and Electronics Engineers, 1975, p 16.

51. MJ LaGasse, RW Schoenlein, JG Fujimoto, PA Schulz. Amplification of femtosecond pulses in Ti:Al_2O_3 using an injection-seeded laser. Opt Lett 14: 1347–1349, 1989.

52. PF Moulton. Tunable paramagnetic-ion lasers. In: M Bass and ML Stitch, eds. Laser Handbook. Amsterdam: Elsevier Science Publishers BV, 1985, p 284.

53. Z Liu, H Ohtake, N Sarukura, MA Dubinskii, RY Abdulsabirov, SL Korableva, AK Naumov, VV Semashko. Subnanosecond tunable ultraviolet pulse generation from a low-Q, short-cavity Ce:LiCAF laser. Jpn J Appl Phys 36: L1384–L1386, 1997.

54. Z Liu, N Sarukura, MA Dubinskii, RY Abdulsabirov, SL Korableva, AK Naumov, VV Semashko. Tunable ultraviolet short-pulse generation from a Ce:LiCAF laser amplifier system and its sum-frequency mixing with an Nd: YAG laser. Jpn J Appl Phys 37: L36–L38, 1998.

55. N Sarukura, Z Liu, H Ohtake, Y Segawa, MA Dubinskii, RY Abdulsabirov, SL Korableva, AK Naumov, VV Semashko. Ultraviolet short pulses from an all-solid-state Ce:LiCAF master oscillator and power amplifier system. Opt Lett 22: 994–996, 1997.

56. N Sarukura, Y Ishida. Ultrashort pulse generation from a passively mode-locked Ti:sapphire laser based system. IEEE J Quantum Electron. 28: 2134–2141, 1992.

57. K Shimamura, N Mujilatu, K Nakano, SL Baldochi, Z Liu, H Ohtake, N Sarukura,

and T Fukuda. Growth and characterization of Ce-doped $LiCaAlF_6$ single crystals. J Crystal Growth 197: 896–900, 1999.

58. Z Liu, S Izumida, H Ohtake, N Sarukura, K Shimamura, N Mujilatu, SL Baldochi, T Fukuda. High-pulse-energy, all-solid-state, ultraviolet laser oscillator using large Czochralski-grown Ce:LiCAF crystal. Jpn J Appl Phys 37: L1318–L1319, 1998.

59. Z Liu, K Shimamura, K Nakano, T Fukuda, T Kozeki, H Ohtake, N Sarukura. High-pulse-energy ultraviolet Ce:LiCAF laser oscillator with newly designed pumping schemes. Jpn J Appl Phys 39: L466–L467, 2000.

60. Z Liu, K Shimamura, K Nakano, N Mujilatu, T Fukuda, T Kozeki, H Ohtake, N Sarukura. Direct generation of 27-mJ, 309-nm pulses from a Ce:LLF oscillator using a large-size Ce:LLF crystal. Jpn J Appl Phys 39: L88–L89, 2000.

61. P Rambaldi, M Douard, J-P Wolf. New UV tunable solid-state lasers for lidar applications. Appl Phys B 61: 117–120, 1995.

62. JF Pinto, L Esterowitz. Distributed-feedback, tunable Ce^{3+}-doped colquiriite lasers. Appl Phys Lett 71: 205–207, 1997.

63. AB Petersen. All solid-state, 228–240 nm source based on Ce:LiCAF. Advanced Solid-State Lasers, OSA Technical Digest, 1997, paper PD2.

64. AJS McGonigle, DW Coutts, CE Webb. 530-mW 7-kHz cerium LiCAF laser pumped by the sum-frequency-mixed output of a copper-vapor laser. Opt Lett 24: 232–234, 1999.

65. DA Hammons, MC Richardson, BHT Chai, M Bass. Spectroscopic properties of Ce^{3+} in orthosilicate, garnet, and fluoride crystals. OSA Trends in Optics and Photonics. Washington DC: Optical Society of America, vol 10, Advanced Solid State Lasers, 1997, p 35.

66. MA Dubinskii, KL Schepler, VV Semashko, R Yu. Abdulsabirov, BM Galjautdinov, SL Korableva, AK Naumov. Ce^{3+}:$LiBaF_3$ as new prospective active material for tunable UV laser with direct UV pumping. OSA Trends in Optics and Photonics. Washington, DC: Optical Society of America, vol 10, Advanced Solid State Lasers, 1997, p 30.

12

Diode-Pumped Picosecond UV Lasers by Nonlinear Frequency Conversions

François Balembois, Frédéric Druon, Patrick Georges, and Alain Brun
Laboratoire Charles Fabry de l'Institut d'Optique, Orsay, France

I. INTRODUCTION

For many applications there is an intense interest in simple, compact, all-solid-state pulsed ultraviolet (UV) laser sources. For example, fluorescence lifetime spectroscopy requires UV pulses with a typical pulse duration of 100 ps, which is shorter than the fluorescence lifetime of a large number of molecules. This application also requires a high repetition rate, typically between 10 kHz and 1 MHz, to be compatible with photon counting detection chains and to ensure a fast acquisition process of the fluorescence decay signal. The average power required is at the milliwatt level. For example, at a 10 kHz repetition rate, the pulse energy required to reach the milliwatt average power level is 0.1 μJ. To answer the need for compactness, efficiency, and reliability, the key sources are semiconductors. However, laser diodes cannot directly produce energetic UV short pulses with high spatial quality. It is necessary to convert the light emitted by laser diodes with laser crystals that can produce energetic picosecond pulses in the near infrared and with frequency converters (nonlinear crystals) in order to reach the UV band.

The laser sources we have developed are based on diode-pumped solid-state laser crystals emitting in the near infrared. They all consist of an oscillator producing picosecond pulses, an amplifier to boost the energy, and nonlinear stages to reach the UV. The sources operate at a high repetition rate between 10

451

and 65 kHz, limited either by the storage lifetime of the crystals or by the amplification process itself.

Three techniques have been used to produce the picosecond pulses: active mode locking of a laser cavity, gain-switching of a laser diode, and passive Q-switching of a microlaser with a very short cavity length.

The first UV laser source we developed is based on a tunable diode-pumped Cr:LiSAF mode-locked oscillator. The output pulses were amplified in an indirectly diode-pumped tunable Ti:sapphire regenerative amplifier. Frequency tripling was achieved by two LBO (LiB_3O_5) crystals to reach the UV band (273–285 nm).

The second laser source used a gain-switched laser diode for seeding the previous Ti:sapphire regenerative amplifier. An LBO crystal and a $\beta B_2 O_4$ (BBO) crystal were used to convert the wavelength to the UV (268–280 nm).

The last source was based on diode-pumped Nd-doped crystals. A Nd:YAG microchip laser passively Q-switched by a Cr:YAG crystal was amplified in a geometrical passive multipass Nd:YVO_4 amplifier. KTP or LBO crystals were used for second harmonic generation. LBO and BBO crystals were used for frequency tripling to 355 nm and frequency quadrupling to 266 nm, respectively.

II. TUNABLE DIODE-PUMPED UV CHAIN BASED ON A MODE-LOCKED Cr:LiSAF OSCILLATOR AND Ti:SAPPHIRE REGENERATIVE AMPLIFIER

This work has been done in collaboration with the University of Limoges in France (IRCOM) [1].

Tunable UV can be very useful to excite specific molecules in their absorption bands. Tunable laser crystals with a large fluorescence bandwidth are not very numerous. In the UV, Ce^{3+}:LiSAF (Ce^{3+}:LiSrAlF$_6$) has proven to be an efficient crystal [2]. However, it needs to be pumped by a frequency-quadrupled Nd:YAG laser operating at a low repetition rate [2], so it is not convenient for our application. In the visible spectrum, praseodymium-doped YLF (Pr:YLiF$_4$) is the only efficient laser crystal but it is not tunable [3]. Moreover, it has to be pumped in the blue by an ionized argon laser. In the near infrared, the Ti:sapphire crystal has a wide fluorescence bandwidth between 680 and 1080 nm. However, it has to be pumped in the blue–green wavelength range where no high-power laser diodes are available. In the same wavelength range, a new crystal from the colquiriite family was developed in the early 1990s; Cr:LiSAF (Cr^{3+}:LiSrAlF$_6$). This presents an absorption bandwidth in the red where high-power InGaAlP laser diodes (up to 4 W) are available. We have investigated Cr:LiSAF lasers in different regimes and under different pumping schemes [4,5]. This crystal appeared to work well at low pump power levels (typically 1 W maximum) but

suffered thermal problems at higher pump power level. Moreover, the small signal gain remains limited in this crystal to 1.1 (typically), which is not enough to
develop an efficient amplifier. These different results have led us to use a Cr:
LiSAF crystal only for a low-power oscillator producing the picosecond pulses.
For that, the only solution is a mode-locked oscillator. Concerning the amplifier,
we decided to use a Ti:sapphire crystal that combines excellent physical properties (high thermal conductivity, high emission cross section) with an emission in
the fluorescence bandwidth of Cr:LiSAF. We pumped it with a diode-pumped
frequency-doubled Nd:YVO$_4$ laser emitting pulses at 532 nm in Q-switched operation.

A. Actively Mode-Locked Cr:LiSAF Oscillator

The oscillator is a diode-pumped actively mode-locked Cr:LiSAF laser (Fig. 1).
The cavity was a Z-fold cavity including the gain medium at Brewster angle, an
acousto-optic modulator for mode-locking, and a Lyot filter to tune the laser
wavelength. The Cr:LiSAF was pumped by a red diode emitting 400 mW at 670
nm. The pumping optics comprised a collimator, a cylindrical telescope, and a

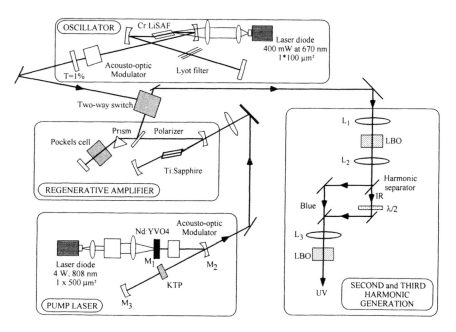

Figure 1 Experimental set up for the Cr:LiSAF, Ti:sapphire laser system used for UV
generation.

doublet focusing the beam inside the Cr:LiSAF crystal. This oscillator produced tunable pulses from 820 to 880 nm at a repetition rate of 100 MHz and with a typical pulse duration of 80 ps. At a wavelength of 850 nm, corresponding to the maximum of the gain curve for Cr:LiSAF, the energy per pulse was 0.15 nJ. The peak power was only 2 W at this wavelength, making any efficient nonlinear conversion (doubling and tripling) impossible.

B. Ti:Sapphire Regenerative Amplifier

In order to get higher peak powers, we increased the energy per pulse in the regenerative amplifier. Because Cr:LiSAF has a long fluorescence lifetime (67μs), it has a good energy storage capability and seems to be well suited for an amplifier. In fact, a diode-pumped Cr:LiSAF regenerative amplifier has already been developed [6]. However, this system required four pumping diodes and also intracavity elements with very low losses, since the double-pass gain in this configuration was low (lower than 1.25). We have shown that the small signal gain in Cr:LiSAF crystals pumped in continuous wave is limited by up-conversion and thermal quenching of fluorescence [7]. These effects all make the use of a highly doped crystal unsuitable. Moreover, the use of a low doped crystal leads to a bad overlap between the pump beam and the cavity beam. Therefore, despite its good storage capability, we concluded that a diode-pumped Cr:LiSAF is not suitable as an amplifier, because the energy extraction is limited by its low gain.

We investigated another possibility with a Ti:sapphire crystal, since it has a common fluorescence band with the Cr:LiSAF crystal. This crystal is well suited to amplification since it has an emission cross section eight times higher than that of Cr:LiSAF. However, its absorption band, located in the blue–green range, makes direct diode-pumping currently impossible. To overcome this problem, one solution is to pump the Ti:sapphire crystal in the green using a diode-pumped frequency-doubled Nd:YVO$_4$ laser. Another problem is that Ti:sapphire has a poor storage capability due to its low fluorescence lifetime (3 μs). One solution is to Q-switch the pump laser. By this means, the pump energy is delivered to the crystal in a duration much shorter than its lifetime (typically 50 ns). Moreover, since the Nd:YVO$_4$ crystal has both good storage capability for pump energy (lifetime of 115 μs) and a very good small signal gain (emission cross section of 30×10^{-19} cm^2), Q-switched operation under cw diode-pumping is very efficient. It can be all the more efficient since high-power and high-brightness laser diodes already exist at the pump wavelength (808 nm).

Following these ideas, we developed a Ti:sapphire regenerative amplifier (Fig. 1). The pump laser used only one laser diode emitting 4 W at 808 nm (SDL 2380). After beam-shaping, the pump light is focused on the Nd:YVO$_4$ crystal. The first mirror of the cavity is directly coated on the laser crystal (M$_1$). The others (M_2 and M_3) are concave mirrors with a radius of curvature of 100 mm. At the waist located between M_2 and M_3, we placed a KTP crystal for second-harmonic generation. All the cavity mirrors are highly reflective at 1064 nm; M_3 was also highly reflective at 532 nm while M_2 has a high transmission at 532 nm. An acousto-optic modulator was introduced near M_1 to Q-switch the laser. This source produced 40 µJ green pulses with a duration of 50 ns at a repetition rate of 10 kHz.

These pulses were used to pump the Ti:sapphire crystal of the regenerative amplifier (Fig. 1). It was a three-mirror cavity with two concave mirrors (100 mm radius of curvature) around the crystal and a plane mirror. All these mirrors supported a high-reflectivity coating between 800 and 900 nm. The Ti:sapphire crystal was 13 mm long and absorbed 70% of the pump power at 532 nm. We recycled the transmitted 30% of the pump power using a concave mirror placed after the second curved mirror of the cavity. For simplicity, this mirror is not shown in Figure 1. A prism was used to tune the amplifier in order to match the operating wavelength of the oscillator, which is shifted from the maximum of the gain curve of Ti:sapphire. A Pockels cell and a polarizer were used to inject a single pulse prior to the amplification and to dump it when the amplification process reached its optimum. A two-way switch consisting of a Faraday rotator, a half-wave plate, and a polarizer was used to separate the ejected beam from the injected one.

At 820 nm, we measured a small signal gain per double pass higher than 1.6. The build-up time to reach the saturation of the gain was relatively low (200 ns corresponding to 50 cavity round trips). As a comparison, the build-up time was 820 ns (corresponding to around 175 round trips) in the case of the Cr:LiSAF regenerative amplifier developed by Mellish et al. [6]. The energy of the amplified pulses (vs. wavelength) is shown in Figure 2. The system (oscillator and amplifier) could be tuned between 820 and 870 nm. The shortest wavelength was imposed by the Cr:LiSAF crystal, whose gain curve is centered at 850 nm. At 820 nm the oscillator was near the oscillation threshold. The longest wavelength was fixed by the Ti:sapphire crystal, whose gain curve is centered at 800 nm. However, the common fluorescence band between these two crystals is sufficient to ensure output pulses with an energy of more than 3 µJ over a bandwidth of 30 nm from our system.

C. Harmonic Generation

The amplified pulses were frequency-converted to the blue and to the UV by harmonic generation. Since the peak power in the infrared remained low, tight focusing of the amplified laser beam was needed to ensure a good conversion efficiency. The small size of the beams in interaction required us to consider the strong influence of the walk-off angles for the choice of the most efficient crystal.

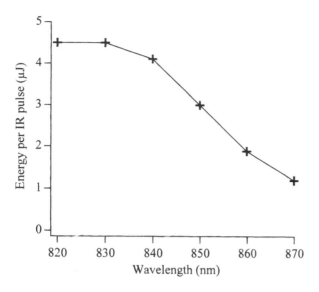

Figure 2 Energy per pulse in the infrared vs. wavelength.

Around 800 nm, three nonlinear crystals provide efficient second-harmonic generation (SHG): LBO, BBO, and LiIO$_3$ [8]. We selected an LBO crystal as described by Nebel and Beigang [9]. Although BBO and LiIO$_3$ can lead to better efficiency [9], they have a walk-off angle significantly larger than LBO. Crystals with a large walk-off produce a harmonic beam with a poor spatial quality, which is a drawback for subsequent nonlinear mixing processes. We chose type I phase matching corresponding to a nonlinear coefficient d_{eff} = 1.25 pm/V (at λ = 830 nm), which is 2.5 times larger than that for type II phase-matching.

For the third-harmonic generation (THG), we selected an LBO crystal with a type I phase matching, as in [10], for the following reasons: (a) to have a good spatial profile of the UV beam for efficient coupling into an optical fiber we rejected the BBO because of its large walk-off (nearly 5 degree) even if a higher conversion efficiency could be expected with this crystal [9]; (b) LiIO$_3$ is not transparent below 300 nm; and (c) type II phase matching is not possible for THG in LBO in this wavelength range.

The set-up for doubling and tripling is shown in Figure 1. A lens L$_1$ (200 mm focal length) focused the infrared beam inside the first LBO crystal. This crystal was cut for type I phase matching at 850 nm (θ = 90 degrees, ϕ = 27 degrees) and its length was 8 mm. We obtained blue pulses with an energy of more than 0.5 μJ over a bandwidth of 10 nm (Fig. 3). By taking a gaussian temporal shape for the infrared (IR) pulse, we calculated that the pulse duration in the blue was 60 ps.

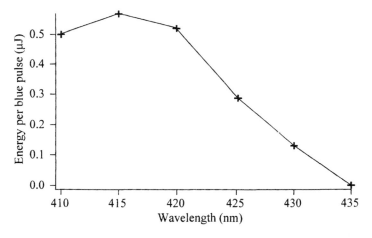

Figure 3 Energy per pulse in the blue vs. wavelength.

The two beams were then collimated by a 100 mm focal-length lens (L_2) and separated by a dichroic mirror. A half wave plate rotated the polarization in the infrared in order to satisfy the type I phase-matching conditions in the second LBO crystal. This crystal was cut in type I phase matching for sum frequency generation of 860 nm and 430 nm ($\theta = 90$ degrees; $\phi = 61$ degrees). Its length was 8 mm. Another dichroic mirror recombined the two pulses and a lens L_3 (90 mm focal length) focused the beam inside the second LBO crystal.

The UV pulses have an estimated duration of 50 ps, calculated by assuming a gaussian temporal shape for the IR pulses. Figure 4 shows the energy per pulse in the UV as a function of the wavelength. We obtained more than 0.1 µJ over

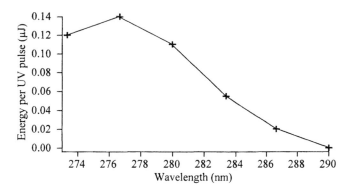

Figure 4 Energy per pulse in the UV vs. wavelength.

a bandwidth of 6 nm between 274 and 280 nm. Since the repetition rate was 10 kHz, this corresponded to more than 1 mW of average power. These performances could be improved by the use of suitably cut crystals, since the experiments reported here were carried out with crystals cut for SHG of 850 nm radiation and SFG of 860 nm and 430 nm wavelengths. Higher UV powers may be obtained with a BBO THG crystal at the expense of a highly elliptical beam shape.

III. GAIN-SWITCHED PICOSECOND LASER DIODE SEEDING A TI:SAPPHIRE REGENERATIVE AMPLIFIER

This work was has been performed in collaboration with the Institut d'Electronique Fondamentale in Orsay (France) [11].

To simplify the source described in Sec. II, we have investigated the possibility of replacing the mode-locked Cr:LiSAF oscillator with a semiconductor laser. In this case, tuning can be achieved, for example, by varying the temperature of the laser diode. For a given semiconductor laser, tunability is on the order of 10 nm in the infrared, for a temperature excursion range of 30°C, but it would be possible to have a few low-cost semiconductor lasers with different emission wavelengths to fulfill the right tuning range.

Compact systems have been realized with a mode-locked external-cavity laser diode producing femtosecond pulses [12] and used to seed a high-energy flashlamp-pumped regenerative amplifier. However, the mode-locking technique has a disadvantage: it produces pulses at a very high repetition rate (100 MHz typically). This pulse rate is then decreased to the slower repetition rate of the amplifier (typically 1–250 kHz). This means that almost all the oscillator power (more than 99.99%) is lost for the amplification process.

Conversely, a laser diode can operate in the gain-switched regime at the right repetition rate and can produce picosecond pulses (below 100 ps), thanks to its high gain and low cavity lifetime. Such very simple pulsed sources have already been used in combination with amplifiers. Seeding an erbium-doped fiber amplifier with gain-switched laser diode pulses has been reported around 1.5 μm [13]. A Ti:sapphire regenerative amplifier, pumped by a flashlamp pump frequency-doubled Nd:YAG laser, and a flashlamp-pumped alexandrite regenerative amplifier were both injected by a picosecond gain-switched laser diode [14,15]. Energies reaching the millijoule level at a few Hertz repetition rate were reported.

In our case, we have used the same indirectly diode-pumped Ti:sapphire regenerative amplifier as described in Sec. II. We replaced the entire Cr:LiSAF mode-locked oscillator by a gain-switched laser diode driven by a pulse generator.

A. Gain-Switched Laser Diode as Seeder
of the Regenerative Amplifier

The laser diode consisted of a separate-confinement single-quantum-well structure with a 170 Å GaAs-active layer in an AlGaAs waveguide. The laser stripe
was 5 μm wide and 2 mm long, and lasing occured on a single transverse mode.
To increase the output efficiency, the front face was antireflection coated (reflection coefficient of 1%) and the back face was high reflection coated (reflection
coefficient of 95%). The gain-switching regime was achieved by pumping the
laser diode with a pulse generator capable of delivering 200 ps current pulses at
repetition rates up to 100 kHz [16]. In these conditions, the laser output spread
over many longitudinal modes. Due to chirping effects on each longitudinal
mode, the pulse spectrum was quasicontinuous and spanned from 800 to 840 nm
with a maximum amplitude at 812 nm (Fig. 5). The pulse duration was 200 ps
and the pulse energy was 22 pJ.

The experimental set up is shown schematically in Figure 6. The picosecond
laser diode output beam was collimated with a high numerical aperture (NA)
objective L_1 (NA = 0.5, focal length = 8 mm). Two cylindrical lenses with focal
lengths of +100 mm and −25.4 mm, respectively, were used to reshape the output
beam and match the beam profile to the regenerative amplifier. The amplifier and
its pump laser have been previously described in Section II. To prevent all feedback from the amplifier to the laser diode, we added a Faraday isolator. The
overall isolation from the amplifier to the picosecond laser diode was then 60 dB.

All the devices were synchronized with the internal clock of the Pockels
cell driver (MEDOX Electro-optics). A first voltage step was applied to the driver
of the acousto-optic modulator to trigger the pump pulse at 532 nm. A second
voltage step was applied to the pulse generator of the picosecond laser diode.
The delay between the two voltage steps was adjusted to synchronize the picosec-

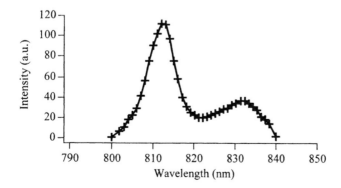

Figure 5   Spectrum of pulses emitted by the gain-switched laser diode.

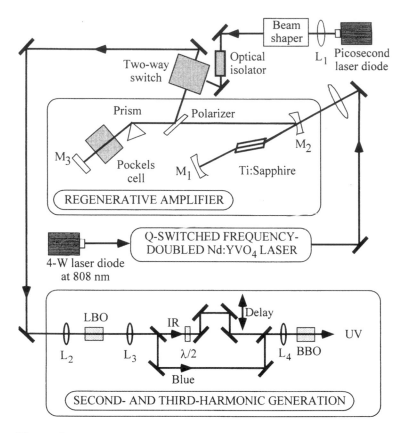

Figure 6 Experimental set up of a Ti:sapphire amplifier seeded by a gain-switched laser diode. Focal lengths of lenses L_1, L_2, L_3, and L_4 are 8 mm, 200 mm, 100 mm, and 90 mm respectively.

ond laser diode pulse injected into the regenerative amplifier with the green pump pulse. Two high-voltage steps were applied to the Pockels cell. The first voltage step was used to trap the picosecond pulse inside the cavity. The second one was used to eject it, once the amplification process had reached its maximum.

B. Performances of the Ti:Sapphire Regenerative Amplifier

Because of the presence of a prism in the regenerative amplifier cavity, only a small window of 5 nm full width at half maximum (FWHM) was selected in the spectrum of the picosecond laser diode. The central wavelength could be adjusted simply by rotating the plane mirror M_3 of the amplifier (Fig. 6). By considering the transmission of the different optics and the shape of the laser diode spectrum (Fig. 5), we estimated that the energy actually injected into the amplifier was 1.6 pJ at 812 nm (i.e., at the laser diode spectrum maximum), whereas it was only 0.3 pJ at 803 nm (at the edges of the laser diode spectrum). This energy level was very low compared to the energy of the pulses coming from the Cr:LiSAF oscillator used in the previous scheme (150 pJ at 830 nm). Figure 7 illustrates the amplification process at 812 nm and 803 nm, obtained with a fast photodiode (risetime 1 ns) monitoring a leakage of one regenerative amplifier mirror. As can be seen, the amplified spontaneous emission (ASE) cannot be neglected during this amplification process. To characterize the amplifier seeded by the gain-switched laser diode, it is important to evaluate the signal-to-background ratio (SBR). The SBR is given by the ratio of the signal energy to the ASE energy emitted during the signal pulse. The SBR can be calculated with the data given in Figure 7, corrected by the the ratio of the FWHM response of the photodiode to the actual pulse width [17]. The SBR was 314 at 812 nm, of the same order of magnitude as those previously obtained by Hariharan et al. [17], showing that the amplifier was correctly seeded. At 803 nm, the actual SBR was 43. Despite its higher sensitivity to alignment at this wavelength, the amplifier could easily be injected. Moreover, once the alignment was achieved, the SBR remained the same during all our experiments (2 hr).

To obtain the injection threshold of our amplifier, we decreased the energy emitted by the laser diode until the measured SBR was equal to 1. We measured the average power emitted by the laser diode at the input of the amplifier. By

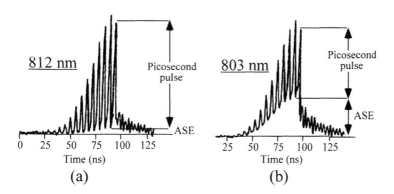

Figure 7 Pulse amplification at (a) 812 nm and (b) 803 nm obtained from a fast photodiode (1 ns risetime) monitoring leakage of one mirror of the regenerative amplifier.

taking into account the laser diode spectrum and the spectral window of 5 nm selected by the amplifier, we estimated that the injection threshold was 0.2 pJ per pulse at 812 nm.

We next studied the ejected amplified pulse. The amplifier output pulse consisted of two parts: an amplified picosecond pulse and a pulse of ASE with a duration of 5 ns that corresponded to one roundtrip in the amplifier cavity. Figure 8 shows that despite ASE, it was possible to generate picosecond pulses at a 10 kHz repetition rate with an energy between 1 and 4 μJ and a tuning range from 803 to 840 nm.

The temporal width of the amplified pulses was measured with a sampling oscilloscope (30 ps rise time) and a high-speed photodiode (25 ps rise time). A value of 70 ps (no deconvolution) was found for the amplified pulsewidth at 812 nm (Fig. 9). As we tuned the amplifier, we observed that the pulsewidth remained between 70 and 100 ps. Note that the amplified pulses were shorter than those emitted by the laser diode (200 ps). This result actually reflects the spectral selectivity of the amplifier: because the different wavelengths are emitted at different times during the laser pulse, selecting a wavelength is just equivalent to selecting a temporal window in the gain-switched pulse.

The sampling oscilloscope was triggered by an electrical signal delivered by the power supply of the picosecond laser diode. The pulse jitter deduced from the trace thickness (Fig. 9) was within the measurement accuracy (30 ps). Actually, timing jitter less than 1 ps can be expected from a gain-switched laser diode system. Present results show that the amplified optical pulses can be precisely

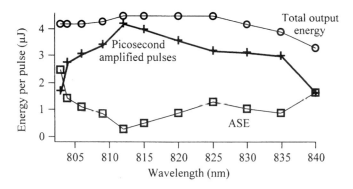

Figure 8 Tunability of infrared pulses around 820 nm. Circles represent the total energy extracted per pulse from the amplifier (including ASE). Crosses represent the energy of the amplified laser pulses. Squares correspond to ASE.

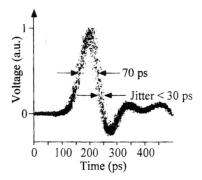

Figure 9 Temporal shape of the amplified optical pulse measured with a sampling oscilloscope.

synchronized with an electrical signal. This feature could be useful in multiple optoelectronic applications.

C. Frequency Conversion to Blue and UV

The amplified pulses were frequency-converted to the blue and the UV by harmonic generation (Fig. 6). In contrast to the set up used for the first source (Section II), we used a BBO for third-harmonic generation (8 mm length, cut in type I for sum frequency generation of 780 nm and 390 nm) and we introduced a delay line to adjust the optical path length of the infrared beam finely to that of the blue beam. The accuracy of the path length in the harmonic generator needed to maximize the UV signal was approximately 2 mm.

Spectral widths of 1 nm and 0.3 nm (FWHM) were measured in the blue and UV, respectively. This may be compared with the 5 nm (FWHM) spectrum of the infrared pulses. Spectral widths were limited by the spectral acceptance of the two nonlinear crystals. Although, the LBO and BBO crystals were not optimized for the operating wavelength range (803–840 nm), we obtained good conversion efficiency: reaching 30% from IR to blue and 12% from IR to UV. We achieved a tunability of more than 15 nm in the blue and of more than 10 nm in the UV (Fig. 10). In both cases, the pulse energy was above 0.1 µJ. For a 10 kHz repetition rate, average powers larger than 11 and 4 mW were achieved at 406 and 271 nm, respectively. As compared with the frequency conversion stage used in Sec. II, we obtained a more powerful UV beam but also more elliptical. This was due to the lower angular acceptance of BBO than LBO and to a higher nonlinear coefficient.

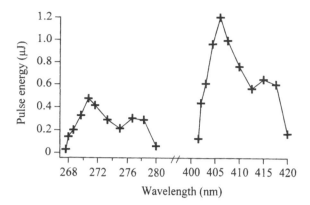

Figure 10 Tuning curves for the second- and third-harmonic generation.

To estimate the maximum contribution of ASE to the UV signal, we operated the amplifier in free running mode (i.e., without seeding) with an output energy per pulse of 4 μJ. We measured a UV energy per pulse lower than 2 nJ; that is, more than two orders of magnitude lower than the energy of picosecond UV pulses. Thus, by means of the nonlinear conversion process, the pulse was removed from its nanosecond ASE background.

The output level in the UV (at 271 nm) was monitored for 2 hr. Over this period, the seeding of the amplifier remained unchanged and the SBR remained at its initial value of 314 (at 812 nm). Peak-to-peak fluctuations were found to be ±2.5%, corresponding to IR fluctuations of less than ±0.8%. Such good stability obtained in spite of a nonlinear conversion process reflects the very good reproducibility of pulses emitted by the gain-switched laser diode as well as the stable pumping of the regenerative amplifier by a continuous wave (CW) laser diode.

The repetition frequency was increased to 25 kHz, which was the maximum value acceptable for our Pockels cell. Because of the long fluorescence lifetime of the Nd^{3+} ion in the yttrium vanadate (YVO_4) matrix (115 μs), the frequency-doubled $Nd:YVO_4$ laser then emitted green pulses of lower energy (18 μJ at 25 kHz instead of 40 μJ at 10 kHz). However, we still obtained 1 mW average power in the UV at 271 nm.

D. Improvement of Spectral Quality

Harmonic conversion efficiency was limited by the wide spectrum of the infrared pulses (5 nm FWHM). Moreover, it could be useful for spectroscopic applications to emit pulses with narrower spectra. For these reasons, we modified the injection setup shown in Figure 11. We used a 2000 groove/mm grating that spatially dispersed the beam of the gain-switched laser diode. Because the injection was highly sensitive to alignment, we could rotate the grating to select finely the wavelength of the pulses seeded into the amplifier. The optical path length between the grating and the input of the regenerative amplifier was 140 cm. With this setup, the amplified spectrum was less than 1 nm FWHM compared with 5 nm FWHM in the former experiment. The pulse duration remained unchanged: between 70 and 100 ps.

At a wavelength of 812 nm corresponding to the maximum of the seeded energy, the SBR was equal to 43. It was lower than previously measured (SBR of 314) because the energy seeded into the amplifier was only 0.3 pJ (compared with 1.6 pJ). A smaller part of the diode spectrum was selected and the transmission of the grating was only 80%.

We also compared the frequency conversion in the two cases with and without the grating. At the same injection level (with an SBR of 43) and at the same wavelength (271 nm), we obtained a UV average power of 3.4 mW with the grating and 1 mW without the grating. As expected, the spectral narrowing of the fundamental wavelength increased the nonlinear conversion efficiencies.

We achieved tunability by simultaneously rotating the grating and the back mirror of the regenerative amplifier (Fig. 6) in order to match the wavelength of the injected beam with the free running wavelength of the amplifier. The system was tunable between 810 nm and 817 nm with SBR between 20 and 43. The UV average power was between 1 mW and 3.4 mW in the spectral range 270–272.3 nm. The performance quality was lower than before because more than the half of the energy ejected from the amplifier was lost in ASE. However, the average powers obtained would be sufficient for time-resolved spectroscopic studies such as those performed on biological media.

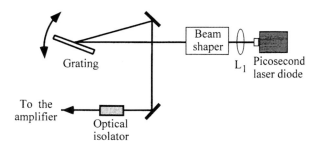

Figure 11   Set up for the injection seeding with a grating.

IV. Q-SWITCHED ND:YAG MICROLASER SEEDING A Nd:YVO₄ AMPLIFIER

This work has been performed in collaboration with Nanolase (France) [18].

The third technique investigated was based on the following observation: widely tunable crystals are generally associated with relatively low emission cross-sections and therefore with low small-signal gain. In the case of our previous Ti:sapphire regenerative amplifier, the small signal gain was approximately 1.6 per double pass. More than 50 passes were needed to reach the microjoule level. The regenerative amplifier with its costly Pockels cell is the only efficient solution for extracting the energy stored inside the Ti:sapphire crystal.

Conversely, the emission cross section is much higher for crystals emitting at fixed wavelengths. For example, the peak emission cross section is 30×10^{-19} cm^2 for Nd:YVO₄ crystals compared with 4.1×10^{-19} cm^2 for Ti:sapphire crystals. This means that the stored energy can be extracted in a few passes in the case of a Nd:YVO₄ amplifier with a simple geometrical multipass amplification scheme. For example, a small signal gain of 10 per pass can be easily achieved in a diode-pumped Nd:YVO₄ crystal: only six passes are needed to increase the pulse energy from the pJ level to the μJ level. For this case, the amplification process is totally passive; no active element such as a Pockels cell limits the repetition rate.

Different seeding sources are available for an Nd-doped amplifier operating at 1.06 μm. If we exclude the solution of a mode-locked oscillator for the reasons mentioned in Sec. III, it is possible to use a gain-swiched semiconductor laser operating at 1.06 μm. However, the energy per pulse is only in the picojoule range, due to the short fluorescence lifetime in semiconductors. The second possibility is to use a passively Q-switched diode-pumped Nd:doped microlaser [19–21]. In this case, thanks to the long fluorescence lifetime of the Nd^{3+} ion (230 μs in YAG), the energy per pulse is six orders of magnitude higher, reaching the microjoule level. The high energy level of the seeded pulses contributes to a good extraction of the stored energy inside the amplifier within a few passes. Moreover, this monolithic device is very simple to use because of its totally passive process and extremely compact due to the small size of the microchip and the diode-pumping system.

However, it is difficult to fulfill the source requirements defined in the introduction (high repetition rate and picosecond pulses) with a passively Q-switched microchip laser in standard use. To overcome this problem, we used a high-brightness laser diode to pump a passively Q-switched Nd:YAG microchip laser. Afterwards, we amplified the output pulses in a Nd:YVO₄ multipass amplifier.

A. Passively Q-Switched Microchip Laser

Our experiment was performed with a microchip laser composed of a 750 μm long piece of gain medium (Nd^{3+}:YAG) on which was grown, by liquid phase epitaxy, a 60 μm layer of saturable absorber (Cr^{4+}:YAG) [22]. The concentration of Nd^{3+} and Cr^{4+} ions was 1.2×10^{20} cm^{-3} and 2.18×10^{18} cm^{-3}, respectively. The output–coupler–mirror transmission was 5%.

We measured a pulse duration of 500 ps with a fast photodiode and a fast sampling oscilloscope. The predicted pulse duration given by the theory [23,24] for this device was 520 ps, which is in good agreement with our experimental measurement.

This kind of short-pulse duration microchip laser is usually pumped with a fiber-coupled CW laser diode and would produce pulses at a relatively low repetition rate. For example, with a 0.85 W, 100 μm core-diameter fiber-coupled diode, the repetition rate would be 5.9 kHz and the average power 15 mW. In fact, for the production of 500 ps pulses, our microchip concept consists in having a short length of gain medium and a saturable absorber with a high ion concentration, both of which are incompatible with a high repetition rate.

To increase the repetition rate, the idea would be to modify the microchip laser structure. To keep the pulse duration short, one should make this modification for the Nd concentration, which requires a different technology. An alternative method of obtaining a higher repetition rate is to optimize the pumping system. We found experimentally that the higher the power density on the laser diode-emitting area, the higher the repetition rate for a given pump power (Table 1). First, we used a standard 1 W, 1×100 μm^2 junction CW laser diode whose beam was reshaped using cylindrical lenses optimized for the repetition rate. This

Table 1 Influence of Pump System Brightness on Microlaser Performance

Pump at 808 nm	0.85 W LD fiber coupled 100 μm diameter	1W, 1×100 μm^2 LD	1W, 1×50 μm^2 LD	2W, 1×100 μm^2 LD	Ti:Sapphire
Incident pump power (W)	0.85	0.9	0.9	1.8	0.67
M^2 (pump source)	18	10	5	10	1
Repetition rate (kHz)	5.9	21	41	45	101
Energy per pulse (μJ)	2.5	0.58	0.37	0.89	0.28
Average power (mW)	15	15	15	40	28

LD, laser diode.

technique allowed, with approximately the same incident power, a repetition rate 3.6 times higher than in the case of the fiber-coupled diode. A second step was to use very high brightness diodes. We used a 1W, 1×50 μm^2 laser diode, which improved the repetition rate by a factor of 6.9. Finally, we pumped with a 2W, 1×100 μm^2 CW laser diode from SDL, Inc. Because its power density was the same as the previous 1W, 1×50 μm^2 diode, the repetition rate was approximately the same, but its power was twice as high, which allowed a higher average output power of 40 mW.

In order to estimate an upper limit to the repetition rate, at a given pump power, we pumped the microchip laser with a diffraction-limited beam at 808 nm from a CW Ti:sapphire laser. The highest repetition rate obtained was 101 kHz at 0.67 W incident pump power. This pump power corresponded to the limit for single-longitudinal-mode operation with a stable pulse train. With a higher incident power, multiple longitudinal modes began to appear, and the pulse train became erratic.

B.   Nd:YVO$_4$ Passive Multipass Amplifier

To increase the energy per pulse, but without reducing the repetition rate, we used a geometric multipass amplifier configuration. The main advantage of a multipass amplifier is its totally passive process, which induces no repetition-rate limitation. For simplicity, efficiency, and compactness we chose a Nd:YVO$_4$ crystal because of its high-emission cross section at the 1064 nm-emitted wavelength of the Nd:YAG/Cr:YAG microchip laser, which favored good energy extraction even in a configuration having few round trips. Moreover, to reduce the thermal effects, we used a composite crystal [25] made with a 1 mm long undoped YVO$_4$ crystal bonded to a 1% doped 3 mm long Nd:YVO$_4$ crystal. Since the pulse seeded in the amplifier had significant energy (fraction of μJ), the energy stored in this crystal could be extracted in a maximum of four passes.

The experimental setup is shown in Figure 12. For the first two passes, the signal beam was focused into the crystal by a lens (L2), reflected on the back of the crystal (which was HR coated at 1.064 μm), and recollimated with the same lens. The beam could be intercepted at this point in order to evaluate the two-pass-amplifier performance. For the four-pass amplifier, the beam was reinjected into the gain medium after a small spatial shift. To optimize the third and fourth passes, and to compensate for the thermal lensing that occurred during the first two passes, the beam propagated through a slightly-detuned 1:1 telescope (L3–L3).

To have a high gain per pass and thus further increase the extraction efficiency of the amplifier, we used a high-brightness 2W, 1×100 μm^2 pump diode. The use of high-brightness diodes seemed to be optimal for an efficient multipass amplifier. For example, after two passes, the gain obtained with a more powerful

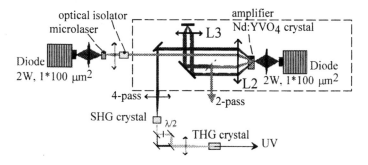

Figure 12 Experimental setup for the Nd:YAG microlaser amplified in an Nd:YVO₄ amplifier.

but less bright diode (4W, 1 × 500-μm²) was 2.5 times lower than the brighter diode. Despite a higher amount of deposited energy in the crystal with the 4W diode, the maximal (after four passes) energy extraction remained almost the same as with the high brightness diode.

With the same objective of having a simple and efficient amplifier, we studied the influence of the injection power (Fig. 13). The energy extraction was naturally higher with a 40 mW injection than the 15 mW one. This explains why a 2W, 1 × 100 μm² diode was also used to pump the microchip laser to obtain

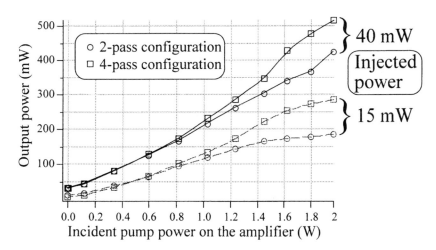

Figure 13 Output power vs. pump power for two different incident signal powers for two passes or four passes inside the amplifier.

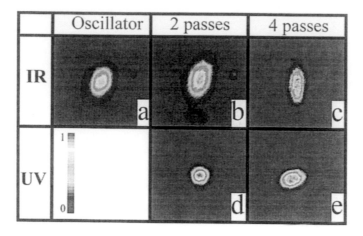

Figure 14   IR and UV beam profiles in the far field. (a) Oscillator beam profile. (b,c) Amplified beam profile (at 1064 nm) after two passes and four passes. (d,e) UV beam profile (355 nm) in the two configurations (two passes and four passes).

a highly efficient amplifier. The average output powers were 408 mW after two passes and 510 mW after four passes at 1064 nm.

We recorded the beam profiles and the average output power for the two-pass and four-pass configurations. We observed that the beam profile shrank in the horizontal direction during the third and fourth passes (Fig. 14 a–c). This could be explained by the fact that the signal beam of the third and fourth passes was amplified on the edges of the gain area. However, the M^2 of the amplified beam was nearly the same in the two-pass and in the four-pass configuration, on the order of 1.2.

The choice between the four-pass configuration, which delivered the maximum output power, and the simpler two-pass configuration, which delivered less energy but had a improved beam profile, was made based on the efficiency of harmonic generation by nonlinear processes.

C.   Nonlinear Conversion Stage

For the nonlinear processes, we first used a 5 mm long type II KTP (KT$_i$OPO$_4$) crystal for second-harmonic generation. The output green power (at 532 nm) was up to 180 mW (4 µJ per pulse, 37% conversion efficiency) in the four-pass configuration, and 120 mW (2.33 µJ per pulse, 31% conversion efficiency) in the two-pass scheme. Subsequently, for optimization of the third harmonic, the IR and the green were separated after the doubling crystal in two different arms

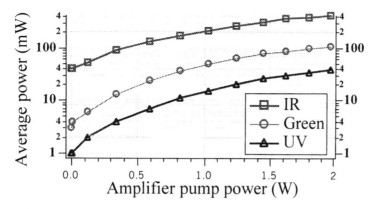

Figure 15 Output power of the source for fundamental and harmonic waves vs. the pump power of the amplifier.

and recombined to give a better overlap in the second nonlinear crystal (8 mm long type II LBO). In order to obtain a circular beam profile in the UV (at 355 nm), we used the angular acceptance and the walk-off of the LBO crystal to reshape the distorted IR beam profile delivered by the multipass amplifier. The UV beam profile was circular for the two-pass amplifier (Fig. 14d) and slightly elliptical for the four-pass amplifier (Fig. 14e), but much less than in the infrared (Fig. 14c). An output power of 38 mW (10% efficiency) using the two-pass configuration was obtained compared to 35 mW (7.2% efficiency) using the four-pass design. Because of its elliptical shape, the output beam after four passes could not allow an efficiency as high as that obtained after two passes. The optimal system for both simplicity and efficiency in the UV was the two-pass amplifier configuration. This UV source delivered 0.84 µJ pulses at 45 kHz corresponding to 40 mW of average power (Fig. 15) with a pulse duration shortened by nonlinear conversion down to 300 ps. The M^2 of the UV beam was 1.34.

We also replaced the LBO crystal used for frequency tripling by a BBO crystal phase matched for frequency doubling the green output. We obtained 0.5 µJ at 266 nm.

D. Towards Higher Power in the UV

Figure 13 demonstrated the strong influence of the input signal average power (for a given repetition rate) on the extraction of the power stored inside the Nd: YVO_4. The upper limit for average power can be obtained from the crystal performance in a cavity in the CW regime. Optical-to-optical conversion efficiency of 50% can be expected from an optimized Nd: YVO_4 laser operating in CW. In our case, the upper limit for the average output power from the amplifier was on the order of 1 W. This suggests that it is possible to extract a higher-output power from the amplifier with a higher-input signal.

Following this idea, we used a different microchip laser producing higher average power than the previous one, at the expense of a slight increase in the pulse duration. The microchip laser consisted of a 1.5 mm long Nd:YAG crystal bonded with a 100 μm long Cr:YAG. The pulse duration was 800 ps (compared to 500 ps for the previous one). The repetition rate was 52 kHz and the average power was a factor of 4 higher (150 mW).

For this case, the power after two-pass-amplification was 800 mW in the infrared. To simplify the experimental setup, we replace the KTP crystal cut in type II by an LBO cut in type I for second-harmonic generation at 532 nm. In this case, frequency tripling can be achieved without separation between the infrared and the green beams. After frequency doubling, the power was 120 mW at 532 nm. The frequency tripling was achieved with the same LBO as used previously. We obtained 100 mW at 355 nm (corresponding to an pulse energy of 1.9 μJ).

V. CONCLUSION

We have presented several diode-pumped sources emitting picosecond pulses in the UV. All the sources we have developed are based on diode-pumped laser crystals emitting in the near infrared and on two nonlinear crystals converting the light to the UV band.

The first two laser sources are based on an indirectly diode-pumped Ti:sapphire regenerative amplifier seeded either by an actively mode-locked diode-pumped Cr:LiSAF picosecond oscillator or a gain-switched picosecond laser diode. Typical performances were 0.1 μJ at 10 kHz (1 mW of average power) between 270 and 285 nm.

The last laser source described was based on a passively Q-switched Nd:YAG laser seeding a two-pass Nd:YVO$_4$ amplifier. We obtained more than 1 μJ at 40 kHz (40 mW of average power) at 266 nm after frequency conversion. The average power obtained at 355 nm reached the 100 mW level. As expected, operation at a fixed wavelength is more simple and more efficient. We are currently working on a tunable version of this system based on nonlinear crystals cut for optical parametric generation and amplification: quasi-phase-matched crystals (periodically poled lithium niobate crystals) can efficiently produce tunable red light from pump pulses at 532 nm [26]. Subsequently, a frequency doubler can be used to reach the UV band.

The total optical pump power for all the systems does not exceed 5 W.

Considering the excellent electrical-to-optical conversion efficiency of the laser diodes (30%), only a few tens of watts of electrical power are needed to supply these sources. These compact systems promise to have many potential applications in the domains of fluorescence spectroscopy, optical sampling, or the semiconductor industry.

REFERENCES

1. F Balembois, M Gaignet, F Louradour, V Couderc, A Barthelemy, P Georges, A Brun. Tunable picosecond UV source at 10 kHz based on an all solid-state diode-pumped laser system. Appl Physics B 65: 255–258, 1997.
2. CD Marshall, SA Payne, JA Speth, WF Krupke. Ultraviolet laser emission properties of Ce^{3+}-doped $LiSrAlF_6$. In TY Fan, BHT Chai, eds. OSA Proceedings on Advanced Solid State Lasers. Washington, DC: Optical Society of America, 1994, pp 389–394.
3. T Sandrock, T Danger, E Heumann, G Huber. Continuous wave laser action of Pr-doped fluorides at room temperature. In TY Fan, BHT Chai, eds. OSA Proceedings on Advanced Solid State Lasers. Washington, DC: Optical Society of America, 1994, pp 357–360.
4. F Falcoz, F Balembois, P Georges, A Brun. Self-starting self mode-locked femtosecond diode pumped Cr:LiSAF laser. Optics Letts 20: 1874–1876, 1995.
5. F Falcoz, F Kerboull, F Druon, F Balembois, P Georges, A Brun. Small signal gain investigations for a continuous-wave diode-pumped Q-switched Cr:LiSAF laser. Optics Letts 21: 1253–1255, 1996.
6. R Mellish, SCW Hyde, NP Barry, R Jones, PMW French, JR Taylor, CJ Van der Poel, A Valster. All-solid-state diode-pumped Cr: LiSAF femtosecond oscillator and regenerative amplifier. Appl Phys B 65: 221–226, 1997.
7. F Balembois, F Falcoz, F Druon, F Kerboull, P Georges, A Brun. Theoretical and experimental investigations of small signal gain for a diode-pumped Q-switched Cr: LiSAF laser. IEEE J Quant Electron 33: 269, 1997.
8. CS Adams, AI Ferguson. Tunable narrow linewidth ultra-violet light generation by frequency doubling of a ring Ti:sapphire laser using lithium tri-borate in an external enhancement cavity. Opt Commun 90: 89, 1992.
9. A Nebel, R Beigang. External frequency conversion of cw mode-locked $Ti:Al_2O_3$ laser radiation. Opt Lett 16: 1729, 1991.
10. L Goldberg, DAV Kliner. Tunable UV generation at 286 nm by frequency tripling of a high-power mode-locked semiconductor laser. Opt Lett 20: 1640, 1995.
11. F Balembois, M Gaignet, P Georges, A Brun, N Stelmakh, JM Lourtioz. Tunable picosecond blue and UV pulses from a diode-pumped laser system seeded by a gain-switched laser diode. Appl Optics 37: 4876–4880, 1998.
12. PJ Deffyett, A Yusim, S Grantham, S Gee, K Kabel, M Richardson, G Alphonse, J Connolly. Ultrafast semiconductor laser-diode-seeded Cr:LiSAF regenerative amplifier. Appl Opt 36: 3375–3380, 1997.

13. A Galvanauskas, ME Fermann, P Blixt, JA Tellefsen, D Harter, Hybrid diode-laser fiber-amplifier source of high-energy ultrashort pulses. Opt Lett 19: 1043–1045, 1994.

14. DJ Harter, O Montoya, J Squier, J Nees, P Bado, G Mourou. Short pulse generation from Ti:doped materials. In: Technical Digest Series vol. 7, Conference on Lasers and Electro-Optics. IEEE/Lasers and Electro-Optics Society and Optical Society of America, Anaheim, 1988, paper PD 6.

15. DJ Harter, P Bado. Wavelength tunable alexandrite regenerative amplifier. Appl Opt, 27: 4392–4395, 1988.

16. N Stelmakh, J-M Lourtioz, G Marquebielle, G Volluet, JP Hirtz. Generation of high-energy (0.3µJ) short pulses (<400 ps) from a gain-switched laser diode stack with sub-ns electrical pump pulses. IEEE J Sel Topics Quant Electron 3: 245–250, 1997.

17. A Hariharan, DJ Harter, TS Sosnowski, S Kane, D Du, TB Norris, J Squier. Injection of ultrafast regenerative amplifiers with low energy femtosecond pulses from an Er-doped fiber laser. Opt Commun 132: 469–473, 1996.

18. F Druon, F Balembois, P Georges, A Brun. High-repetition-rate 300-ps pulsed ultra-violet source with a passively Q-switch microchip laser and a multipass amplifier. Optics Letts. 24: 499–501, 1999.

19. JJ Zayhowski, C Dill III. Diode-pumped passively Q-switched picosecond microchip lasers. Opt Lett 19: 1427, 1994.

20. JJ Zayhowski. Nonlinear frequency conversion with passively Q-switched microchip lasers. CLEO 96 Tech Dig 9: CWA6, 236, 1996.

21. S Zhou, KK Lee, YC Chen, S Li. Monolithic self-Q-switched Cr,Nd:YAG laser. Opt Lett 18: 511, 1993.

22. L Fulbert, J Marty, B Ferrand, E Molva. Passively Q-switched monolithic microchip laser. CLEO 95 Tech Dig 15: CWC5, 176, 1995.

23. JJ Degnan. Optimization of passively Q-switched lasers. IEEE J Quant Electron 31: 1890, 1995.

24. C Larat, M Schwarz, JP Pocholle, G Feugnet, MR Papuchon. High repetition rate solid-state laser for space communications. Proc SPIE 2381: 256, 1995.

25. M Tsunekane, N Taguchi, H Inaba. High power operation of diode-end pumped Nd:YVO4 laser using composite rod with undoped end. Electron Letts 1 (32): 40, 1996.

26. U Bäder, JP Meyn, J Bartschke, T Weber, A Borsutszy, R Wallenstein, RG Batchko, MM Fejer, RL Byer. Visible nanosecond PPLN optical parametric generator pumped by a passively Q-switched single frequency Nd:YAG laser. CLEO 99 Tech Dig CThB4, 1999.

13
Atmospheric Ultraviolet Spectroscopy

Kelly Chance and Laurence S. Rothman
Harvard–Smithsonian Center for Astrophysics, Cambridge, Massachusetts

I. INTRODUCTION

Ultraviolet (UV) spectroscopy is a rich source of information about the Earth's atmosphere. In this chapter we will survey the field of ultraviolet remote sensing measurements of the Earth's atmosphere as currently practiced and as planned for the next decade, with emphasis on satellite-based measurements. We will review the underlying physics, particularly as it differs from that contributing to atmospheric measurements at longer wavelengths, and describe the techniques of data analysis used to process raw spectra into atmospheric quantities. We will describe progress in constructing and maintaining the extension of the HITRAN database into the ultraviolet, to serve as the standard repository for absorption cross sections for the analysis of atmospheric spectra. Finally, we will review the heritage of satellite UV spectrometers for atmospheric measurements and the current and planned instruments.

Ground-based (and balloon-based and aircraft-based) measurements of the atmosphere using ultraviolet spectroscopy have a rich history and have, naturally, provided much of the heritage for the satellite measurements discussed here, as well as substantial validation of the satellite measurements. They are not specifically discussed in this chapter, but much of the discussion of underlying physics, measurement techniques, and database consideration applies equally to them.

II. PHYSICS CONSIDERATIONS

The physical characteristics of the observing environment (temperature, pressure, interfering absorption, and scattering properties) are particularly important to the viability of UV atmospheric measurements.

The average temperature at the Earth's surface (the bottom of the "troposphere") is about 285 K, at a pressure of 1 atm. At increasing altitude the pressure and temperature fall, with the pressure decreasing at an exponential scale height of about 8 km and the temperature decreasing at roughly the adiabatic lapse rate, that of cooling by adiabatic thermodynamic expansion (the dry adiabatic lapse rate is ~ 10 K km^{-1}), until one reaches the "tropopause," which is the boundary between the troposphere and stratosphere. The tropopause is at ~ 15 km, 0.1 atm, and 215 K (although this is highly variable with latitude, season, and dynamical conditions). Above the tropopause the pressure continues to fall, but the temperature rises, due to photochemical heating induced by the absorption of UV radiation by O_2 in the Herzberg continuum, producing oxygen atoms, the associated photochemistry of "odd oxygen" (oxygen atoms and ozone), and the further absorption of UV radiation by ozone. In very simplified terms, the oxygen photochemistry (the "Chapman" chemistry [Chapman, 1930; Brasseur and Solomon, 1986]) can be summarized as:

$$O_2 + h\nu(\lambda \leq 242nm) \rightarrow 2O \tag{1}$$

$$O + O + M \rightarrow O_2 + M \tag{2}$$

$$O + O_2 + M \rightarrow O_3 + M \tag{3}$$

$$O + O_3 \rightarrow 2O_2 \tag{4}$$

$$O_3 + h\nu(\lambda \leq 310nm) \rightarrow O_2 + O \tag{5}$$

where M is any third body (usually an N_2 or O_2 molecule) needed in order to provide momentum conservation for binary recombination. Most of the stratospheric heating is from absorption of O_3 in the Hartley band, from 242 to 310 nm, which also is responsible for much of the atmospheric opacity to UV radiation. Some additional contribution comes from absorption, with weak dissociation, in the O_3 Huggins bands, from 310 to 400 nm. (The long-wavelength limit is arbitrarily selected by intensity; there is no physical band cutoff.) The departures from this simplified scheme, particularly those affected by anthropogenic actions, occupy much of the effort of stratospheric ozone chemistry studies. For present purposes we are concerned with the physical background conditions for measurements, and these are well described by Eqs. (1)–(5). The relative heating reaches a maximum at the "stratopause" (~ 50 km, 0.0008 atm, 270 K). Above this, in the "mesosphere" (which reaches an altitude of 80 km) polyatomic constituents, including ozone, become increasingly rarer, and the atmosphere again cools.

Atmospheric UV measurements are dominated by absorptions in the stratosphere, with significant contributions from the troposphere and, to a lesser extent, the mesosphere. Conditions of local thermodynamic equilibrium (LTE), where populations are adequately described by a single electronic, vibrational, rotational, and translational temperature, normally apply to the physical processes encountered in UV atmospheric spectroscopy. Non-LTE conditions are occasionally encountered, particularly at higher altitudes, in the mesosphere and above, but they will not be addressed in this chapter, where we concentrate on the UV spectroscopy applicable to the troposphere and stratosphere.

Molecular absorption in the ultraviolet wavelength range is thus dominated by absorption of ozone. Much of the additional contribution to spectral absorption is due to small, often chemically reactive species. Many of the more complex species that enrich the infrared atmospheric spectrum lack electronic transitions in the ultraviolet, or have transitions that are too weak to be (as yet) useful diagnostics of the atmospheric state (NO_2 is a notable exception). The strong presence of absorptions due to reactive species is in part the result of the physics that dictates that, for example, small oxygen-containing free radicals have transitions in the portion of the UV atmospheric spectrum readily accessible to measurement. It is also due to the highly oxidized, photochemically driven state of the upper portions of the atmosphere.

Rayleigh scattering is the other major contributor to the UV atmospheric spectrum. Rayleigh scattering (also known as "molecular scattering") is dipole scattering in which electric dipole moments are induced in the air molecules (primarily N_2 and O_2) by the interaction of their polarizabilities with the electric field of the radiation. Rayleigh scattering has been thoroughly reviewed by Bates [1984]. The well-known approximate λ^{-4} intensity of the Rayleigh scattering cross section is described more exactly by the relationship

$$Q_R \times 10^{24} (\text{cm}^2) = \frac{3.9993 \times 10^{-4} \sigma^4}{1 - 1.069 \times 10^{-2} \sigma^2 - 6.681 \times 10^{-5} \sigma^4} \tag{6}$$

where σ (μm^{-1}) $= 1/\lambda$ (μm), as derived by Chance and Spurr [1997] from an analysis of the data of Bates [1984]. This expression is valid from 0.2–1 μm (200–1000 nm). In addition to being a significant competitor to absorption in the modulation of transmitted UV solar radiation, Rayleigh scattering is the major source of light for measurements in limb scattering in the UV and visible (to about 1000 nm) and for backscattered light measurements below about 310 nm. It provides the backscatter signal for the SBUV measurements of ozone profiles in the stratosphere by differential penetration into the O_3 Hartley band versus wavelength (see sections on BUV, SBUV, and SSBUV instruments, below).

Atmospheric UV spectroscopy is thus largely the measurement of O_3 plus the measurement of small absorptions in the presence of O_3 absorption and Rayleigh scattering. A further complicating effect must be considered, particularly for small absorptions (absorption optical depths as small as several $\times 10^{-4}$ may provide useful measurements, as discussed in the section on GOME, below): Since N_2 and O_2 are not spherically symmetrical, their molecular polarizabilities are anisotropic. The polarizability anisotropies introduce an inelastic component to the Rayleigh scattering, where a portion of the Rayleigh-scattered light (3.7% of Rayleigh scattering at 350 nm, for example) is rotationally Raman-scattered (vibrational Raman scattering also contributes, but to a much lesser extent). This effect was first described by Grainger and Ring [1962], who observed the filling in (broadening and reduction in depth) of solar Fraunhofer lines when they were viewed from the ground in scattered sunlight. The effect was firmly established as being primarily due to rotational Raman scattering by Kattawar et al. [1981] and has since been fully quantified by Chance and Spurr [1997] using the data provided by Bates [1984].

Since the Ring effect spreads out several percent of the Fraunhofer spectrum in wavelength, by convolving it with the rotational Raman spectra of N_2 and O_2, it constitutes a major interference source for the analysis of spectra: a "non-Beer's law" contribution. It can usually be included in spectrum fitting as a small additional pseudoabsorption source, where methods of determining the correction include measuring the polarization of scattered sunlight [Solomon et al., 1987] and modeling the effect directly from molecular scattering processes [Fish and Jones, 1995; Joiner et al., 1995; Chance and Spurr, 1997].

Contributors to the UV atmospheric spectrum include transitions with discrete line spectra as well as broader spectra exhibiting significant dissociation broadening. The cataloging of spectra is generally organized into line parameters (i.e., for line-by-line description of spectra) and absorption cross sections, as detailed in the review of HITRAN database, below. For the range of temperatures and pressures encountered in atmospheric UV spectroscopy, pressure broadening of discrete lines is usually negligible in comparison to Doppler broadening. The narrowest discrete lines will have widths (full-width at $1/e$ intensity) of less than 0.001 nm, setting a natural limit to the resolution for atmospheric spectrometers that is rarely approached in practice. A more practical limit, as deduced from GOME studies for a variety of molecules, is that spectral resolution capable of fully resolving the highly temperature-dependent and weakly predissociating O_3 Huggins bands (0.2–0.3 nm) is sufficient for practically all detailed spectroscopic measurements of the atmosphere in the UV. (In contrast, higher resolution in the visible and near infrared is capable of producing extra information on cloud properties [Kuze and Chance, 1994] and on the vertical distribution of H_2O [Maurellis et al., 2000].)

A. Measurement Geometries

1. Backscatter Measurements

These measurements are performed by looking from the satellite in the direction of the Earth's surface. The surface itself is often masked: by ozone absorption in the wavelength range of the Hartley band, by Rayleigh scattering that limits penetration, or by clouds, so that spectral measurements sample volumes representative of the wavelength-dependent penetration. The BUV method, described below, uses wavelength-dependent penetration into the atmosphere in the ozone Hartley band to produce profiles of stratospheric ozone. This method may be combined with measurements in the temperature-dependent Huggins bands at sufficient spectral resolution to resolve the vibrational hot-band structure, taking advantage of temperature decrease with altitude in the troposphere, to produce altitude profiles down to the ground (or cloud-top scattering layer) [Chance et al., 1991, 1997; Munro et al., 1992, 1998]. In general, because of the strength of Rayleigh scattering, UV backscatter measurements are more sensitive to stratospheric gases than tropospheric gases on a per-molecule basis. Backscatter measurements are useful for a number of gas total-column measurements in the UV, including BrO, OClO, NO_2, SO_2, and HCHO.

Spectral Analysis. The GOME satellite has fulfilled initial expectations of being able to measure absorption optical depths as small as several $\times 10^{-4}$ ($3^{1}/_{2}$ orders of magnitude below the full-scale backscattered radiance). Several methods have been used for analyzing the spectra obtained from GOME backscatter measurements and proposed for use in later instruments. These generally fall into two categories: those in which line-of-sight column abundances (slant columns) are first fitted, and then adjusted to give vertical column abundances; and those in which vertical abundances, and sometimes altitude profile information, are fitted directly. For occultation and limb-scattering measurements, series of spectra measured at different elevation angles must be analyzed together to give vertical profile information.

The most fundamental method of spectral analysis for slant columns consists of synthesis of the satellite radiance measurements, beginning with a satellite-measured irradiance, and adjusting the parameters describing the synthesis by iteration until they agree with the measurements in the least squares sense. This is the long-standing technique used in generations of ground-based, airborne, and satellite (e.g., infrared) measurements. Parameters for the analysis of GOME spectra are typical of what will be needed for other planned satellite instruments:

absorption cross sections, Ring effect correction, albedo, wavelength calibration correction, a correction for undersampling of the atmospheric spectrum by the GOME instrument (radiance and irradiance measurements are taken at slightly different Doppler shifts; resampling to construct a synthetic radiance spectrum introduces aliasing since the GOME instrument does not fully Nyquist sample the spectrum [Goldman, 1953; Chance, 1998; Slijkhuis et al., 1999]), and low-frequency closure terms to account for albedo variation and Mie scattering.

Differential optical absorption spectroscopy (DOAS) is a commonly used variant of the procedure for determination of slant column abundances [Platt, 1994]. In the DOAS technique, the radiance is divided by the irradiance and the logarithm taken, to linearize Beer's law contributions. The data are then high-pass filtered to give resulting variations about zero. The result is fitted using linear least-squares analysis to a sum of high-pass filtered cross sections times abundances, a high-pass-filtered version of the Ring effect spectrum divided by the Fraunhofer spectrum, and low-frequency closure terms. If wavelength adjustment is needed, this is accomplished by alternating the linear fitting steps with nonlinear steps in wavelength scale adjustments.

Although DOAS was principally designed for situations where the irradiance is unstable (i.e., a varying lamp source) or cannot be precisely measured (as in ground-based atmospheric measurements), it has proved quite useful in the analysis of GOME satellite data, and is used in the current operational processing for O_3 and NO_2 column abundance measurements [Loyola et al., 1997]. However, because it involves discarding part of the spectral information and introducing an approximation in the treatment of the Ring effect, it is inherently less accurate than the basic fitting procedure.

For either method of slant column determination, it is necessary to divide the measured slant column abundance by an air mass factor (AMF), given by

$$\text{AMF} = \frac{\text{slant column}}{\text{vertical column}} = \frac{\tau_s}{\tau} \qquad (7)$$

where τ_s and τ are the optical depths for the slant and vertical columns. In practice, AMFs are determined from radiative transfer calculation with assumption about the shape of the species' vertical distribution usually obtained from modeling. In UV measurements, it is chiefly Rayleigh scattering that affects the atmospheric sampling so that determinations of AMFs are not simply geometric calculations (as noted earlier, UV backscatter measurements are usually more sensitive to stratospheric gases than tropospheric gases on a per-molecule basis).

GOME has successfully demonstrated the technique of determining ozone profiles and tropospheric ozone abundances by combining the backscatter information on the stratospheric profile [Singer and Wentworth, 1957] with spectrally resolved measurements in the temperature-dependent Huggins bands. Because of the altitude-dependent temperature structure of the atmosphere, the Huggins bands provide spectral fingerprints for the tropospheric ozone distribution [Chance et al., 1991, 1997; Munro et al., 1992, 1998].

Optical absorption coefficient spectroscopy (OACS) is a novel technique for applying the opacity coefficient method (OCM) to fit GOME spectra in regions where the absorptions are discrete and highly structured [Maurellis et al., 2000]. OCM is a refinement over exponential sum-fitting techniques [Kato et al., 1999] for sampling average transmittance over detector pixels. OACS is successful in determining H_2O column abundances from GOME spectra in the visible, and is a promising technique for determining height-resolved (profile) information from backscatter measurements of highly structured spectra of H_2O and other species.

Cloud Parameter Determination. Information on cloud parameters present in backscatter data includes absolute radiances levels and their spatial variation; absorption depth in the O_2 A band and the O_2–O_2 collision-induced absorption; and the depth of the Ring effect, particularly for the Ca II H and K lines near 395 nm. The GOME operational data processor employs an initial cloud-fitting algorithm (ICFA) based on cloud-top height climatology from the ISCCP [Rossow et al., 1996] and absorption in the visible O_2 A band [Kuze and Chance, 1994; Loyola et al., 1997]. This method is being upgraded to the cloud retrieval algorithm for GOME (CRAG), which employs spatial variations of the albedo from the GOME sensor in combination with O_2 A band absorption depth and absolute radiance level to determine cloud height, fractional coverage, and optical depth simultaneously [Kurosu et al., 1997, 1998, 1999]. Note that the ODUS instrument will be flown simultaneously with an O_2 A band camera for cloud coverage measurements on the GCOM-1A satellite. Joiner and Bhartia [1995] have proposed the use of the Ring effect for cloud-top pressure measurements, and tested the method on Nimbus 7 TOMS and SBUV data. They indicate that for an instrument with 0.2 nm spectrum resolution (e.g., GOME) this method should be capable of cloud-top pressure determination to 30 mbar precision for fully cloudy scenes. The use of absorption depth due to the O_2–O_2 collision complex as an indicator of cloud coverage has also been proposed as an indicator of cloud coverage and used with some success with GOME data [Chance et al., 1991; Wagner and Platt, 1998].

2. Occultation Measurements

Solar occultation is the most common of this type of measurement, although lunar occultations are planned (for example, by SAGE III) to permit measurements of species, including NO_3 and OClO, that are easily photolyzed and are observable principally at night. Occultations have the inherent advantages of high vertical resolution and high signal due to the direct beam source, and simplified restitution of measurements to location on the limb in comparison to limb-scattering and limb-emission measurements. (Limb pointing is, in general, quite difficult; the availability of ephemeris data for occultation measurements greatly simplifies the geolocation, although it may still be complicated by refraction.) A disadvantage of occultation measurements is that occultation opportunities occur infrequently (e.g. one sunrise and one sunset per orbit) and at locations that are not freely selectable. The use of stellar occultations, to be employed by GOMOS, overcomes this by making use of a number of stars of sufficient magnitude. An additional advantage of stellar occultations is that many of the important atmospheric species have concentrations with strong diurnal variation. Thus a sunrise occultation is not equivalent to a sunset occultation and the different altitudes encountered during a solar occultation are observed at different solar times, requiring photochemical modeling to correct the data [Roscoe and Pyle, 1987]. Solar occultation measurements have by far the greatest heritage in satellite as well as balloon-borne measurements. In general, they are capable of measuring to below the tropopause, to altitudes determined by cloud and aerosol interference. In the longer-wave UV and into the visible, they are capable of measuring to an altitude as low as 5 km a significant fraction of the time.

The spectral analysis of occultation spectra will normally be done either by fitting to obtain line-of-sight column abundances and then determining profile information sequentially, using spectra at successively lower measurement tangent heights (the "onion-peeling" method) or by simultaneous fitting of a set of occultation spectra to a selected altitude grid, with appropriate constraints on interpolation (the "global fitting" method [Carlotti, 1988]).

3. Limb-Scatter Measurements

Measurements in limb-scattered radiation use Rayleigh scattering as the light source. They trade the advantages of occultation measurements for the ability to measure at any geolocation (except for the direct illumination of an occultation source). Because of the weaker line-of-sight signal, limb-scatter measurements suffer from more interference by multiply scattered light arising from lower in the atmosphere than the tangent height of observation, requiring more sophisticated retrieval algorithms. This interference and the weaker source may make limb-scatter measurements more difficult at lower altitudes in the troposphere than for occultations. Altitude registration is more difficult than for occultations, but significant progress has been made in using the intensity of Rayleigh scattering outside of molecular absorption for this purpose, for example by the Rayleigh scattering attitude sensor (RSAS) [Janz et al., 1996], which flew on Space Shuttle flight STS-72 in 1996. Limb-scatter measurements have substantial balloon-borne heritage [McElroy, 1988]. The technique is planned for SCIAMACHY, SAGE III, OSIRIS, and OMPS.

Spectral analysis is similar to that required for occultation measurements, except that radiative transfer modeling is generally more important to correct for multiply scattered radiation from below the tangent height, and limb registration is more difficult due to the lack of ephemeris data.

B. HITRAN Database

The HITRAN spectroscopic database was created as a compilation of spectroscopic parameters that could serve to provide the basic information to computer programs modeling atmospheric attenuation due to molecular gaseous absorption in the infrared. The first edition [McClatchey et al., 1973] was made public in 1973. The work had evolved from several monographs produced in the 1960s at the National Bureau of Standards [Gates et al., 1964; Calfee and Benedict, 1966; Benedict and Calfee, 1967].

HITRAN is the international standard molecular spectroscopic database and has developed over time with respect to several key aspects:

The spectral range now spans millimeter through ultraviolet transitions (the database employs the units of wavenumber in cm^{-1} throughout, which is proportional to energy and very convenient to most modeling codes). Most of the transitions included in HITRAN are due to the electric dipole moment of the molecules, although there are also magnetic dipole and electric quadrupole transitions (which, although weak, are of interest in the case of very abundant species such as O_2 and N_2).

The HITRAN compilation includes absorption cross sections in addition to the traditional line-by-line spectroscopic parameters. The cross sections are used where the density of spectral features is very crowded (as is the case for "heavy" molecular species in the infrared such as the chlorofluorocarbons and especially for most bands in the UV). The cross sections are given at different pressure–temperature pairs, which facilitates modeling of transmission or radiance in nonisothermal paths. The file structure and format for the cross section files are summarized in Table 1.

The number of gases covered has expanded from the original seven that were the principal atmospheric absorbers in the infrared to about 60 different molecules along with their isotopomers when significant to terrestrial atmospheric remote-sensing problems.

The quantity of parameters for individual transitions in the line-by-line portion of the HITRAN compilation has substantially increased. The original database contained information for the line position (transition frequency in vacuum wavenumbers), the intensity of the transition, the Lorentzian halfwidth (broadening of the spectral line due to collisions

Table 1   File Structure and Format (in FORTRAN) for Cross-Section Files

Molecule	$\nu_{1,\ init}$ cm^{-1}	$\nu_{1,\ final}$ cm^{-1}	# of points	T_1 K	P_1 torr	σ_{max} cm^2	Reference Resolution, etc.
A10	F10.4	F10.4	I10	F10.4	F10.4	E10.3	3A10

Cross sections (10 per line) in cm^2/molecule T_1, P_1: 10E10.3

Molecule	$\nu_{1,\ init}$ cm^{-1}	$\nu_{1,\ final}$ cm^{-1}	# of points	T_2 K	P_2 torr	σ_{max} cm^2	Reference Resolution, etc.
A10	F10.4	F10.4	I10	F10.4	F10.4	E10.3	3A10

Cross sections (10 per line) in cm^2/molecule T_2, P_2: 10E10.3

Molecule	$\nu_{2,\ init}$ cm^{-1}	$\nu_{2,\ final}$ cm^{-1}	# of points	T_1 K	P_1 torr	σ_{max} cm^2	Reference Resolution, etc.
A10	F10.4	F10.4	I10	F10.4	F10.4	E10.3	3A10

Cross sections (10 per line) in cm^2/molecule T_1, P_1: 10E10.3

Molecule	$\nu_{2,\ init}$ cm^{-1}	$\nu_{2,\ final}$ cm^{-1}	# of points	T_2 K	P_2 torr	σ_{max} cm^2	Reference Resolution, etc.
A10	F10.4	F10.4	I10	F10.4	F10.4	E10.3	3A10

Cross sections (10 per line) in cm^2/molecule T_2, P_2: 10E10.3

of the molecule with air), the lower-state energy of the transition, and the unique quantum number identification of each transition. These parameters were considered minimal for the calculation of molecular attenuation via the Lambert–Beers law. With the increased sophistication of high-resolution instrumentation, and the advances in retrieval algorithms, more fundamental parameters have been added, including transition probabilities, self-broadened halfwidths, temperature dependence coefficients for the halfwidth, and pressure–shift coefficients.

The media and software for HITRAN have also evolved. Originally released on magnetic tape, it is currently available on CD-ROM and more and more data are becoming accessible on the Internet. The CD-ROM has certainly emphasized the archival nature of the database. HITRAN is not a program (although the MODTRAN band-model code has been supplied with the 1996 edition of HITRAN). However, software has been included that allows various database functions, such as filtering, sorting, plotting the data, and access to relevant references.

Current information concerning HITRAN, as well as links to related spectroscopic databases, can be found at the HITRAN web-site (http://CfA—www.Harvard.edu/HITRAN).

1. UV Data in HITRAN

The edition of the HITRAN compilation of 1996 [Rothman et al., 1998] extended into the ultraviolet region for the first time with the inclusion of UV cross sections for N_2O at 296 K in the wavelength region 170–222 nm, and SO_2 at 213 K in the wavelength region 172–240 nm. The cross-section data employ the identical format as adopted for the IR cross sections, namely a table of cross sections at equal wavenumber interval. These sets of cross-sections are introduced with a standardized header providing essential information about each pressure–temperature set (see Table 1). High-resolution data for the O_2 Schumann–Runge system were also added to the compilation with line-transition parameters similar in format to the main body of HITRAN. The UV data are kept in a separate directory in the HITRAN compilation and further separated by two subdirectories, one for the cross-section data and the other for line parameter data.

For the upcoming edition of the database, HITRAN2001, we anticipate adding high-resolution cross sections for several molecular band systems: O_3, OClO, BrO, H_2CO, ClO, and NO_2. In addition, there will be line-by-line data for O_2, NO, and OH. The spectra of these gases have applications to a number of programs, including GOME and the NASA Earth observing system (EOS) program [King et al., 1995]. While the choices of cross sections used in analysis of field measurements are currently evolving, HITRAN2001 includes spectra that have been demonstrated to provide reasonable fitting for GOME spectra. Tables 2 and 3 summarize the cross-section and line-by-line UV data of HITRAN2001.

UV Line-by-Line Parameters. The HITRAN line list for the O_2 Schumann-Runge system contains 11,020 lines covering a spectral range of 44,606.676 to 57,027.590 cm^{-1} (224–175 nm). This system represents transitions between the $B\,^3\Sigma_u^-$ electronic state and the $X\,^3\Sigma_g^-$ ground electronic state. Nearly all of the line parameters are based on high-resolution absorption measurements from two groups: one from the Australian National University [Lewis et al., 1986] and the other from the Harvard–Smithsonian Center for Astrophysics [Yoshino et al., 1983; Yoshino et al., 1992]. A description of this line compilation is given in Minschwaner et al. [1992], which also contains relevant citations for measured quantities. The file in HITRAN lists all principal branch triplets over the range $v' = 0–19$, $v'' = 0–2$, $N'' = 1–51$. Satellite branches are included for $v' = 0–19$, $v'' = 0$, $N'' = 1–15$. Principal branches for the $^{16}O^{18}O$ isotopomer are listed for $v' = 2–16$, $v'' = 0$, $N'' = 1–24$. (A single prime (') designates the upper state of a transition, and double primes ('') denote the lower state.)

Table 2 UV Cross Sections in HITRAN

Molecule	Band system	Spectral range (cm^{-1})	Temperatures (K)
N_2O	$(^1A''\,^1\Sigma^-,\,^1\Delta)? \leftarrow \tilde{X}^1\Sigma^-$	44925–58956	296
SO_2	$\tilde{C}^1B_2 \leftarrow \tilde{X}^1A_1$	41691–58452	213
		45455–50505	295
O_3	$^1B_2 \leftarrow \tilde{X}^1A_1$ (Hartley-Huggins)	24570–33311	203–293
	$^1B_1 \leftarrow \tilde{X}^1A_1$ (Chappuis)		
	(Wulf)		
HCHO	$^1A'' \leftarrow \tilde{X}^1A_1$	27391–33311	223, 293
BrO	$A^2\Pi_{3/2} \leftarrow X^2\Pi_{3/2}$	25756–32013	228, 298
OClO	$\tilde{A}^2A_2 \leftarrow \tilde{X}^2B_1$	20992–41228	228
ClO	$A^2\Pi_{3/2} \leftarrow X^2\Pi_{3/2}$	32000–37700	220
NO_2	$A^2B_1 \leftarrow \tilde{X}^2A_1$	17540–32260	213, 298

There are two caveats for the user associated with these data. First, line positions of high N'' are not reliable. Second, there is an error in the indexing of the lower "global" quantum state, in that the $B^3\Sigma_u^-$ electronic state was referred to rather than the $X^3\Sigma_g^-$.

Line positions were calculated from energy level differences based on measured molecular constants. Line intensities were obtained from measured band oscillator strengths normalized according to Hönl-London factors, assuming a Boltzmann distribution of energies at 296 K. The tabulated widths are measured predissociation widths at zero pressure. Pressure-broadened widths are not listed because this effect is comparable to predissociation, broadening only for pressures larger than 1 atm. (The self-broadening coefficient is on the order of 0.20 $\text{cm}^{-1}/\text{atm}$ [Lewis et al., 1988].) Use of a Voigt line shape, composed of a thermal-broadened Doppler profile and a predissociation-broadened Lorentz profile, is

Table 3 UV Line-by-Line Data in HITRAN

Molecule	Band system	Spectral range (cm^{-1})
O_2	$B^1\Sigma_u^- \leftarrow X^3\Sigma_g^-$ (Schumann-Runge)	44604–57028
	$A^3\Sigma_u^+ \leftarrow X^3\Sigma_g^-$ (Herzberg I)	37000–41600
	$c^1\Sigma_u^- \leftarrow X^3\Sigma_g^-$ (Herzberg II)	36400–41600
	$A'^3\Delta_u \leftarrow X^3\Sigma_g^-$ (Herzberg III)	36400–41600
OH	$A^2\Sigma^+ \leftarrow X^2\Pi_i$	31000–36000
NO	$A^2\Sigma^+,\, B^2\Pi_r,\, C^2\Pi,\, D^2\Sigma^+ \leftarrow X^2\Pi_i$ (γ, β, δ, ϵ bands)	44300–62500

adequate for most atmospheric applications. Temperature-dependent parameters listed in the line-by-line portion of the HITRAN compilation are given at a standard reference temperature of 296 K. The appropriate scaling of these quantities is given in the Appendix of Rothman et al. [1998]. Line intensities at temperatures other than 296 K can be obtained using the energy of the lower quantum state in conjunction with the temperature dependence of the total internal partition sum. The latter has been calculated over the range 70–400 K, and is reproduced to within 1% from the following expressions [Gamache et al., 1990]:

$$^{16}O^{16}O : Q(T) = 0.604 + 0.731\ T - 4.00 \times 10^{-5}T^2 + 8.71 \times 10^{-8}T^3$$

$$^{16}O^{16}O : Q(T) = -1.04 + 1.51\ T - 3.44 \times 10^{-4}T^2 - 1.09 \times 10^{-6}T^3$$

The summation for $^{16}O^{16}O$ covers sufficient range of vibrational and rotational energy levels to ensure accuracy over the temperature range 70–400 K. For $^{16}O^{18}O$, however, a small error near 400 K may be expected as a result of neglecting the higher vibrational and rotational states.

UV Cross Sections.   Cross sections for two species, nitrous oxide and sulfur dioxide, had been placed in the 1996 edition of HITRAN. These cross sections have been cast into the same format as the IR cross-section data described in Rothman et al. [1998]; transforming the UV data into the equal-interval wavenumber scale required interpolation from measured wavelength scales employed in the UV.

High-resolution cross-section measurements of N_2O at 295–299 K have been performed in the wavelength region 170–222 nm with a 6.65 m scanning spectrometer of sufficient resolution to yield cross sections independent of the instrumental function [Yoshino et al., 1984]. The measured cross sections are available throughout the region 44,925–58,955 cm^{-1} at intervals of 0.1–0.2 cm^{-1}. Previously unresolved details of the banded structure, which is superimposed on the continuous absorption in the region 174–190 nm, are observed.

Laboratory measurements at high resolution of the absorption cross section of SO_2 at the temperature 213 K have been performed in the wavelength region 172–240 nm with a 6.65 m scanning spectrometer operated at an instrumental width of 0.002 nm [Freeman et al., 1984]. The measured cross sections are available throughout the region 172–240 nm at wavenumber intervals of 0.4–0.1 cm^{-1}. The measured cross sections, which are relevant to the photochemistry of planetary atmospheres, possess significantly more spectroscopic structure, and are more accurate than previous measurements made at lower resolution. However, values at peak cross sections could be affected by the instrumental widths even at 0.002 nm, which is larger than the Doppler widths.

Since these cross-section data in HITRAN are only given at a single temperature (296K in the case of N_2O and 213K in the case of SO_2), they are insufficient for simulation of optical paths that are not isothermal.

III. SATELLITE SPECTROMETERS

A. Heritage

A number of instruments have provided valuable atmospheric measurements using discrete photometer bands, while demonstrating the utility and providing the heritage for the more comprehensive spectroscopic measurements of the atmosphere that are now in progress and planned for the future. Most of the previous measurements are, not surprisingly, of ozone (O_3), with nitrogen dioxide (NO_2) accounting for most of the remaining gas measurements. A comprehensive review of satellite ozone measurements is given by Krueger et al. [1990]. Several excellent review articles on the instruments mentioned here and their measurements have been published in the special issue of *Planetary and Space Science*, particularly by Miller [1989], Krueger [1989], and McCormick et al. [1989]. Of particular note among photometer instruments are the following.

1. BUV, SBUV, and SSBUV Instruments

This series of instruments, which began with the Nimbus 4 BUV (1970 launch) measures solar radiation backscattered in the nadir viewing of the Earth's atmosphere at a series of discrete wavelengths from 255 to 340 nm, with 1 nm bandpass (resolution). These are actually scanning monochromator (i.e., spectrometer) instruments, but the taking of data from specific bands, and subsequent analysis, mimics the function of photometer measurements. In addition to total ozone (determined by discrete band measurements between 312 and 340 nm in the highly structured Huggins bands), they measure stratospheric ozone profiles, following on the technique first proposed by Singer and Wentworth [1957]. This technique determines profile information for stratospheric ozone from the differential penetration depth of solar radiation into the strong ultraviolet Hartley band (242–310 nm). (See Heath et al. [1973] for further information on the Nimbus 4 BUV instrument, Fleig et al. [1990] for details of the Nimbus 7 SBUV instrument, and Hilsenrath et al. [1995] for details of the SSBUV instrument.)

2. Total Ozone Mapping Spectrometer Instruments

The TOMS series of instruments, which began with the Nimbus 7 TOMS (1978 launch, Heath et al. [1975]) measures total ozone using six 1 nm bandpass wavelength bands from 312 to 380 nm (308–360 nm in later versions). Ozone is determined from differences in absorption in the Huggins bands.

3. Stratospheric Gas and Aerosol Experiment Instrument

This series (SAGE I and II) includes sun photometers that measure solar occultations in selected bands in the ultraviolet, visible, and infrared [McCormick et al.,

1989]. SAGE I, launched in 1979 aboard the AEM-2 satellite, measured O_3, NO_2, and aerosols using bands at 380, 450, 600, and 1000 nm. SAGE II, launched in 1984 aboard the ERBS satellite (and still operating in 2001), measures O_3, NO_2, H_2O, and aerosols using bands at 385, 453, 448, 525, 600, 940, and 1020 nm. A linear CCD array spectrometer version, SAGE III (described below), is to be launched in the near future.

4. Polar Ozone and Aerosol Measurement

The POAM II instrument is a sun photometer that measures solar occultations at nine spectral bands between 350 and 1060 nm [Glaccum et al., 1996]. POAM II was launched in 1993 on board the SPOT-3 satellite, and measures atmospheric profiles of O_3, NO_2, H_2O, aerosols, and temperature.

B. Current and Planned Instruments

The Global Ozone Monitoring Experiment [GOME; European Space Agency, 1995] is the first space-based spectrometer to utilize a broad range of the ultraviolet spectrum at continuous wavelength coverage and moderate spectral resolution for the purpose of measuring the composition of the Earth's atmosphere. It was not, however, the first planned, since it is a scaled-back version of the SCIAMACHY instrument, which was first proposed in 1985 [Burrows and Chance, 1991; Bovensmann et al., 1999; Noël et al., 1999]. The utility of measurements in a wide range of the ultraviolet, visible, and near-infrared for determining abundances of a large fraction of species that are of importance in current atmospheric science is now widely recognized. Table 4 gives an overview of instruments that are planned over the next decade involving this region. Measurements include various viewing geometries, including nadir, limb, solar (and lunar) occultation, each with its own advantages for focusing on the various spatial and temporal aspects of atmospheric composition and variability. They also utilize varying spectral and spatial resolutions and wavelength coverages, generally optimized for various measurement aspects in tradeoffs that compromise among global coverage, cost, complexity, and availability of satellite and support resources (e.g., telemetry bandwidth).

1. GOME

The first satellite instrument to measure UV spectra of the Earth's atmosphere, GOME was launched in polar, sun-synchronous orbit on the ERS-2 satellite in April 1995 [European Space Agency, 1995], and is still performing normally as of September, 2001. GOME obtains about 30,000 Earth radiance spectra each day, covering the ultraviolet (237–405 nm at 0.2 nm resolution) and the visible (407–794 nm at 0.4 nm resolution), measured with silicon diode-array detectors.

Table 4   Selected UV/Visible Satellite Spectrometers

Instrument	Wavelengths (nm)	Geometry	Gases[a]
GOME (GOME-2)	240–790	Nadir	O_3, NO_2, BrO, OClO, SO_2, HCHO, H_2O
SCIAMACHY[b]	240–2400	Nadir, limb, occultation	O_3, NO_2, BrO, OClO, SO_2, HCHO, H_2O
OMI	270–500	Nadir	O_3, NO_2, BrO, OClO, SO_2, HCHO
OMPS	290–1000	Nadir, limb	O_3, NO_2, BrO, OClO, SO_2, HCHO, H_2O
SAGE III	280–1040	Occultation, limb	O_3, NO_2, BrO, OClO, H_2O, NO_3
GOMOS	250–952	Stellar occultation	O_3, NO_2, NO_3, OClO
Odin/OSIRIS	280–800	Limb	O_3, NO_2, BrO, OClO, SO_2, HCHO, H_2O
ACE/MAESTRO	285–1030	Occultation, nadir	O_3, NO_2, BrO, OClO, SO_2, HCHO, H_2O
ODUS	306–420	Nadir	O_3, NO_2, BrO, OClO, SO_2, HCHO

[a] H_2O is measured in the visible and infrared.
[b] Additional species are measured in the infrared.

GOME also employs a polarization measurement device (PMD) that measures two polarizations of backscattered light at low spectral resolution, but at 16× higher spatial resolution than the spectral channels. The PMD is primarily for correction of instrument polarization sensitivity, but it also serves as a higher resolution sensor for albedo variations. GOME observes the entire Earth with 3 day coverage at the equator. Current operational data products available for GOME include total column abundances of O_3 and NO_2, and cloud fractional coverage. It was demonstrated that GOME should be able to measure a large set of atmospheric gases, including ozone profiles and tropospheric ozone, and atmospheric columns of NO_2, H_2O, BrO, OClO, SO_2 (e.g., from volcanic eruptions), and HCHO, to scientifically useful sensitivity [Chance et al., 1991]. This has now been shown to be the case, and numerous publications on these molecules and the geophysical processes they illuminate are being produced. Among the highlights of GOME results are the first direct measurements of tropospheric ozone from space [Munro et al., 1998]; the global measurement of BrO, including tropospheric enhancements in the polar springtime [Richter et al., 1998; Chance, 1998; Wagner and Platt, 1998]; and measurement of tropospheric formaldehyde from biogenic sources including biomass burning and isoprene production by forests [Thomas et al., 1998; Chance et al., 2000; Palmer et al., 2001]. Analysis of GOME data has shown that fitting of spectra to absorption optical depths of several $\times 10^{-4}$ of the full-scale radiance values is achievable in practice in favorable cases, given proper attention to the underlying spectroscopy, the fitting method, and the instrument characterization. In addition to gaseous components of the Earth's atmosphere, GOME measures aerosols using the TOMS-type absorbing aerosol index [Hsu et al., 1996; Herman et al., 1997; Hsu et al., 1999; Torres et al., 1999; van Oss, 1999] and cloud parameters, using a combination of information from the visible $O_2 A$ band, the PMD, and the radiance measurements [Loyola et al., 1997; Kurosu et al., 1997, 1998, 1999]. GOME instruments (the "GOME-2" series) have been selected as the operational ozone monitors on the Eumetsat Metop polar platforms beginning in 2003.

2. Scanning Imaging Absorption Spectrometer for Atmospheric Chartography

The SCIAMACHY is scheduled for launch into polar, sun-synchronous orbit on the ESA Envisat satellite in 2001. SCIAMACHY is the original concept from which the GOME instrument was derived, as a smaller instrument with particular emphasis on atmospheric ozone [Burrows and Chance, 1991; Chance et al., 1991; Noël et al., 1999; Bovensmann et al., 1999]. SCIAMACHY will observe the Earth's atmosphere in nadir viewing, in limb-scattered radiation, and in solar occultation. The wavelength range 240–1700 nm is measured fully; there are additional wavelength regions in the infrared at 1940–1040 nm and 2265–2380 nm, principally for the measurement of CO, N_2O, CH_4, and CO_2. The ultraviolet species measurements from SCIAMACHY will be the same as those for GOME. Addition of the limb and occultation measurement geometries will permit the determination of altitude profiles for most species, down to the limit of limb observation. Sequential limb and nadir viewing may help to determine budgets of tropospheric components that are difficult to obtain from nadir-only (e.g., GOME) measurements because of the overlying stratospheric loading (this is particularly relevant for O_3 and NO_2).

3. Ozone Monitoring Instrument

The OMI is a contribution of the Netherlands and Finland to the NASA EOS AURA (formerly CHEM) mission, scheduled for launch in 2003 in polar, sun-synchronous orbit. OMI measures in the nadir at higher spatial resolution than GOME (13×24 km^2 nominal instantaneous field of view) and with a wider orbital swath: OMI provides full global coverage each day. To accommodate the higher spatial sampling, OMI measures at lower spectral resolution than GOME. OMI measures 270–500 nm with resolution of 0.45–0.64 nm. OMI will provide measurements of O_3 (columns, plus vertical profile information at vertical resolution between that of SBUV and GOME); column abundances of the gases NO_2, BrO, OClO, SO_2, and HCHO; cloud height and fractional coverage; and a TOMS-type absorbing aerosol index.

4. Ozone Mapping and Profiler Suite

The OMPS has been selected as the operational ozone instrument for the US National Polar Orbiting Environmental Satellite System (NPOESS) satellites. The first of these sun-synchronous satellites will be launched in 2008. OMPS replaces the SBUV instruments as the US operational ozone-monitoring instruments. OMPS provides nadir measurements from 250 to 380 nm at 1 nm resolution and limb scattering measurements, employing a prism-based monochromator, from 290–1000 nm in 16 total bands, with bandpass resolutions selected from 1.5–40 nm. (These specifications are subject to minor changes as the detailed design is accomplished.) OMPS UV measurements include ozone total columns and profiles; the gases NO_2, BrO, OClO, SO_2, HCHO; and cloud and aerosol information. OMPS provides full global coverage each day at spatial resolutions from $46 \times 50 \text{ km}^2$ (nadir) to $215 \times 50 \text{ km}^2$ (limit of across-track swath).

5. SAGE III

The SAGE III instruments are planned for three missions, the first scheduled for launch aboard the Russian METEOR 3-M satellite in 2001. SAGE III is the fourth generation in this series of instruments, after the Stratospheric Aerosol Measurement II (SAM II), SAGE I, and SAGE II. In addition to the solar occultation measurements of the past instruments, SAGE III will measure lunar occultations (primarily for NO_3 and OClO) and make measurements in limb-scattered radiation. The linear array CCD detector provides continuous coverage from 280 to 1030 nm at 1 nm resolution. An additional (discrete photodiode) detector provides 1550 nm measurements to extend aerosol diagnostic capability. SAGE III measures O_3, NO_2, BrO, OClO, H_2O, NO_3, and aerosols.

6. Global Ozone Monitoring by Occultation of Stars

The GOMOS instrument will fly on the ESA Envisat satellite in 2001. It is designed to combine the high vertical resolution inherent in occultation measurements (better than 1.7 km in this case) with day- and night-side capability by performing stellar occultation on a selection of several hundred stars. With the Envisat orbital coverage, this will provide typically more than 600 profile measurements each day. GOMOS measures from 250 to 952 nm, with 0.7–0.9 nm resolution in the UV and visible. Additionally, two photometers measure from

466 to 528 nm and 644 to 705 nm. The GOMOS spectrometer measures profiles of ozone, NO_2, NO_3, OClO, H_2O, and temperature.

7. Odin/Optical Spectrograph and InfraRed Imaging System

Odin is a Swedish-led joint astronomy/aeronomy mission that includes participation from Canada, Finland, and France. Odin includes a UV/visible/IR spectrometer, the OSIRIS, and a microwave radiometer. The spectrometer will measure from 280 to 800 nm at 1 nm resolution, with an additional 10 nm wide radiometer channel at 1270 nm. It was launched in 2001 into a sun-synchronous orbit. The UV/visible/IR atmospheric measurements are made by limb scanning in scattered sunlight. OSIRIS measures O_3, NO_2, BrO, OClO, SO_2, HCHO, and H_2O.

8. Atmospheric Chemistry Experiment/Measurements of Aerosol Extinction in the Stratosphere and Troposphere Retrieved by Occultation

The ACE is a Canadian project to be launched aboard the SCISAT-2 satellite in 2002. It includes an infrared Fourier transform spectrometer, operating in solar occultation, and the UV/visible/IR MAESTRO instrument. MAESTRO operates from 285 to 1030 nm with a resolution of 0.6 nm in the UV and 1.0 nm in the visible/IR. It is primarily an occultation instrument, but will make nadir, backscatter measurements subject to available satellite resources. In addition to aerosols, MAESTRO will measure O_3, NO_2, BrO, OClO, SO_2, HCHO, and H_2O.

9. Ozone Dynamics Ultraviolet Spectrometer

The ODUS is scheduled for launch on the Japanese GCOM-A1 satellite in 2005, in a 70 degree inclination orbit. An additional instrument will fly on the GCOM-A2 satellite in 2008. ODUS measures backscattered radiation from 306 to 420 nm at 0.5 nm resolution. The nadir field-of-view has 20 km spatial resolution, degraded off axis in the 120 degree total field-of-view. ODUS will measure O_3, NO_2, BrO, OClO, SO_2, and HCHO.

REFERENCES

Bates DR. Rayleigh scattering by air. Planet Space Sci 32: 785–790, 1984.
Benedict WS, RF Calfee. Line Parameters for the 1.9 and 6.3 Micron Water Vapor Bands. U.S. Dept of Commerce, ESSA paper 2, 1967.
Bovensmann H, JP Burrows, M Buchwitz, J Frerick, S Noël, VV Rozanov, KV Chance, APH Goed. SCIAMACHY: mission objectives and measurement modes. J Atmos Sci 56: 127–150, 1999.

Brasseur G, S Solomon. Aeronomy of the Middle Atmosphere. Dordrecht, The Netherlands: D Reidel, 1986.

Burrows JP, KV Chance. Scanning imaging absorption spectrometer for atmospheric chartography. Proc SPIE, Future European and Japanese Remote Sensing Sensors and Programs, 1490: 146–154, 1991.

Calfee RF, WS Benedict. Carbon dioxide spectral line positions and intensities calculated for the 2.05 and 2.7 micron regions. NBS Technical Note 332, 1966.

Carlotti M. Global-fit approach to the analysis of limb-scanning atmospheric measurements. Appl Opt 27: 3250–3254, 1988.

Chance KV, JP Burrows, W Schneider. Retrieval and molecule sensitivity studies for the global ozone monitoring experiment and the scanning imaging absorption spectrometer for atmospheric chartography. Proc SPIE, Remote Sensing of Atmospheric Chemistry, 1491: 151–165, 1991.

Chance KV, JP Burrows, D Perner, W Schneider. Satellite measurements of atmospheric ozone profiles, including tropospheric ozone, from UV/visible measurements in the nadir geometry: a potential method to retrieve tropospheric ozone. J Quant Spectrosc Radiat Transfer 57: 467–476, 1997.

Chance K, RJD Spurr. Ring effect studies: Rayleigh scattering, including molecular parameters for rotational Raman scattering, and the Fraunhofer spectrum. Appl Opt 36: 5224–5230, 1997.

Chance K. Analysis of BrO measurements from the global ozone monitoring experiment Geophys Res Lett 25: 3335–3338, 1998.

Chance K, PI Palmer, RJD Spurr, RV Martin, TP Kurosu, DJ Jacob. Satellite observations of formaldehyde over North America from GOME. Geophys Res Lett 27: 3461–3464, 2000.

Chapman S. On ozone and atomic oxygen in the upper atmosphere. Phil Mag 10: 369, 1930.

European Space Agency, The GOME Users Manual, F. Bednarz, ed. European Space Agency Publication SP-1182, Noordwijk, The Netherlands: ESA Publications Division, ESTEC, 1995.

Fish DF, RL Jones. Rotational Raman scattering and the Ring effect in zenith-sky spectra. Geophys Res Lett 22: 811–814, 1995.

Fleig AJ, RD McPeters, PK Bhartia, BM Schlesinger, RP Cebula, KF Klenk, SL Taylor, DF Heath. Nimbus 7 Solar Backscatter Ultraviolet (SBUV) Ozone Products User's Guide. National Aeronautics and Space Administration, NASA RP-1234, Washington, D.C., 1990.

Freeman DE, K Yoshino, JR Esmond, WH Parkinson. High resolution absorption cross section measurements of SO_2 at 213°K in the wavelength region 172–240 nm. Planet Space Sci 32: 1125–1134, 1984.

Gamache RR, RL Hawkins, LS Rothman. Total internal partition sums in the temperature range 70–3000K: atmospheric linear molecules. J Mol Spectrosc 142: 205–219, 1990.

Gates DM, RF Calfee, DW Hansen, WS Benedict. Line parameters and computed spectra for water vapor bands at 2.7 μm. NBS Monograph 71, 1964.

Glaccum W, R Lucke, RM Bevilacqua, EP Shettle, JS Hornstein, DT Chen, JD Lumpe, SS Krigman, DJ Debrestian, MD Fromm, F Dalaudier, E Chassefiere, C Deniel,

CE Randall, DW Rusch, JJ Olivero, C Brogniez, J Lenoble, R Kremer. The polar ozone and aerosol measurement (POAM II) instrument. J Geophys Res 101: 14,479–14,487, 1996.

Goldman S, Information Theory. New York: Prentice-Hall, 1953.

Grainger JF, J Ring. Anomalous Fraunhofer line profiles. Nature 193: 762, 1962.

Heath DF, CL Mateer, AJ Krueger. The nimbus-4 backscatter ultraviolet (BUV) atmospheric ozone experiment—two years' operation. Pure Appl Geophys 106/108: 1238–1253, 1973.

Heath DF, AJ Krueger, HA Roeder, BD Henderson. The solar backscatter ultraviolet and total ozone mapping spectrometer (SBUV/TOMS) for Nimbus G. Opt Eng 14: 323–331, 1975.

Herman JR, PK Bhartia, O Torres, C Hsu, C Seftor, E Celarier. Global distribution of UV-absorbing aerosols from Nimbus 7/TOMS data. J Geophys Res 102: 16,911–16,922, 1997.

Hilsenrath E, RP Cebula, MT Deland, K Laamann, S Taylor, C Wellemeyer, PK Bhartia. Calibration of the NOAA-11 solar backscatter ultraviolet (SBUV/2) ozone data set from 1989 to 1993 using in-flight calibration data and SSBUV. J Geophys Res 100: 1351–1366, 1995.

Hsu NC, JR Herman, PK Bhartia, CJ Seftor, O Torres, AM Thompson, JF Gleason, TF Eck, BN Holben. Detection of biomass burning smoke from TOMS measurements. Geophys Res Lett 23: 745–748, 1996.

Hsu NC, JR Herman, O Torres, BN Hoben, D Tanre, TF Eck, A Smirnov, B Chatenet, F Lavenu. Comparisons of TOMS aerosol index with sun-photometer aerosol optical thickness: Results and applications. J Geophys Res 104: 6269–6279, 1999.

Janz SJ, E Hilsenrath, D Flittner, D Heath. Rayleigh scattering attitude sensor. Proc SPIE 2831: 146–153, 1996.

Joiner J, PK Bhartia, RP Cebula, E Hilsenrath, RD McPeters, H Park. Rotational Raman scattering (Ring effect) in satellite backscatter ultraviolet measurements. Appl Opt 34: 4513–4525, 1995.

Joiner J, PK Bhartia. The determination of cloud pressures from rotational Raman scattering in satellite backscatter ultraviolet measurements. J Geophys Res 100: 23,019–23,026, 1995.

Kato S, TP Ackerman, JH Mather, EE Clothiaux. The k-distribution method and correlated-k approximation for a shortwave radiative transfer model. J Quant Spectrosc Radiat Transfer 62: 109–121, 1999.

Kattawar GW, AT Young, TJ Humphreys. Inelastic scattering in planetary atmospheres. I. The Ring effect, without aerosols. Astrophys J 243: 1049–1057, 1981.

King MD, DD Hering, DJ Diner. The Earth observing system (EOS): a space-based program for assessing mankind's impact on the global environment. Opt Photon News 6: 34–39, 1995.

Krueger AJ, B Guenther, AJ Fleig, DF Heath, E Hilsenrath, R McPeters, C Prabhakara. Satellite ozone measurements. Phil Trans R Soc Lond A 296: 191–204, 1990.

Krueger AJ. The global distribution of total ozone: TOMS satellite measurements. Planet Space Sci 37: 1555–1565, 1989.

Kurosu T, VV Rozanov, JP Burrows. Parameterization schemes for terrestrial water clouds in the radiative transfer model GOMETRAN. J Geophys Res 102: 21,809–21,823, 1997.

Kurosu TP, KV Chance, RJD Spurr. Cloud retrieval algorithm for the European Space Agency's Global Ozone Monitoring Experiment. Proc SPIE, Satellite Remote Sensing of Clouds and the Atmosphere III 3495: 17–26, 1998.

Kurosu TP, KV Chance, RJD Spurr. CRAG—cloud retrieval algorithm for ESA's global ozone monitoring experiment. Proc 1999 European Symposium on Atmospheric Measurements from Space, ESA WPP-161: 513–521, 1999.

Kuze A, KV Chance. Analysis of cloud-top height and cloud coverage from satellites using the O_2 A and B bands. J Geophys Res 99: 14,481–14,491, 1994.

Lewis BR, L Berzins, JH Carver. Oscillator strengths for the Schumann-Runge bands of O_2. J Quant Spectrosc Radiat Transfer 36: 209–232, 1986.

Lewis BR, L Berzins, CJ Dedman, TT Scholz, JH Carver. Pressure broadening in the Schumann-Runge system of molecular oxygen. J Quant Spectrosc Radiat Transfer 39: 271–282, 1988.

Loyola D, W Balzer, B Aberle, M Bittner, K Kretschel, E Mikusch, H Mühle, T Ruppert, C Schmid, S Slijkhuis, R Spurr, W Thomas, T Wieland, M Wolfmüller, Ground segment for ERS-2 GOME sensor at the German D-PAF. Proc 3rd ERS Symposium, Space at the Service of Our Environment, ESA SP-414: 591–596, 1997.

Maurellis AN, R Lange, W van der Zande, I Aben, W Ubachs. Precipitable water column retrieval from GOME data. Geophys Res Lett 27: 903–906, 2000.

McClatchey RA, WS Benedict, SA Clough, DE Burch, RF Calfee, K Fox, LS Rothman, JS Garing. AFCRL atmospheric absorption line parameters compilation. AFCRL-TR-0096, 1973.

McCormick MP, JM Zawodny, RE Veiga, JC Larsen, PH Wang. An overview of SAGE I and SAGE II ozone measurements. Planet Space Sci 37: 1567–1586, 1989.

McElroy CT. Stratospheric nitrogen dioxide concentrations as determined from limb brightness measurements made on June 17, 1983. J Geophys Res 93: 7075–7083, 1988.

Miller AJ, A review of satellite observations of atmospheric ozone. Planet Space Sci 37: 1539–1554, 1989.

Minschwaner K, GP Anderson, LA Hall, K Yoshino. Polynomial coefficients for calculating O_2 Schumann-Runge cross sections at 0.5 cm^{-1} resolution. J Geophys Res 97: 10,103–10,108, 1992.

Munro R, BJ Kerridge, JP Burrows, K Chance. Ozone profile retrievals from the ESA GOME instrument. 1992 Quadrennial Ozone Symposium, 1992.

Munro R, R Siddans, WJ Reburn, BJ Kerridge. Direct measurements of tropospheric ozone distributions from space. Nature 392: 168–171, 1998.

Noël S, H Bovensmann, JP Burrows, J Frerick, KV Chance, APH Goede. Global atmospheric monitoring with SCIAMACHY. Physics Chem Earth 24: 427–434, 1999.

Palmer PI, DJ Jacob, K Chance, RV Martin, RJD Spurr, TP Kurosu, I Bey, R Yantosca, A Fiore, Q Li. Air-mass factor formulation for spectroscopic measurements from satellites: application to formaldehyde retrievals from the global ozone monitoring experiment. J Geophys Res 106: 14,539–14,550, 2001.

Platt U, Differential optical absorption spectroscopy (DOAS). In: MW Sigrist, ed. Air

Monitoring by Spectroscopic Techniques. New York: John Wiley & Sons, 1994, pp 27–84.

Richter A, F Wittrock, M Eisinger, J Burrows. GOME observations of tropospheric BrO in northern hemisphere spring and summer 1997. Geophys Res Lett 25: 2683–2686, 1998.

Roscoe HK, JA Pyle. Measurements of solar occultation: the error in a naive retrieval if the constituent's concentration changes. J Atmos Chem 5: 323–341, 1987.

Rossow WB, AW Walker, DE Beuschel, MD Roiter. International satellite cloud climatology project (ISCCP) documentation of new cloud datasets. WMO/TD-No. 737, World Meteorological Organization, 1996.

Rothman LS, CP Rinsland, A Goldman, ST Massie, DP Edwards, J-M Flaud, A Perrin, C Camy-Peyret, V Dana, J-Y Mandin, J Schroeder, A McCann, RR Gamache, RB Wattson, K Yoshino, KV Chance, KW Jucks, LR Brown, V Nemtchinov, P Varanasi. The HITRAN Molecular Spectroscopic Database and HAWKS (HITRAN Atmospheric Workstation), 1996 edition. J Quant Spectrosc Radiat Transfer 60: 665–710, 1998.

Singer SF, RC Wentworth. A method for the determination of the vertical ozone distribution from a satellite. J Geophys Res 62: 299–308, 1957.

Slijkhuis S, A von Bargen, W Thomas, K Chance. Calculation of undersampling correction spectra for DOAS spectral fitting. Proc ESAMS'99—European Symposium on Atmospheric Measurements from Space, 1999, pp 563–569.

Solomon S, AL Schmeltekopf, RW Sanders. On the interpretation of zenith sky absorption measurements. J Geophys Res 92: 8311–8319, 1987.

Thomas W, E Hegels, S Slijkhuis, R Spurr, K Chance. Detection of biomass burning combustion products in Southeast Asia from backscatter data taken by the GOME spectrometer. Geophys Res Lett 25: 1317–1320, 1998.

Torres O, PK Bhartia, JR Herman, Z Ahmad, J Gleason. Derivation of aerosol properties from satellite measurements of backscattered ultraviolet radiation: theoretical basis. J Geophys Res 103: 17,099–17,110, 1999.

van Oss R. Aerosol retrieval with GOME. Proc 1999 European Symposium on Atmospheric Measurements from Space, ESA WPP-161, 1999, pp 581–585.

Wagner T, U Platt. Satellite mapping of enhanced BrO concentrations in the troposphere. Nature 395: 486–490, 1998.

Yoshino K, DE Freeman, JR Esmond, WH Parkinson. High resolution absorption cross section measurements and band oscillator strengths of the (1,0)–(12,0) Schumann-Runge bands of O_2. Planet Space Sci 31: 339–353, 1983.

Yoshino K, DE Freeman, WH Parkinson. High resolution absorption cross section measurements of N_2O at 295–299K in the wavelength region 170–222 nm. Planet Space Sci 32: 1219–1222, 1984.

Yoshino K, JR Esmond, AS-C Cheung, DE Freeman, WH Parkinson. High resolution absorption cross sections in the transmission window region of the Schumann-Runge bands and Herzberg continuum of O_2. Planet Space Sci 40: 185–192, 1992.

14

Ultraviolet Spectroscopy in Astronomy

George R. Carruthers
Naval Research Laboratory, Washington, D.C.

I. ULTRAVIOLET ASTRONOMY: AN INTRODUCTION

A. Why Ultraviolet Astronomy?

Astronomy, to a much greater degree than most other fields of science, is dependent on remote sensing measurements, because most of the objects under study are inaccessible to direct, in situ measurements or sample returns. Until the advent of space-based observations, astronomical observations (including planetary astronomy and remote-sensing measurements of Earth's upper atmosphere) were limited to the wavelength ranges of the electromagnetic spectrum that can penetrate Earth's lower atmosphere, which is opaque to radiations having wavelengths below about 300 nm = 3000 Å (see Fig. 1). This precluded observations of nearly all of the ultraviolet, x-ray, and gamma ray regions of the spectrum.

Spectroscopy is the most detailed and quantitative remote sensing technique for determination of the compositions and temperatures of the objects under study. By observing radiation emitted or absorbed at specific wavelengths of the spectrum, one can determine, to varying degrees (depending on the wavelength and spectral resolution selected), both the composition and the temperature of an astronomical object. In some cases, one can also determine the velocity (in the line of sight) of the object or gas cloud, using the Doppler effect.

The most sensitive and quantitative measurements of the compositions of gaseous objects are obtained from the study of resonance transitions: spectral transitions involving the ground state of an atom or molecule. This is due to the fact that in most cases, even at relatively high temperatures, nearly all of the atoms or molecules of common substances reside in the ground electronic state.

499

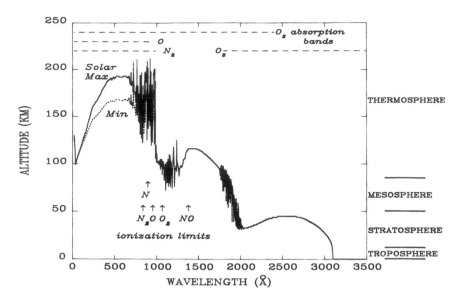

Figure 1   Graph of the altitude at which unit optical depth (attenuation to $1/e$, or 37%) for radiation vertically incoming to Earth's atmosphere is reached, versus wavelength in the ultraviolet spectral range. The atmospheric constituents contributing to this absorption, and significant ionization limits, are shown. (From R Meier Naval Research Laboratory [NRL].)

Therefore, particularly for the most common atoms and molecules present in astronomical objects or their atmospheres, observations of resonance spectral transitions provide the highest sensitivity of detection, and (especially for resonance absorption spectroscopy) the greatest quantitative accuracy.

The resonance transitions of the most abundant atoms and molecules in astronomical objects, such as those of H, H_2, He, C, CO, O, O_2, N, N_2, and Ar, all lie in the far-ultraviolet region of the electromagnetic spectrum, which is inaccessible to ground-based observatories (except for very high-redshift objects, such as quasars).

The temperatures of objects can be determined from observations of both spectral line transitions (of gases) and the continuous spectrum (of stars and cooler, high-density objects such as planets). As is well known, the peak of the continuous spectrum of a "black body" radiator shifts toward shorter wavelengths in inverse proportion to temperature, and the total emitted radiation increases as the fourth power of temperature. Although stellar spectra show only crude resemblance to the spectra of black body radiators, it is nevertheless true that the peak of the continuous spectrum shifts toward shorter wavelengths as the "effective" temperature of the star (the temperature of a perfect black-body radiator having the same total luminosity and surface area as the star) increases.

For stars much hotter than the Sun, most of the emitted radiation falls in the ultraviolet (UV) wavelength range below the transmission limit of Earth's atmosphere (about 300 nm), and hence it is difficult to make accurate determinations of the total stellar luminosity and effective temperature from ground-based observations alone. Figure 2 is a graph of theoretical model atmosphere flux vs. wavelength curves, for stars with a range of effective temperatures, showing that

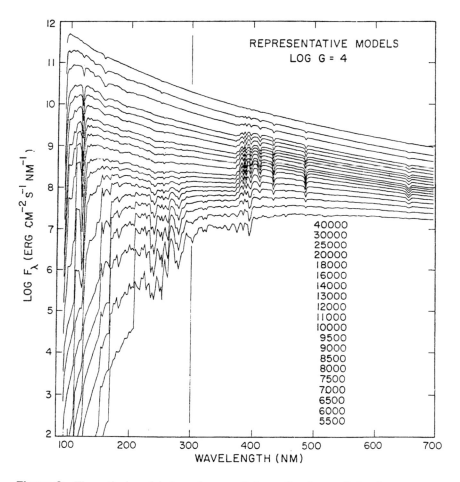

Figure 2 Theoretical model atmosphere predictions of surface radiation flux vs. wavelength, for a range of effective temperatures. Only the portion of the spectra longward or 300 nm are accessible to ground-based observatories. (From Kurucz, 1979.)

the sensitivity of emitted flux to temperature is much greater near or shortward of the peak of maximum emission than far longward of that peak (Kurucz, 1979).

Prior to the advent of space-based far-UV observations of hot stars, the only directly observable evidence concerning their high temperatures (besides their visible-light spectra) was the fluorescence of interstellar gas (emission nebulas, also known as ionized hydrogen, or H II, regions) induced by the far-UV emissions from these stars. Far-UV radiation of wavelengths less than 91.2 nm, the series limit of the Lyman (resonance) line series of atomic hydrogen, is capable of ionizing hydrogen. When the hydrogen ions recombine with electrons, spectral line emission in the visible (Balmer) series, as well as in other line series, of atomic hydrogen is produced. The integrated intensity of the nebular emission, when ratioed to the observed visible brightness of the star, provides an indication (but not an accurate measure) of the temperature of the star.

B. Difficulties of Ultraviolet Astronomy

The primary difficulty, which prevented ultraviolet astronomy until the early 1950s, is that atmospheric absorption prevents observations in the UV wavelength range below 300 nm from ground-based observatories. Therefore, the advent of UV astronomy had to await the advent of the "space age" to allow space-based observations. Figure 1 shows the altitudes to which the instrumentation must be carried to observe 1/e (36.8%) of the total incoming radiation from an (overhead) astronomical object, vs. wavelength. As seen, although significant improvements can be made in observations of infrared and microwave wavelengths, at altitudes accessible by aircraft and balloons, only rocket vehicles can reach altitudes high enough to enable observations in the UV and x-ray wavelength ranges.

The first UV spectra of an astronomical object, the Sun, were obtained by Naval Research Laboratory scientists using instruments flown on V-2 rockets brought to the United States from Germany after World War II (see Fig. 3). Similar observations of the (much fainter) other stars were made significantly later, beginning in the early 1960s, in part because of the requirement for accurate pointing systems on the rocket vehicles used for the observations.

As is true of most other scientific investigations requiring space flight, the cost and required reliability of the instrumentation are much higher than for equivalent instrumentation in ground-based observatories. Also, opportunities for making space-based observations (except from spacecraft already in orbit) are much fewer and more limited in observing flexibility.

Another difficulty is that the nature of instrumentation usable for ultraviolet measurements is somewhat different than for equivalent instrumentation operating in visible and near-visible wavelengths (see Samson, 1967). In particular, the reflectivities of commonly used mirror coatings and the transmissions of re-

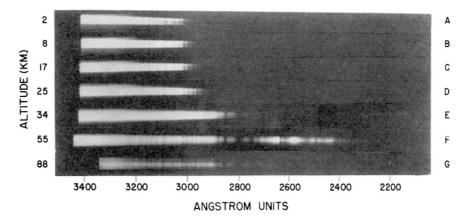

Figure 3 Spectra of the Sun obtained with a spectrograph flown on a V-2 rocket in 1946, vs. altitude, showing (at the higher altitudes) the UV wavelength range below 300 nm (3000 A) for the first time. (From R Tousey, NRL.)

fractive optics made of commonly used glasses are very low in the ultraviolet, particularly at the far-UV wavelengths (below 200 nm). Also, window and mirror coating materials that are usable in far-UV wavelengths may be degraded by exposure to moisture and other forms of contamination to a greater degree than materials used at visible and near-visible wavelengths. These considerations result in a much more limited choice of refractive materials and mirror coatings, and may require minimizing, to the extent possible, the total number of reflective and/or refractive elements in the optical system.

At the very shortest (extreme-UV; EUV) wavelengths (below 100 nm), the limitations on reflectivities and transmissions of optical materials are even more severe than in the far-UV (100–200 nm) wavelength range. This requires, in general, use of grazing incidence reflective optics, and/or limiting the instrument to a single reflection. No known materials usable for windows or refractive optics transmit below about 105 nm, which is the transmission limit of lithium fluoride (LiF). However, very thin films of metals such as Al and In can be used to transmit selected wavelength ranges in the EUV, and recently developed multilayer interference coatings can provide high normal-incidence reflectivities in selected, relatively narrow wavelength ranges.

As is true in other wavelengths of the electromagnetic spectrum, there is a tradeoff between spectral resolution and sensitivity: in general, the higher the resolution, the lower the sensitivity, because the same (or similar) amount of radiation is spread over a larger number of detector elements. Therefore, to an even greater extent than in visible wavelengths, the ultraviolet spectrograph must be optimized for the primary scientific objective it is to address, and it is difficult to build a practical "multipurpose" instrument.

II. ULTRAVIOLET SPECTROSCOPIC INSTRUMENTATION

A. Characteristics of Materials for Transmission Optics

Most glasses used in ground-based astronomical instruments do not transmit, or transmit poorly, in UV wavelengths shortward of about 300 nm. The most commonly used UV transmissive materials for refractive optics and windows, in order of decreasing short-wavelength transmission cutoff, are as follows:

UV-grade fused silica (SiO_2)	165 nm
Sapphire (Al_2O_3)	145 nm
Barium fluoride (BaF_2)	135 nm
Strontium fluoride (SrF_2)	128 nm
Calcium fluoride (CaF_2)	123 nm
Magnesium fluoride (MgF_2)	115 nm
Lithium fluoride (LiF)	105 nm

It is noteworthy that the fluorides have short-wavelength limits that are temperature sensitive: they shift to longer wavelengths when heated above room temperature, and to shorter wavelengths when cooled below room temperature.

This temperature dependence can be used to "fine tune" the short-wavelength cutoffs. For example, we have used LiF windows cooled to liquid nitrogen temperatures to detect hydrogen Lyman-β (102.5 nm) radiation (Carruthers, 1971), and SrF_2 heated to 100°C to transmit atomic oxygen 135.6 nm radiation while mostly excluding atomic oxygen 130.4 nm radiation (Carruthers and Seeley, 1996).

Since the fluorides are crystalline materials, they are more difficult to work with in generating optical surfaces than are fused silica and most glasses. Also, they have relatively high coefficients of thermal expansion and hence are susceptible to thermal shock. They are also (to varying degrees) water soluble, and their transmission characteristics may be degraded by exposure to moisture or humid air.

Another difficulty in using these materials for refractive optics is that the index of refraction varies with wavelength more rapidly toward shorter wavelengths (and the curve is very steep near the short-wavelength transmission limit). This means that refractive elements such as lenses and Schmidt correctors must be designed for specific wavelengths, or relatively narrow-wavelength ranges, of interest. On the other hand, the high indices of refraction near the short-wavelength cutoffs mean that flatter prisms and/or less steeply curved lens surfaces can be used, which minimizes transmission losses in these wavelength ranges.

B. Characteristics of Coatings for Reflective Optics

In the visible and near-UV spectral regions, aluminum is usually the coating of choice for mirrors and reflection gratings. However, at shorter wavelengths the naturally occurring oxidation of bare aluminum coatings causes the reflectivity to fall off markedly in the far-UV range.

This problem can be circumvented by deposition of a thin film of a UV-transmissive material, such as MgF_2, on the mirror surface immediately after aluminization, while still in vacuum. As seen in Figure 4, optimal coating thicknesses of MgF_2 can provide reflectivities as high as 70% down to about 115 nm wavelength. With the use of LiF coatings, the useful range of reflectivity can be extended to below 102 nm (although with the disadvantages of somewhat lower maximum reflectivity, and of higher susceptivity to degradation by atmospheric humidity) (see Bradford et al., 1969).

For wavelengths below 100 nm, no known materials can provide high reflectivities (greater than 50%) over significant wavelength ranges. The best materials to use include the heavy metals platinum, osmium, and tungsten, and also silicon carbide (SiC). Figure 5 shows typical reflectance vs. wavelength curves for these materials.

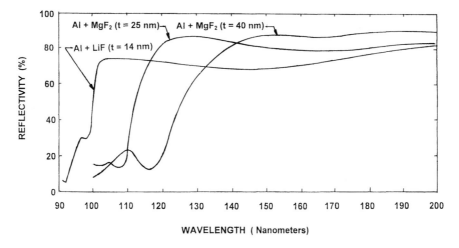

Figure 4 Reflectance vs. wavelength for mirror coatings useful in the far-UV spectral range. (From Bradford et al., 1969.)

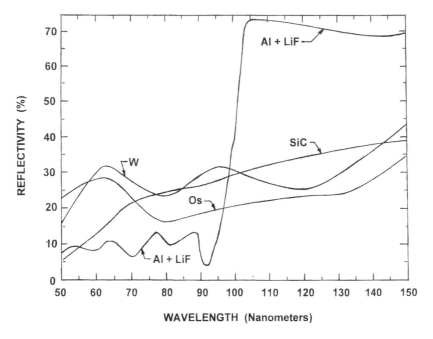

Figure 5   Reflectance vs. wavelength for mirror coatings useful in the wavelength range below 100 nm. (From W Hunter, NRL; Keski-Kuha et al., 1988.)

In recent years, multilayer interference coatings have been developed that can provide very high reflectances in selected, narrow-wavelength ranges in the extreme ultraviolet (EUV). These have been particularly useful for monochromatic imaging of the sun in selected EUV and soft x-ray wavelengths, usually corresponding to known emission-line wavelengths.

C.   Minimizing the Number of Optical Elements to Maximize Throughput

The throughput of spectrographs can be maximized, although sometimes with some compromise in the optical quality, by using a single reflective or transmissive element for more than one purpose. The best example is the use of a concave grating, which both disperses the radiation and brings it to a focus, instead of using a plane reflectance grating plus a focusing mirror or lens. Likewise, a telescope with a single, long-focal-length primary mirror may be used to feed a spectrograph instead of a more compact, two-mirror Cassegrain or Ritchey–Chretien telescope of the same collecting aperture.

As mentioned, in the EUV spectral region, reflectivities of broad-band normal-incidence mirror coatings are quite low in comparison to shorter wavelengths, and grazing-incidence optics may be required to obtain acceptable throughputs. This may cause the optical system to be considerably longer (and hence more difficult to accommodate in a spacecraft) than an equivalent-aperture normal-incidence-reflection optical system.

D. Typical Spectrograph Configurations

1. Objective Grating Spectrographs

The objective grating, or so-called slitless, spectrograph is the simplest type of spectrograph, consisting of only a grating and a camera (or only the grating, if it is a concave grating), plus detector. Instruments of this type were among the first ones used in ultraviolet astronomy, largely because of (a) the relatively poor pointing accuracies achievable in early sounding-rocket experiments; and (b) the desire to survey large regions of the sky, as distinct from targeting only a few selected objects. Since the field of view can be large compared to that achievable with a slit-type spectrograph, the spectra of several objects can be recorded simultaneously (using a two-dimensional imaging detector or film).

A disadvantage of this technique is that there is no intrinsic wavelength reference, since the wavelength recorded at a particular point on the detector format depends on both the actual wavelength and the position of the object in the camera field of view. However, it is often the case that the positions of particular spectral lines or features known or expected to be present in the spectrum can be used to calibrate a wavelength scale for each detected object.

Another disadvantage is that diffuse sources of large angular size cannot be studied with as high spectral resolution as pointlike sources, such as stars. Also, any diffuse sky glow (due to upper atmospheric or near-Earth space emissions) will not be spectrally resolved, but is also recorded (as a diffuse background), and hence may degrade the detectivity or signal-to-noise ratio of the spectra of objects of interest.

Objective Plane Grating Spectrographs.   Figure 6 shows diagrams of two types of plane grating objective spectrographs. This type of instrument was used in sounding rocket flights, beginning in the early 1960s, by groups at Princeton University and the Naval Research Laboratory, to observe the spectra of hot stars extending down to wavelengths as short as 100 nm (see Morton, 1967; Carruthers, 1970, Carruthers et al., 1984). Among the results of these experiments were the first detections of interstellar atomic hydrogen, via its Lyman-α (121.6 nm) resonance absorption line (Princeton), and of interstellar molecular hydrogen via its Lyman band resonance absorption in the 100–115 nm wavelength range (NRL).

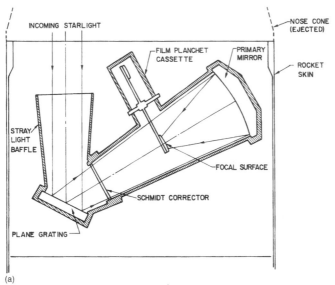

INCOMING STARLIGHT

NOSE CONE
(EJECTED)

FILM PLANCHET
CASSETTE

PRIMARY
MIRROR

ROCKET
SKIN

STRAY
LIGHT
BAFFLE

FOCAL SURFACE

SCHMIDT CORRECTOR

PLANE GRATING

(a)

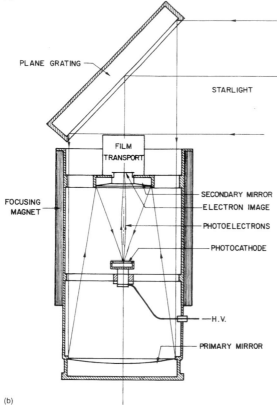

PLANE GRATING

STARLIGHT

FILM
TRANSPORT

FOCUSING
MAGNET

SECONDARY MIRROR

ELECTRON IMAGE

PHOTOELECTRONS

PHOTOCATHODE

H.V.

PRIMARY MIRROR

(b)

Inverse Wadsworth Concave Grating. This is the simplest of all spectrographs (see Fig. 7), consisting of only a concave grating and a detector. Parallel light rays are incident on the concave grating, which then projects a focused spectral image on the detector. A collimator is usually necessary on the input side to avoid contamination of the spectrum by off-axis radiation.

Because of the typically larger focal ratios (f/numbers) of concave gratings used in this mode than with the use of a plane grating with a Schmidt or other small-focal-ratio camera, the field of view is usually smaller, but the useful spectral resolution is usually higher. This type of instrument has been used in sounding rocket flights in which a single star was the target, rather than a wide field of stars, but where the pointing uncertainties were too large to make use of a slit spectrograph, and/or where the number of optical surfaces needed to be minimized in order to maximize sensitivity (see Smith, 1969, 1972).

2. Rowland Circle Spectrographs

The Rowland-circle spectrograph (see Fig. 8) is the simplest slit spectrograph design based on the use of a concave grating. Unlike in the case of objective grating spectrographs, the spectral resolution is independent of the angular size of the source. It can be used alone to observe the spectra of diffuse sources whose angular extent is larger than the field of view of the grating, as projected through the slit. However, for observations of stars and other discrete objects, an external optical element or telescope must be used to focus an image of the object onto the slit, and a very accurate and stable pointing system is needed to keep the object fixed on the slit.

The spectral resolution is determined by (a) the radius of curvature of the grating, (b) the number of lines per millimeter of the grating ruling, and (c) the width of the entrance slit. Resolution increases in direct proportion to variables (a) and (b), and in inverse proportion to (c). As one might guess, sensitivity varies in the opposite direction from resolution for all three variables.

Another important consideration in spectrograph design is the wavelength range to be covered in a single observation. Spherical concave gratings are subject to increasing aberrations with increasing angle from the grating normal (as projected on the detector), although special figuring of the grating (e.g., toroidal

Figure 6 (a) An objective grating spectrograph, using a Schmidt camera and photographic film, flown on sounding rockets by Princeton University to obtain far-ultraviolet spectra of hot stars. (b) An objective grating spectrograph, using a two-mirror, all-reflecting camera with electrographic recording on film, flown on sounding rockets by the Naval Research Laboratory to observe the far-UV spectra of hot stars.

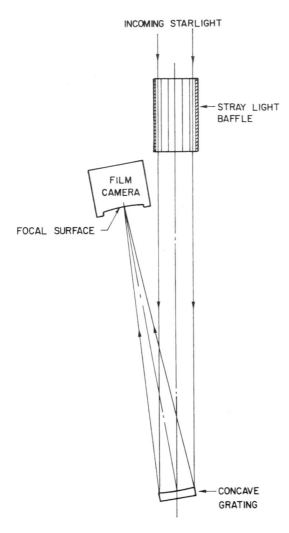

Figure 7 An inverse-Wadsworth spectrograph, a type of which was used in sounding rocket flights by NASA Goddard Space Flight Center to observe the far-UV spectra of hot stars.

figuring) can give significantly improved useful wavelength ranges. Therefore, it is usually easier to increase spectral resolution (over a given wavelength range of simultaneous observation) by increasing the focal length of the grating, rather than increasing the number of lines per millimeter of its ruling; however, this also makes the total instrument larger (which is a disadvantage for space flight

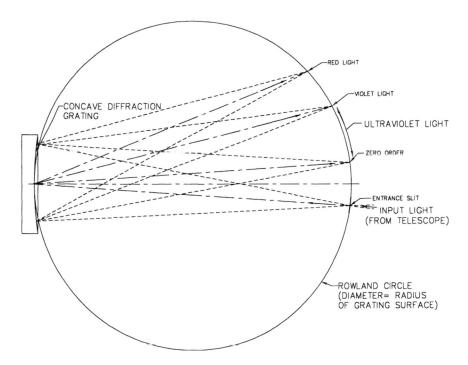

RED LIGHT

VIOLET LIGHT

CONCAVE DIFFRACTION
GRATING

ULTRAVIOLET LIGHT

ZERO ORDER

ENTRANCE SLIT

INPUT LIGHT
(FROM TELESCOPE)

ROWLAND CIRCLE
(DIAMETER= RADIUS
OF GRATING SURFACE)

Figure 8 Diagram of a Rowland-circle spectrograph, using a concave diffraction
grating.

experiments) and increases the effective focal ratio (which decreases the sensitivity to diffuse sources).

3.   Other Spectrograph Designs

There have been numerous other types of spectrographs and spectrometers used in UV space astronomy, as is true also in ground-based astronomy and laboratory spectroscopy. Among these are devices using combinations of prisms and diffraction gratings, and ones incorporating two (rather than one) diffraction gratings.

For very high spectral resolution studies, a combination of two diffraction gratings whose rulings are oriented at right angles to each other (i.e., so-called cross-dispersed arrangements) can be used. A very high dispersion (i.e., finely ruled) grating produces high spectral resolution, but only a small portion of the spectral range can be recorded in the direction of dispersion. However, if a more coarsely ruled echelle grating is placed ahead of this high-dispersion grating with its dispersion in the transverse direction, multiple segments of the spectrum can be recorded simultaneously, on a two-dimensional format detector (or film). This type of spectrograph has been used in several UV astronomy space instruments, initially for observations of the Sun, and later for stellar observations. Instruments of the latter type include the International Ultraviolet Explorer (IUE) satellite, launched in 1978 and remaining in operation for more than 10 years (see Kondo et al., 1989); spectrographs used in the Hubble Space Telescope (HST) (see Harms et al., 1982; Brandt et al., 1982); and the Interstellar Medium Absorption Profile Spectrometer (IMAPS) (Jenkins et al., 1988). The latter three instruments are discussed in more detail below.

E. Ultraviolet Detectors for Astronomical UV Spectroscopy

The earliest ultraviolet astronomy experiments, through the mid-1960s, used one of two sensor types, similar to those then used in ground-based astronomy: (a) photographic film, or (b) photomultiplier tubes (or other equivalent single-element detectors). Although adequate for many first-look investigations in UV astronomy, the limitations of these detectors are such that they have now been almost totally replaced by two-dimensional electronic imaging detectors.

1. Photographic Film in Ultraviolet Astronomy

Photographic film was the first detection technology used in UV space astronomy, such as the pioneering observations of the Sun in the far-UV obtained by NRL in V-2 rocket flights beginning in the late 1940s (see Fig. 3), and objective-grating spectroscopy of stars in the UV obtained from sounding rockets and a Gemini manned space mission in the mid-1960s. At the present time, however, photography has been almost totally replaced by electronic media, particularly for long-duration space missions.

The most important disadvantages of photographic film in UV astronomy are: (a) the film has to be specially prepared to provide useful sensitivity in the far UV, because the absorption of UV by the gelatin in photographic emulsions greatly reduces the amount of UV that can activate the silver halide grains that are the active ingredients of the film; (b) the film cannot be made sensitive *only* in the UV and insensitive to visible light, which is a problem particularly for observations of the Sun and other objects that are much brighter in visible light than in the UV; and (c) it is difficult to obtain quantitative photometric information from film images, because of the nonlinear, highly variable, and exposure-time-dependent responses of photographic emulsions. Other potential problems include degradation of the photographic emulsion with time, in long-duration space missions, due to the high vacuum and/or energetic charged particle environments in space.

2. Photomultipliers and Channel Electron Multipliers

Photomultipliers, and the closely related channel electron multipliers, are the simplest of the purely electronic detectors. Unlike film, they have linear and easily calibrated responses to electromagnetic radiation and can be made to have high sensitivity in the UV, and little or no visible light sensitivity, by appropriate choice of the photocathode and window materials (including, in space, totally windowless operation). They also can have much higher detection efficiencies than film (counts per incident photon, or quantum efficiency). Their output electronic signals can be sent back to Earth by radio telemetry, eliminating the need for film return and hence making possible long-duration and/or deep-space missions.

Their major disadvantage is that they can record only one wavelength or spatial resolution element at one time. To record a full spectrum of an astronomical object, the detector must be scanned over the desired wavelength range, taking a reading at each spectral resolution increment along the spectrum. Not only does this greatly increase the time required to record a spectrum, but it also does not differentiate time variations of spectral intensity (e.g., due to pointing oscillations or changes in atmospheric absorption) from changes of spectral intensity with wavelength.

Both of these devices make use of photoelectric emission, from a photosensitive surface (photocathode). By the use of appropriate photocathode materials and window materials, the sensitivity of the device can be tailored to the wavelength range of interest (Sommer, 1968). For example, photocathode materials that are highly sensitive in the middle- and far-UV ranges but are totally blind to visible light can be used. Windowless photomultipliers and channel multipliers can be made sensitive in the EUV, as well as far-UV, wavelength ranges.

Both devices make use of the process of secondary emission to amplify the charge corresponding to a single photoelectron to produce output signals of 10^6–10^8 output electrons, which is adequate to produce a readily detectable signal, or count, in the output electronic signal. A device that can count the individual photoelectrons released from the front-end photosensitive surface provides the maximum possible signal-to-noise ratio. This ratio is limited by statistics: $S/N = N/\sqrt{N} = \sqrt{N}$, where N is the total number of photoelectron counts.

In a photomultiplier, the secondary amplification is achieved using a series of secondary-emitting dynodes (see Fig. 9). In the channel electron multiplier, a continuous tube of semiconducting glass is used to amplify photoelectrons produced at the front end of the device, which makes possible a simpler and more compact device than the discrete-dynode photomultiplier (see Fig. 10).

A major UV astronomy mission that used photomultiplier detectors was the Princeton University's instrument on the third Orbiting Astronomical Observatory (OAO-3, re-named Copernicus) launched by the National Aeronautics and

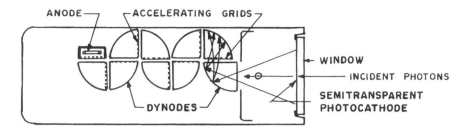

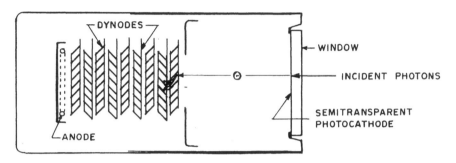

Figure 9 Diagrams of two photomultiplier tubes, using different types of dynodes for amplification (by secondary emission) of electrons produced by photoemission from semitransparent photocathodes.

Space Administration (NASA) in 1972. The primary scientific objective of Copernicus was to study the composition and spatial distribution of interstellar gas, via its far-UV absorption spectra. This instrument (see Fig. 11) used a concave grating spectrograph and an 80 cm aperture telescope mirror. Windowless photomultipliers were physically scanned over the Rowland circle of the spectrograph, providing moderate (0.02 nm) and high (0.005 nm) spectral resolutions in the 95–160 nm wavelength range.

3. Electronic Imaging Devices

Television Camera Tubes. Television camera tubes, of the type called secondary electron conduction (SEC) vidicons, were used in two major UV astronomy missions. The first was for far-UV direct imagery of stars, in the second Orbiting Astronomical Observatory (OAO-2) launched in 1968. The second was for the spectrographs on board the International Ultraviolet Explorer (IUE) launched in 1978. These devices circumvented most of the disadvantages of film;

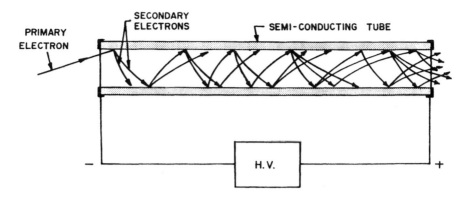

Figure 10 Diagram of a channel electron multiplier, which uses a continuous secondary-emitting tube surface to amplify photoelectron input to, or produced at, the front end of the tube.

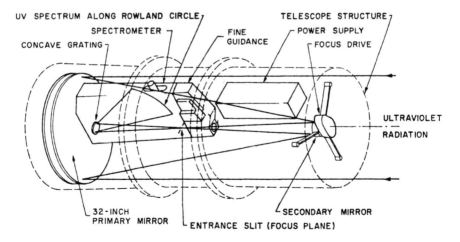

Figure 11 Diagram of the instrumentation of the third Orbiting Astronomical Observatory (OAO-3), named Copernicus, launched into orbit in 1972. The instrument, provided by Princeton University, used a reflecting telescope and concave-grating spectrograph, and windowless photomultiplier tubes which were scanned along the Rowland circle, to record the far-UV spectra of stars. (Courtesy of Princeton University Observatory.)

in particular, since they generated electronic signals that could be radioed to ground stations, they were not subject to the limitations on observing time set by an initial supply of film, or to degradation of film in the space environment. In particular, the IUE's ability to take long time exposures with its two-dimensional, integrating SEC vidicon detectors enabled it to observe much fainter stars than was possible with Copernicus, which had a much larger telescope aperture (although with lower spectral resolution, and limited to wavelengths longward of 115 nm).

Microchannel Plate-Based Detectors. The microchannel plate (MCP) is an extension of the channel electron multiplier technology to provide a two-dimensional array of individual, miniaturized channel multipliers that can operate simultaneously, but with independently recorded signal outputs (see Fig. 12). This provides, as a result, the major advantages of photographic film (simultaneous recording of a two-dimensional field of view) with the much-improved sensitivity, quantitative accuracy, and wavelength range selectability of a photoelectric

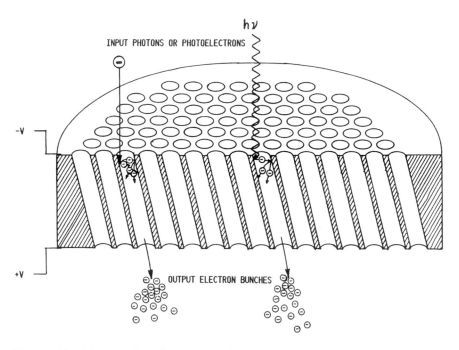

Figure 12 Diagram of a microchannel plate, which is based on the principle of the channel electron multiplier (Fig. 10). Here, many channel multipliers are fused together into a continuous plate, which has the ability to detect simultaneously the input radiation at all points over a two-dimensional image.

detector. When the MCP is used with various photocathode materials, deposited either on the front face of the MCP or on a separate photocathode substrate, the wavelength range of response can be varied to meet the investigation's objectives.

In comparison to UV-sensitive television camera tubes, the MCP detectors have the advantage of being much more compact and lighter in weight, and can count individual photoelectron events rather than producing an analog video signal. This results in both better sensitivity and better quantitative accuracy.

Charge-Coupled Device-Based Detectors. The charge-coupled device (CCD) is a solid-state electronic imaging device that can provide large-format, two-dimensional images in visible and near-infrared radiation by the principle of photoconduction. However, CCDs can also be used for detection of ultraviolet radiation by one of three techniques: (a) the CCD can be made sensitive to UV radiation, as well as visible and UV radiation, by special processing; (b) the CCD can be coupled to an image-intensifier tube with a UV-sensitive photocathode, to provide UV images without contamination by visible and near-IR radiation; and (c) photoelectrons emitted by a UV-sensitive photocathode can be made to impact the CCD directly, thereby producing an electronic output signal (electron-bombarded CCD). All three of these approaches have been used in UV astronomical investigations.

Film Recording Electronic Imaging Devices. Many of the disadvantages of film can be circumvented by using it in conjunction with electronic imaging devices. Image intensifiers or image converters are devices with photoelectric emission surfaces (photocathodes) of the same types usable with photomultipliers and channel electron multipliers, but that form so-called electron images on a phosphor screen, which then gives off visible light (as in a cathode-ray tube). These devices may also include microchannel plates to amplify the electron image before it reaches the output phosphor screen. The light from the phosphor screen can then be transferred to the recording photographic film, using either lenses or (more efficiently) a fiber-optic rear faceplate (with the phosphor screen deposited on one side and the film contacting the other side).

Another variation on this theme is the electrographic or electronographic detector (see Fig. 6b), in which film is used to record the electron image of the image tube directly. This has the advantage of greater simplicity, for detectors operating in the vacuum ultraviolet, provided that photocathode materials (such as the alkali halides) relatively insensitive to degradation by outgassing of the recording emulsion are used (Carruthers, 1965). The electrographic technique is more quantitative than direct or image-intensified photography, because energetic electrons striking the emulsion directly leave discrete spots, or tracks, which can be individually counted, and (unlike ordinary photography) the optical density of the processed emulsion is directly proportional to the total number of photo-electron hits per unit of area.

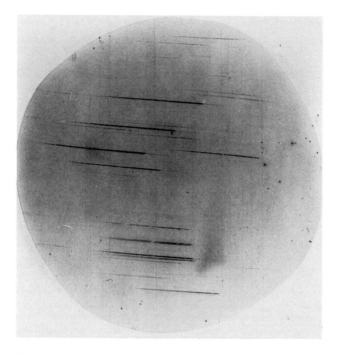

Figure 13   An objective-grating spectrogram of the region of the Belt of Orion, obtained by NRL in a 1982 sounding rocket flight, using an all-reflecting objective-grating spectrograph and electrographic recording. The field of view is 10 degrees in diameter, and the spectral range is 90–160 nm (shorter wavelengths to the right in each spectrum). The spectral resolution is about 0.3 nm.

The electrographic recording technique has been used by NRL in several sounding rocket flights of objective-grating spectrographs to record the spectra of hot stars, in some cases down to the Lyman limit of 91.2 nm (see Fig. 13).

III.   EXAMPLES OF CURRENT UV SPECTROGRAPHIC INSTRUMENTS AND APPLICATIONS TO UV SPACE ASTRONOMY

A.   Solar Ultraviolet Spectroscopy Space Missions

As mentioned previously, the Sun was the first target of UV space astronomy, both because of its practical importance to us here on Earth and because it is many orders of magnitude brighter than any other star, even in the far UV. Most of the spectrograph types and detector technologies described in the preceding section were used first in solar physics studies, before they were used to study objects in deep space. Solar UV astronomy began with the first V-2 rocket flights from White Sands Missile Range in the late 1940s, and followed with other sounding rocket flights (continuing to the present time). The first dedicated solar ultraviolet astronomy satellite missions were the Orbiting Solar Observatories (OSOs), beginning in the 1960s. Later missions included the Skylab space station launched in 1973, the Solar Maximum Mission (NASA SMM, 1987), launched on a Delta rocket in 1980 (and later repaired by a space shuttle crew in 1984), and the Spacelab-2 mission launched in 1985.

Because the Sun is many orders of magnitude brighter in the visible than in the far UV, spectrograph and detector technologies must include special precautions to avoid contamination of the far-UV spectra by stray visible light. These include use of cross dispersion (see Fig. 14), in which a low-dispersion diffraction

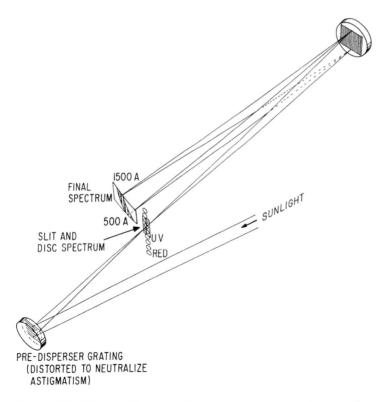

Figure 14 Diagram of a solar ultraviolet spectrograph, using a predisperser grating to minimize the input of visible solar radiation into the main spectrograph.

grating, ahead of the main spectrograph entrance slit and oriented perpendicularly to the primary spectrograph grating, predisperses the radiation before it enters the main spectrograph slit. This effectively prevents most of the visible light from entering the main spectrograph slit. Other methods include the use of solar-blind photoelectric detectors, which are sensitive in the far UV but insensitive to visible and infrared radiation.

An example of a high-resolution solar UV spectroscopy instrument is the Naval Research Laboratory's High-Resolution Telescope and Spectrometer (HRTS) (Bartoe and Brueckner, 1975), which was flown in several sounding rocket flights and as part of the Spacelab-2 Shuttle mission in 1985 (see Fig. 15). This long-slit imaging spectrograph was used to measure both the spatial and spectral distributions of solar far-UV radiation at various locations on the solar surface (including, for example, sunspots and other active regions) and variations with solar activity (by multiple sounding rocket reflights).

Another complementary instrument developed by NRL was the Solar Ultraviolet Spectral Irradiance Monitor (SUSIM) (Brueckner et al., (1991). This (see Fig. 16) was intended to monitor the total solar UV radiation output, with lower spectral resolution than provided by HRTS, but emphasizing high photometric accuracy and time stability. This instrument flew, along with HRTS, on the Spacelab-2 shuttle mission. SUSIM also was flown on three later Shuttle-based Atmospheric Laboratory for Applications and Science (ATLAS) missions, the first launched in 1992. It also flew on a much longer duration unmanned satellite mission, the Upper Atmosphere Research Satellite (UARS), launched in 1991 (and still in operation at this writing). Because solar activity varies in an 11 year cycle, long-duration monitoring with highly stable and accurately calibrated instrumentation is essential.

The overlap between the ATLAS and UARS missions allowed direct, simultaneous comparisons of the results of two SUSIM instruments. The Shuttle-based instrument had the advantage of being calibrated immediately before and after its relatively short-duration missions, so that it could track any changes in the calibration of the long-duration UARS instrument.

B. Hubble Space Telescope

The Hubble Space Telescope (HST) is the most powerful and best known (to the general public, as well as the scientific community) of all nonsolar space astronomy projects (NASA NP-126) (see Fig. 17). Although its primary objective is not specifically UV space astronomy, this is nevertheless an important part of the HST's total suite of observation programs and complement of scientific instruments. HST's large aperture, high spatial resolving power, and long duration on orbit make it by far the best single example illustrating the instrumental technologies, and their evolution in time, of UV space astronomy.

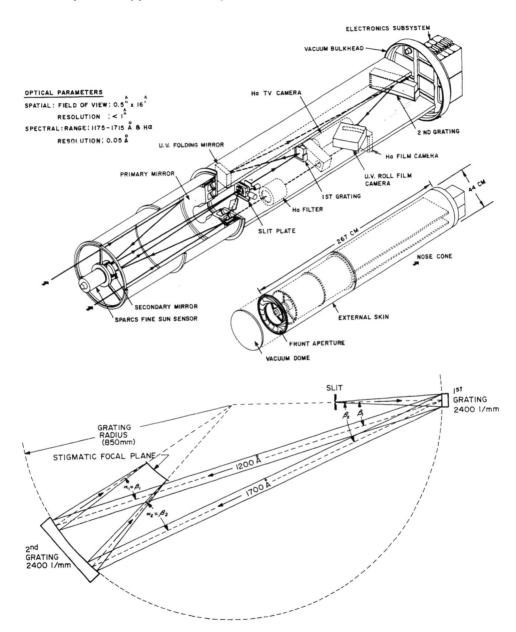

Figure 15 Diagram of the Naval Research Laboratory's High Resolution Telescope and Spectrograph (HRTS), which was flown on several sounding rockets and the Spacelab-2 Shuttle mission to obtain high-resolution, spatially resolved far-ultraviolet spectra of the Sun.

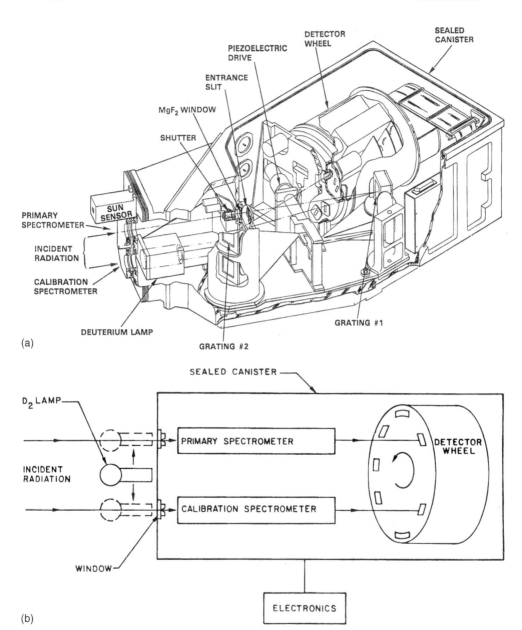

Figure 16   (a) Diagram of the Naval Research Laboratory's SUSIM, which was intended for highly accurate measurements of the intensity and time variations of the solar UV spectrum. (b) Diagram of the use of two identical spectrometers and a calibration lamp, which ensure the highest possible accuracy of measurements by SUSIM of the solar UV spectrum and its potential time variations.

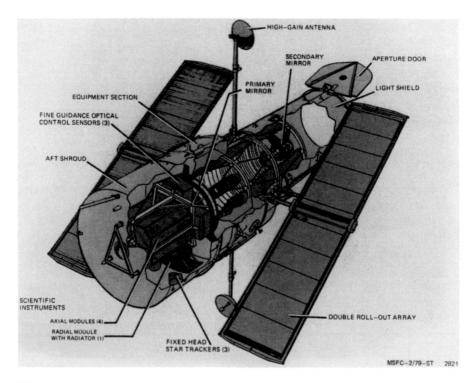

Figure 17 Diagram of the Hubble Space Telescope (HST) shows the main telescope optics and the scientific instrument modules. (From NASA NP-126.)

When the HST and its first-generation suite of scientific instruments were approved for development and flight, in the latter 1970s (NASA CP-2244, 1982), the state of the art in UV astronomy instrumentation, and specifically detector technology, was far from what it is at present. However, since HST was designed for servicing by astronauts in space shuttle refurbishment missions, this has allowed the initially installed instruments to be upgraded or replaced with newer ones over the duration of HST's mission.

1. Faint Object Spectrograph and High-Resolution Spectrograph

When HST was launched into orbit in 1990, it contained (among several other instruments) a faint object spectrograph (FOS; Harms et al., 1982) and a high-resolution spectrograph (HRS; Brandt et al., 1982). Both had sensitivities extending into the far UV, down to wavelengths as short as 115 nm. As their names imply, the two instruments had different tradeoffs between spectral resolution and sensitivity, although both instruments individually provided wide ranges of selectable spectral resolutions.

The FOS (see Fig. 18) had two separate but similar spectrographic channels, each incorporating its own detector, emphasizing different but overlapping wavelength ranges. The "red" detector was sensitive in the 180–850 nm spectral range, whereas the "blue" detector covered the 115–500 nm spectral range. The FOS also had a selection of gratings (and a prism) providing resolving powers ($\lambda/\Delta\lambda$) in the range 100–1200.

The HRS (see Fig. 19) also had two separate detector channels, covering the wavelength ranges 110–170 nm and 115–320 nm, and selections of gratings providing spectral resolving powers in the range 2000–100,000. The latter, high-resolution modes utilized cross dispersion gratings in conjunction with first-order gratings to provide this capability.

Both the FOS and HRS utilized one-dimensional electronic imaging detectors, called digicons, which were similar in principle to the current electron-bombarded CCDs described previously, except that they had only one-dimensional arrays of 512 detector elements each. This meant that they could simultaneously image 512 spectral elements, but only one spatial element (transverse to the dis-

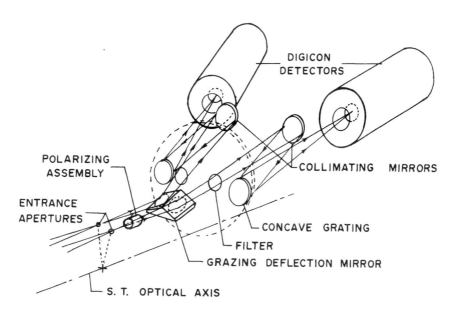

Figure 18   Diagram of the FOS, which was one of the first set of scientific instruments on HST when it was first launched in 1990. As shown, two separate optical systems and detectors were used to observe in the UV and visible spectral ranges. (From NASA CP-2244.)

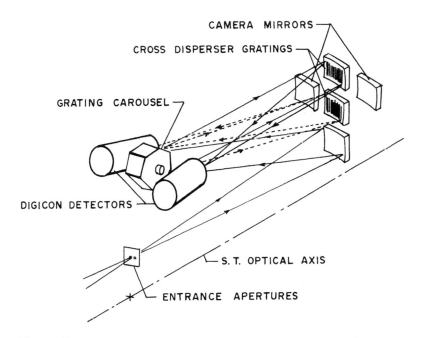

Figure 19 Diagram of the HRS, one of the original complement of HST scientific instruments. The HRS used two detectors and various combinations of mirrors and gratings, to cover the UV spectral range with various spectral resolutions and wavelength ranges. (From NASA CP-2244.)

persion direction), at a time. Nevertheless, the capabilities of these instruments for UV astronomy (in the wavelength ranges to which they were sensitive) far exceeded those of previous space experiments, particularly following the installation of the spherical-aberration-correcting optics in the first Shuttle HST servicing mission in 1993.

However, NASA also solicited proposals for more advanced scientific instruments, to be installed in HST during later servicing missions. The first of these with far-UV spectroscopic capabilities was the Space Telescope Imaging Spectrograph (STIS), which replaced both the FOS and HRS in the 1997 shuttle-servicing mission. Planned for a near-future servicing mission is another far-UV spectrograph, the Cosmic Origins Spectrograph (COS).

2. Space Telescope Imaging Spectrograph

The STIS (Woodgate et al., 1986) was one of several second-generation scientific instruments replacing or supplementing originally installed scientific instruments in space shuttle servicing missions. STIS was designed to replace both the FOS

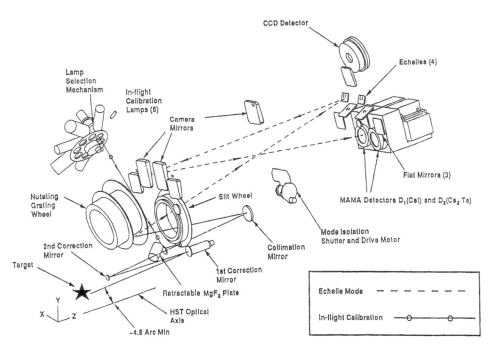

Figure 20   Diagram of the STIS, which replaced both the FOS and HRS during the 1996 Shuttle refurbishment mission to HST. By the use of two-dimensional imaging detectors, STIS was able to increase greatly the efficiency of observations, especially those of extended sources. (From B Woodgate, NASA Goddard Space Flight Center.)

and the HRS, and was installed in the second space shuttle servicing mission in 1997.

The major improvement in the STIS (see Fig. 20) over the previous FOS and HRS was provided by its use of two-dimensional-format electronic imaging devices, in place of the one-dimensional digicons used in the first-generation instruments. This made possible the concept of an imaging spectrograph, in which spectral imaging takes place along one dimension of the two-dimensional array, and spatial imaging (in the along-slit direction) takes place in the transverse dimension. In effect, this allows the variation in intensity of incoming radiation at each wavelength along the spectrum to be measured in the direction perpendicular to that of spectral dispersion. For "extended" objects (i.e., other than stars or other pointlike sources), such as planets, nebulae, and galaxies, this provides much higher sensitivity and/or provides a much more efficient capability for "mapping" the brightness distribution over the extended object (see Table 1).

This has proven very effective in, for example, verifying the presence of massive black holes in the centers of active galaxies.

The STIS uses two types of detectors: (a) a microchannel-plate-based detector, known as the multianode microchannel array (MAMA), used for the middle- and far-UV spectral ranges (110–300 nm); and (b) large-format CCD arrays for the near-UV, visible, and near-infrared spectral range.

3. Cosmic Origins Spectrograph

The COS is a new, third-generation instrument to be installed in HST in a future Shuttle servicing mission (Green et al., 1999). Its scientific objectives require studies of some of the faintest and most distant objects yet observed in the far-UV. This requires it to have the maximum possible sensitivity in the far-UV spectral range, which is achieved by minimizing the number of reflections in the optical system and by other technological improvements vs. the current multipurpose STIS instrument.

C. Hopkins Ultraviolet Telescope

The Hopkins Ultraviolet Telescope (HUT) (Davidsen et al., 1993) was one of three UV astronomy instruments which was part of the Astro space shuttle payload, flown on two shuttle Spacelab missions, in 1990 and 1995 (see Fig. 21). It was optimized for the very far ultraviolet wavelength range, below 110 nm, which is inaccessible to HST and most previous UV astronomy space missions (except Copernicus). This is achieved by using iridium coatings on the telescope mirror and concave grating, and using a windowless microchannel-plate detector.

The HUT made the first high-sensitivity, moderate-resolution spectrographic studies in the 91–110 nm wavelength range, which included studies of quasars and other extragalactic objects, as well as stars in our galaxy, planets in our solar system, and our Moon.

D. Interstellar Medium-Absorption Profile Spectrometer

The IMAPS instrument is a small, special-purpose instrument developed by Princeton University, which was first flown on sounding rockets and then on two Shuttle AstroSPAS missions (Jenkins et al., 1988) (see Fig. 22). It was specifically designed to provide extremely high spectral resolution ($\lambda/\Delta\lambda = 2 \times 10^5$) at very short far-UV wavelengths (95–115 nm), which are not accessible with HST due to the low reflectance of its mirror coatings in this wavelength range. Species of the interstellar medium of special interest in this wavelength range include H_2, HD, D, and O^{+5} (O VI). As shown in Figure 22, IMAPS uses a cross-dispersed optical system, in which the echelle grating is operated as an objective grating.

Table 1 STIS Modes Summary

Detector (photocathode)/window	MAMA (CsI)/MgF$_2$	MAMA (Cs$_2$Te)/MgF$_2$	CCD MgF$_2$ window	
Wavelength (nm)	115–170	165–310	305–555	550–1000
Spectral band	1	2	3	4
Pixel angular size (arc sec)	0.025	0.025	0.050	0.050
Low-resolution spectral imaging (first order)				
Mode number	1.1[a]	2.1[a]	3.1[b]	4.1[b]
Resolving power	960–1400	525–985	560–1010	505–971
Slit length (X, arc sec)	25	25	50	50
Effective area (cm^2)	500	940	6500	5200
Medium-resolution spectral imaging (scanning)				
Mode number	1.2	2.2	3.2	4.2
Resolving power ($\times 10^4$)	1.00–1.51	0.93–1.8	0.54–1.0	0.49–0.90
Slit length (X, arc sec)	30	30	50	50
Scan angle (degree)	5.92	9.55	9.06	8.10
Effective area (cm^2)	360	740	7100	5900
No. exposures/band	11	18	10	9
Medium-resolution echelle				
Mode number	1.3	2.3	N/A	N/A
Resolving power ($\times 10^4$)	2.3	2.3		
Slit length (X, arc sec)	0.43	0.27		
Effective area (cm^2)	320	660		
No. exposures/band	1	2		

High-resolution echelle	Mode number	1.4	2.4	N/A	N/A
	Resolving power ($\times 10^5$)	1.05	1.05		
	Slit length (X, arc sec)	0.12	0.1		
	Effective area (cm^2)	230	480		
	No. exposures/band	3	5		
Objective spectroscopy	Mode number	N/A	2.5	N/A	N/A
	Dispersing element		Prism		
	Resolving power		26–200		
	Field height (Y, arc sec)		25		
	Field width (X, arc sec)		25		
	Effective area (cm^2)		1200		
Camera	Mode number	1.6[a]	2.6[a]	3.6[b]	4.6[b]
	Field height (Y, arc sec)	25	25	50	50
	Field width (X, arc sec)	25	25	50	50
	Effective area (cm^2)	830	1600	9400	7200

All six-reflection configurations, except where noted.
Effective spectral resolution element: 50 μm for MAMA detectors (2 pixels); 42 μm for CCD detector (2 pixels).
Resolving power values use 0.05 arc sec wide slit for bands 1 and 2 and 0.10 arc sec wide slit for bands 3 and 4.
[a] Four-reflection configuration.
[b] Lyot stop with grating wheel element.

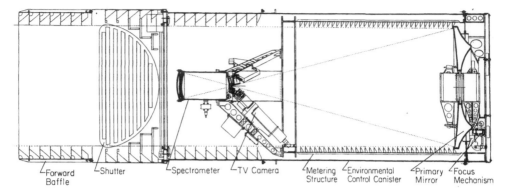

Figure 21   Diagram of the Hopkins Ultraviolet Telescope (HUT), which was flown on two Shuttle Astro Spacelab missions, in 1990 and 1995. HUT was a telescope/spectrograph combination optimized for very short UV wavelengths (down to or below 90 nm). (From Davidsen, 1993.)

E.   Far-Ultraviolet Spectroscopic Explorer

The objectives of FUSE, like Copernicus, HUT, and IMAPS, are to study astronomical objects in the very far UV wavelength range (below the 115 nm limit of HST), and to do so with high sensitivity (to reach faint objects) with both high and moderate spectral resolutions (Sahnow et al., 1996). However, it has the advantage of being a long-duration, dedicated mission (rather than a limited-duration space shuttle investigation). FUSE, which was launched in June, 1999, actually consists of four separate but coaligned telescope/spectrograph combinations, with combinations of two wavelength ranges and two spectral resolution capabilities (see Fig. 23). Two of the instruments use silicon carbide mirror coatings, which provide the best reflectivity in the 90–100 nm wavelength range, whereas the other two use LiF-overcoated aluminum coatings, which provide the best reflectance in the 100–110 nm range. The actual wavelength ranges provide overlapping coverage of the 90.5–118.7 nm wavelength range, with about 0.005 nm spectral resolution.

The use of two-dimensional, photon-counting microchannel plate detectors and other technical improvements makes FUSE far more sensitive than the Copernicus instrument, which had similar wavelength range and spectral resolution capabilities.

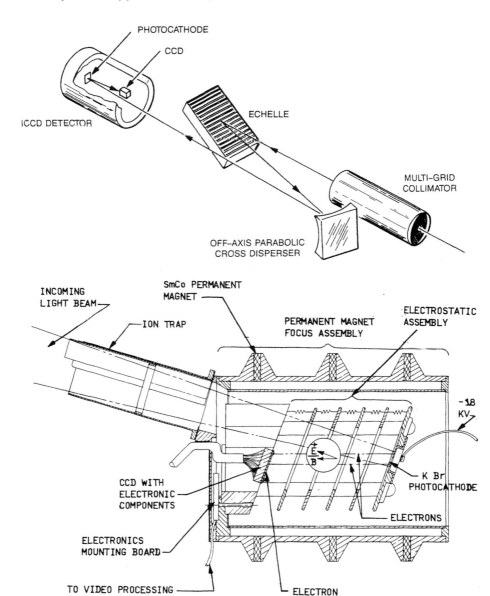

Figure 22 Diagram of Princeton University's IMAPS and its detector system, flown on sounding rockets and on two AstroSPAS Shuttle missions (in 1993 and 1996). (From Jenkins et al., 1988.)

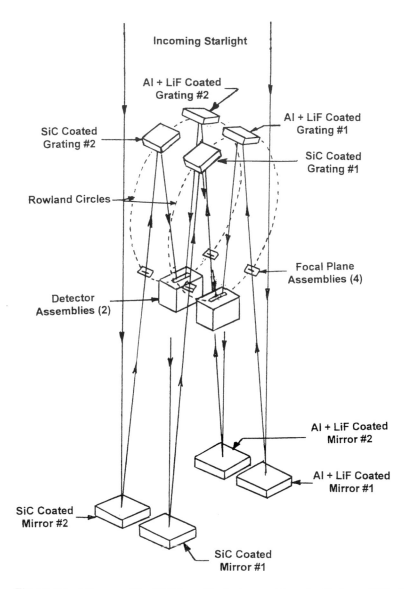

Figure 23   Diagram of the FUSE satellite instrument, launched in June, 1999. As shown, FUSE consists of four nearly identical telescope/spectrometer assemblies, with two pairs having different mirror/grating coatings to optimize different portions of the 90–187 nm spectral range. (Adapted from Moos et al. Ap J Lett 538(1): L1, 2000.)

F. Extreme Ultraviolet Spectroscopy Missions

The extreme-UV spectral range (below about 90 nm wavelength) requires the use of different optical technologies than do the longer UV wavelength ranges. This is due to the extremely low normal-incidence reflectivities of most mirror coatings. Therefore, as is also true for the x-ray spectral range, EUV systems make use of grazing incidence reflective optical systems (see Fig. 24). However, relatively recently, multilayer interference mirror coatings have been developed that provide relatively high reflectivities with normal-incidence reflecting optics, but only for very narrow, specifically selected wavelength ranges.

There have been numerous missions with instruments devoted, primarily or in part, to EUV spectroscopy of the Sun. These have been primarily to observe the very-high-temperature regions of the solar atmosphere; the corona, and active regions (associated with solar flares). Temperature diagnostics can be obtained from comparison of the spectral emissions due to different stages of ionization of a single element (such as Fe or Si; see Fig. 25) but it is also important to have imaging spectrographic capability, to map out variations with position on the Sun or in its atmosphere. Observations in the EUV spectral range complement those in the far-UV and x-ray wavelength ranges, by covering an intermediate range of temperature.

The EUV is also characterized, for observations of objects outside the solar system, by the fact that atomic hydrogen in the interstellar medium had been thought to absorb nearly all of the radiation from celestial objects below the

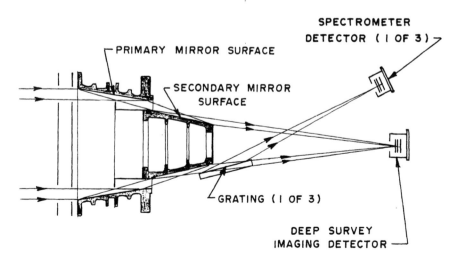

Figure 24 Diagram of an extreme-ultraviolet (below 90 nm) telescope and spectrometer flown on the EUVE satellite, launched in 1991.

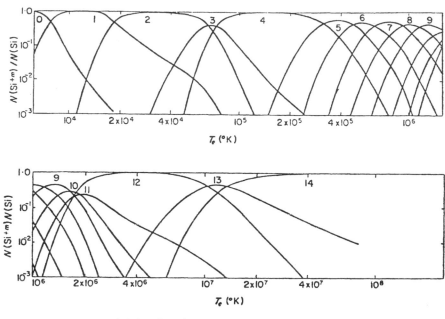

(a) Ionization equilibrium of Si ions

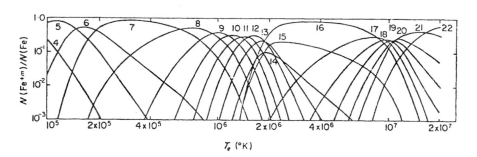

(b) Ionization equilibrium of Fe ions

Figure 25 Variation of the relative degree of ionization with temperature, for (a) silicon and (b) iron.

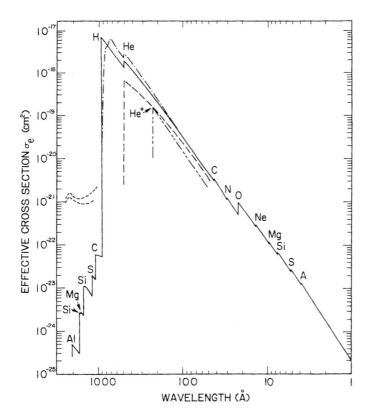

Figure 26 Diagram of the expected opacity of the interstellar medium vs. wavelength, in the far- and extreme-UV spectral ranges, with contributions of the predominant atoms or ions indicated. (From R Cruddace, NRL.)

hydrogen ionization threshold wavelength (91.2 nm) (see Fig. 26). For this reason, it was originally thought that there would be little to be seen in this wavelength range from objects beyond our solar system.

However, early exploratory investigations, including an EUV instrument carried on the Apollo-Soyuz mission in 1975 (Lampton et al., 1976), revealed that in directions in space with exceptionally low hydrogen column densities, very hot objects could be detected at EUV wavelengths. This led to the development and launch of the Extreme Ultraviolet Explorer (EUVE) in 1992 (Bowyer, 1994). This satellite had instruments for both direct imaging and spectroscopy of objects during the mission, which had an all-sky EUV survey as a major objective.

The EUVE sky survey verified that the distribution of absorbing interstellar material was far less uniform than thought previously, and showed that observations in some directions could extend to very great distances, including some extragalactic objects. Objects detected included not only very hot, white dwarf stars, but also many late-type stars (similar to our Sun in surface temperature, or even cooler) from which chromospheric and coronal EUV emissions (also similar to those of our Sun) were observed. EUVE also observed solar system planetary atmospheric emissions, and the Moon's reflectance of solar EUV radiation.

IV. EXAMPLES OF SCIENTIFIC RESULTS FROM UV SPACE ASTRONOMY MISSIONS

We summarize here some of the new and unique results of UV space astronomy, categorized by the type of object or measurement (rather than by mission or instrument type). Although UV astronomical measurements have only been possible in the most recent half-century, compared to the four centuries since the invention of the astronomical telescope, it is clear that the contributions to astronomical knowledge from observations in this part of the electromagnetic spectrum have been significant.

A. Measurements of the Solar Atmosphere and Solar Activity

The benefits of the availability of far- and extreme-ultraviolet spectroscopy in solar physics include the ability to detect and measure the very-high-temperature regions of the solar atmosphere, without need for a solar eclipse (which, in any case, would allow observations of only those portions of the solar atmosphere not covered by the Moon or a coronagraph disk).

In general, the characteristic temperatures of observable ionic species increase as one goes toward shorter wavelengths, since the photons are more energetic. However, there is also the advantage that emissions from the outer regions of the solar atmosphere (the chromosphere and corona) are not obscured by the (in the visible) far more intense radiation from the solar photosphere. The relative intensities of different ionization stages of the same element (e.g., Fe^+, Fe^{2+}, and Fe^{3+}) can be used to determine the temperature, since the degree of ionization increases with temperature in a known fashion, as shown in Figure 25. In addition, even in cooler regions (such as sunspots), far-UV spectroscopy can detect species not observable at longer wavelengths; for example, the NRL HRTS instrument first detected H_2 emission lines in sunspots in a 1975 sounding rocket flight.

The Doppler shifts of emission lines of known rest wavelengths can be used to determine the line-of-sight velocities of gases being ejected from (or falling back into) active regions on the solar surface. This, in combination with methods for measuring the direction and magnitude of the local magnetic field, has verified that motions of ionized gases near sunspots and other regions of solar activity are guided by magnetic field lines in the solar atmosphere. Hence, variations in the far-UV and EUV radiation from the Sun can serve to trace variations in the magnetic field strength and configuration.

Although the far-UV and x-ray regions of the solar spectrum constitute only a small percentage of the Sun's total energy output, this portion (and its time variation) has practical importance because it is responsible for the creation of the ionosphere, the ionized component of Earth's upper atmosphere that facilitates long-distance radio communication. It also is responsible for heating of Earth's upper neutral atmosphere. Variations in the upper atmospheric temperature produce corresponding changes in the atmospheric density at high altitudes (which, in turn, affects atmospheric drag on satellites in low Earth orbit). Therefore, it is of high practical importance to have instruments in space to monitor the solar far-UV and x-ray outputs, and accurately measure any variations on both short- and long-term time scales.

Figure 27 shows examples of solar far-UV spectra obtained with HRTS. Since this is an imaging spectrograph with high spectral resolution, it can measure, simultaneously, variations in spectral intensity with wavelength and with position on the solar surface. It also can measure velocities of emitting gas clouds, by the Doppler shifts in their spectral images.

Figure 28 is a graph of the middle- and far-UV spectrum of the Sun obtained with SUSIM. As mentioned previously, the short-wavelength parts of the solar spectrum are much less bright than the near-UV and visible portions, but also are much more variable with solar activity.

B. Measurements of the Temperatures and Luminosities of Hot Stars

Before the advent of UV space astronomy, the temperatures of stars much hotter than our Sun could be determined only by (a) comparing the spectral distribution of the star's (ground accessible) continuous spectrum with those of classical "black body" radiators or theoretical model atmosphere flux distributions (see Fig. 2), after correcting for "reddening" of the starlight due to interstellar dust extinction (if known); or (b) comparing the visible brightnesses of the star and of any associated emission nebula (ionized hydrogen, or H II region) associated with the star (the ionization of the nebula being attributed to extreme-UV emission from the star, at wavelengths shortward of the atomic hydrogen ionization

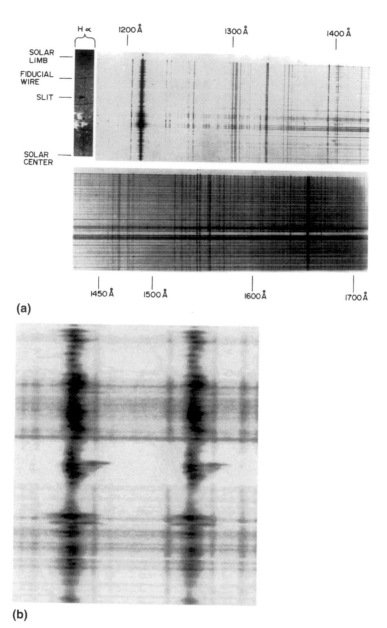

(a)

(b)

Figure 27   (a) A far-UV spectral image (and direct image, in Hα, of the slit field of view) of the Sun, obtained by NRL's HRTS in a sounding rocket flight. (b) A more detailed view of the region of the C IV doublet (154.8–155 nm) shows redshift (downflow) in a sunspot region, near vertical center.

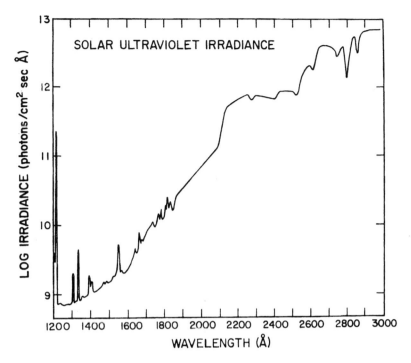

Figure 28   A full-disk solar ultraviolet spectrum, obtained with NRL's Solar Ultraviolet Spectral Irradiance Monitor, during the Spacelab-2 Shuttle mission.

threshold of 91.2 nm). Such estimates were fraught with many uncertainties, which could be circumvented only by more complete observations, particularly of the ground-inaccessible far-UV photospheric spectrum.

Accurate determination of the total energy outputs of hot stars is important because this determines the lifetimes of these stars, which are initially much more massive than our Sun. This is also important for a better understanding of the total galactic radiation environment, and its effects on the interstellar medium.

Ultraviolet spectroscopy from space has provided much new information about the atmospheres and total luminosities of stars of all types (including cool stars, with temperatures similar to or less than that of our Sun). The OAO-2, Copernicus, and IUE satellites, with their long observing times, built upon early sounding rocket investigations to increase greatly our knowledge of stellar atmospheres and stellar evolution.

One complication in such measurements is that for stars lying behind significant amounts of interstellar dust, the extinction due to this dust, and its wavelength dependence, has to be taken into account before the UV brightnesses of the stars can be accurately determined. As discussed in the next section, UV measurement of dust extinction is an astrophysical research objective in its own right. The technique used is to compare two or more stars of spectral types as nearly the same as possible, one of which is known (from ground-based observations) to lie behind significant amounts of interstellar dust, and the other behind a negligible or very small column density, of dust.

One of the early results from rocket flights (Morton 1967; Carruthers et al., 1984) (see Fig. 29), which was confirmed in more detail by the following long-duration satellite missions, is that very hot and luminous stars have intense "stellar winds" in which matter is being ejected at much higher rates than the "solar wind"—in some cases, at rates exceeding 10^{-6} solar masses per year. This is revealed by the so-called "P-Cygni profile" of some of the strong spectral lines (named for the star in which this type of line profile was first observed in the ground-accessible wavelength range). It consists of "blue-shifted" absorption

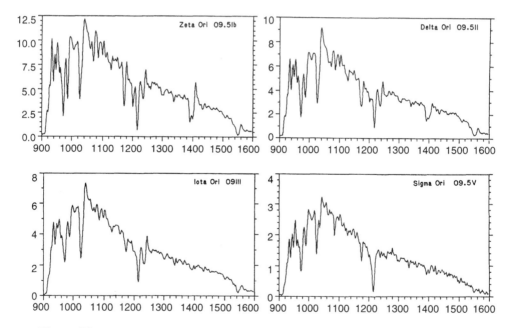

Figure 29 Microdensitometer tracings of some of the stellar far-ultraviolet spectra obtained by NRL in a sounding rocket flight (Fig. 13), showing absorption lines, and "P Cygni" profiles (O VI, 1035 A; N V, 1235 A; Si IV, 1400 A, and C IV, 1550 A). Note that the latter are prominent in the supergiant star ζ Ori, less prominent in the giant stars δ Ori and ι Ori, and absent (replaced by pure absorption lines) in the main-sequence star σ Ori. The Lyman-α (121.6 nm) interstellar absorption line due to atomic hydrogen is prominent in all of these stellar spectra.

features in combination with "red-shifted" emission lines. The interpretation of this profile, as shown in Figure 30, is that the star is ejecting mass at high velocity in all directions. Gas in the line of sight between us and the star is moving toward us while also absorbing starlight, whereas gas moving away or at right angles to the line of sight is responsible for the emission features. It is found that, for a given spectral type of star, the mass loss is more extreme for highly luminous giant or supergiant stars than for main-sequence ("dwarf") stars and, for a given luminosity classification, increases with surface temperature. Resonance features that most prominently exhibit this P-Cygni profile, in order of increasing temperature of the ejected gas, are C IV (154.8–155.1 nm), Si IV (139.4–140.3 nm), N V (123.9–124.3 nm), and O VI (103.2–103.8 nm). These indicate that even for these very hot stars (30,000–40,000 K effective temperatures), their outer atmospheres must be very much hotter still (150,000–300,000 K). Also, it is clear that these intense stellar winds significantly affect the evolution and total lifetimes of these initially very massive stars (10 to more than 40 times the Sun's mass).

Figure 31 shows Copernicus spectra of two hot stars, ζ Ophiuchi (spectral type O9.5) and ζ Puppis (spectral type O5) (Morton, 1976) showing different

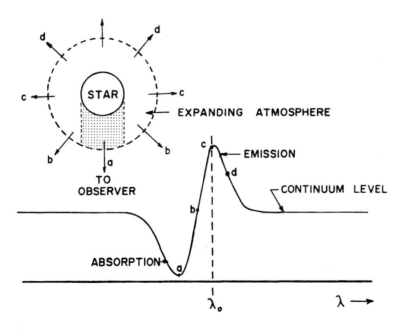

Figure 30 Diagram illustrates the geometry of "stellar winds" that produce the P-Cygni profiles of spectral features. The gas moving toward the observer produces the "blue-shifted" absorption line component; the gas moving away from the observer (but not hidden by the star itself) produces the "red-shifted" emission line component.

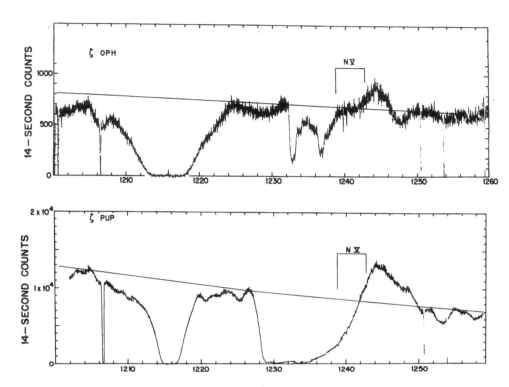

Figure 31   Copernicus spectra, with 0.2 Å (0.02 nm) resolution, of two stars in the 1200–1260 Å (120–126 nm) spectral range show in detail the "P Cygni" profile of the N V spectral feature. The star ζ Puppis, of spectral type O5, shows a more prominent outflow than the star ζ Ophiuchi, of spectral type O9.5. However, the interstellar absorption line of atomic hydrogen at 1216 Å (121.6 nm) is stronger in the latter star, which is observed through a larger column density of interstellar gas. (From Morton, 1976.)

degrees of mass loss in the "P-Cygni profile" of the four-times-ionized nitrogen (N V) spectral feature. The "rest" wavelengths of the two components of this doublet resonance feature are indicated.

C.   Measurements of the Composition and Properties of the Interstellar Medium

Measurements of the interstellar medium, including both gaseous and solid (dust) components, have benefitted greatly from the advent of UV astronomical measurements. In particular, the resonance absorption lines of most of the common gaseous components of interstellar gas, including atomic and molecular hydrogen, helium, oxygen, nitrogen, and carbon, lie in the far UV. As mentioned previously, since the vast majority of these atoms or molecules reside in the ground states (even in relatively "hot" regions, such as in emission nebulae), resonance absorption measurements provide much more accurate measurements of the relative abundances and total column densities of these constituents of the interstellar gas in the lines of sight to UV-bright stars. Also, access to shorter wavelengths in the UV allows access to resonance lines of a given element in higher stages of ionization.

Figure 32 shows some of the spectral line properties used in quantitative measurements of interstellar gas column abundances. As shown, for very weak absorption lines (low optical depth at line center), the spectral line equivalent width is proportional to the column abundance. At the other extreme, where the starlight is totally absorbed within the Doppler width of the line, the equivalent width of the absorption line (including wings) is proportional to the square root of the column density. For intermediate cases, the analysis is more complex, and often involves comparing two or more spectral lines of the same specie having different intrinsic absorption strengths.

The Doppler half-intensity width of an "optically thin" spectral line is given by

$$\Delta\lambda_{1/2} = 7.16 \times 10^{-8}\lambda\sqrt{\frac{T}{M}}$$

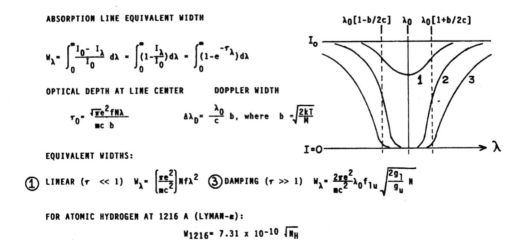

ABSORPTION LINE EQUIVALENT WIDTH

$$W_\lambda = \int_0^\infty \frac{I_0 - I_\lambda}{I_0}\, d\lambda = \int_0^\infty (1 - \frac{I_\lambda}{I_0})\, d\lambda = \int_0^\infty (1 - e^{-\tau_\lambda})\, d\lambda$$

OPTICAL DEPTH AT LINE CENTER   DOPPLER WIDTH

$$\tau_0 = \frac{\sqrt{\pi} e^2 f N \lambda}{m c\, b} \qquad \Delta\lambda_D = \frac{\lambda_0}{c}\, b, \text{ where } b = \sqrt{\frac{2kT}{M}}$$

EQUIVALENT WIDTHS:

① LINEAR ($\tau \ll 1$) $W_\lambda = \left[\frac{\pi e^2}{m c^2}\right] N f \lambda^2$ ③ DAMPING ($\tau \gg 1$) $W_\lambda = \frac{2\pi e^2}{m c^2}\lambda_0 f_{1u}\sqrt{\frac{2g_1}{g_u}}\, N$

FOR ATOMIC HYDROGEN AT 1216 Å (LYMAN-α):

$$W_{1216} = 7.31 \times 10^{-10} \sqrt{N_H}$$

Figure 32 Some of the considerations used in quantitative measurements of column densities of gas, from absorption line spectroscopy.

where T is the temperature (in kelvins (K)), M is the atomic or molecular mass, and the wavelength λ is in nanometers.

For the case of atomic hydrogen, by far the most abundant constituent of the interstellar gas, the square root proportionality of equivalent width to column density, N, usually applies with high accuracy. The line equivalent width vs. column density relationship (Morton, 1967) is given by:

$$W_\lambda = \frac{2\pi e^2}{m_e c^2}\lambda f \sqrt{\left(\frac{2g_1}{g_2} N\right)} = 7.31 \times 10^{-11}\sqrt{N}\, \text{nm}$$

where W_λ is the equivalent width, f is the oscillator strength of the transition, and $g_{1,2}$ are the statistical weights of the lower and upper levels.

In addition to measurements of the column abundances of various atoms, molecules, and ions, high-resolution UV spectroscopy can also be used to determine the temperatures and velocities of interstellar gas clouds, by measuring the Doppler broadening and Doppler shifts of the spectral lines (see Fig. 33).

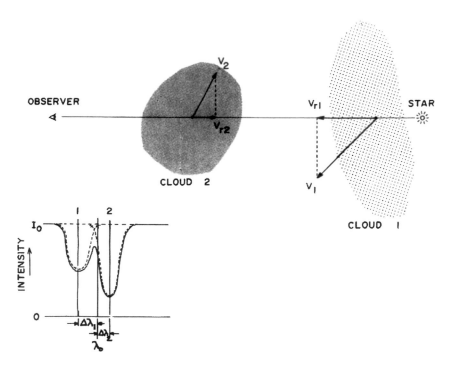

Figure 33 Illustration of how high-resolution spectroscopy can be used to determine the line-of-sight velocities, as well as column densities, of interstellar gas clouds.

The first far-UV measurements of atomic hydrogen, by Princeton University (Morton, 1967), and of molecular hydrogen, by NRL (Carruthers, 1970), were obtained in sounding rocket flights. Among the early results of rocket UV astronomy were the findings that, in many regions of space, the column densities of atomic hydrogen were much lower than estimates from ground-based measurements, such as atomic hydrogen 21 cm radio radiation.

These early results were followed up in much more detail with the launch of the Copernicus satellite in 1972. Figure 31, the Copernicus spectra of 0.02 nm resolution, shows that the Lyman-α absorption line of atomic hydrogen at 121.6 nm in the spectrum of ζ Ophiuchi (which is known to be behind a significant amount of interstellar material) is considerably stronger than that in the spectrum of the star ζ Puppis.

The Copernicus observations of interstellar hydrogen Lyman-α absorption could also be extended to some much cooler stars (in which cases the interstellar Lyman-α was seen superimposed on the stellar chromospheric Lyman-α emission line (DuPree et al., 1977; see Fig. 34)). Observations of these relatively nearby stars indicate that the distribution of atomic hydrogen in the near-solar neighborhood is unexpectedly nonuniform.

Figure 35 shows a higher-resolution Copernicus spectrum of ζ Ophiuchi, at somewhat shorter wavelengths (Spitzer and Jenkins, 1975). The spectrum, with a resolution of 0.005 nm, reveals interstellar absorption lines of molecular hydro-

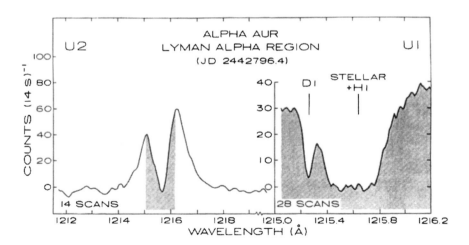

Figure 34 Observations with the Copernicus satellite of the Lyman-α emission line of the star α Aurigae (Capella). The spectra, with resolutions of 0.02 and 0.005 nm, show the absorptions due to interstellar atomic hydrogen and deuterium superimposed on the stellar line. (From DuPree et al., 1977.)

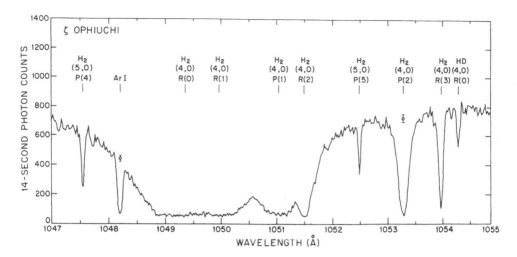

Figure 35 A high-resolution (0.005 nm) Copernicus spectrum of the star ζ Ophiuchi shows absorption features due to interstellar molecular hydrogen. Also present are absorptions due to interstellar HD and Ar. (From Spitzer and Jenkins, 1975.)

gen. Analysis of these spectra indicates that the amounts of atomic and molecular hydrogen in the line of sight to this star are very similar. By intercomparing the individual H_2 lines (which indicate the degree of rotational excitation), one can also determine the temperature of the gas, which in this case was found to be about 80 K.

The Copernicus satellite made the first observations of deuterium ("heavy hydrogen") in the interstellar gas, in both the atomic form (Rogerson and York, 1976, DuPree et al., 1977; see Figs. 34, 36), and in the form of the molecule HD (Fig. 35). The ratio of D to H in interstellar space is important to theories of the origin and evolution of the universe, since D can be destroyed, but not created, by thermonuclear reactions in stars.

Another unanticipated discovery was that five-times-ionized oxygen (O VI), with resonance lines at 103.2 and 103.6 nm, is a significant component of the low-density regions of the interstellar medium (Spitzer and Jenkins, 1975). With a characteristic temperature of more than 2×10^6 K, this "coronal" gas is thought to be heated by supernova explosions. The IUE satellite also observed this "hot" phase of the interstellar medium (Kondo et al., 1989), detecting absorption lines of C IV, Si IV, and N V, not only toward stars in our galaxy but also in the Magellanic Clouds. Intercomparisons of the strengths of these lines in the same line of sight can provide more accurate measurements of the gas temperature than measurements of only one line.

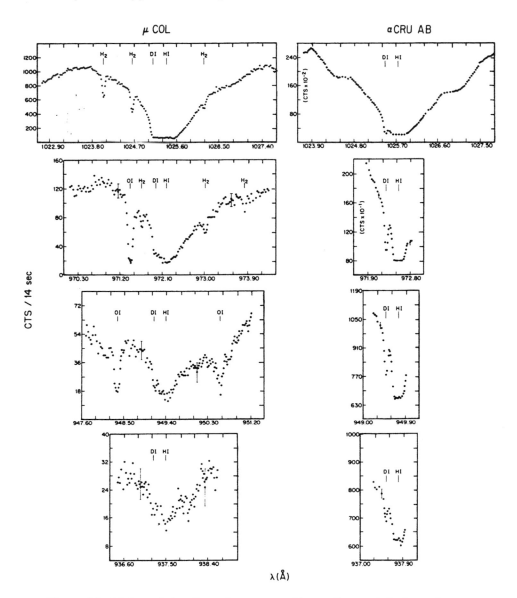

Figure 36 High-resolution Copernicus spectra of interstellar atomic hydrogen Lyman-β, γ, δ, and ε absorption lines toward two stars show the corresponding features due to interstellar deuterium. (From York and J. Rogerson, 1976.)

The HST STIS instrument has extended studies of O VI absorption to intergalactic space; at large distances from our galaxy, the red-shift of spectral lines due to expansion of the universe is sufficient to shift the O VI lines into the wavelength range accessible by HST (Tripp et al., 2000). It was found that the strengths of the O VI lines were not well correlated with atomic hydrogen (Lyman-α or Lyman-β) absorptions at the same redshifts. Since the O VI can only be excited by collisions with other hot gas particles (mostly ionized hydrogen), this shows that there is a significant amount of (otherwise invisible) ionized hydrogen in intergalactic space.

Currently, the FUSE satellite is extending the observations shortward of 115 nm made by Copernicus to include much fainter and more distant stars (and external galaxies), and observations made by HST to shorter wavelengths, for studies of the interstellar and intergalactic media (see FUSE, 2000). Figure 37 is a spectrum of the central star of a planetary nebula, observed by FUSE, in the 91–99 nm wavelength range, with about 0.005 nm resolution. Among other early results of FUSE are measurements of O VI absorption in the galactic halo (using distant active galactic nuclei and quasistellar objects as light sources), and much more extensive mapping of the distributions of H_2 and HD in our galaxy and in the Magellanic Clouds.

The IMAPS instrument, although it has a relatively small collecting aperture, has the highest spectral resolution (0.001 nm) of any instrument yet used in the 95–110 nm spectral range. This has given it the ability to measure the profiles and central wavelengths of interstellar absorption lines toward bright stars

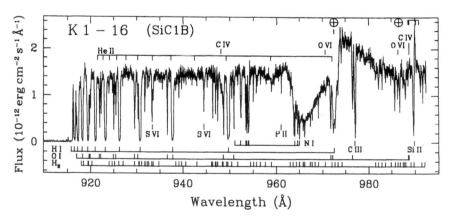

Figure 37 A FUSE spectrum of the central star of a planetary nebula, K1-16, in the 91.2–99.2 nm spectral range. Numerous stellar and interstellar absorption lines, as well as the P-Cygni O VI profile, are apparent. (From HW Moos et al., Ap J Lett 538 (1): L1, 2000.)

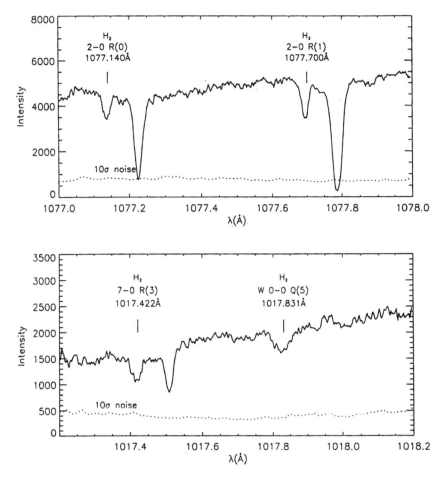

Figure 38 IMAPS spectra of interstellar molecular hydrogen absorption lines toward the star ζ Orionis, obtained in the AstroSPAS-1 shuttle mission. Three of the four marked features are accompanied by stronger features due to the same transition, but redshifted by 25 km/s. This indicates that two separate and discrete clouds with different temperatures are responsible for the two component features. (From Jenkins and Peimbert, 1997.)

with high accuracy, from which temperature and velocity information, as well as abundance information, on interstellar clouds in the line of sight can be obtained (see Fig. 33, Fig. 38). It has been used to measure, with high accuracy, the ratio of D to H in the interstellar medium (from higher members of the Lyman line series), and has indicated that this ratio is variable (Jenkins et al., 1999).

Ultraviolet spectroscopy also gave new insights into the compositions and properties of interstellar dust particles. Ground-based observations had shown that interstellar dust "reddens" the light of stars seen through it: its effectiveness in absorbing or scattering starlight increases toward shorter wavelengths. However, no spectral features gave clues to the compositions of the dust particles. The advent of space-based UV measurements, beginning with a NASA Goddard Space Flight Center sounding rocket flight (Stecher, 1965), revealed that interstellar dust extinction vs. wavelength reached a maximum near 220 nm, beyond which the extinction decreased with decreasing wavelength, down to about 160 nm, below which the extinction began to increase again toward shorter wavelengths (see Fig. 39).

Observations of interstellar dust extinction, beginning with the OAO-2 satellite (Code, 1972) and greatly extended by the IUE satellite, allowed many more stars (including some in the Magellanic Clouds; see Koorneef and Code, 1981) to be observed, and showed that the extinction vs. wavelength curve was highly variable (Witt et al., 1984). It was also found that the variability of the extinction curve with direction in space was much more marked at the shorter wavelengths than in the wavelength range longward of the "220 nm peak," as also indicated in Figure 39. Theoretical and laboratory studies initially led to the "220 nm bump" being attributed to graphite particles, but later studies indicated that this is unlikely.

Observations of ultraviolet starlight reflected or scattered by interstellar dust complement the observations of extinction (which has both true absorption and scattering components) in providing more information about the nature, and potentially the composition, of interstellar dust particles. It has been found that, in at least some cases, the scattering efficiency (albedo) of dust grains has a minimum near 220 nm and then rises toward shorter wavelengths (Lillie and Witt 1976; see Fig. 40).

Currently, there is still no consensus regarding the identity of the material responsible for the 220 nm extinction peak, but it is generally presumed to be carbon (or carbonaceous compounds) in some form. Both carbon and silicates may be responsible, in varying proportions, for the far-UV extinction below 200 nm.

Measurements of the gaseous component of the interstellar medium also shed light on the likely compositions of dust particles. It has been observed, beginning with Copernicus observations and extended with HST HRS and STIS observations, that in regions where there are significant concentrations of dust and molecular hydrogen, the amounts of some of the gaseous components of the interstellar medium (relative to the total hydrogen column density and relative abundances in the solar atmosphere) are reduced. The elements most depleted from the gas are those with high "condensation temperatures," such as iron, silicon, calcium, and titanium. That is, those most likely to form solid particles

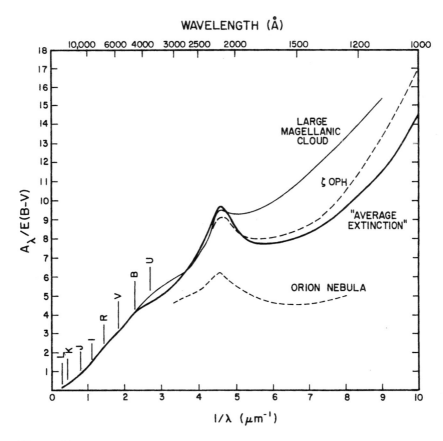

Figure 39 Typical extinction curves due to interstellar dust, covering the infrared through ultraviolet spectral ranges, normalized to a "reddening" or color excess, E(B-V), of one magnitude. Note the prominent peak in the middle-UV near 2200 A (220 nm), and the marked variations for different lines of sight in the far-UV range (below 200 nm).

at the low temperatures of the interstellar clouds (Savage and Sembach 1996; see Fig. 41).

D.   Studies of Extragalactic Objects and Cosmology

Observations of external galaxies in the ultraviolet allow comparisons of the distributions of hot stars with those of the much more common (and, in the visible, in total much brighter) cooler stars. They also allow more accurate measurements of the effective temperatures and total luminosities of individual objects, includ-

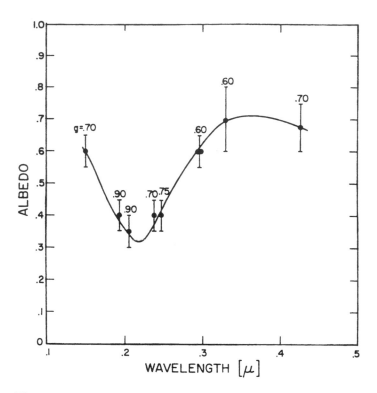

Figure 40 Measurements of the reflectance, or albedo, and scattering phase function asymmetry g, of interstellar dust particles, obtained with the Orbiting Astronomical Observatory-2 (OAO-2) narrow-band photometers. (From Lillie and Witt, 1976.)

ing active galactic nuclei and quasistellar objects (quasars). For sufficiently luminous individual stars or star groupings, the compositions, column densities, and spatial distributions of interstellar gas (within the external galaxy) and of any intergalactic gas in the line of sight to the galaxy can also be determined.

Since galaxies and quasars at very large distances from our galaxy display redshifts of their radiation due to the expansion of the universe, it is possible to observe radiation (with any given instrument) to shorter rest wavelengths in more distant objects. This allows, for example, determinations of the distributions in space of intergalactic gas clouds (for example, atomic hydrogen clouds in intergalactic space give rise to a "Lyman-α forest" of absorption lines, each localized in redshift by its apparent wavelength). With a far-UV spectrograph (as on HUT and FUSE), it is possible to observe neutral helium (resonance line at 58.4 nm) and even ionized helium (30.4 nm resonance line rest wavelength) in clouds of sufficiently high redshift.

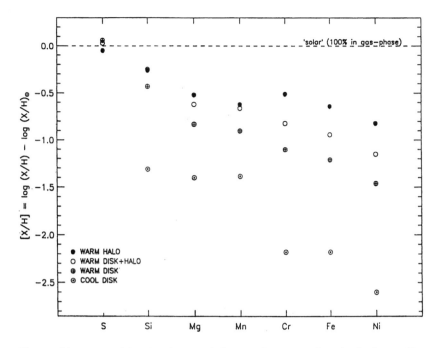

Figure 41 Ratio of the abundances of elements in gaseous form in the interstellar medium to that of hydrogen, and relative to the same ratio in the atmosphere of the Sun, as observed with the HST HRS. Note that elements expected to be major constituents of interstellar dust particles are under-represented in the interstellar gas, as they have been depleted in the process of dust particle production. (From Savage and Sembach, 1996.)

The STIS instrument on HST, normally limited to wavelengths longward of about 115 nm, has observed strong, redshifted O VI absorption lines (rest wavelengths 103.2–103.7 nm) toward distant quasars (Tripp et al., 2000; see Fig. 42). In comparing these with hydrogen Lyman-α and Lyman-β absorption features at these same redshifts, it was found that much of the O VI absorption was occurring in regions of space with little corresponding H absorption. From this it is inferred that (invisible) ionized hydrogen must be of higher than previously known abundance in these very distant regions of intergalactic space, assuming the O/H abundance ratio is uniform. Also, collisions with energetic protons are the only practical means for creating the observed highly ionized oxygen.

Current and future observations with STIS and COS are likely to prove valuable complements of observations with large ground-based telescopes, and space-based infrared telescopes, in furthering our understanding of the very distant regions of our universe.

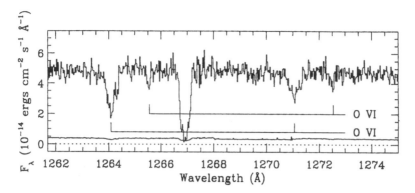

Figure 42   HST STIS high-resolution spectra of redshifted O VI absorption lines, toward the quasistellar object H1821 + 643. The two pairs of O VI lines are produced in intergalactic space, in clouds with redshifts less than that of the QSO. The amount of O VI in this line of sight, not correlated with hydrogen Lyman α or Lyman β absorption, implies a large amount of unseen, ionized hydrogen in intergalactic space. (From Tripp et al., 2000.)

REFERENCES

Bartoe JDF, Brueckner GE. New stigmatic, coma-free, concave-grating spectrograph. J Opt Soc Am 65: 13, 1975.

Bohlin RC. Copernicus observations of interstellar absorption in Lyman-Alpha. Astrophys J 200: 402, 1975.

Bowyer CS. Astronomy and the Extreme Ultraviolet Explorer Satellite. Science 263: 55, 1994.

Bradford AP, Hass G, Osantowski JF, Toft AR. Preparation of mirror coatings for the vacuum ultraviolet in a 2-m evaporator. Appl Optics 8: 1183, 1969.

Brandt JC, HRS Team. The high resolution spectrograph for the Space Telescope. In: The Space Telescope Observatory, NASA CP-2244, 1982, p 76.

Brueckner GE, Bartoe JDF, Sandlin GD, VanHoosier ME. Lines of H_2 in extreme-ultraviolet solar spectra. Nature 270: 326, 1977.

Brueckner G, Lean J, VanHoosier M. SUSIM—Solar Ultraviolet Spectral Irradiance Monitor. Naval Research Laboratory report, 1991.

Carruthers GR. Magnetically focused image converters with internal reflecting optics. Rep NRL Progr July: 7, 1965.

Carruthers GR. Rocket observation of interstellar molecular hydrogen. Astrophys J Lett, 161: L81, 1970.

Carruthers GR. Narrow-band filter for the Lyman-β wavelength region. Appl Optics 10: 1461, 1971.

Carruthers GR, Heckathorn HM, Opal CB. Stellar ultraviolet flux distributions in the 912–

1200 A wavelength range. In: Future of Ultraviolet Astronomy Based on Six Years of IUE Research. NASA Conference Publication 2349, 1984, p 534.

Carruthers GR, Seeley TD. Global Imaging Monitor of the Ionosphere (GIMI): a far-ultraviolet imaging experiment on ARGOS. In: Proc SPIE, Ultraviolet Atmospheric and Space Remote Sensing: Methods and Instrumentation 2831: 65, 1996.

Code AD, ed. The Scientific Results From the Orbiting Astronomical Observatory (OAO-2). NASA SP-310, 1972.

Davidsen AF, ed. The Hopkins Ultraviolet Telescope: Collected Scientific Papers, 1991–1992. Baltimore: Johns Hopkins University, 1993.

Dupree AK, Baliunas SL, Shipman HL. Deuterium and hydrogen in the local interstellar medium. Astrophys J 218: 361, 1977.

Far Ultraviolet Spectroscopic Explorer (FUSE) Special Issue. Astrophys J Lett, July 20: 538, 2000.

Green JC, Morse JA, Andrews JP, Wilkinson E, Siegmund OHW, Ebbetts, D. Performance of the cosmic origins spectrograph for the Hubble Space Telescope. In Ultraviolet-Optical Space Astronomy Beyond HST, Astronomical Society of the Pacific Conference Series, 164: 176, 1999.

Harms RJ, FOS Teams. Astronomical capabilities of the Faint Object Spectrograph on Space Telescope. In: The Space Telescope Observatory, NASA CP-2244, 1982, p 55.

Jenkins EB, Joseph CL, Long D, Zucchino PM, Carruthers GR, Bottema M, Delamere WA. IMAPS: high resolution echelle spectrograph to record far-UV spectra of stars from sounding rockets. In: Proc SPIE, Ultraviolet Technology II. 932: 213, 1988.

Jenkins EB, Peimbert A. Molecular hydrogen in the direction of ζ Orionis A. Astrophys J 477: 265, 1997.

Jenkins EB, Tripp TM, Wozniak PR, Sofia UJ, Sonneborn G. Spatial variability in the ratio of interstellar atomic deuterium to hydrogen. I. Observations toward δ Orionis by the Interstellar Medium Absorption Profile Spectrograph. Astrophys J 520: 182, 1999.

Keski-Kuha RAM, Osantowski JF, Herzig H, Gum JF, Toft AR. Normal incidence reflectance of ion beam deposited SiC films in the EUV. Appl Optics 27: 2815, 1988.

Kondo Y, Boggess A, Maran SP. Astrophysical contributions of the International Ultraviolet Explorer. Annu Rev Astron Astrophys 27: 397, 1989.

Koorneef J, Code AD. Ultraviolet interstellar extinction in the large Magellanic Cloud using observations with the International Ultraviolet Explorer satellite. Astrophys J, 247: 860, 1981.

Kurucz RL. Model atmospheres for G, F, A, B, and O stars. Astrophys J Suppl 40: 1, 1979.

Lampton M, Margon B, Paresce F, Stern R, Bowyer S. Discovery of a nonsolar extreme-ultraviolet source. Astrophys J Lett 203: L71, 1976.

Lillie CF, Witt AN. Ultraviolet photometry from the Orbiting Astronomical Observatory. XXV. Diffuse galactic light in the 1500–4200 A region and the scattering properties of interstellar dust grains. Astrophys J 208: 64, 1976.

Morton DC. The far-ultraviolet spectra of six stars in Orion. Astrophys J, 147: 1017, 1967.

Morton DC. P Cygni profiles in Zeta Ophiuchi and Zeta Puppis. Astrophys J, 203: 386, 1976.

NASA NP-126. Exploring the Universe with the Hubble Space Telescope. NASA–US Government Printing Office.

Sahnow DJ, Friedman SD, Oegerle WR, Moos HW, Green JC, Siegmund OHW. Design and predicted performance of the Far Ultraviolet Spectroscopic Explorer (FUSE). Proc SPIE 2807, 1996.

Samson JAR. Techniques of Vacuum Ultraviolet Spectroscopy. New York: John Wiley & Sons, 1967.

Savage BD, Sembach KR. Interstellar abundances from absorption-line observations with the Hubble Space Telescope. Annu Rev Astron Astrophys 34: 279, 1996.

Shull JM, Beckwith S. Interstellar molecular hydrogen. Annu Rev Astron Astrophys 20: 163, 1982.

Smith AM. Rocket spectrographic observations of α Virginis. Astrophys J 156: 93, 1969.

Smith, AM. Rocket spectroscopy of ζ Orionis. Astrophys J 172: 129, 1972.

Sommer AH. Photoemissive Materials: Preparation, Properties, and Uses. New York: John Wiley & Sons, 1968.

Spitzer L, Jenkins EB. Ultraviolet studies of the interstellar gas. Annu Rev Astron Astrophys 13: 133, 1975.

Stecher TP. Interstellar extinction in the ultraviolet. Astrophys J 142: 1683, 1965.

Tripp TM, Savage BD, Jenkins EB. Intervening O VI quasar absorption systems at low redshift: a significant baryon reservoir. Astrophys J Lett 534: L1, 2000.

Witt AN, Bohlin RC, Stecher TP. The variation of galactic interstellar extinction in the ultraviolet. Astrophys J 279: 698, 1984.

Woodgate BE, Boggess A, Gull TR, Heap SR, Krueger VL, Maran SP, Melcher RW, Rebar FJ, Vitagliano HD, Green RF, Wolff SC, Hutchings JB, Jenkins EB, Linsky JL, Moos HW, Roesler F, Shine RA, Timothy JG, Weistrop DE, Bottema M, Meyer W. Second generation spectrograph for the Hubble Space Telescope. In: Proc SPIE, Instrumentation in Astronomy VI. 627: 350, 1986.

York DG, Rogerson JB. The abundance of deuterium relative to hydrogen in interstellar space. Astrophys J 203: 378, 1976.

Index

Printed and bound by CPI Group (UK) Ltd, Croydon, CR0 4YY

23/10/2024

01778267-0001